# BIG DATA

## ALGORITHMS, ANALYTICS, AND APPLICATIONS

# Chapman & Hall/CRC
# Big Data Series

**SERIES EDITOR**
**Sanjay Ranka**

## AIMS AND SCOPE

This series aims to present new research and applications in Big Data, along with the computational tools and techniques currently in development. The inclusion of concrete examples and applications is highly encouraged. The scope of the series includes, but is not limited to, titles in the areas of social networks, sensor networks, data-centric computing, astronomy, genomics, medical data analytics, large-scale e-commerce, and other relevant topics that may be proposed by potential contributors.

## PUBLISHED TITLES

**BIG DATA : ALGORITHMS, ANALYTICS, AND APPLICATIONS**
**Kuan-Ching Li, Hai Jiang, Laurence T. Yang, and Alfredo Cuzzocrea**

Chapman & Hall/CRC
Big Data Series

# BIG DATA

## ALGORITHMS, ANALYTICS, AND APPLICATIONS

EDITED BY

KUAN-CHING LI
PROVIDENCE UNIVERSITY
TAIWAN

HAI JIANG
ARKANSAS STATE UNIVERSITY
USA

LAURENCE T. YANG
ST. FRANCIS XAVIER UNIVERSITY
CANADA

ALFREDO CUZZOCREA
ICAR -CNR & UNIVERSITY OF CALABRIA
ITALY

CRC Press is an imprint of the
Taylor & Francis Group, an informa business
A CHAPMAN & HALL BOOK

CRC Press
Taylor & Francis Group
6000 Broken Sound Parkway NW, Suite 300
Boca Raton, FL 33487-2742



Printed on acid-free paper
Version Date: 20141210

International Standard Book Number-13: 978-1-4822-4055-9 (Hardback)



**Library of Congress Cataloging-in-Publication Data**

Big data (CRC Press)
Big data : algorithms, analytics, and applications / edited by Kuan-Ching Li [and 3 others].
pages cm. -- (Chapman & Hall/CRC big data series)
Includes bibliographical references and index.
ISBN 978-1-4822-4055-9 (hardback)
1. Big data. 2. Database management. 3. Data mining. 4. Machine theory. I. Li, Kuan-Ching, editor. II. Title.

QA76.9.D343B545 2015
005.7--dc23 2014043108

**Visit the Taylor & Francis Web site at**
**http://www.taylorandfrancis.com**

**and the CRC Press Web site at**
**http://www.crcpress.com**

# Contents

## SECTION II Big Data Processing



## SECTION III Big Data Stream Techniques and Algorithms

# Foreword by Jack Dongarra

It is apparent that in the era of *Big Data*, when every major field of science, engineering, business, and finance is producing and needs to (repeatedly) process truly extraordinary amounts of data, the many unsolved problems *of when, where, and how all those data are to be produced, transformed, and analyzed* have taken center stage. This book, *Big Data: Algorithms, Analytics, and Applications*, addresses and examines important areas such as management, processing, data stream techniques, privacy, and applications.

The collection presented in the book covers fundamental and realistic issues about *Big Data*, including efficient algorithmic methods to process data, better analytical strategies to digest data, and representative applications in diverse fields such as medicine, science, and engineering, seeking to bridge the gap between huge amounts of data and appropriate computational methods for scientific and social discovery, and to bring together technologies for media/data communication, elastic media/data storage, cross-network media/data fusion, SaaS, and others. It also aims at interesting applications related to Big Data.

This timely book edited by Kuan-Ching Li, Hai Jiang, Laurence T. Yang, and Alfredo Cuzzocrea gives a quick introduction to the understanding and use of Big Data and provides examples and insights into the process of handling and analyzing problems. It presents introductory concepts as well as complex issues. The book has five major sections: management, processing, streaming techniques and algorithms, privacy, and applications.

Throughout the book, examples and practical approaches for dealing with Big Data are provided to help reinforce the concepts that have been presented in the chapters.

This book is a required understanding for anyone working in a major field of science, engineering, business, and financing. It explores Big Data in depth and provides different angles on just how to approach Big Data and how they are currently being handled today.

I have enjoyed and learned from this book, and I feel confident that you will as well.

**Jack Dongarra**
*University of Tennessee, Knoxville*

# Foreword by Jack Dongarra

# Foreword by Dr. Yi Pan

In 1945, mathematician and physicist John von Neumann and his colleagues wrote an article, "First Draft of a Report on the EDVAC," to describe their new machine EDVAC based on some ideas from J. Presper Eckert and John Mauchly. The proposed concept of stored-program computer, known as *von Neumann machine*, has been widely adopted in modern computer architectures. Both the program and data for an application are saved in computer memory for execution flexibility. Computers were not dedicated to single jobs anymore.

In 1965, Intel's cofounder Gordon Moore proposed the so-called Moore's law to predict that the number of transistors doubles every 18 months. Computer hardware roughly sticks to Moore's law over these years. However, to ensure that the computability doubles every 18 months, computer hardware has to work with system software and application software closely. For each computer job, both computations and data aspects should be handled properly for better performance.

To scale up computing systems, multiple levels of parallelism, including instruction-level, thread-level, and task-level parallelism, have been exploited. Multicore and many-core processors as well as symmetric multiprocessing (SMP) have been used in hardware. Data parallelism is explored through byte-level parallelization, vectorization, and SIMD architectures. To tolerate the latency of data access in memory, memory hierarchy with multiple levels of cache is utilized to help overcome the *memory wall* issue.

To scale out, multicomputer systems such as computer clusters, peer-to-peer systems, grids, and clouds are adopted. Although the aggregated processing and data storage capabilities have been increased, the communication overhead and data locality remain as the major issues. Still, scaled-up and scaled-out systems have been deployed to support computer applications over the years.

Recently, *Big Data* has become a buzzword in the computer world, as the size of data increases exponentially and has reached the petabyte and terabyte ranges. The representative Big Data sets come from some fields in science, engineering, government, private sector, and daily life. *Data digestion* has taken over the previous *data generation* as the leading challenge for the current computer world. Several leading countries and their governments have noticed this new trend and released new policies for further investigation and research activities. As distributed systems and clouds become popular, compared to the traditional data in common applications, Big Data exhibit some distinguished characteristics, such as volume, variety, velocity, variability, veracity, value, and complexity. The manners for data collection, data storage, data representation, data fusion, data processing, and visualization have to be changed. Since modern computers still follow *von Neumann architecture* and other potential ones such as biocomputers and quantum computers are still in infancy, the algorithmic and software adjustments are still the main consideration for Big Data.

Li, Jiang, Yang, and Cuzocrea's timely book addresses the Big Data issue. These editors are active researchers and have done a lot of work in the area of Big Data. They assembled a group of outstanding authors. The book's content mainly focuses on algorithm, analytics, and application aspects in five separate sections: Big Data management, Big Data processing, Big Data stream techniques and algorithms, Big Data privacy, and Big Data applications. Each section contains several case studies to demonstrate how the related issues are addressed.

Several Big Data management strategies such as indexing and clustering are introduced to illustrate how to organize data sets. Some processing and scheduling schemes as well as representative frameworks such as MapReduce and Hadoop are also included to show how to speed up Big Data applications. Particularly, stream technique is explained in detail due to its importance in Big Data processing. Privacy is still a concern in Big Data and proper attention is necessary. Finally, the text includes several actual Big Data applications in finance, media, biometrics, geoscience, and the social sector. They help demonstrate how the Big Data issue has been addressed in various fields.

I hope that the publication of this book will help university students, researchers, and professionals understand the major issues in Big Data applications as well as the corresponding strategies to tackle them. This book might also stimulate research activities in all aspects of Big Data. As the size of data still increases dramatically every day, the Big Data issue might become obviously more challenging than people expect. However, this may also be a good opportunity for new discoveries in this exciting field. I highly recommend this timely and valuable book. I believe that it will benefit many readers and contribute to the further development of Big Data research.

**Dr. Yi Pan**
*Distinguished University Professor of Computer Science*
*Interim Associate Dean of Arts and Science*
*Georgia State University, Atlanta*

# Foreword by D. Frank Hsu

Due to instrumentation and interconnection in the past decades, the Internet and the World Wide Web have transformed our society into a complex cyber–physical–social ecosystem where everyone is connected to everything (the Internet of Everything [IOE]) and everybody is an information user as well as an information provider. More recently, *Big Data* with high volume and wide variety have been generated at a rapid velocity. Everyone agrees that Big Data have great potential to transform the way human beings work, live, and behave. However, their true value to the individual, organization, society, planet Earth, and intergalactic universe hinges on understanding and solving fundamental and realistic issues about Big Data. These include efficient algorithms to process data, intelligent informatics to digest and analyze data, wide applications to academic disciplines in the arts and sciences, and operational practices in the public and private sectors. The current book, *Big Data: Algorithms, Analytics, and Applications* (BDA$^3$), comes at the right time with the right purpose.

In his presentation to the Information and Telecommunication Board of the National Research Council (http://www.TheFourthParadigm.com), Jim Gray describes the long history of the scientific inquiry process and knowledge discovery method as having gone through three paradigms: from empirical (thousands of years ago), to theoretical (in the last few hundred years, since the 1600s), and then to modeling (in the last few decades). More recently, the complexity of Big Data (structure vs. unstructured, spatial vs. temporal, logical vs. cognitive, data-driven vs. hypothesis-driven, etc.) posts great challenges not only in application domains but also in analytic method. These domain applications include climate change, environmental issues, health, legal affairs, and other critical infrastructures such as banking, education, finance, energy, healthcare, information and communication technology, manufacturing, and transportation. In the scientific discovery process, efficient and effective methods for analyzing data are needed in a variety of disciplines, including STEM (astronomy, biomedicine, chemistry, ecology, engineering, geology, computer science, information technology, and mathematics) and other professional studies (architecture, business, design, education, journalism, law, medicine, etc.). The current book, BDA$^3$, covers many of the application areas and academic disciplines.

Big Data phenomenon has altered the way we analyze data using statistics and computing. Traditional statistical problems tend to have many observations but few parameters with a small number of hypotheses. More recently, problems such as analyzing fMRI data set and social networks have diverted our attention to the situation with a small number of observations, a large number of variables (such as cues or features), and a relatively bigger number of hypotheses (also see D.J. Spiegelhalter, *Science*, 345: 264–265, 2014). In the Big Data environment, the use of statistical significance (the *P* value) may not always be appropriate. In analytics terms, correlation is not equivalent to causality and normal distribution may not be that normal. An ensemble of multiple models is often used to improve forecasting, prediction, or decision making. Traditional computing problems use static data base, take input from logical and structured data, and run deterministic algorithms. In the Big Data era, a relational data base has to be supplemented with other structures due to scalability issue. Moreover, input data are often unstructured and illogical (due to acquisition through cognition, speech, or perception). Due to rapid streaming of incoming data, it is necessary to bring computing to the data acquisition point. Intelligent informatics will use data mining and semisupervised machine learning techniques to deal with the uncertainty factor of the complex Big Data environment. The current book, BDA$^3$, has included many of these data-centric methods and analyzing techniques.

The editors have assembled an impressive book consisting of 22 chapters, written by 57 authors, from 12 countries across America, Europe, and Asia. The chapters are properly divided into five sections on Big Data: management, processing, stream technologies and algorithms, privacy, and applications. Although the authors come from different disciplines and subfields, their journey is the same: to discover and analyze Big Data and to create value for them, for their organizations and society, and for the whole world. The chapters are well written by various authors who are active researchers or practical experts in the area related to or in Big Data. BDA$^3$ will contribute tremendously to the emerging new paradigm (the fourth paradigm) of the scientific discovery process and will help generate many new research fields and disciplines such as those in computational x and x-informatics (x can be biology, neuroscience, social science, or history), as Jim Gray envisioned. On the other hand, it will stimulate technology innovation and possibly inspire entrepreneurship. In addition, it will have a great impact on cyber security, cloud computing, and mobility management for public and private sectors.

I would like to thank and congratulate the four editors of BDA$^3$—Kuan-Ching Li, Hai Jiang, Laurence T. Yang, and Alfredo Cuzzocrea—for their energy and dedication in putting together this significant volume. In the Big Data era, many institutions and enterprises in the public and private sectors have launched their Big Data strategy and platform. The current book, BDA$^3$, is different from those strategies and platforms and focuses on essential Big Data issues, such as management, processing, streaming technologies, privacy, and applications. This book has great potential to provide fundamental insight and privacy to individuals, long-lasting value to organizations, and security and sustainability to the cyber–physical–social ecosystem on the planet.

**D. Frank Hsu**
*Fordham University, New York*

# Preface

As data sets are being generated at an exponential rate all over the world, *Big Data* has become an indispensable issue. While organizations are capturing exponentially larger amounts of data than ever these days, they have to rethink and figure out how to digest it. The implicit meaning of data can be interpreted in reality through novel and evolving algorithms, analytics techniques, and innovative and effective use of hardware and software platforms so that organizations can harness the data, discover hidden patterns, and use newly acquired knowledge to act meaningfully for competitive advantages.

This challenging vision has attracted a great deal of attention from the research community, which has reacted with a number of proposals focusing on *fundamental issues*, such as *managing Big Data*, *querying and mining Big Data*, *making Big Data privacy-preserving*, designing and running *sophisticated analytics over Big Data*, and *critical applications*, which span over a large family of cases, from *biomedical (Big) Data* to *graph (Big) Data*, from *social networks* to *sensor* and *spatiotemporal stream networks*, and so forth.

A conceptually relevant point of result that inspired our research is recognizing that classical managing, query, and mining algorithms, even developed with very large data sets, are not suitable to cope with Big Data due to both methodological and performance issues. As a consequence, there is an emerging need for devising innovative models, algorithms, and techniques capable of managing and mining Big Data while dealing with their inherent properties, such as *volume*, *variety*, and *velocity*.

Inspired by this challenging paradigm, this book covers *fundamental and realistic issues about Big Data*, including efficient algorithmic methods to process data, better analytical strategies to digest data, and representative applications in diverse fields such as medicine, science, and engineering, seeking to bridge the gap between huge amounts of data and appropriate computational methods for scientific and social discovery and to bring technologies for media/data communication, elastic media/data storage, cross-network media/data fusion, Software as a Service (SaaS), and others together. It also aims at interesting applications involving Big Data.

According to this methodological vision, this book is organized into five main sections:

- "Big Data Management," which focuses on research issues related to the effective and efficient management of Big Data, including *indexing* and *scalability* aspects.
- "Big Data Processing," which moves the attention to the problem of processing Big Data in a widespread collection of resource-intensive computational settings, for

example, those determined by *MapReduce environments*, *commodity clusters*, and *data-preponderant networks*.

- "Big Data Stream Techniques and Algorithms," which explores research issues concerning the management and mining of Big Data in *streaming environments*, a typical scenario where Big Data show their most problematic drawbacks to deal with—here, the focus is on how to manage Big Data *on the fly*, with limited resources and approximate computations.
- "Big Data Privacy," which focuses on models, techniques, and algorithms that aim at making Big Data *privacy-preserving*, that is, protecting them against *privacy breaches* that may prevent the anonymity of Big Data in conventional settings (e.g., *cloud environments*).
- "Big Data Applications," which, finally, addresses a rich collection of practical applications of Big Data in several domains, ranging from *finance applications* to *multimedia tools*, from *biometrics applications* to *satellite (Big) Data processing*, and so forth.

In the following, we will provide a description of the chapters contained in the book, according to the previous five sections.

The first section (i.e., "Big Data Management") is organized into the following chapters.

Chapter 1, "Scalable Indexing for Big Data Processing," by Hisham Mohamed and Stéphane Marchand-Maillet, focuses on the *K-nearest neighbor* (*K*-NN) search problem, which is the way to find and predict the most closest and similar objects to a given query. It finds many applications for information retrieval and visualization, machine learning, and data mining. The context of Big Data imposes the finding of approximate solutions. Permutation-based indexing is one of the most recent techniques for approximate similarity search in large-scale domains. Data objects are represented by a list of references (pivots), which are ordered with respect to their distances from the object. In this context, the authors show different distributed algorithms for efficient indexing and searching based on permutation-based indexing and evaluate them on big high-dimensional data sets.

Chapter 2, "Scalability and Cost Evaluation of Incremental Data Processing Using Amazon's Hadoop Service," by Xing Wu, Yan Liu, and Ian Gorton, considers the case of *Hadoop* that, based on the *MapReduce* model and *Hadoop Distributed File System* (HDFS), enables the distributed processing of large data sets across clusters with scalability and fault tolerance. Many data-intensive applications involve continuous and incremental updates of data. Understanding the scalability and cost of a Hadoop platform to handle small and independent updates of data sets sheds light on the design of scalable and cost-effective data-intensive applications. With these ideas in mind, the authors introduce a motivating movie recommendation application implemented in the MapReduce model and deployed on *Amazon Elastic MapReduce* (EMR), a Hadoop service provided by Amazon. In particular, the authors present the deployment architecture with implementation details of the Hadoop application. With metrics collected by *Amazon CloudWatch*, they present an empirical scalability and cost evaluation of the *Amazon Hadoop* service on processing continuous and incremental data streams. The evaluation result highlights the potential of autoscaling for cost reduction on Hadoop services.

Chapter 3, "Singular Value Decomposition, Clustering, and Indexing for Similarity Search for Large Data Sets in High-Dimensional Spaces," by Alexander Thomasian, addresses a popular paradigm, that is, representing objects such as images by their feature vectors and searching for similarity according to the distances of the points representing them in high-dimensional space via *K-nearest neighbors* (*K*-NNs) to a target image. The authors discuss a combination of *singular value decomposition* (SVD), clustering, and indexing to reduce the cost of processing *K*-NN queries for large data sets with high-dimensional data. They first review dimensionality reduction methods with emphasis on SVD and related methods, followed by a survey of clustering and indexing methods for high-dimensional numerical data. The authors describe combining SVD and clustering as a framework and the main memory-resident *ordered partition* (*OP*)*-tree index* to speed up *K*-NN queries. Finally, they discuss techniques to save the OP-tree on disk and specify the *stepwise dimensionality increasing* (SDI) index suited for *K*-NN queries on dimensionally reduced data.

Chapter 4, "Multiple Sequence Alignment and Clustering with Dot Matrices, Entropy, and Genetic Algorithms," by John Tsiligaridis, presents a set of algorithms and their efficiency for *Multiple Sequence Alignment* (MSA) and *clustering problems*, including also solutions in distributive environments with Hadoop. The strength, the adaptability, and the effectiveness of the *genetic algorithms* (GAs) for both problems are pointed out. MSA is among the most important tasks in computational biology. In biological sequence comparison, emphasis is given to the simultaneous alignment of several sequences. GAs are stochastic approaches for efficient and robust search that can play a significant role for MSA and clustering. The *divide-and-conquer* principle ensures undisturbed consistency during vertical sequences' segmentations. Indeed, the *divide-and-conquer method* (DCGA) can provide a solution for MSA utilizing appropriate cut points. As far as clustering is concerned, the aim is to divide the objects into clusters so that the validity inside clusters is minimized. As an internal measure for cluster validity, the *sum of squared error* (SSE) is used. A *clustering genetic algorithm with the SSE criterion* (CGA_SSE), a hybrid approach, using the most popular algorithm, the *K-means*, is presented. The CGA_SSE combines local and global search procedures. Comparison of the *K*-means and CGA_SSE is provided in terms of the accuracy and quality of the solution for clusters of different sizes and densities. The complexity of all proposed algorithms is examined. The Hadoop for the distributed environment provides an alternate solution to the CGA_SSE, following the MapReduce paradigm. Simulation results are provided.

The second section (i.e., "Big Data Processing") is organized into the following chapters.

Chapter 5, "Approaches for High-Performance Big Data Processing: Applications and Challenges," by Ouidad Achahbar, Mohamed Riduan Abid, Mohamed Bakhouya, Chaker El Amrani, Jaafar Gaber, Mohammed Essaaidi, and Tarek A. El Ghazawi, puts emphasis on social media websites, such as *Facebook*, *Twitter*, and *YouTube*, and job posting websites like *LinkedIn* and *CareerBuilder*, which involve a huge amount of data that are very useful for economy assessment and society development. These sites provide sentiments and interests of people connected to web communities and a lot of other information. The Big Data collected from the web is considered an unprecedented source to fuel data processing and business intelligence. However, collecting, storing, analyzing, and processing these Big Data as quickly as possible creates new challenges for both scientists and analytics. For example, analyzing Big Data from social media is now widely accepted by many companies as a way of testing the acceptance of their products and services based on customers' opinions. *Opinion mining* or *sentiment analysis methods* have been recently proposed for extracting positive/negative words from Big Data. However, highly accurate and timely processing and analysis of the huge amount of data to extract their meaning requires new processing techniques. More precisely, a technology is needed to deal with the massive amounts of unstructured and semistructured information in order to understand hidden user behavior. Existing solutions are time consuming given the increase in data volume and complexity. It is possible to use high-performance computing technology to accelerate data processing through MapReduce ported to cloud computing. This will allow companies to deliver more business value to their end customers in the dynamic and changing business environment. This chapter discusses approaches proposed in literature and their use in the cloud for Big Data analysis and processing.

Chapter 6, "The Art of Scheduling for Big Data Science," by Florin Pop and Valentin Cristea, moves the attention to applications that generate Big Data, like social networking and social influence programs, cloud applications, public websites, scientific experiments and simulations, data warehouses, monitoring platforms, and e-government services. Data grow rapidly, since applications produce continuously increasing volumes of both unstructured and structured data. The impact on data processing, transfer, and storage is the need to reevaluate the approaches and solutions to better answer user needs. In this context, scheduling models and algorithms have an important role. A large variety of solutions for specific applications and platforms exist, so a thorough and systematic analysis of existing solutions for *scheduling models*, *methods*, and *algorithms* used in Big Data processing and storage environments has high importance. This chapter presents the best of existing solutions and creates an overview of current and near-future trends. It highlights, from a research perspective, the performance and limitations of existing solutions and offers an overview of the current situation in the area of scheduling and resource management related to Big Data processing.

Chapter 7, "Time–Space Scheduling in the MapReduce Framework," by Zhuo Tang, Ling Qi, Lingang Jiang, Kenli Li, and Keqin Li, focuses on *the significance of Big Data*, that is, analyzing people's behavior, intentions, and preferences in the growing and popular social networks and, in addition to this, processing data with nontraditional structures and exploring their meanings. Big Data is often used to describe a company's large amount of unstructured and semistructured data. Using analysis to create these data in a relational database for downloading will require too much time and money. Big Data analysis and cloud computing are often linked together because real-time analysis of large data requires a framework similar to MapReduce to assign work to hundreds or even thousands of computers. After several years of criticism, questioning, discussion, and speculation, Big Data finally ushered in the era belonging to it. Hadoop presents MapReduce as an analytics engine, and under the hood, it uses a distributed storage layer referred to as the Hadoop Distributed File System (HDFS). As an open-source implementation of MapReduce, Hadoop is, so far, one of the most successful realizations of large-scale data-intensive cloud computing platforms. It has been realized that when and where to start the reduce tasks are the key problems in enhancing MapReduce performance. In this so-delineated context, the chapter proposes a framework for supporting *time–space scheduling* in MapReduce. For *time scheduling*, a *self-adaptive reduce task scheduling policy* for reduce tasks' start times in the Hadoop platform is proposed. It can decide the start time point of each reduce task dynamically according to each job context, including the task completion time and the size of the map output. For *space scheduling*, suitable algorithms are released, which synthesize the network locations and sizes of reducers' partitions in their scheduling decisions in order to mitigate network traffic and improve MapReduce performance, thus achieving several ways to avoid scheduling delay, scheduling skew, poor system utilization, and low degree of parallelism.

Chapter 8, "GEMS: Graph Database Engine for Multithreaded Systems," by Alessandro Morari, Vito Giovanni Castellana, Oreste Villa, Jesse Weaver, Greg Williams, David Haglin, Antonino Tumeo, and John Feo, considers the specific case of organizing, managing, and analyzing massive amounts of data in several contexts, like social network analysis, financial risk management, threat detection in complex network systems, and medical and biomedical databases. For these areas, there is a problem not only in terms of size but also in terms of performance, because the processing should happen sufficiently fast to be useful. *Graph databases* appear to be a good candidate to manage these data: They provide an efficient data structure for heterogeneous data or data that are not themselves rigidly structured. However, exploring large-scale graphs on modern high-performance machines is challenging. These systems include processors and networks optimized for regular, floating-point intensive computations and large, batched data transfers. At the opposite, exploring graphs generates fine-grained, unpredictable memory and network accesses, is mostly memory bound, and is synchronization intensive. Furthermore, graphs often are difficult to partition, making their processing prone to load unbalance. Following this evidence, the chapter describes *Graph Engine for Multithreaded Systems* (GEMS), a full software stack that implements a graph database on a commodity cluster and enables scaling in data set size while maintaining a constant query throughput when adding more cluster nodes. The GEMS software stack comprises a SPARQL-to-data parallel C++ compiler, a library of distributed data structures, and a custom, multithreaded, runtime system. Also, an evaluation of GEMS on a typical SPARQL benchmark and on a Resource Description Format (RDF) data set is proposed.

Chapter 9, "KSC-net: Community Detection for Big Data Networks," by Raghvendra Mall and Johan A.K. Suykens, demonstrates the applicability of the *kernel spectral clustering* (KSC) method for community detection in Big Data networks, also providing a practical exposition of the KSC method on large-scale synthetic and real-world networks with up to $10^6$ nodes and $10^7$ edges. The KSC method uses a primal–dual framework to construct a model on a smaller subset of the Big Data network. The original large-scale kernel matrix cannot fit in memory. So smaller subgraphs using a *fast and unique representative subset* (FURS) selection technique is selected. These subsets are used for training and validation, respectively, to build the model and obtain the model parameters. It results in a powerful out-of-sample extensions property, which allows inferring of the community affiliation for unseen nodes. The KSC model requires a *kernel function*, which can have kernel parameters and what is needed to identify the number of clusters *k* in the network. A memory-efficient and computationally efficient model selection technique named *balanced angular fitting* (BAF) based on angular similarity in the eigenspace was proposed in the literature. Another parameter-free KSC model was proposed as well. Here, the model selection technique exploits the structure of projections in eigenspace to automatically identify the number of clusters and suggests that a normalized linear kernel is sufficient for networks with millions of nodes. This model selection technique uses the concept of entropy and balanced clusters for identifying the number of clusters *k*. In the scope of this context literature, the chapter describes the software *KSC-net*, which obtains the representative subset by FURS, builds the KSC model, performs one of the two (BAF and parameter-free) model selection techniques, and uses out-of-sample extensions for community affiliation for the Big Data network.

Chapter 10, "Making Big Data Transparent to the Software Developers' Community," by Yu Wu, Jessica Kropczynski, and John M. Carroll, investigates the *open-source software* (OSS) development community, which has allowed technology to progress at a rapid pace around the globe through shared knowledge, expertise, and collaborations. The broad-reaching open-source movement bases itself on a share-alike principle that allows anybody to use or modify software, and upon completion of a project, its source code is made publically available. Programmers who are a part of this community contribute by voluntarily writing and exchanging code through a collaborative development process in order to produce high-quality software. This method has led to the creation of popular software products including *Mozilla Firefox*, *Linux*, and *Android*. Most OSS development activities are carried out online through formalized platforms (such as *GitHub*), incidentally creating a vast amount of interaction data across an ecosystem of platforms that can be used not only to characterize open-source development work activity more broadly but also to create Big Data awareness resources for OSS developers. The intention of these awareness resources is to enhance the ability to seek out much-needed information necessary to produce high-quality software in this unique environment that is not conducive to ownership or profits. Currently, it is problematic that interconnected resources are archived across stand-alone websites. Along these research lines, this chapter describes the process through which these resources can be *more conspicuous* through Big Data in *three interrelated sections* about the context and issues of the collaborating process in online space and a fourth section on how *Big Data can be obtained and utilized.*

The third section (i.e., "Big Data Stream Techniques and Algorithms") is organized into the following chapters.

Chapter 11, "Key Technologies for Big Data Stream Computing," by Dawei Sun, Guangyan Zhang, Weimin Zheng, and Keqin Li, focuses on the two main mechanisms for *Big Data computing*, that is, *Big Data stream computing* (*BDSC*) and *Big Data batch computing*. BDSC is a model of straight-through computing, such as *Twitter Storm* and *Yahoo! S4*, which does for stream computing what Hadoop does for batch computing, while Big Data batch computing is a model of storing and then computing, such as the MapReduce framework, open-sourced by the Hadoop implementation. Essentially, Big Data batch computing is not sufficient for many real-time application scenarios, where a data stream changes frequently over time and the latest data are the most important and most valuable. For example, when analyzing data from real-time transactions (e.g., financial trades, e-mail messages, user search requests, sensor data tracking), a data stream grows monotonically over time as more transactions take place. Ideally, a real-time application environment can be supported by BDSC. In this specific applicative setting, this chapter introduces data stream graphs and the system architecture for BDSC and key technologies for BDSC systems. Among other contributions, the authors present the system architecture and key technologies of four popular example BDSC systems, that is, *Twitter Storm*, *Yahoo! S4*, *Microsoft TimeStream*, and *Microsoft Naiad*.

Chapter 12, "Streaming Algorithms for Big Data Processing on Multicore Architecture," by Marat Zhanikeev, studies Hadoop and MapReduce, which are well known as *de facto standards* in Big Data processing today. Although they are two separate technologies, they form a single package as far as *Big Data processing*—not just storage—is concerned. This chapter treats them as one package, according to the depicted vision. Today, Hadoop and/or MapReduce lacks popular alternatives. Hadoop solves the practical problem of not being able to store Big Data on a single machine by distributing the storage over *multiple nodes*. MapReduce is a framework on which one can run *jobs that process the contents of the storage*—also in a distributed manner—and generate statistical summaries. The chapter shows that performance improvements mostly target MapReduce. There are several fundamental problems with MapReduce. First, the *map* and *reduce* operators are restricted to key–value hashes (data type, not hash function), which places a cap on usability. For example, while the *data streaming* is a good alternative for Big Data processing, MapReduce fails to accommodate the necessary data types. Second, MapReduce jobs create heterogeneous environments where jobs compete for the same resource with no guarantee of fairness. Finally, MapReduce jobs lack *time awareness*, while some algorithms might need to process data in their time sequence or using a time window. The core premise of this chapter is to replace MapReduce with a *time-aware storage and processing logic*. Big Data is replayed along the timeline, and all the jobs get the time-ordered sequence of data items. The major difference here is that the new method collects all the jobs in one place—the node that replays data—while MapReduce sends jobs to remote nodes so that data can be processed locally. This architecture is chosen for the sole purpose of accommodating a wide range of data streaming algorithms and the data types they create.

Chapter 13, "Organic Streams: A Unified Framework for Personal Big Data Integration and Organization Towards Social Sharing and Individualized Sustainable Use," by Xiaokang Zhou and Qun Jin, moves the attention to the rapid development of *emerging computing paradigms*, which are often applied to our dynamically changed work, life, playing and learning in the highly developed information society, a kind of seamless integration of the real physical world and cyber digital space. More and more people have been accustomed to sharing their personal contents across the social networks due to the high accessibility of social media along with the increasingly widespread adoption of wireless mobile computing devices. User-generated information has spread more widely and quickly and provided people with opportunities to obtain more knowledge and information than ever before, which leads to an explosive increase of data scale, containing big potential value for individual, business, domestic, and national economy development. Thus, it has become an increasingly important issue to sustainably manage and utilize *personal Big Data* in order to mine useful insight and real value to better support information seeking and knowledge discovery. To deal with this situation in the Big Data era, a *unified approach to aggregation and integration of personal Big Data from life logs* in accordance with individual needs is considered essential and effective, which can benefit the sustainable information sharing and utilization process in the social networking environment. Based on this main consideration, this chapter introduces and defines a new concept of *organic stream*, which is designed as a flexibly extensible data carrier, to provide a simple but efficient means to formulate, organize, and represent personal Big Data. As an abstract data type, organic streams can be regarded as a *logic metaphor*, which aims to meaningfully process the raw stream data into an associatively and methodically organized form, but no concrete implementation for physical data structure and storage is defined. Under the conceptual model of organic streams, a heuristic method is proposed and applied to extract diversified individual needs from the tremendous amount of social stream data through social media. And an integrated mechanism is developed to aggregate and integrate the relevant data together based on individual needs in a meaningful way, and thus personal data can be physically stored and distributed in private personal clouds and logically represented and processed by a set of newly introduced, metaphors named heuristic stone, associative drop, and associative ripple. The architecture of the system with the foundational modules is described, and the prototype implementation with the experiment's result is presented to demonstrate the usability and effectiveness of the framework and system.

Chapter 14, "Managing Big Trajectory Data: Online Processing of Positional Streams," by Kostas Patroumpas and Timos Sellis, considers *location-based services*, which have become all the more important in social networking, mobile applications, advertising, traffic monitoring, and many other domains, following the proliferation of smartphones and GPS-enabled devices. Managing the locations and trajectories of numerous people, vehicles, vessels, commodities, and so forth must be efficient and robust since this information must be processed online and should provide answers to users' requests in real time. In this *geo-streaming* context, such long-running continuous queries must be repeatedly evaluated against the most recent positions relayed by moving objects, for instance, reporting which people are now moving in a specific area or finding friends closest to the current location of a mobile user. In essence, modern processing engines must cope with huge amounts of streaming, transient, uncertain, and heterogeneous spatiotemporal data, which can be characterized as *big trajectory data*. Inspired by this methodological trend, this chapter examines Big Data processing techniques over *frequently updated locations* and *trajectories of moving objects*. Indeed, the Big Data issues regarding *volume*, *velocity*, *variety*, and *veracity* also arise in this case. Thus, authors foster a *close synergy* between the established stream processing paradigm and spatiotemporal properties inherent in motion features. Taking advantage of the spatial locality and temporal timeliness that characterize each trajectory, the authors present methods and heuristics that address such problems.

The fourth section (i.e., "Big Data Privacy") is organized into the following chapters.

Chapter 15, "Personal Data Protection Aspects of Big Data," by Paolo Balboni, focuses on *specific legal aspects of managing and processing Big Data* by also providing a relevant vision on European privacy and data protection laws. In particular, the analysis considers applicable EU data protection provisions and their impact on both businesses and consumers/data subjects, and it introduces and conceptually assesses a methodology to determine whether (1) data protection law applies and (2) personal Big Data can be (further) processed (e.g., by way of analytic software programs). Looking into more detail, this chapter deals with diverse aspects of data protection, providing an understanding of Big Data from the perspective of personal data protection using the *Organization for Economic Co-operation and Development*'s four-step life cycle of personal data along the value chain, paying special attention to the concept of compatible use. Also, the author sheds light on the development of the concept of *personal data* and its relevance in terms of data processing. Further focus is placed on aspects such as *pseudoanonymization*, *anonymous data*, and *reidentification*. Finally, conclusions and recommendations that focus on the privacy implications of Big Data processing and the importance of strategic data protection compliance management are illustrated.

Chapter 16, "Privacy-Preserving Big Data Management: The Case of OLAP," by Alfredo Cuzzocrea, highlights the *security and privacy of Big Data repositories* as among the most challenging topics in *Big Data research*. As a relevant instance, the author considers the case of *cloud systems*, which are very popular now. Here, cloud nodes are likely to exchange data very often. Therefore, the *privacy breach risk* arises, as distributed data repositories can be accessed from a node to another one, and hence, *sensitive information* can be inferred. Another relevant data management context for Big Data research is represented by the issue of effectively and efficiently supporting *data warehousing and online analytical processing (OLAP) over Big Data*, as multidimensional data analysis paradigms are likely to become an "enabling technology" for *analytics over Big Data*, a collection of models, algorithms, and techniques oriented to extract useful knowledge from cloud-based Big Data repositories for decision-making and analysis purposes. At the convergence of the three axioms introduced (i.e., security and privacy of Big Data, data warehousing and OLAP over Big Data, analytics over Big Data), a critical research challenge is represented by the issue of *effectively and efficiently computing privacy-preserving OLAP data cubes over Big Data*. It is easy to foresee that this problem will become more and more important in future years as it not only involves relevant theoretical and methodological aspects, not all explored by actual literature, but also regards significant modern scientific applications. Inspired by these clear and evident trends, this chapter moves the attention to privacy-preserving OLAP data cubes over Big Data and provides two kinds of contributions: (1) a complete survey of privacy-preserving OLAP approaches available in literature, with respect to both *centralized* and *distributed environments*, and (2) an innovative framework that relies on *flexible sampling-based data cube compression techniques for computing privacy-preserving OLAP aggregations on data cubes*.

The fifth section (i.e., "Big Data Applications") is organized into the following chapters.

Chapter 17, "Big Data in Finance," by Taruna Seth and Vipin Chaudhary, addresses Big Data in the context of the *financial industry*, which has always been driven by data. Today, Big Data is prevalent at various levels in this field, ranging from the financial services sector to capital markets. The availability in Big Data in this domain has opened up new avenues for innovation and has offered immense opportunities for growth and sustainability. At the same time, it has presented several new challenges that must be overcome to gain the maximum value from it. Indeed, in recent years, the financial industry has seen an upsurge of interest in Big Data. This comes as no surprise to finance experts, who understand the potential value of data in this field and are aware that no industry can benefit more from Big Data than the financial services industry. After all, the industry not only is driven by data but also thrives on data. Today, the data, characterized by the four *V*s, which refer to volume, variety, velocity, and veracity, are prevalent at various levels of this field, ranging from capital markets to the financial services industry. Also, capital markets have gone through an unprecedented change, resulting in the generation of massive amounts of high-velocity and heterogeneous data. For instance, about 70% of US equity trades today are generated by high-frequency trades (HFTs) and are machine driven. In the so-delineated context, this chapter considers the impact and applications of Big Data in the financial domain. It examines some of the key advancements and transformations driven by Big Data in this field. The chapter also highlights important Big Data challenges that remain to be addressed in the financial domain.

Chapter 18, "Semantic-Based Heterogeneous Multimedia Big Data Retrieval," by Kehua Guo and Jianhua Ma, considers *multimedia retrieval*, an important technology in many applications such as web-scale multimedia search engines, mobile multimedia search, remote video surveillance, automation creation, and e-government. With the widespread use of multimedia documents, our world will be swamped with multimedia content such as massive images, videos, audio, and other content. Therefore, traditional multimedia retrieval has been switching into a *Big Data environment*, and the research into solving some problems according to the features of *multimedia Big Data retrieval* attracts considerable attention. Having as reference the so-delineated application setting, this chapter proposes a heterogeneous multimedia Big Data retrieval framework that can achieve good retrieval accuracy and performance. The authors begin by addressing the particularity of heterogeneous multimedia retrieval in a Big Data environment and introducing the background of the topic. Then, literature related to current multimedia retrieval approaches is briefly reviewed, and the general concept of the proposed framework is introduced briefly. The authors provide in detail a description of this framework, including semantic information extraction, representation, storage, and multimedia Big Data retrieval. Finally, the proposed framework's performance is experimentally evaluated against several multimedia data sets.

Chapter 19, "Topic Modeling for Large-Scale Multimedia Analysis and Retrieval," by Juan Hu, Yi Fang, Nam Ling, and Li Song, similarly puts emphasis on the *exponential growth of multimedia data* that occurred in recent years, with the arrival of the Big Data era and thanks to the rapid increase in processor speed, cheaper data storage, prevalence of digital content capture devices, as well as the flooding of social media like *Facebook* and *YouTube*. New data generated each day have reached 2.5 quintillion bytes as of 2012. Particularly, more than 10 h of videos are uploaded onto *YouTube* every minute, and millions of photos are available online every week. The explosion of multimedia data in social media raises a great demand for developing effective and efficient *computational tools* to facilitate producing, analyzing, and retrieving large-scale multimedia content. *Probabilistic topic models* prove to be an effective way to organize large volumes of text documents, while much fewer related models are proposed for other types of unstructured data such as multimedia content, partly due to the high computational cost. With the emergence of cloud computing, *topic models* are expected to become increasingly applicable to multimedia data. Furthermore, the growing demand for a deep understanding of multimedia data on the web drives the development of sophisticated machine learning methods. Thus, it is greatly desirable to develop topic modeling approaches to multimedia applications that are consistently effective, highly efficient, and easily scalable. Following this methodological scheme, this chapter presents a review of topic models for large-scale multimedia analysis and shows the current challenges from various perspectives by presenting a comprehensive overview of related work that addresses these challenges. Finally, the chapter discusses several research directions in the field.

Chapter 20, "Big Data Biometrics Processing: A Case Study of an Iris Matching Algorithm on Intel Xeon Phi," by Xueyan Li and Chen Liu, investigates the applicative setting of *Big Data biometrics repositories*. Indeed, the authors recognize that, with the drive towards achieving higher computation capability, the most advanced computing systems have been adopting alternatives from the traditional *general purpose processors* (GPPs) as their main components to better prepare for Big Data processing. *NVIDIA's graphic processing units* (GPUs) have powered many of the top-ranked supercomputer systems since 2008. In the latest list published by *Top500.org*, two systems with *Intel Xeon Phi* coprocessors have claimed positions 1 and 7. While it is clear that the need to improve efficiency for Big Data processing will continuously drive changes in hardware, it is important to understand that these new systems have their own advantages as well as limitations. The required effort from the researchers to port their codes onto the new platforms is also of great significance. Unlike other coprocessors and accelerators, the *Intel Xeon Phi* coprocessor does not require learning a new programming language or new parallelization techniques. It presents an opportunity for the researchers to share parallel programming with the GPP. This platform follows the standard parallel programming model, which is familiar to developers who already work with *x86*-based parallel systems. From another perspective, with the rapidly expanded biometric data collected by various sources for identification and verification purposes, how to manage and process such Big Data draws great concern. On the one hand, biometric applications normally involve comparing a huge amount of samples and templates, which has strict requirements on the computational capability of the underlying hardware platform. On the other hand, the number of cores and associated threads that hardware can support has increased greatly; an example is the newly released *Intel Xeon Phi* coprocessor. Hence, Big Data biometrics processing demands the execution of the applications at a higher parallelism level. Taking an *iris matching algorithm* as a case study, the authors propose an *OpenMP* version of the algorithm to examine its performance on the *Intel Xeon Phi* coprocessor. Their target is to evaluate their parallelization approach and the influence from the optimal number of threads, the impact of thread-to-core affinity, and the built-in vector engine. This does not mean that achieving good performance on this platform is simple. The hardware, while presenting many similarities with other existing multicore systems, has its own characteristics and unique features. In order to port the code in an efficient way, those aspects are fully discussed in the chapter.

Chapter 21, "Storing, Managing, and Analyzing Big Satellite Data: Experiences and Lessons Learned from a Real-World Application," by Ziliang Zong, realizes how Big Data has shown great capability in yielding extremely useful information and extraordinary potential in revolutionizing scientific discoveries and traditional commercial models. Indeed, numerous corporations have started to utilize Big Data to understand their customers' behavior at a fine-grained level, rethink their business process work flow, and increase their productivity and competitiveness. Scientists are using Big Data to make new discoveries that were not possible before. As the volume, velocity, variety, and veracity of Big Data keep increasing, significant challenges with respect to innovative Big Data management, efficient Big Data analytics, and low-cost Big Data storage solutions arise. This chapter provides a case study on how the *big satellite data* (at the petabyte level) of the world's largest satellite imagery distribution system is captured, stored, and managed by the *National Aeronautics and Space Administration* (NASA) and the *US Geological Survey* (USGS) and gives a unique example of how a changed policy could significantly affect the traditional ways of Big Data storage and distribution, which will be quite different from typical commercial cases driven by sales. Also, the chapter discusses how the USGS *Earth Resources Observation and Science* (EROS) center swiftly overcomes the challenges from serving few government users to hundreds of thousands of global users and how *data visualization* and *data mining* techniques are used to analyze the characteristics of millions of requests and how they can be used to improve the performance, cost, and energy efficiency of the EROS system. Finally, the chapter summarizes the experiences and lessons learned from conducting the target Big Data project in the past 4 years.

Chapter 22, "Barriers to the Adoption of Big Data Applications in the Social Sector," by Elena Strange, focuses on the *social aspects of dealing with Big Data*. The author recognizes that effectively working with and leveraging Big Data has the potential to change the world. Indeed, if there is a ceiling on realizing the benefits of *Big Data algorithms, applications, and techniques*, we have not yet reached it. The research field is maturing rapidly. No longer are we seeking to understand what "Big Data" is and whether it is useful. No longer is Big Data processing the province of niche computer science research. Rather, the concept of Big Data has been widely accepted as important and inexorable, and the buzzwords "Big Data" have found their way beyond computer science into the essential tools of business, government, and media. Tools and algorithms to leverage Big Data have been increasingly democratized over the last 10 years. By 2010, over 100 organizations reported using the distributed file system and framework *Hadoop*. Early adopters leveraged *Hadoop* on in-house *Beowulf clusters* to process tremendous amounts of data. Today, well over 1000 organizations use *Hadoop*. That number is climbing and now includes companies with a range of technical competencies and those with and without access to internal clusters and other tools. Yet the benefits of Big Data have not been fully realized by businesses, governments, and particularly the social sector. In this so-delineated background, this chapter describes the impact of this gap on the social sector and the broader implications engendered by the sector in a broader context. Also, the chapter highlights the opportunity gap—the unrealized potential of Big Data in the social sector—and explores the historical limitations and context that have led up to the current state of Big Data. Finally, it describes the current perceptions of and reactions to Big Data algorithms and applications in the social sector and offers some recommendations to accelerate the adoption of Big Data.

Overall, this book represents a solid research contribution to state-of-the-art studies and practical achievements in algorithms, analytics, and applications on Big Data, and sets the basis for further efforts in this challenging scientific field that will, more and more, play a leading role in next-generation database, data warehousing, data mining, and cloud computing research. The editors are confident that this book will represent an authoritative milestone in this very challenging scientific road.

# Editors

**Kuan-Ching Li** is a professor in the Department of Computer Science and Information Engineering at Providence University, Taiwan. Dr. Li is a recipient of awards from NVIDIA, the Ministry of Education (MOE)/Taiwan, and the Ministry of Science and Technology (MOST)/Taiwan. He also received guest professorships at universities in China, including Xiamen University (XMU), Huazhong University of Science and Technology (HUST), Lanzhou University (LZU), Shanghai University (SHU), Anhui University of Science and Technology (AUST), and Lanzhou Jiaotong University (LZJTU). He has been involved actively in conferences and workshops as a program/general/steering conference chairman and in numerous conferences and workshops as a program committee member, and he has organized numerous conferences related to high-performance computing and computational science and engineering.

Dr. Li is the Editor in Chief of the technical publications *International Journal of Computational Science and Engineering (IJCSE)*, *International Journal of Embedded Systems (IJES)*, and *International Journal of High Performance Computing and Networking (IJHPCN)*, all published by Interscience. He also serves on a number of journals' editorial boards and guest editorships. In addition, he has been acting as editor/coeditor of several technical professional books published by CRC Press and IGI Global. His topics of interest include networked computing, GPU computing, parallel software design, and performance evaluation and benchmarking. Dr. Li is a member of the Taiwan Association of Cloud Computing (TACC), a senior member of the IEEE, and a fellow of the IET.

**Hai Jiang** is an associate professor in the Department of Computer Science at Arkansas State University, United States. He earned a BS at Beijing University of Posts and Telecommunications, China, and MA and PhD degrees at Wayne State University. His research interests include parallel and distributed systems, computer and network security, high-performance computing and communication, Big Data, and modeling and simulation. He has published one book and several research papers in major international journals and conference proceedings. He has served as a US National Science Foundation proposal review panelist and a US Department of Energy (DoE) Smart Grid Investment Grant (SGIG) reviewer multiple times. He serves as an editor for the *International Journal of High Performance Computing and Networking* (*IJHPCN*); a regional editor for the *International Journal of Computational Science and Engineering* (*IJCSE*) as well as the *International Journal of Embedded Systems* (*IJES*); an editorial board member for the *International Journal of Big Data Intelligence* (*IJBDI*), the *Scientific World Journal* (*TSWJ*), the *Open Journal of Internet of Things* (*OJIOT*), and the *GSTF Journal on Social Computing* (*JSC*); and a guest editor for the *IEEE Systems Journal*, *International Journal of Ad Hoc and Ubiquitous Computing*, *Cluster Computing*, and *The Scientific World Journal* for multiple special issues. He has also served as a general chair or program chair for some major conferences/workshops and involved in 90 conferences and workshops as a session chair or as a program committee member. He has reviewed six cloud computing–related books (*Distributed and Cloud Computing*, *Virtual Machines*, *Cloud Computing: Theory and Practice*, *Virtualized Infrastructure and Cloud Services Management*, *Cloud Computing: Technologies and Applications Programming*, *The Basics of Cloud Computing*) for publishers such as Morgan Kaufmann, Elsevier, and Wiley. He serves as a review board member for a large number of international journals. He is a professional member of ACM and the IEEE Computer Society. Locally, he serves as US NSF XSEDE (Extreme Science and Engineering Discovery Environment) Campus Champion for Arkansas State University.

**Professor Laurence T. Yang** is with Department of Computer Science at St. Francis Xavier University, Canada. His research includes parallel and distributed computing, embedded and ubiquitous/pervasive computing, cyber–physical–social systems, and Big Data.

He has published 200+ refereed international journal papers in the above areas; about 40% are in IEEE/ACM transactions/journals and the rest mostly are in Elsevier, Springer, and Wiley journals. He has been involved in conferences and workshops as a program/general/steering conference chair and as a program committee member. He served as the vice chair of the IEEE Technical Committee of Supercomputing Applications (2001–2004), the chair of the IEEE Technical Committee of Scalable Computing (2008–2011), and the chair of the IEEE Task Force on Ubiquitous Computing and Intelligence (2009–2013). He was on the steering committee of the IEEE/ACM Supercomputing Conference series (2008–2011) and on the National Resource Allocation Committee (NRAC) of Compute Canada (2009–2013).

In addition, he is the editor in chief and editor of several international journals. He is the author/coauthor or editor/coeditor of more than 25 books. *Mobile Intelligence* (Wiley 2010) received an Honorable Mention by the American Publishers Awards for Professional and Scholarly Excellence (the PROSE Awards). He has won several best paper awards (including IEEE best and outstanding conference awards, such as the IEEE 20th International Conference on Advanced Information Networking and Applications [IEEE AINA-06]), one best paper nomination, the Distinguished Achievement Award (2005, 2011), and the Canada Foundation for Innovation Award (2003). He has given 30 keynote talks at various international conferences and symposia.

**Alfredo Cuzzocrea** is a senior researcher at the Institute of High Performance Computing and Networking of the Italian National Research Council, Italy, and an adjunct professor at the University of Calabria, Italy. He is habilitated as an associate professor in computer science engineering by the Italian National Scientific Habilitation of the Italian Ministry of Education, University and Research (MIUR). He also obtained habilitation as an associate professor in computer science by the Aalborg University, Denmark, and habilitation as an associate professor in computer science by the University of Rome Tre, Italy. He is an adjunct professor at the University of Catanzaro "Magna Graecia," Italy, the University of Messina, Italy, and the University of Naples "Federico II," Italy. Previously, he was an adjunct professor at the University of Naples "Parthenope," Italy. He holds 35 visiting professor positions worldwide (Europe, United States, Asia, and Australia). He serves as a Springer Fellow Editor and as an Elsevier Ambassador. He holds several roles in international scientific societies, steering committees for international conferences, and international panels, some of them having directional responsibility. He serves as a panel leader and moderator in international conferences. He was an invited speaker in several international conferences worldwide (Europe, United States, and Asia). He is a member of scientific boards of several PhD programs worldwide (Europe and Australia). He serves as an editor for the Springer series *Communications in Computer and Information Science*. He covers a large number of roles in international journals, such as editor in chief, associate editor, and special issue editor (including *JCSS, IS, KAIS, FGCS, DKE, INS,* and *Big Data Research*). He has edited more than 30 international books and conference proceedings. He is a member of the editorial advisory boards of several international books. He covers a large number of roles in international conferences, such as general chair, program chair, workshop chair, local chair, liaison chair, and publicity chair (including ODBASE, DaWaK, DOLAP, ICA3PP, ICEIS, APWeb, SSTDM, IDEAS, and IDEAL). He served as the session chair in a large number of international conferences (including EDBT, CIKM, DaWaK, DOLAP, and ADBIS). He serves as a review board member for a large number of international journals (including *TODS, TKDE, TKDD, TSC, TIST, TSMC, THMS, JCSS, IS, KAIS, FGCS, DKE,* and *INS*). He also serves as a review board member in a large number of international books and as a program committee member for a large number of international conferences (including VLDB, ICDE, EDBT, CIKM, IJCAI, KDD, ICDM, PKDD, and SDM). His research interests include multidimensional data modeling and querying, data stream modeling and querying, data warehousing and OLAP, OLAM, XML data management, web information systems modeling and engineering, knowledge representation and management models and techniques, Grid and P2P computing, privacy and security of very large databases and OLAP data cubes, models and algorithms for managing uncertain and imprecise information and knowledge, models and algorithms for managing complex data on the web, and models and algorithms for high-performance distributed computing and architectures. He is the author or coauthor of more than 330 papers in international conferences (including EDBT, CIKM, SSDBM, MDM, DaWaK, and DOLAP), international journals (including *JCSS, IS, KAIS, DKE,* and *INS*), and international books (mostly edited by Springer). He is also involved in several national and international research projects, where he also covers responsibility roles.

# Contributors

**Mohamed Riduan Abid**
Al Akhawayn University
Ifrane, Morocco

**Ouidad Achahbar**
Al Akhawayn University
Ifrane, Morocco

**Mohamed Bakhouya**
International University of Rabat
Sala el Jadida, Morocco

**Paolo Balboni**
ICT Legal Consulting
Milan, Italy

and

European Privacy Association
Brussels, Belgium

**John M. Carroll**
College of Information Sciences and Technology
Pennsylvania State University
State College, Pennsylvania

**Vito Giovanni Castellana**
Pacific Northwest National Laboratory
Richland, Washington

**Vipin Chaudhary**
Department of Computer Science and Engineering
University at Buffalo (SUNY)
Buffalo, New York

**Valentin Cristea**
Computer Science Department
Faculty of Automatic Control and Computers
University Politehnica of Bucharest
Bucharest, Romania

**Alfredo Cuzzocrea**
ICAR-CNR
and
University of Calabria
Rende, Italy

**Chaker El Amrani**
Université Abdelmalek Essaadi, Tangier
Tangier, Morocco

**Tarek A. El Ghazawi**
George Washington University
Washington, DC

**Mohammed Essaaidi**
Ecole Nationale Supérieure d'Informatique et d'Analyse des Systemes
Agdal Rabat, Morocco

**Yi Fang**
Department of Computer Engineering
Santa Clara University
Santa Clara, California

**John Feo**
Pacific Northwest National Laboratory
Richland, Washington

**Jaafar Gaber**
Universite de Technologie de Belfort-Montneliard
Belfort, France

**Ian Gorton**
Software Engineering Institute
Carnegie Mellon University
Pittsburgh, Pennsylvania

**Kehua Guo**
School of Information Science and Engineering
Central South University
Changsha, China

**David Haglin**
Pacific Northwest National Laboratory
Richland, Washington

**Juan Hu**
Department of Computer Engineering
Santa Clara University
Santa Clara, California

**Lingang Jiang**
College of Information Science and Engineering
Hunan University
Changsha, China

**Qun Jin**
Faculty of Human Sciences
Waseda University
Tokorozawa-shi, Japan

**Jessica Kropczynski**
College of Information Sciences and Technology
Pennsylvania State University
State College, Pennsylvania

**Kenli Li**
College of Information Science and Engineering
Hunan University
Changsha, China

**Keqin Li**
College of Information Science and Engineering
Hunan University
Changsha, China

and

Department of Computer Science
State University of New York
New Paltz, New York

**Xueyan Li**
Department of Electrical and Computer Engineering
Clarkson University
Potsdam, New York

**Nam Ling**
Department of Computer Engineering
Santa Clara University
Santa Clara, California

**Chen Liu**
Department of Electrical and Computer Engineering
Clarkson University
Potsdam, New York

**Yan Liu**
Faculty of Computer Science and Engineering
Concordia University
Montreal, Canada

**Jianhua Ma**
Faculty of Computer and Information Sciences
Hosei University
Tokyo, Japan

**Raghvendra Mall**
KU Leuven–ESAT/STADIUS
Leuven, Belgium

**Stéphane Marchand-Maillet**
Viper Group, Computer Science Department
University of Geneva
Geneva, Switzerland

**Hisham Mohamed**
Viper Group, Computer Science Department
University of Geneva
Geneva, Switzerland

**Alessandro Morari**
Pacific Northwest National Laboratory
Richland, Washington

**Kostas Patroumpas**
National Technical University of Athens
Athens, Greece

**Florin Pop**
Computer Science Department
Faculty of Automatic Control and Computers
University Politehnica of Bucharest
Bucharest, Romania

**Ling Qi**
College of Information Science and Engineering
Hunan University
Changsha, China

**Timos Sellis**
RMIT University
Melbourne, Australia

**Taruna Seth**
Department of Computer Science and Engineering
University at Buffalo (SUNY)
Buffalo, New York

**Li Song**
Institute of Image Communication and Information Processing
Shanghai Jiao Tong University
Shanghai, China

**Elena Strange**
University of Memphis
Memphis, Tennessee

**Dawei Sun**
Department of Computer Science and Technology
Tsinghua University
Beijing, China

**Johan A.K. Suykens**
KU Leuven–ESAT/STADIUS
Leuven, Belgium

**Zhuo Tang**
College of Information Science and Engineering
Hunan University
Changsha, China

**Alexander Thomasian**
Thomasian & Associates
Pleasantville, New York

**John Tsiligaridis**
Math and Computer Science Department
Heritage University
Toppenish, Washington

**Antonino Tumeo**
Pacific Northwest National Laboratory
Richland, Washington

**Oreste Villa**
NVIDIA Research
Santa Clara, California

**Jesse Weaver**
Pacific Northwest National Laboratory
Richland, Washington

**Greg Williams**
Rensselaer Polytechnic Institute
Troy, New York

**Xing Wu**
Department of Electrical and Computer Engineering
Concordia University
Montreal, Canada

**Yu Wu**
College of Information Sciences and Technology
Pennsylvania State University
State College, Pennsylvania

**Guangyan Zhang**
Department of Computer Science and Technology
Tsinghua University
Beijing, China

**Marat Zhanikeev**
Department of Artificial Intelligence
Computer Science and Systems Engineering
Kyushu Institute of Technology
Fukuoka, Japan

**Weimin Zheng**
Department of Computer Science and Technology
Tsinghua University
Beijing, China

**Xiaokang Zhou**
Faculty of Human Sciences
Waseda University
Tokorozawa-shi, Japan

**Ziliang Zong**
Texas State University
San Marcos, Texas

# I

## Big Data Management

CHAPTER 1

# Scalable Indexing for Big Data Processing

Hisham Mohamed and Stéphane Marchand-Maillet

CONTENTS

## ABSTRACT

The $K$-nearest neighbor ($K$-NN) search problem is the way to find and predict the closest and most similar objects to a given query. It finds many applications for information retrieval and visualization, machine learning, and data mining. The context of Big Data imposes the finding of approximate solutions. Permutation-based indexing is one of the most recent techniques for approximate similarity search in large-scale domains. Data objects are represented by a list of references (pivots), which are ordered with respect to their distances from the object. In this chapter, we show different distributed algorithms for efficient indexing and searching based on permutation-based indexing and evaluate them on big high-dimensional data sets.

## 1.1 INTRODUCTION

Similarity search [1] aims to extract the most similar objects to a given query. It is a fundamental operation for many applications, such as genome comparison to find all the similarities between one or more genes and multimedia retrieval to find the most similar picture or video to a given example.

Similarity search is known to be achievable in at least three ways. The first technique is called *exhaustive search*. For a given query, the distance between the query and the database object is calculated while keeping track of which objects are similar (near) to the query. The main problem with this technique is that it does not scale with large collections. The second technique is *exact search*. Using space decomposition techniques, the number of objects that are compared to the query is reduced. This technique is not efficient for high-dimensional data sets, due to the curse of dimensionality [2]. As the dimensions increase, a large part of the database needs to be scanned. Hence, the performance becomes similar to that of exhaustive search. The third technique is *approximate search*. It reduces the amount of data that needs to be accessed by using some *space partitioning* (generally done using pivots) [3–5] or *space transformation* techniques [6–10]. It provides fast and scalable response time while accepting some imprecision in the results. The most common techniques for that are the locality-sensitive hashing (LSH) [7], FastMap [6], and M-tree [8]. A survey can be found in Reference 11. Recently, a new technique based on permutations is proposed, which is called *permutation-based indexing* (PBI) [12].

The idea of PBI was first proposed in Reference 12. It relies on the selection and ordering of a set of reference objects (pivots) by their distance to every database object. Order permutations are used as encoding to estimate the real ordering of distance values between the submitted query and the objects in the database (more details in Section 1.2). The technique is efficient; however, with the exponential increase of data, parallel and distributed computations are needed.

In this chapter, we review the basics of PBI and show its efficiency on a distributed environment. Section 1.2 introduces the PBI model and its basic implementation. Section 1.3 shows a review of the related data structures that are based on PBI. Section 1.4 introduces indexing and searching algorithms for PBI on distributed architectures. Finally, we present our evaluation on large-scale data sets in Section 1.5 and conclude in Section 1.6.

## 1.2 PERMUTATION-BASED INDEXING

### 1.2.1 Indexing Model

**Definition 1**

*Given a set of N objects $o_i$, $D = \{o_1,\ldots,o_N\}$ in m-dimensional space, a set of reference objects $R = \{r_1,\ldots,r_n\} \subset D$, and a distance function that follows the metric space postulates, we define the ordered list of R relative to $o \in D$, $L(o,R)$, as the ordering of elements in R with respect to their increasing distance from o:*

$$L(o,R)=\{r_{i_1},\ldots,r_{i_n}\} \text{ such that } d(o,r_{i_j}) \le d(o,r_{i_{j+1}}) \forall j=1,\ldots,n-1$$

*Then, for any $r \in R$, $L(o,R)_{|r}$ indicates the position of r in $L(o,R)$. In other words, $L(o,R)_{|r} = j$ such that $r_{i_j} = r$. Further, given $\tilde{n} > 0$, $\tilde{L}(o,R)$ is the* pruned ordered list *of the $\tilde{n}$ first elements of $L(o,R)$.*

Figure 1.1b and c gives the ordered lists $L(o,R)$ ($n = 4$) and the pruned ordered lists $\tilde{L}(o,R)$ ($\tilde{n} = 2$) for $D$ and $R$ illustrated in Figure 1.1a.

In $K$-nearest neighbor ($K$-NN) similarity queries, we are interested in ranking objects (to extract the $K$ first elements) and not in the actual interobject distance values. PBI relaxes distance calculations by assuming that they will be approximated in terms of their ordering when comparing the ordered lists of objects. Spearman footrule distance ($d_{SFD}$) is considered to define the similarity between ordered lists. Formally,

$$d(q,o_i) \overset{\text{rank}}{\simeq} d_{SFD}(q,o_i) = \sum_{j=1}^{n} \left| \tilde{L}(q,R)_{|r_j} - \tilde{L}(o_i,R)_{|r_j} \right|. \tag{1.1}$$

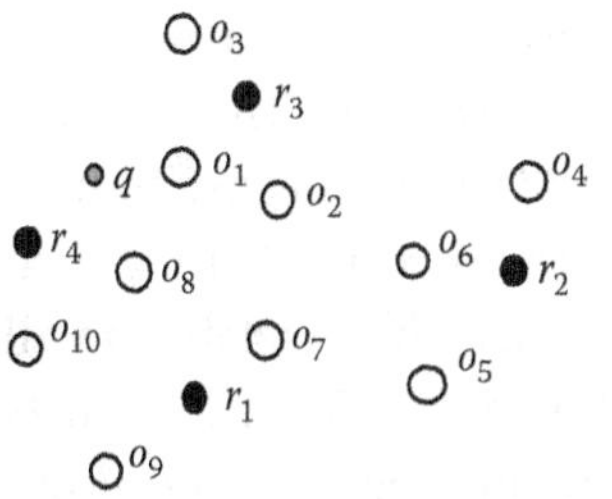

(a)

(b)

$L(o_1,R) = (r_3,r_4,r_1,r_2)$ $L(o_2,R) = (r_3,r_2,r_4,r_1)$
$L(o_3,R) = (r_3,r_4,r_1,r_2)$ $L(o_4,R) = (r_2,r_3,r_1,r_4)$
$L(o_5,R) = (r_2,r_1,r_3,r_4)$ $L(o_6,R) = (r_2,r_3,r_1,r_4)$
$L(o_7,R) = (r_1,r_2,r_4,r_3)$ $L(o_8,R) = (r_4,r_1,r_3,r_2)$
$L(o_9,R) = (r_1,r_4,r_2,r_3)$ $L(o_{10},R) = (r_4,r_1,r_3,r_2)$
$L(q,R) = (r_4,r_3,r_1,r_2)$

(c)

$L(o_1,R) = (r_3,r_4)$ $L(o_2,R) = (r_3,r_2)$
$L(o_3,R) = (r_3,r_4)$ $L(o_4,R) = (r_2,r_3)$
$L(o_5,R) = (r_2,r_1)$ $L(o_6,R) = (r_2,r_3)$
$L(o_7,R) = (r_1,r_2)$ $L(o_8,R) = (r_4,r_1)$
$L(o_9,R) = (r_1,r_4)$ $L(o_{10},R) = (r_4,r_1)$
$L(q,R) = (r_4,r_3)$

FIGURE 1.1 (a) White circles are data objects $o_i$; black circles are reference objects $r_j$; the gray circle is the query object $q$. (b) Ordered lists $L(o_i,R)$, $n = 4$. (c) Pruned ordered lists $\tilde{L}(o_i,R)$, $\tilde{n} = 2$.

To efficiently answer users' queries, the authors of References 13 through 17 proposed several strategies to decrease the complexity of the $d_{SFD}$, which we discuss in detail in Section 1.3.

### 1.2.2 Technical Implementation

Algorithm 1 details the basic indexing process for PBI. For each object in $D$, the distance $d(o_i,r_j)$ with all the references in the reference set $R$ is calculated (lines 1–4). After sorting the distances in increasing order using the suitable sorting algorithm (*quicksort* is used), the full ordered list $L(o_i,R)$ is created (line 5). In line 6, partial lists $\tilde{L}(o_i,R)$ are generated by choosing the top $\tilde{n}$ references from $L(o_i,R)$. In line 7, $\tilde{L}(o_i,R)$ are stored in the appropriate data structure [13–17]. Theoretically, the sorting complexity is $O(n\log n)$, which leads to $O(N(n + n\log n))$ indexing complexity.

**Algorithm 1: Permutation-Based Indexing**

IN: $D$ of size $N$, $R$ of $n$ and $\tilde{n} \leq n$
OUT: $\tilde{L}(o_i,R) \forall i = 1,\ldots,N-1$
1. For $o_i \in D$
2.     For $r_j \in R$
3.         $b[j].dis = d(o_i,r_j)$
4.         $b[j].indx = j$
5.     $L(o_i, R) = quicksort(b, n)$
6.     $\tilde{L}(o_i,R) = partiallist(L(o_i,R),\tilde{n})$
7.     Store the ID $i$ of $o_i$ and its $\tilde{L}(o_i,R)$ for other processing.

## 1.3 RELATED DATA STRUCTURES

### 1.3.1 Metric Inverted Files

Amato and Savino [14] and Amato et al. [17] introduced the metric inverted file (MIF) as a disk-based data structure to store the permutations in an inverted file [18]. Inverted files are the backbone of all current text search engines. They are composed of two main components, which are the *vocabulary* and the *inverted list*. The *vocabulary* contains all the unique words in a text corpus. For each vocabulary, there is an *inverted list*. The inverted list contains all the positions in which this vocabulary word existed in the corpus. To map that to the PBI, the reference objects act as the vocabulary, and the inverted list for each reference object stores the object ID and the ranking of this reference object relative to all objects in the database $\left(o_i,\tilde{L}(o_i,R)_{|r_j}\right)$. Figure 1.2a shows the MIF for the partial ordered lists in Figure 1.1c. SFD (Equation 1.1) is used to estimate the similarity between the query and the database objects. For a given query $q$, an accumulator is assigned to each object $o_i$ and initialized to zero. The inverted list for each reference point is accessed, and the accumulator is updated by adding the difference between the position of the current reference point in the ordered list of the query and the position of the reference point in the ordered

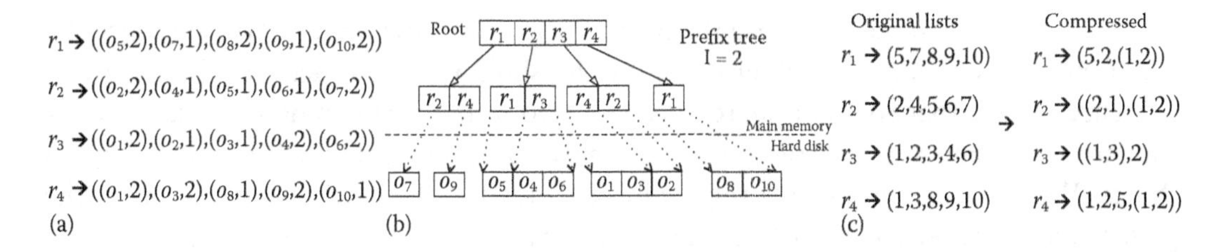


FIGURE 1.2 (a) Metric inverted file. (b) Prefix permutation index. (c) Neighborhood approximation.

list of the objects (saved in the inverted list). After checking the inverted lists of all the reference points, the objects are sorted based on their accumulator value. Objects with a small accumulator value are more similar to the query. Later, the authors of Reference 19 proposed a distributed implementation of the MIF.

### 1.3.2 Brief Permutation Index

To decrease the memory usage, the authors of Reference 20 proposed the brief permutation index. The algorithm proposes a smaller representation of the ordered lists in order to save memory. After generating the ordered lists $L(o_i,R)$, each list is encoded using a fixed-size bit string. Once a query is received, its order list is also encoded in the same way and compared with the vectors using the binary Hamming distances [21]. In Reference 22, the authors proposed a version with LSH, but the recall drops when the speed increases.

### 1.3.3 Prefix Permutation Index

In Reference 23, the authors proposed a new technique called the *prefix permutation index* (PP-index). The algorithm aims to decrease the memory usage by reducing the number of references and performing some direct distance calculation (DDC) with a portion from the database. PP-index only stores a prefix of the permutations of length $l$ (where $l < n$) in the main memory in a trie tree, and the data objects are saved in blocks on the hard disk. Objects that share the same prefix are saved close to each other. Figure 1.2b shows the PP-index for the partial ordered lists in Figure 1.1c. Once a query is submitted, all the objects in the block sharing the same prefix are ranked based on their distance to the query. The PP-index requires less memory than the previous data structures and uses the permutation prefix in order to quickly retrieve the relevant objects, in a hash-like manner. The proposed technique initially gives a lower recall value. An improvement is made using multiple queries. When a query is received, the permutation list for the query is computed. Different combinations of the query permutation list are then submitted to the tree as new queries.

### 1.3.4 Neighborhood Approximation

In Reference 16, the authors proposed *neighborhood approximation* (NAPP). The main idea of the NAPP is to get the objects that are located in the intersection region of reference objects shared between the query and database objects. Hence, if there is a query $q$, the similar objects to the query are the objects where $\tilde{L}(o_i,R) \cap \tilde{L}(q,R) \neq \phi$. To reduce the complexity, they proposed another term, the threshold value $t$. It defines the number of reference objects that could be shared between two ordered lists in order to define their similarity. The data structure related to this algorithm is similar to the MIF except for two things. The first difference is that the $t$ value is defined in the indexing, and it affects the performance. The second difference is the data compression on the inverted lists for saving storage. The indexing time of this technique is high especially due to the compressing time. The searching time and efficiency are based on the indexing technique and the threshold value. Figure 1.2c shows the NAPP for the partial ordered lists in Figure 1.1c.

### 1.3.5 Metric Suffix Array

In References 24 and 25, the authors further proposed the metric suffix array (MSA) as a memory-based fast and effective data structure for storing the permutations. An MSA $\Psi$ acts like a suffix array [26–28] over $S$. More specifically, $\Psi$ is a set of $M$ integers corresponding to the permutation induced by the lexical ordering of all $M$ suffixes in $S\left(\left\{r_{i_k},\ldots,r_{i_M}\right\}\forall k \leq M\right)$. The MSA is sorted into buckets. A bucket $b_j$ contains the positions of all the suffixes of $S$ of the form $\left\{r_j,\ldots,r_{i_M}\right\}$, that is, where reference point $r_j$ appears first. Figure 1.3a shows the MSA for the partial ordered lists in Figure 1.1c. At query time, $\tilde{L}(q,R)$ is computed, and buckets $b_j$ are scanned for all $r_j \in \tilde{L}(q,R)$. From each suffix value $\Psi_k$ in a bucket $b_j$, the object ID $i = \left\lfloor \frac{\Psi_k}{\tilde{n}} \right\rfloor + 1$ and the location of the reference point in the ordered list of the object are identified, $\tilde{L}(o_i,R)_{|r_j} = (\Psi_k \bmod \tilde{n}) + 1$. Similar to the MIF, the objects are ranked based on their accumulator value using Equation 1.1.

### 1.3.6 Metric Permutation Table

In Reference 13, the authors proposed the metric permutation table (MPT). The main idea is to simplify further the distance approximation by quantizing the ranks within the ordered lists into $B \leq \tilde{n}$ intervals (buckets). A bucket $B$ is defined

$$b_{ij} = \left\lceil \frac{B}{\tilde{n}} \cdot \tilde{L}(o_i,R)_{|r_j} \right\rceil \quad \text{and} \quad b_{qj} = \left\lceil \frac{B}{\tilde{n}} \cdot \tilde{L}(q,R)_{|r_j} \right\rceil \tag{1.2}$$

| | $o_1$ | | $o_2$ | | $o_3$ | | $o_4$ | | $o_5$ | | $o_6$ | | $o_7$ | | $o_8$ | | $o_9$ | | $o_{10}$ | |
|---|---|---|---|---|---|---|---|---|---|---|---|---|---|---|---|---|---|---|---|---|
| $\tilde{L}(o_i,R)_{\mid rj}$ | 1 | 2 | 1 | 2 | 1 | 2 | 1 | 2 | 1 | 2 | 1 | 2 | 1 | 2 | 1 | 2 | 1 | 2 | 1 | 2 |
| S: | $r_3$ | $r_4$ | $r_3$ | $r_2$ | $r_3$ | $r_4$ | $r_2$ | $r_3$ | $r_2$ | $r_1$ | $r_2$ | $r_3$ | $r_1$ | $r_2$ | $r_4$ | $r_1$ | $r_1$ | $r_4$ | $r_4$ | $r_1$ |
| Index: | 1 | 2 | 3 | 4 | 5 | 6 | 7 | 8 | 9 | 10 | 11 | 12 | 13 | 14 | 15 | 16 | 17 | 18 | 19 | 20 |

| | $\text{buk}_{r1}$ | | | | | $\text{buk}_{r2}$ | | | | | $\text{buk}_{r3}$ | | | | | $\text{buk}_{r4}$ | | | | |
|---|---|---|---|---|---|---|---|---|---|---|---|---|---|---|---|---|---|---|---|---|
| MSA $\Psi$: | 10 | 13 | 16 | 17 | 20 | 4 | 7 | 9 | 11 | 14 | 1 | 3 | 5 | 8 | 12 | 2 | 6 | 15 | 18 | 19 |

(a)

| | $r_1$ | | $r_2$ | | $r_3$ | | $r_4$ | | ← $R$ |
|---|---|---|---|---|---|---|---|---|---|
| Memory | 1 | 2 | 1 | 2 | 1 | 2 | 1 | 2 | ← Buckets |
| Hard disk | $o_7$ $o_9$ | $o_5$ $o_8$ $o_{10}$ | $o_4$ $o_5$ $o_6$ | $o_2$ $o_7$ | $o_1$ $o_2$ $o_3$ | $o_4$ $o_6$ | $o_8$ $o_1$ 0 | $o_1$ $o_3$ $o_9$ | |

(b)

FIGURE 1.3 (a) Metric suffix array. (b) Metric permutation table.

as a regular quantization scheme. Equation 1.1 now transforms into

$$d(q,o_i) \overset{\text{rank}}{\simeq} \sum_{\substack{r_j \in \tilde{L}(q,R) \\ r_j \in \tilde{L}(o_i,R)}} |b_{qj} - b_{ij}|. \tag{1.3}$$

For each reference, $B$ buckets are allocated. Each bucket contains a list of objects $z$, which contains the IDs of the objects that are located in the same bucket. Figure 1.3b shows the MIF for the partial ordered lists in Figure 1.1c. Then, these lists are compressed using *delta encoding* [29].

For searching, Equation 1.3 counts the discrepancy in quantized ranking from common reference objects between each object and the query. That is, each object $o_i$ scores

$$s_i = \left|\left\{r_j \in \tilde{L}(q,R) \text{ such that } (r_j \in \tilde{L}(o_i,R) \text{ and } |b_{ij} - b_{qj}| \leq 1)\right\}\right|. \tag{1.4}$$

Objects $o_i$ are then sorted according to their decreasing $s_i$ scores. This approximate ranking can be improved by DDC. For a $K$-NN query, we use a DDC on the $K_c = \Delta.K$ first objects in our sorted candidate list and call $\Delta > 1$ the *DDC factor*.

## 1.4 DISTRIBUTED INDEXING

In this section, we show three different strategies to handle the PBI on a distributed environment. The first strategy is based on data division (Section 1.4.1), the second strategy is based on reference division (Section 1.4.2), and the third strategy is based on data and reference division (Section 1.4.3).

### 1.4.1 Data Based

#### *1.4.1.1 Indexing*

Given a data domain $D$ of $N$ objects, a reference set $R$ of $n$ points, and $P$ processes, the reference set $R$ is shared between all the processes. On the other hand, the data domain $D$ is randomly divided into subdomains $D_1 \ldots D_p$ of equal sizes $\frac{N}{P}$ among the processes. Each process $p$ builds the partial ordered lists $\tilde{L}(o_i, R)$ for all $o_i \in D_p$.

The ordered lists for these objects are saved locally in the appropriate data structure $DP_p$ for each process. Accordingly, the workload of indexing is divided equally between the processes. Using this procedure, the complexity of indexing is reduced from $O(N(n + n\log n))$ to $O\left(\frac{N(n+n\log n)}{P}\right)$.

#### *1.4.1.2 Searching*

The data are equally divided among all the processes. Accordingly, when a query $q$ is submitted, all the processes have to respond to this query. Algorithm 2 shows the searching process. There is a broker process (line 1). This broker process is responsible for handling users' requests. Once a query is received, the query is broadcasted to all the processes, which we call the *workers* (line 2). Each worker creates the $\tilde{L}(q,R)$ and searches through its local partial data structure $DS_p$ (lines 3–6). Once the search is finished, the list of the objects from each process $OL_p$ is sent back to the broker (line 7). The broker organizes the received results lists $OL_1,\ldots,OL_P$ and sends them to the user $OL$ (lines 8–11). The organization of the results is simple. The similarity between the query and the objects in any data domain is identified using the full same reference list. Hence, the organization of the lists consists of merging the lists and sorting them based on the similarity score. This similarity score is defined based on the searching algorithm that is used.

**Algorithm 2: Data-Based Approach: Searching**

IN: Data structure $DS_p$, references $R$, and query: $q$
OUT: Ranked output list $OL$
1. if(*broker*):
2. Broadcast($q$)
3. else:
4. Recv. $q$ from the *broker*
5. Create the query partial ordered list $\tilde{L}(q,R)$
6. $OL_p$ = search in $DS_p$ using $\tilde{L}(q,R)$
7. Send $OL_p$ to the *broker*
8. if(*broker*)
9. For $p \in P$
10. Recv($OL_p$)
11. Create $OL$ and send the results to the user.

### 1.4.2 Reference Based

#### *1.4.2.1 Indexing*

Similar to the previous technique, the data domain is divided into $P$ subdomains $D_1\ldots D_P$. On the other hand, the references have a different distribution depending on the role of the processes. Each process has two roles (*indexer* and *coordinator*). When the processes are in the indexing role, they share all the reference information $R$. When they are in the coordinating role, the reference set is divided between the processes. Each process has a subset of the references $R_p \subset R$ and $|R_p| = \frac{n}{P}$. Algorithm 3 shows the indexing process. All the processes start in the indexing role. For each object $o_i \in D_p$, the partial ordered list $\tilde{L}(o_i,R)$ is created using the full reference set $R$ (lines 1–7). Afterwards, the locations of the closest reference points in the partial ordered list of each object $\tilde{L}(o_i,R)_{|r_j}$, the reference ID $j$, and the related object global ID are sent to the corresponding coordinating process (line 8–12). A corresponding coordinating process is a process $p$, where $r_j \in R_p$. Accordingly, each process $p$ is responsible for a group of objects, which have the related set of references $R_p$ as the closest references. Formally, a set of objects $o_i$ belonging to process $p$ is characterized as

$$\{o_i \text{ such that } r_j \in \tilde{L}(o_i, R) \text{ and } r_j \in R_p\}.$$

That means that the data are divided between the processes based on the references, not randomly as in the previous technique. The complexity of indexing is $O\left(\frac{N(n+n\log n)}{P}\right)+t_1$, where $t_1$ is the time needed to transfer the object information from the indexing process to the coordinating process.

**Algorithm 3: Reference-Based Approach: Indexing**

IN: $D_p$ of size $\frac{N}{P}$, $R$ of $n$, $R_p$ of $\frac{n}{P}$, $\tilde{n}$
OUT: Data structure $DS_p$
1. if(*indexer*)
2. For $o_i \in D_p$
3. For $r_j \in R$
4. $b[j].dis = d(o_i, r_j)$
5. $b[j].indx = j$
6. $L(o_i,R) = quicksort(b,n)$
7. $\tilde{L}(o_i,R) = partiallist(L(o_i,R),\tilde{n})$
8. For $r_c \in \tilde{L}(o_i,R)$
9. Send the global object ID, the $\tilde{L}(o_i,R)|_{r_c}$, and the reference ID $r_c$ to coordinating process.
10. if(*coordinator*)
11. Recv. data from any indexer.
12. Store the received data in $DS_p$.

*1.4.2.2 Searching*

Unlike in the previous search scenario, the processes that participate to answer the query are the processes that have the references, which are located in $\tilde{L}(q,R)$. Algorithm 4 shows the searching process. Once a broker process receives a query, the partial ordered list is created $\tilde{L}(q,R)$ (line 2). An active process is a process whose references are located in $\tilde{L}(q,R)$. More formally, a process $p$ is defined as active if there exists $r_j \in R_p$, such that $r_j \in \tilde{L}(q,R)$. The broker notifies the active processes and sends the position of the related references $\tilde{L}(q,R)_{r_j}$ to them (lines 3–4). Each active process searches in its local partial data structure and send the results back to the broker (lines 5–9). The broker organizes the received results and sends them to the user (lines 10–13). The organization of the results is not simple. In this approach, each process defines the similarity between the query and the related objects based on a partial reference list $R_p$. That means that the same object can be seen by different processes, so the object can be found in each output list by each process with a partial similarity score. Hence, the broker has to add these partial similarity scores that are related to each object and then sort the objects based on the summed similarity score.

**Algorithm 4: Reference-Based Approach: Searching**

IN: Data structure *DS*, references *R* and query: $q$
OUT: Output list *OL*
1. if(*broker*):
2. Create the query partial ordered list $\tilde{L}(q,R)$.
3. Notify the active processes: $r_j \in \tilde{L}(q,R)$ *and* $r_j \in R_p$
4. Send $\tilde{L}(q,R)_{r_j}$ to the active processes $p$
5. if(*active*):
6. Recv. $\tilde{L}(q,R)$
7. For $r_j \in \tilde{L}(q,R)$ *and* $r_j \in R_p$
8. $OL_p$ = search in $DS_p$.
9. Send $OL_p$ to the *broker*
10. if(*broker*):
11. For $p \in P$
12. Recv($OL_p$).
13. Create *OL* and send the results to the user.

### 1.4.3 Index Based

#### *1.4.3.1 Indexing*

Similar to the previous techniques, the data domain is divided into $P$ subdomains $D_1 \ldots D_p$. It can be divided randomly or based on a clustering algorithm. The references are selected independently from each subdomain and not shared with the other processes $R_p$. Once the references are defined for each portion, the processes create the partial ordered lists $\tilde{L}\left(o_{i_p}, R_p\right)$ for their own local objects $D_p$ based on their local reference sets $R_p$. Hence, with the approach, the processes do not need to communicate to each other while indexing. The indexing complexity is reduced to $O\left(\frac{N\left(n+n\log\frac{n}{P}\right)}{P^2}\right)$. The main reason for this reduction is the division of the references between all the processes.

#### *1.4.3.2 Searching*

Algorithm 5 shows the searching process. The broker process has information about all the references $R = \{R_1 \cup R_2 \cup \ldots \cup R_p\}$. Hence, once a query $q$ is submitted, the partial ordered list $\tilde{L}(q,R)$ is created using the full reference list. If the query is located within a certain data domain $D_p$, the reference points that were selected from that data domain $R_p$ are located as the closest references to the query. Hence, the searching is only done through the process that is responsible for the corresponding data domain (lines 1–4). If the query is located between two or more data domains, the closest references are shared among these different domains. Formally, a process $p$ is an active process if there exists $r_j \in R_p$, such that $r_j \in \tilde{L}(q,R)$. After searching, the results from each process cannot be sent to the broker directly. The similarity between the query and each object for each process is defined by different sets of reference points $R_p$ that are not related to each other. To combine and rank them, each process computes the distances between the submitted query and the $K_{cp} = \Delta \times k$ first objects in the candidate list of each process. We call $\Delta \geq 1$ the DDC factor. Then, this distance information with the related object IDs is sent to the broker process (lines 5–10). Finally, the broker ranks these objects based on their distances and sends the results back to the user. For example, if $\Delta = 2$, for *K-NNquery(q,30)*, each process returns the most similar $k = 30$ objects to the query. Each process searches within its local data structure and gets the top set candidate list $K_{c_p} = 2 \times 30 = 60$. Then, a DDC is performed between the query and this top $K_{c_p} = 60$ by each process. If the references of the query ordered list are located within the sets of 3 processes, the broker receives $3 \times 60 = 180$ distances. These 180 distances are sorted, and the top 30 objects are sent back to the user.

**Algorithm 5: Index-Based Approach: Searching**

IN: Data structure *DS*, references *R*, and query: $q$
OUT: Output list *OL*
1. if (*broker*):
2. Create the query partial ordered list $\tilde{L}(q,R)$.
3. Notify the active processes.
4. Send $q$ to the active processes.
5. if (*active*):
6. Recv. $q$ and create $\tilde{L}(q,R_p)$.
7. For $r_j \in \tilde{L}(q,R_p)$.
8. $OL_p$ = search in $DS_p$.
9. $d_p$ = calculate_distance($q$, $2k$, $OL_p$).
10. Send the distances $d_p$ to the broker.

## 1.5 EVALUATION

Large-scale experiments were conducted to evaluate the efficiency and the scalability of the three approaches in terms of recall (RE), position error (PE), indexing time, and searching time. The three approaches were implemented using message passing interface (MPI) [30]. The results section is divided into three subsections.

In Section 1.5.1, we measure the average RE and PE for the three distributed approaches. The experiments were performed using a data set consisting of 5 million objects. The data set is composed of visual shape features (21 dimensional) and was extracted from the 12-million ImageNet corpus [31]. Three sets of references of sizes 1000, 2000, and 3000 are selected randomly from the data set. The average RE and PE are based on 250 different queries, which are selected randomly from the data set.

In Section 1.5.2, we measure the scalability of the three approaches using the same data set in terms of searching and indexing time. We compare the implementation of the three approaches using MPI in terms of indexing and searching time. The average searching time is based on 250 different queries, which are selected randomly from the data set. For the previous experiments, the MSA data structure was used. All the evaluation including indexing and searching time is performed using the full permutation list, as we consider using the full ordered list as the worst-case scenario. That means that whatever the approach used, the searching is done by all the processes.

In Section 1.5.3, we perform large-scale experiments using the full content-based photo image retrieval (CoPhIR) data set [32] and the best distributed approach. The CoPhIR [32] data set consists of 106 million MPEG-7 features. Each MPEG-7 feature contains five descriptors (scalable color, color structure, color layout, edge histogram, and homogeneous texture). These five features are combined to build a 280 dimensional vector. The database size is about 75 GB. The evaluation was performed using a set of references of size 2000. The number of nearest points $\tilde{n}$ is 10.

MPICH2 1.4.1p1 is installed on a Linux cluster of 20 dual-core computers (40 cores in total) each holding 8 GB of memory and 512 GB of local disk storage, led by a master eight-core computer holding 32 GB of memory and a terabyte storage capacity.

### 1.5.1 Recall and Position Error

Given a query $q$, the RE and PE are defined as [11]

$$RE = \frac{|S \cap S_A|}{|S|}, \tag{1.5}$$

$$PE = \frac{\sum_{o \in S_A} |P(X,o) - P(S_A,o)|}{|S_A| \cdot N}, \tag{1.6}$$

where $S$ and $S_A$ are the ordering of the top-$K$ ranked objects to $q$ for exact similarity search and approximate similarity search, respectively. $X$ is the ordering of items in the data set $D$ with respect to their distance from $q$, $P(X,o)$ indicates the position of $o$ in $X$, and $P(S_A,o)$ indicates the position of $o$ in $S_A$. An RE = 0.5 indicates that the output results of the searching algorithm contain 50% of the $k$ best objects retrieved by the exact search. A PE = 0.001 indicates that the average shifting of the best $K$ objects ranked by the searching algorithm, with respect to the rank obtained by the exact search, is 0.1% of the data set.

Figure 1.4 shows the average RE and PE for 10 $K$-NNs using 10 and 20 processes. The figure represents four different strategies, which are as follows. The first strategy is *sequential* (*Seq*). The second strategy is the *data-based* (*DB*) approach. The third strategy is the *reference-based* (*RB*) approach. The last strategy is the *index-based* (*IB*) approach. The *DDC* factor for the *IB* approach is equal to 1 for a fair comparison.

Figure 1.4a and b show that the *Seq, DB*, and *RB* approaches give the same RE and PE. Also, the number of processes does not affect the performance. This is expected as all the

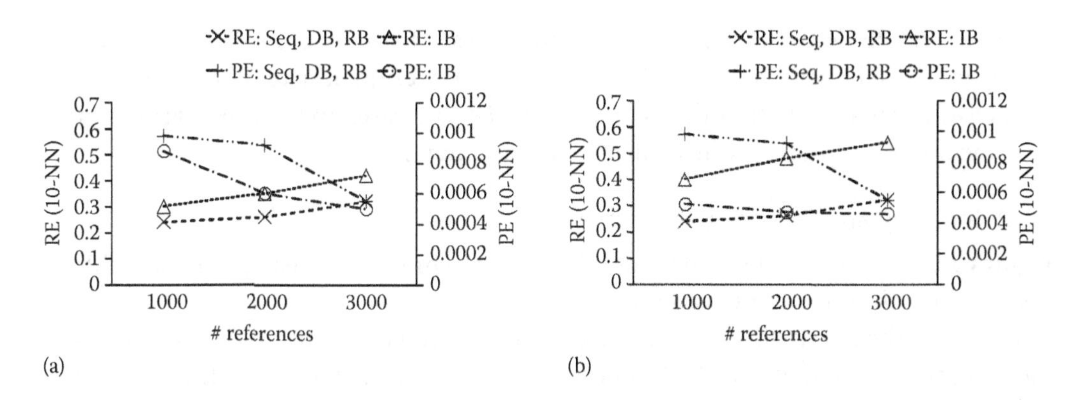


FIGURE 1.4 Average RE and PE (21 dimensional data set, MSA, 250 queries) for 10 *K*-NNs using different sets of reference points (1000, 2000, and 3000) using 10 and 20 processes. (a) RE and PE (1 DDC, 10 nodes). (b) RE and PE (1 DDC, 20 nodes).

objects, whatever their domain, are indexed using the same set of references. Hence, the *DB* and *RB* approaches give the same RE and PE as the sequential algorithm, whatever the number of processes.

The *IB* approach gives a better RE and PE compared to the other three approaches, *Seq*, *DB*, and *RB*. In addition, its efficiency increases when the number of processes increases. There are two reasons for that. First, the data domain with respect to each process is independent, which helps to decrease the search space for each process. That identifies the objects in a better way using the local references. The second reason is the DDC, between the query and the candidate list from each process. Hence, the *IB* approach gives a better recall and position error than the sequential and the other distributed implementations. In addition, a better performance is achieved when the number of processes increases. As the number of processes increases, the number of objects that are assigned to each process decreases, which improves the locality and provides a better representation of the objects, which leads to a better performance.

Figure 1.5 shows the effect of the *DDC* factor used in the *IB* approach for 10 *K*-NNs, using 20 processes and the same sets of references. It is clear from the figure that the RE

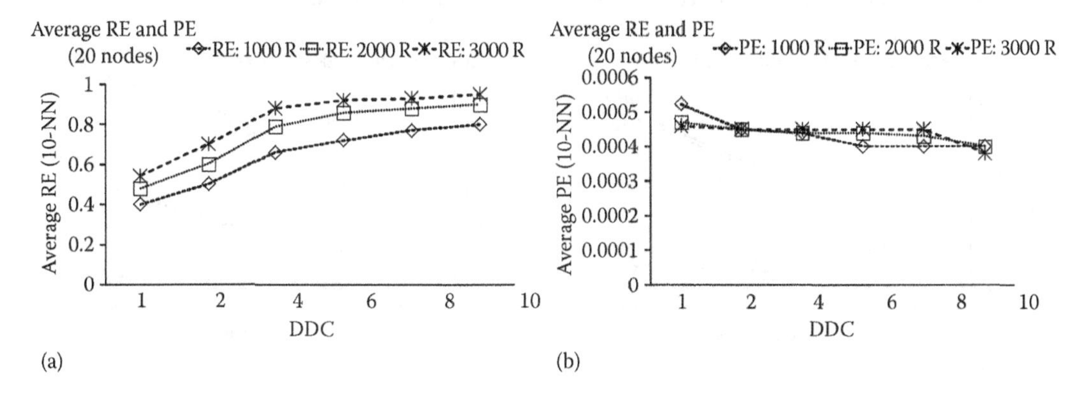


FIGURE 1.5 Effect of changing DDC for 10 *K*-NNs using 20 processes (21 dimensional data set, MSA, 250 queries). (a) Average recall. (b) Average position error.

increases and PE decreases when the *DDC* increases. Also, due to the data distribution, a better recall can be achieved with a low *DDC* value. For example, a recall of 0.9 is achieved with *DDC* equal to 4. On the other hand, when *DDC* increases, more objects are compared with the query per process, which affects the searching time.

### 1.5.2 Indexing and Searching Performance

Figures 1.6, 1.7, and 1.8 show the indexing and searching times using 5, 10, and 20 processes for the three distributed approaches.

*Indexing.* The *x*-axis shows the number of reference points, and the *y*-axis shows the indexing time in seconds for the full ordered list $L(o_i,R)$. For the three algorithms, when the number of cores increases, the indexing time decreases. Also, when the number of reference points increases, the indexing time increases. The *RB* gives a longer indexing time due to the communication required between the processes in order to exchange the data. Using five processes, the *RB* approach is slower than the sequential approach, since most of the time is consumed in exchanging the data between the processes. Exchanging object by object to the corresponding process is an expensive process. Hence, the communication is performed by chunks of size 1 MB. For the *IB* approach, there is a big drop in the indexing time compared to the sequential time. There are two reasons for that. The first reason is

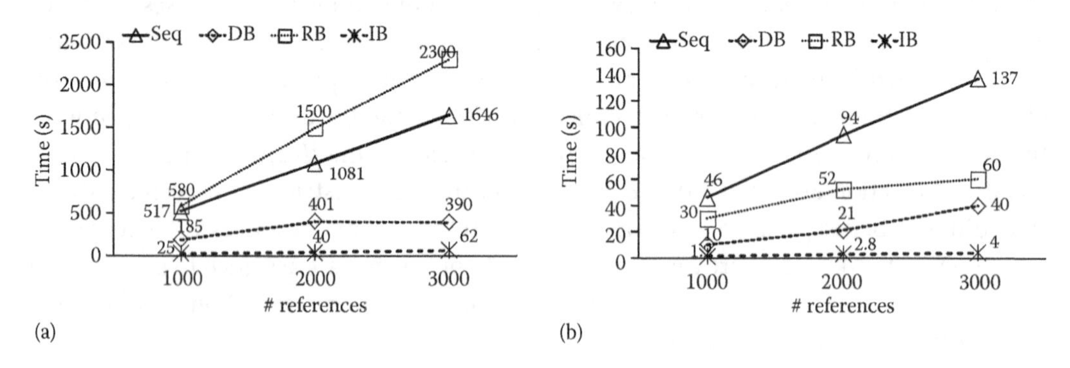


FIGURE 1.6 Indexing and searching time (21 dimensional data set, MSA) in seconds using five processes. (a) Indexing time: 5 processes. (b) Searching time: 5 processes.

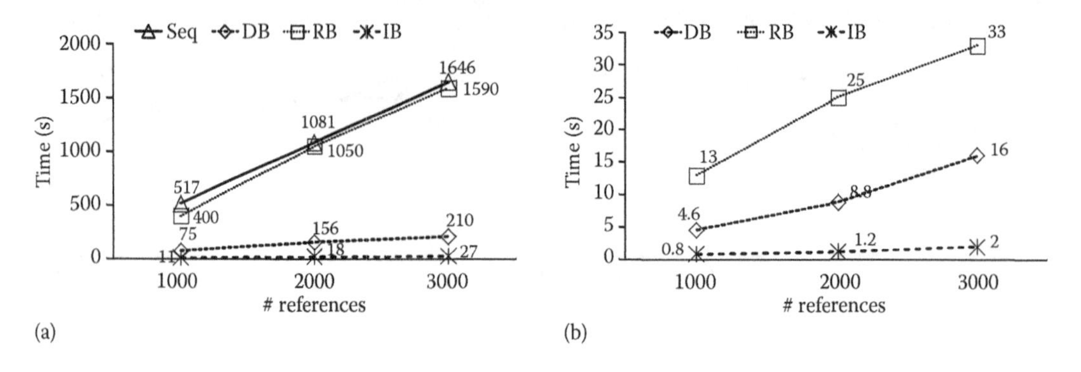


FIGURE 1.7 Indexing and searching time (21 dimensional data set, MSA) in seconds using 10 processes. (a) Indexing time: 10 processes. (b) Searching time: 10 processes.

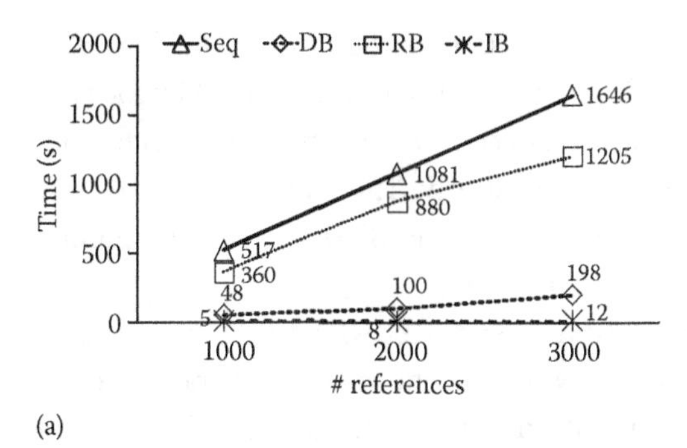


(a)

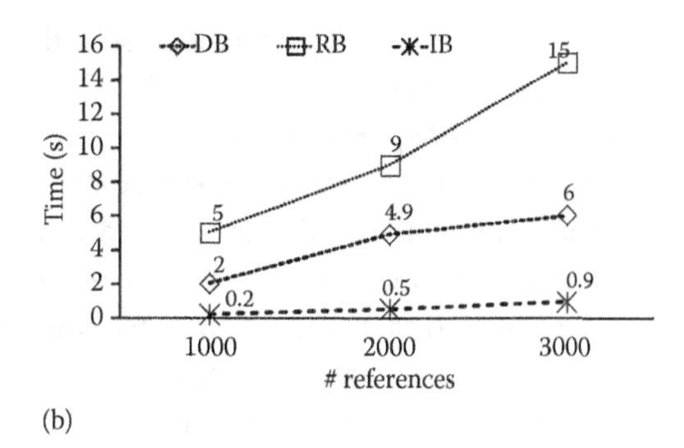


(b)

FIGURE 1.8 Indexing and searching time (21 dimensional data set, MSA) in seconds using 20 processes. (a) Indexing time: 20 processes. (b) Searching time: 20 processes.

that there is no communication time between the processes. The second reason is the division of the data and the references. That makes the indexing process within each process fast compared to the sequential one.

*Searching.* The $x$-axis shows the number of reference points and the $y$-axis shows the searching time in seconds using the full ordered list $L(o_i,R)$. For the three algorithms, when the number of processes increases, the searching time decreases, with the same RE and PE for *DB* and *RB* approaches and with a better RE and PE for the *IB* approach. Also, when the number of reference points increases, the searching time increases. For the *RB* approach, searching is very slow compared to the other two approaches. The main reason for that is the gathering process. In *RB*, all the nodes participate to answer a user's query, and each node has $\frac{n}{P}$ references, where each reference is responsible for $N$ objects as we use the full ordered list. If we have 20 processes and 20 reference points, the broker process receives 20 lists of size $N$. After that, these lists are combined and sorted. The combination process is difficult as the similarity with respect to the objects is done using a portion of the references, not the full references. Hence, this reduces the running time, which means that most of the time for the *RB* is consumed in combining the results from the processes to give the final output to the user.

The *IB* searching algorithm gave the best running time, although there is some overhead time due to the hard disk access and the DDC between the query and the top $K_{cp}$ objects.

### 1.5.3 Big Data Indexing and Searching

Table 1.1 shows the average RE, the indexing time, and the average searching time for top 10 and 100 $K$-NNs for the full CoPhIR [32] data set (106 million objects, 280 dimensional).

TABLE 1.1 Indexing and Searching the CoPhIR Data Set Using the MPT Data Structure

| **2000** | **Sequential** | | **Distributed** | |
|---|---|---|---|---|
| $K$-NN | 10 | 100 | 10 | 100 |
| RE | 0.64 | 0.7 | 0.76 | 0.86 |
| Indexing | 25 h | | 10 min | |
| Searching | 6 s | 6.5 s | 1.8 s | 2 s |

The experiments were performed using the MPT data structure (Section 1.3.6). The table shows the results for the sequential and the distributed implementation using 34 processes based on the *IB* approach. The number of nearest references is $\tilde{n} = 50$ out of 2000 reference points that were selected from each database portion randomly. The DDC factor is $\Delta = 40$ for the sequential and the distributed implementations. For the distributed implementation, the number of reference points for each processor is 60.

The table shows that the distributed implementation gives better RE and PE compared to sequential implementation for the top 10 and 100 *K*-NNs due to the locality and better representation of each object in the distributed environment. In terms of indexing, the distributed implementation is much faster than the sequential implementation due to the division of data and references. For searching, the distributed implementation did not give up high speed like the indexing, due to the number of DDCs that need to be done by each process.

## 1.6 CONCLUSION

PBI is one of the most recent techniques for approximate *K*-NN similarity queries in large-scale domains. It aims to predict the similarity between the data objects based on how they see the surroundings using some reference objects. With the exponential increase of data, parallel and distributed computations are needed. In this chapter, we showed three different distributed algorithms for efficient indexing and searching using PBI. The three algorithms are divided based on the distribution of the data or the references or both of them. We evaluated them on big high-dimensional data sets and showed that the technique, which is based on distributing the data and the references, gives the best performance due to the better representations of the objects by the local references and the low number of objects for each process.

## ACKNOWLEDGMENT

This work is jointly supported by the Swiss National Science Foundation (SNSF) via the Swiss National Center of Competence in Research (NCCR) on Interactive Multimodal Information Management (IM2) and the European Cooperation in Science and Technology (COST) Action on Multilingual and Multifaceted Interactive Information Access (MUMIA) via the Swiss State Secretariat for Education and Research (SER).

## REFERENCES

1. H. V. Jagadish, A. O. Mendelzon and T. Milo. Similarity-based queries. In *Proceedings of the Fourteenth ACM SIGACT-SIGMOD-SIGART Symposium on Principles of Database Systems*, PODS '95, pages 36–45, New York, 1995. ACM.
2. H. Samet. *Foundations of Multidimensional and Metric Data Structures*. The Morgan Kaufmann Series in *Computer Graphics and Geometric Modeling*. Morgan Kaufmann, San Francisco, CA, 2006.
3. B. Bustos, O. Pedreira and N. Brisaboa. A dynamic pivot selection technique for similarity search. In *Proceedings of the First International Workshop on Similarity Search and Applications*, pages 105–112, Washington, DC, 2008. IEEE Computer Society.
4. L. G. Ares, N. R. Brisaboa, M. F. Esteller, Ó. Pedreira and Á. S. Places. Optimal pivots to minimize the index size for metric access methods. In *Proceedings of the 2009 Second International Workshop on Similarity Search and Applications*, pages 74–80, Washington, DC, 2009. IEEE Computer Society.

5. B. Bustos, G. Navarro and E. Chávez. Pivot selection techniques for proximity searching in metric spaces. *Pattern Recognition Letters,* 24(14):2357–2366, 2003.
6. C. Faloutsos and K.-I. Lin. Fastmap: A fast algorithm for indexing, data-mining and visualization of traditional and multimedia datasets. *SIGMOD Record,* 24(2):163–174, 1995.
7. A. Andoni and P. Indyk. Near-optimal hashing algorithms for approximate nearest neighbor in high dimensions. *Communications of the ACM,* 51(1):117–122, 2008.
8. P. Ciaccia, M. Patella and P. Zezula. M-tree: An efficient access method for similarity search in metric spaces. In M. Jarke, M. J. Carey, K. R. Dittrich, F. H. Lochovsky, P. Loucopoulos and M. A. Jeusfeld, editors, *VLDB '97, Proceedings of 23rd International Conference on Very Large Data Bases, August 25–29, 1997, Athens, Greece,* pages 426–435. Morgan Kaufmann, San Francisco, CA, 1997.
9. O. Egecioglu, H. Ferhatosmanoglu and U. Ogras. Dimensionality reduction and similarity computation by inner-product approximations. *Knowledge and Data Engineering, IEEE Transactions on,* 16(6):714–726, 2004.
10. Ü. Y. Ogras and H. Ferhatosmanoglu. Dimensionality reduction using magnitude and shape approximations. In *Proceedings of the Twelfth International Conference on Information and Knowledge Management,* CIKM '03, pages 99–107, New York, 2003. ACM.
11. P. Zezula, G. Amato, V. Dohnal and M. Batko. *Similarity Search: The Metric Space Approach,* volume 32 of *Advances in Database Systems.* Springer, Secaucus, NJ, 2006.
12. E. C. Gonzalez, K. Figueroa and G. Navarro. Effective proximity retrieval by ordering permutations. *IEEE Transactions, on Pattern Analysis and Machine Intelligence,* 30(9):1647–1658, 2008.
13. H. Mohamed and S. Marchand-Maillet. Quantized ranking for permutation-based indexing. In N. Brisaboa, O. Pedreira and P. Zezula, editors, *Similarity Search and Applications,* volume 8199 of *Lecture Notes in Computer Science,* pages 103–114. Springer, Berlin, 2013.
14. G. Amato and P. Savino. Approximate similarity search in metric spaces using inverted files. In *Proceedings of the 3rd International Conference on Scalable Information Systems,* InfoScale '08, pages 28:1–28:10, ICST, Brussels, Belgium, 2008. ICST.
15. A. Esuli. Mipai: Using the PP-index to build an efficient and scalable similarity search system. In *Proceedings of the 2009 Second International Workshop on Similarity Search and Applications,* pages 146–148, Washington, DC, 2009. IEEE Computer Society.
16. E. S. Tellez, E. Chavez and G. Navarro. Succinct nearest neighbor search. *Information Systems,* 38(7):1019–1030, 2013.
17. G. Amato, C. Gennaro and P. Savino. Mi-file: Using inverted files for scalable approximate similarity search. *Multimedia Tools and Applications,* 71(3):1333–1362, 2014.
18. J. Zobel and A. Moffat. Inverted files for text search engines. *ACM Computing Surveys,* 38(2):6, 2006.
19. H. Mohamed and S. Marchand-Maillet. Parallel approaches to permutation-based indexing using inverted files. In *5th International Conference on Similarity Search and Applications (SISAP),* Toronto, CA, August 2012.
20. E. S. Tellez, E. Chavez and A. Camarena-Ibarrola. A brief index for proximity searching. In E. Bayro-Corrochano and J.-O. Eklundh, editors, *Progress in Pattern Recognition, Image Analysis, Computer Vision, and Applications,* volume 5856 *of Lecture Notes in Computer Science,* pages 529–536. Springer, Berlin, 2009.
21. R. W. Hamming. Error detecting and error correcting codes. *Bell System Technical Journal,* 29:147–160, 1950.
22. E. S. Tellez and E. Chavez. On locality sensitive hashing in metric spaces. In *Proceedings of the Third International Conference on SImilarity Search and APplications,* SISAP '10, pages 67–74, New York, 2010. ACM.
23. A. Esuli. PP-index: Using permutation prefixes for efficient and scalable approximate similarity search. *Proceedings of LSDSIR 2009,* i(July):1–48, 2009.

24. H. Mohamed and S. Marchand-Maillet. Metric suffix array for large-scale similarity search. In *ACM WSDM 2013 Workshop on Large Scale and Distributed Systems for Information Retrieval*, Rome, IT, February 2013.
25. H. Mohamed and S. Marchand-Maillet. Permutation-based pruning for approximate *K*-NN search. In *DEXA Proceedings, Part I*, Prague, Czech Republic, August 26–29, 2013, pages 40–47, 2013.
26. U. Manber and E. W. Myers. Suffix arrays: A new method for on-line string searches. *SIAM Journal on Computing*, 22(5):935–948, 1993.
27. K.-B. Schumann and J. Stoye. An incomplex algorithm for fast suffix array construction. *Software: Practice and Experience*, 37(3):309–329, 2007.
28. H. Mohamed and M. Abouelhoda. Parallel suffix sorting based on bucket pointer refinement. In *Biomedical Engineering Conference (CIBEC), 2010 5th Cairo International*, Cairo, Egypt, pages 98–102, December 2010.
29. S. W. Smith. *The Scientist and Engineer's Guide to Digital Signal Processing*. California Technical Publishing, San Diego, CA, 1997.
30. The MPI Forum. MPI: A message passing interface, 1993. Available at http://www.mpi-forum.org.
31. J. Deng, W. Dong, R. Socher, L.-J. Li, K. Li and L. Fei-Fei. ImageNet: A large-scale hierarchical image database. In *IEEE Computer Vision and Pattern Recognition (CVPR)*, 2009.
32. P. Bolettieri, A. Esuli, F. Falchi, C. Lucchese, R. Perego, T. Piccioli and F. Rabitti. CoPhIR: A test collection for content-based image retrieval. *CoRR*, abs/0905.4627v2, 2009.

CHAPTER 2

# Scalability and Cost Evaluation of Incremental Data Processing Using Amazon's Hadoop Service

Xing Wu, Yan Liu, and Ian Gorton

CONTENTS



ABSTRACT

Based on the MapReduce model and Hadoop Distributed File System (HDFS), Hadoop enables the distributed processing of large data sets across clusters with scalability and fault tolerance. Many data-intensive applications involve continuous and incremental updates of data. Understanding the scalability and cost of a Hadoop platform to handle small and independent updates of data sets sheds light on the design of scalable and cost-effective data-intensive applications. In this chapter, we introduce a motivating movie recommendation application implemented in the MapReduce model and deployed on Amazon Elastic MapReduce (EMR), a Hadoop service provided by Amazon. We present the deployment architecture with implementation details of the Hadoop application. With metrics collected by Amazon CloudWatch, we present an empirical scalability and cost evaluation of the Amazon Hadoop service on processing continuous and incremental data streams. The evaluation result highlights the potential of autoscaling for cost reduction on Hadoop services.

## 2.1 INTRODUCTION

Processing large-scale data is an increasing common and important problem for many domains [1]. The *de facto* standard programming model MapReduce [2] and the associated run-time systems were originally adopted by Google. Subsequently, an open-source platform named Hadoop [3] that supports the same programming model has gained tremendous popularity. However, MapReduce was not designed to efficiently process small and independent updates to existing data sets. This means the MapReduce jobs must be run again over both the newly updated data and the old data. Given enough computing resources, MapReduce's scalability makes this approach feasible. However, reprocessing the entire data discards the work done in earlier runs and makes latency proportional to the size of the entire data, rather than the size of an update [4].

In this chapter, we present an empirical scalability and cost evaluation of Hadoop on processing continuously and incrementally updated data streams. We first introduce the programming model of MapReduce for designing software applications. We select an application for finding the most mutual fans of a movie for recommendations using Netflix data [5]. We then discuss the implementation and deployment options available in Amazon Web Services (AWS), a public cloud environment. We design experiments to evaluate the scalability and resource usage cost of running this Hadoop application and explore the insights of the empirical results using monitoring data at both the system and the platform level.

## 2.2 INTRODUCTION OF MapReduce AND APACHE HADOOP

MapReduce is a programming model designed to process large volumes of data in parallel by dividing the work into a set of independent tasks [2]. In general, a MapReduce program processes data with three phases: *map*, *shuffle*, and *reduce*. We use a word count example to demonstrate these three phases. This MapReduce application counts the number of occurrences of each word in a large collection of documents.

Assume two text files map.txt and reduce.txt and their contents as shown in Figure 2.1. The processing of the word count application in the MapReduce model is shown in Figure 2.2. The *map* phase takes the input data and produces intermediate data tuples. Each tuple consists of a key and a value. In this example, the word occurrences of each file are counted in the *map* phase. Then in the *shuffle* phase, these data tuples are ordered and distributed to reducers by their keys. The *shuffle* phase ensures that the same reducer can process all the data tuples with the same key. Finally, during the *reduce* phase, the values of the data tuples with the same key are merged together for the final result.

Apache Hadoop is an open-source implementation of the MapReduce model [3]. Apache Hadoop implements a scalable fault-tolerant distributed platform for MapReduce programs. With the Hadoop Distributed File System (HDFS), which provides high-throughput

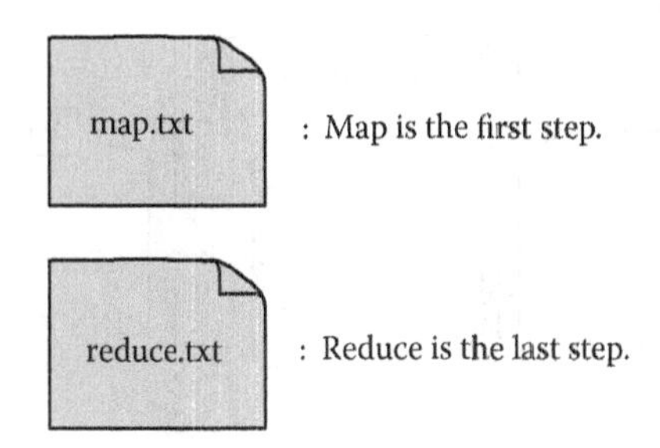


FIGURE 2.1 Word count input text files.

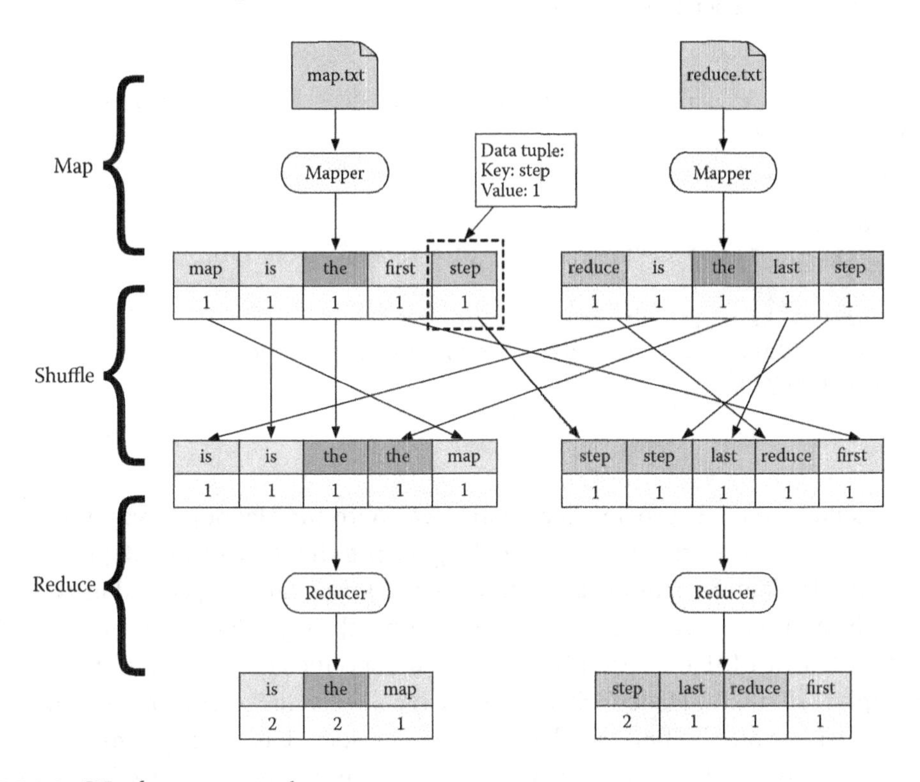


FIGURE 2.2 Word count example.

access to application data, Hadoop is reliable and efficient for Big Data analysis on large clusters.

The interface for application developers using Apache Hadoop is simple and convenient. Developers are only required to implement two interfaces: *mapper* and *reducer*. As Source Code 1 shows, the two void functions *map* and *reduce* are corresponding to the *map* and *reduce* phases of the MapReduce model. Hadoop takes care of the other operations on the data such as shuffling, sorting, dividing input data, and so on.

**Source Code 1: Mapper and Reducer in Java**

```
@Deprecated
publicinterfaceMapper<K1, V1, K2, V2>extendsJobConfigurable, Closeable {
```

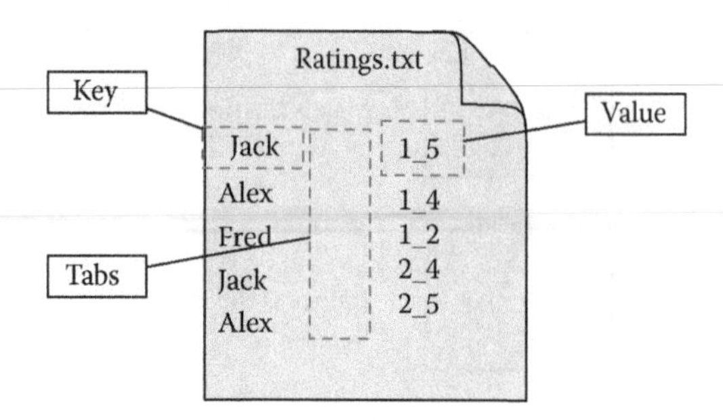


FIGURE 2.3 How Hadoop Streaming works.

```
voidmap(K1 key, V1 value,OutputCollector<K2, V2> output, Reporter reporter)
throwsIOException;
}

@Deprecated
publicinterfaceReducer<K2, V2, K3, V3>extendsJobConfigurable, Closeable {

voidreduce(K2 key, Iterator<V2> values,
      OutputCollector<K3, V3> output, Reporter reporter)
throwsIOException;

}
```

By default, Hadoop is programmed with Java. With the Hadoop Streaming interface, developers can choose any programming languages to implement the MapReduce functions. Using Hadoop Streaming, any executable or script can be specified as the mapper or the reducer. When map/reduce tasks are running, the input file is converted into lines. Then these lines are fed to the standard input of the specified mapper/reducer. Figure 2.3 is an example to show how Hadoop Streaming extracts key/value pairs from lines. By default, the characters before the first tab character are considered as the key, while the rest are values. If there is no tab character in a line, then the entire line is the key, and the value is null. The standard output of the mapper/reducer is collected, and the key/value pairs are extracted from each line in the same way. In this chapter, we choose Python to implement mappers and reducers using the Hadoop Streaming interface.

## 2.3 A MOTIVATING APPLICATION: MOVIE RATINGS FROM NETFLIX PRIZE

Recommendation systems implement personalized information filtering technologies. They are designed to solve the problem of information overload on individuals [6], which refers to the difficulty a person may face when understanding an issue and then making decisions due to the presence of too much information. Many popular commercial web services and social networking websites implement recommendation systems and provide users with enhanced information awareness. For example, movie recommendation improves the user experience by providing a list of movies that most likely cover the movies the user may like. Collaborative filtering (CF) is a widely used algorithm for recommendation systems [7]. It contains two major forms, namely, user-based and item-based CF. The user-based CF aims to recommend a user movies that other users like [8,9], while the item-based CF recommends a user movies similar to the user's watched list or high-rating movies [10–12].

As users keep rating and watching movies, both algorithms cannot avoid the problem of incremental data processing, which means analyzing the new ratings and most recent watched histories and updating the recommendation lists. As the numbers of users and movies grow, incremental data processing can impose immense demands on computation and memory usage. Take the item-based CF algorithm in Reference 12 as an example. Assuming there are $M$ users and $N$ movies, the time complexity to compute the similarity between two movies is $O(M)$. For the final recommendation result, the similarities for all the possible movie pairs must be computed, which has a complexity of $O(M*N^2)$. With millions of users and tens of thousands of movies, the complexity will be extremely large, and scalability becomes a serious problem. Implementing the CF algorithms with the MapReduce model using scalable platforms is a reasonable solution. Hence, a movie recommendation application on real-world data sets provides an excellent motivating scenario to evaluate the scalability and cost of the Hadoop platform in a cloud environment.

For this evaluation, we use the sample data sets from Netflix Prize [5]. The size is approximately 2 GB in total. The data set contains 17,770 files. The MovieIDs range from 1 to 17,770 sequentially with one file per movie. Each file contains three fields, namely, UserID, Rating, and Date. The UserID ranges from 1 to 2,649,429 with gaps. There are, in total, 480,189 users. Ratings are on a five-star scale from 1 to 5. Dates have the format YYYY-MM-DD. We merged all the 17,770 files into one file, with each line in the format of "UserID MovieID Rating Date," ordered by ascending date, to create the input data for our application.

In the item-based CF algorithm, we define a *movie fan* as a user who rates a particular movie higher than three stars. *Mutual fans* of two movies are the users who have rated both of the movies higher than three stars. Then we measure the similarity between two movies by the number of their mutual fans. As a result, we output a recommendation list that contains top-$N$ most similar movies to each movie. Following this application logic, we present the implementation in Hadoop and highlight key artifacts in the MapReduce programming model.

## 2.4 IMPLEMENTATION IN HADOOP

With the MapReduce model in Hadoop, the algorithm consists of three rounds of MapReduce processing as follows:

**Round 1.** Map and sort the *user–movie* pairs. Each pair implies that the user is a fan of the movie. A reducer is not needed in this round. Figure 2.4 shows an example of the input and output.

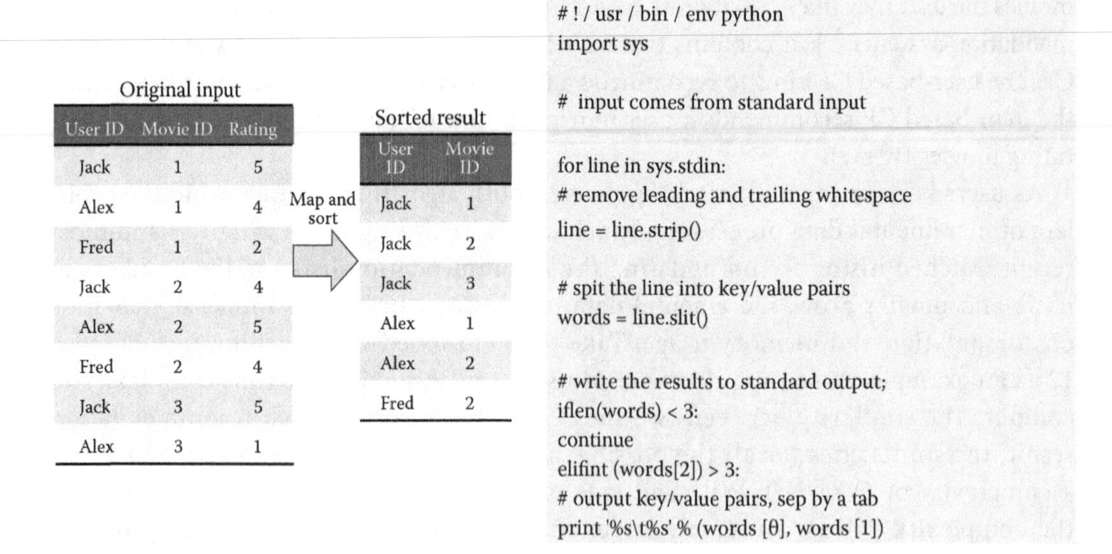


FIGURE 2.4 An example and sample code for round 1.

**Round 2.** Calculate the number of mutual fans of each movie. Figure 2.5 demonstrates the processing with an example. Assume Jack is a fan of movies 1, 2, and 3; then movies 1 and 2 have one mutual fan as Jack. Likewise, movies 1 and 3 and movies 2 and 3 also have one mutual fan as Jack. A mapper finds all the mutual fans of each movie and outputs a line for every mutual fan. Then the reducer aggregates the result based on the movie pair that has one mutual fan and counts the number of mutual fans. The sample codes of the mapper and reducer are shown in Appendix A.

**Round 3.** Extract the movie pairs in the result of round 2 and find the top-*N* movies that have the most mutual fans of each movie. The mapper extracts the movie IDs from the movie pairs, and the reducer aggregates the result from the mapper and orders the recommended movies based on the numbers of their mutual fans. As Figure 2.6 demonstrates, movie pair 1–2 has a mutual fan count of two, and movie pair 1–3 has one mutual fan. Therefore, movie ID 1 has mutual fans with movie IDs 2 and 3.

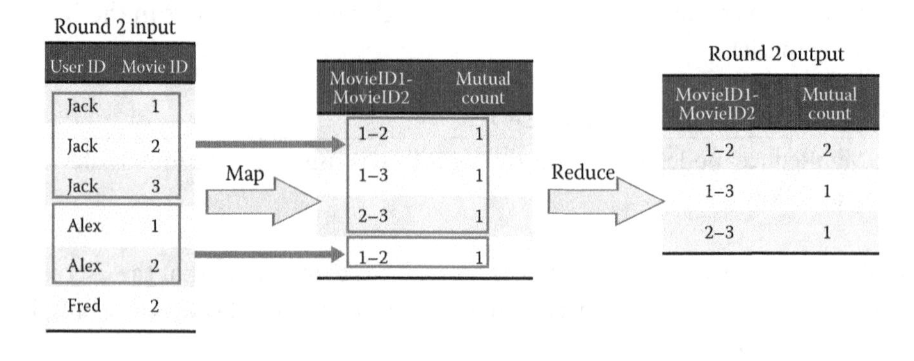


FIGURE 2.5 An example of round 2.

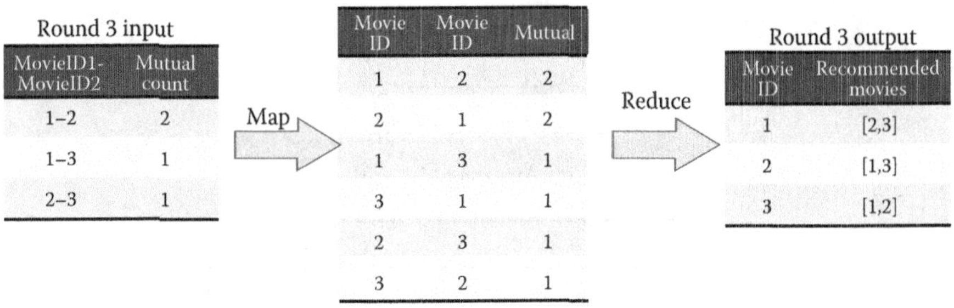

Round 3 input

| MovieID1-MovieID2 | Mutual count |
|---|---|
| 1–2 | 2 |
| 1–3 | 1 |
| 2–3 | 1 |

Map

| Movie ID | Movie ID | Mutual |
|---|---|---|
| 1 | 2 | 2 |
| 2 | 1 | 2 |
| 1 | 3 | 1 |
| 3 | 1 | 1 |
| 2 | 3 | 1 |
| 3 | 2 | 1 |

Reduce

Round 3 output

| Movie ID | Recommended movies |
|---|---|
| 1 | [2,3] |
| 2 | [1,3] |
| 3 | [1,2] |

FIGURE 2.6 An example of round 3.

The recommended movies for movie ID 1 are movies [2,4] in descending order of the count of mutual fans. The sample codes of the mapper and reducer are shown in Appendix A.

## 2.5 DEPLOYMENT ARCHITECTURE

Our evaluation is performed on AWS, which provides the essential infrastructure components for setting up the Hadoop platform. The infrastructure components are listed in Table 2.1. AWS allows flexible choices on virtual machine images, capacities of virtual machine instances, and associated services (such as storage, monitoring, and cost billing).

The deployment of the Hadoop architecture uses AWS Elastic MapReduce (EMR), which is a Hadoop platform across a resizable cluster of Amazon Elastic Compute Cloud (EC2) instances. The deployment architecture is shown in Figure 2.7. In the EMR configuration, we set up 1 master instance and 10 core instances to run the Hadoop implementation of the movie recommendation application. The master instance manages the Hadoop cluster. It assigns tasks and distributes data to core instances and task instances. Core instances run map/reduce tasks and store data using HDFS. All instances are in the type of *m1*.small, which has 1.7 GB of memory, one virtual core of an EC2 Compute Unit, and 160 GB of local instance storage.

We upload the rating file of Netflix data to Amazon Simple Storage Service (S3), which is a storage service that provides an interface to store and retrieve any amount of data. EMR reads data files on S3 as the input file of Hadoop jobs.

We configure the Hadoop cluster and jobs in Amazon EMR. The entire configuration page includes cluster configuration, tags, software configuration, hardware configuration, security and access, bootstrap actions, and steps. For our experiments, we have set up

TABLE 2.1 Infrastructure Components of Movie Recommendation App

| **Application** | **Movie Recommendation** |
|---|---|
| Platform | Apache Hadoop |
| Amazon Web Services | Simple Storage Service (S3) |
| | Elastic MapReduce (EMR) |
| Monitoring tools | Amazon CloudWatch |

Movie recommendation job
Amazon EMR
Handoop cluster
Master
EC2 instance
Core
EC2 instance
10
Amazon S3
Netflix data set
Result

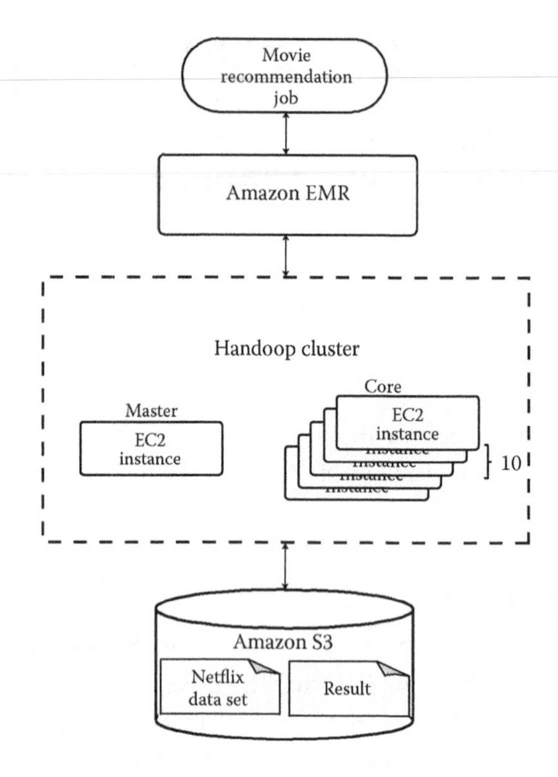

FIGURE 2.7 Deployment architectures on AWS.

cluster configuration, hardware configuration, and steps. We use the default settings for the other sections.

In cluster configuration, we set up our cluster name and the log folder on S3 as Figure 2.8 shows.

In hardware configuration, as Figure 2.9 shows, we leave the network and EC2 subnet as default and change the EC2 instance type and instance count to what we need.

As Figure 2.10 shows, the steps section is where we configure the Hadoop jobs. Since the recommendation application requires three rounds of MapReduce operations, we add

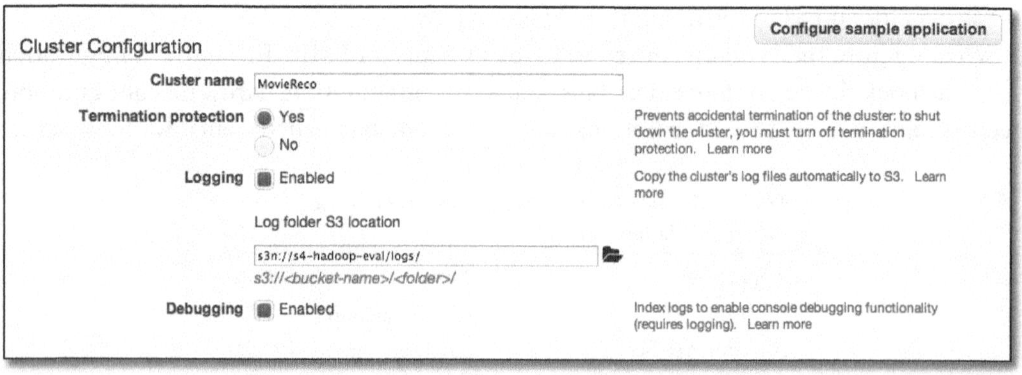

Cluster Configuration
Configure sample application
Cluster name
MovieReco
Termination protection
Yes
No
Prevents accidental termination of the cluster: to shut down the cluster, you must turn off termination protection. Learn more
Logging
Enabled
Copy the cluster's log files automatically to S3. Learn more
Log folder S3 location
s3n://s4-hadoop-eval/logs/
s3://<bucket-name>/<folder>/
Debugging
Enabled
Index logs to enable console debugging functionality (requires logging). Learn more

FIGURE 2.8 Cluster configuration.

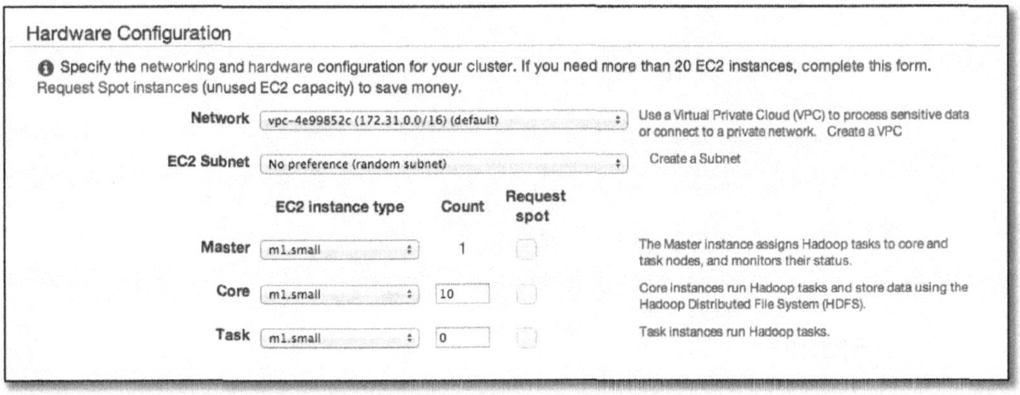

Hardware Configuration

Specify the networking and hardware configuration for your cluster. If you need more than 20 EC2 instances, complete this form. Request Spot instances (unused EC2 capacity) to save money.

Network: vpc-4e99852c (172.31.0.0/16) (default) — Use a Virtual Private Cloud (VPC) to process sensitive data or connect to a private network. Create a VPC

EC2 Subnet: No preference (random subnet) — Create a Subnet

| | EC2 instance type | Count | Request spot | |
|---|---|---|---|---|
| Master | m1.small | 1 | | The Master instance assigns Hadoop tasks to core and task nodes, and monitors their status. |
| Core | m1.small | 10 | | Core instances run Hadoop tasks and store data using the Hadoop Distributed File System (HDFS). |
| Task | m1.small | 0 | | Task instances run Hadoop tasks. |

FIGURE 2.9 Hardware configuration.

Steps

A step is a unit of work you submit to the cluster. A step might contain one or more Hadoop jobs, or contain instructions to install or configure an application. You can submit up to 256 steps to a cluster. Learn more

| Name | Action on failure | JAR S3 location | Arguments |
|---|---|---|---|
| R1 | Terminate cluster | /home/hadoop/contrib/streaming/hadoop-streaming.jar | -mapper s3://s4-hadoop-eval/job/map_r1.py -reducer s3://s4-hadoop-eval/job/reduce_r1.py -input s3://s4-hadoop-eval/data/rating_17m.txt -output s3://s4-hadoop-eval/result/17m1/ |
| R2 | Terminate cluster | /home/hadoop/contrib/streaming/hadoop-streaming.jar | -mapper s3://s4-hadoop-eval/job/map_r2.py -reducer s3://s4-hadoop-eval/job/reduce_r2.py -input s3://s4-hadoop-eval/result/17m1/part-* -output s3://s4-hadoop-eval/result/17m2/ |
| | | | -mapper s3://s4-hadoop-eval/job/map_r3.py -reducer |

Add step: Select a step

Configure and add

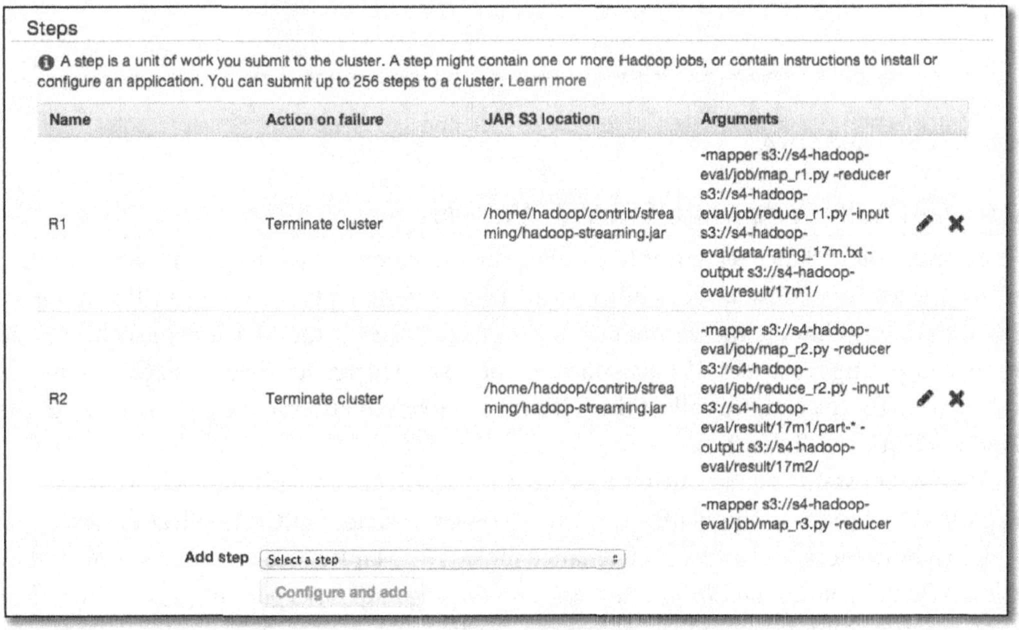

FIGURE 2.10 Steps configuration.

three steps here. Each step is a MapReduce job. The specification of S3 location of the mapper, the reducer, the input files, and the output folder for each step is shown in Figure 2.11.

We use Amazon CloudWatch [13] to monitor the status of Hadoop jobs and tasks and EC2 instances. For EC2 instances, CloudWatch provides a free service named Basic Monitoring, which includes metrics for CPU utilization, data transfer, and disk usage at a 5 min frequency. For the EMR service, CloudWatch has metrics about HDFS, S3, nodes, and map/reduce tasks. All these built-in EC2 and EMR metrics are automatically collected and reported. AWS users can access these metrics in real time on the CloudWatch website or through the CloudWatch application programming interfaces (APIs).

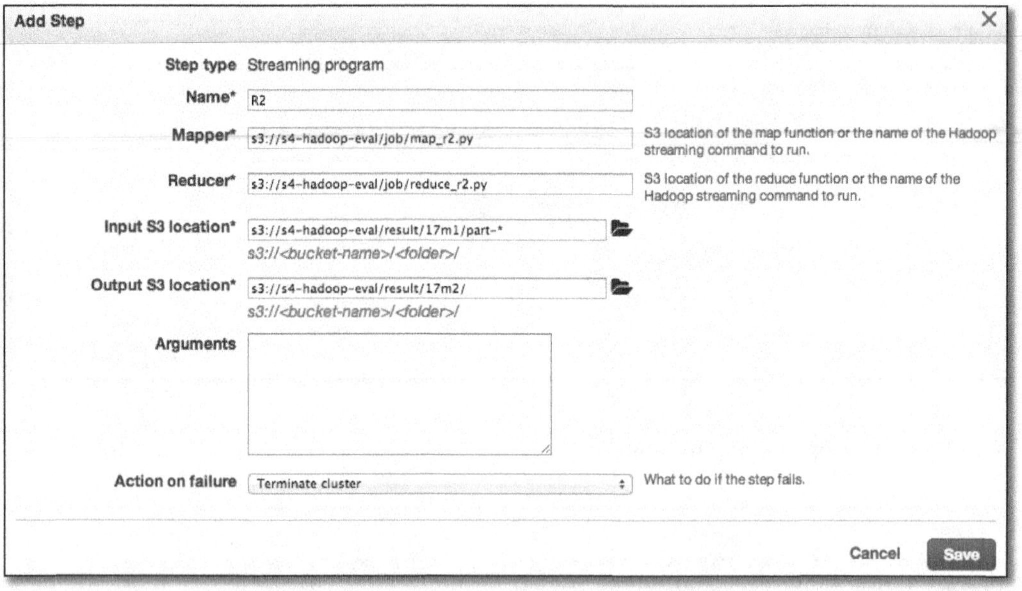

FIGURE 2.11 Add a step.

## 2.6 SCALABILITY AND COST EVALUATION

The evaluation includes scenarios to emulate the incremental updates of data sets, in terms of data sizes. We consider scalability as to what extent a platform can utilize computing resources to handle increasing data sizes or data rates before a platform scales out on more computing resources. We also observe the cost incurred as the data sizes and update rates grow. To this end, we collect the following metrics to observe the performance of the Hadoop platform (Table 2.2).

The system status metrics help to identify any bottleneck that limits performance. We explore the metrics of the platform status to observe factors at the application level that link to a bottleneck and affect performance most. The MapReduce task status provided by CloudWatch includes *remaining map tasks*, *remaining map tasks per slot*, *remaining reduce tasks*, *running map tasks*, and *running reduce tasks.*

In the Hadoop implementation of the movie recommendation application, the rating data are constantly increasing as users keep rating movies on the website. Therefore, Hadoop needs to process the whole set of rating data available as the size increases over time. To evaluate the scalability and cost of Hadoop processing incremental data, we create

TABLE 2.2 Collected Metrics

| Platform | Apache Hadoop |
|---|---|
| System status | CPU utilization |
| | HDFS I/O, S3 I/O |
| | Network I/O |
| Platform status | MapReduce task status |

*Note:* I/O, input/output.

experiments that process different-sized rating files. First, we order the Netflix data set by time. Assuming 200 ratings made by users every second, the data set is of *200 ratings per second * 24 (hours) * 3600 (seconds) * N* records on the *N*th day. We run our movie recommendation jobs over the period of days as $N = 1, 3, 6$ (Figure 2.12) and compare the results.

Understanding how the minimal time interval or frequency of data processing varies according to the data update sizes helps to make scaling decisions on the optimal resource provision settings. Table 2.3 shows the elapsed time of Hadoop jobs with different sizes of input data. The elapsed time implies, given a certain data size, the minimal time required to run the Hadoop application that updates the movie recommendation list. For example, the first entry in Table 2.3 means that given the current capacity, it is only possible to rerun the Hadoop movie recommendation application with a frequency of once per hour for processing rating data updated in 1 day. For any shorter frequencies or larger data sets, more EMR instances need to be provisioned. Further analysis in Figure 2.13 shows the linear trend of the elapsed time of this Hadoop application according to the data sizes.

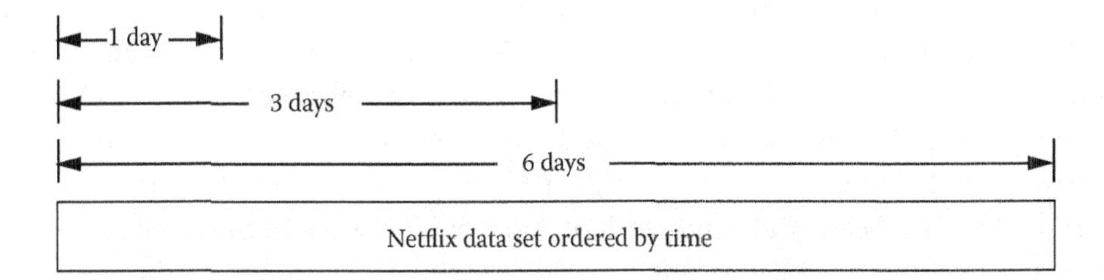


FIGURE 2.12 Input rating file.

TABLE 2.3 Elapsed Time of Movie Recommendation Jobs

| Scenario | Data File Size | Records | Elapsed Time |
|---|---|---|---|
| 1 day | 428 MB | 17,280,000 | 56 min |
| 3 day | 1.2 GB | 51,840,000 | 177 min |
| 6 day | 2.4 GB | 103,680,000 | 407 min |

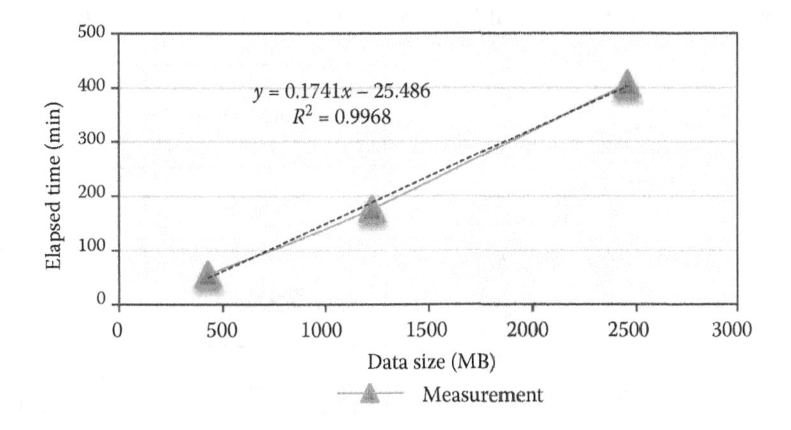


FIGURE 2.13 Linear trend between elapsed time and data sizes.

From the platform status, Figure 2.14 shows the number of running tasks comparing the results of the 1-day input and the 3-day input. The evaluation environment has 10 core instances. Each instance has two mappers and one reducer. The running map and reduce tasks are at their maximum most of the time. In addition, all the mappers and reducers finish their job approximately at the same time, which implies that there is no data skew problem here. In Section 2.4, "Implementation in Hadoop," we present three rounds of the MapReduce jobs of this movie recommendation application. The time running each round is listed in Figure 2.14.

Figure 2.15 shows the system status comparing 1-day data input and 3-day data input. As Figure 2.15c shows, in both experiments, there are no data written to HDFS, and the data read from HDFS total less than 5 KB/s all the time. This is mainly due to the fact that EMR uses S3 for input and output instead of HDFS. In the EMR services, the mappers read from S3 and hold intermediate results in a local disk of EC2 instances. Then in the shuffling stage, these intermediate results are sent through the network to corresponding reducers that may sit on other instances. In the end, the reducers write the final result to S3.

In Figure 2.15a, the experiment with 3-day input has the CPU utilization at 100% most of the time, except at three low points at 100, 125, and 155 min, which are the times when the shuffling or writing results of reduce tasks occur. The average network I/O measurement in Figure 2.15b shows spikes at around the same time points, 100, 125, and 155 min, respectively, while other times, the average network I/O stays quite low. The highest network I/O rate is still below the bandwidth of Amazon's network in the same region, which is 1 Gbps to our best knowledge. The experiment with 1-day input has the same pattern with different time frames. From these observations, we can infer that CPU utilization would be the bottleneck for processing a higher frequency of data.

Table 2.4 shows the cost of the EMR service under different terms of data sizes and updating frequencies. For example, with a once-per-3 h updating frequency and 1.2G input files, we need to run the Hadoop recommendation jobs 8 times a day. Each time, it takes 2 h and 57 min. The EMR price on AWS is 0.08 USD instance per hour. For a sub-hour's usage, the billing system rounds it to an hour. As we have used 11 instances in total, it costs *$0.08*11*3*8 = $21.12* per day. The *NA* (not applicable) in the table means that the processing time for data at this size is longer than the updating interval so that more instances need to be provisioned.

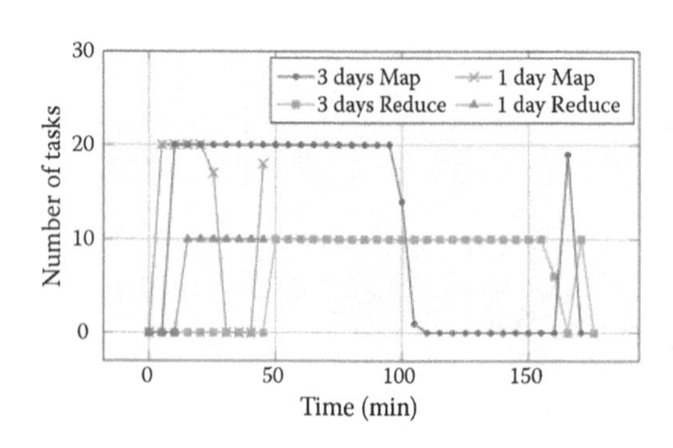


| Round # of MapReduce | 1 day | 3 days |
|---|---|---|
| Round 1 | 4 min | 5 min |
| Round 2 | 41 min | 155 min |
| Round 3 | 4 min | 10 min |

FIGURE 2.14 Number of running MapReduce tasks and time spent.

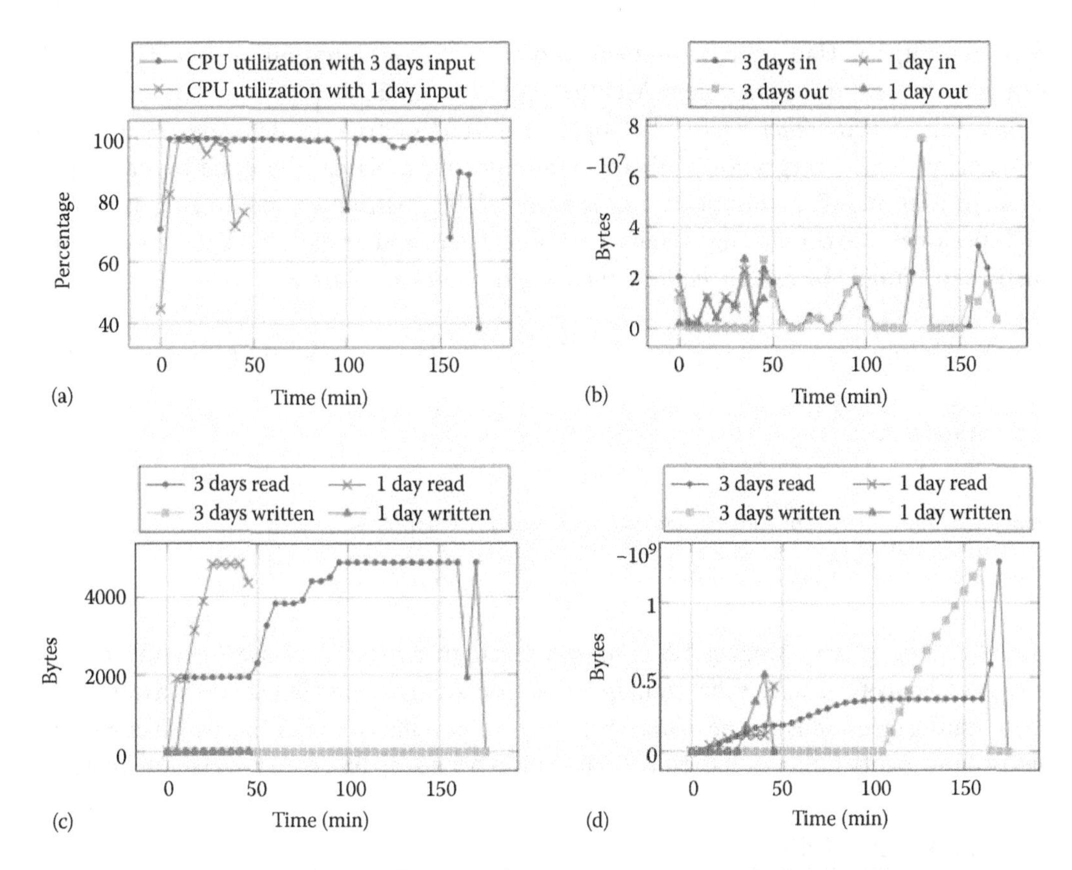


FIGURE 2.15 System status of Hadoop cluster. (a) Core instance average CPU utilization. (b) Core instance average network in/out. (c) HDFS bytes read/written. (d) S3 bytes read/written.

TABLE 2.4 Daily Cost Table of Hadoop Approach

| **Update Frequency** | **Once per Day** | | | **Once Every 3 Hours** | | | **Once per Hour** | | |
|---|---|---|---|---|---|---|---|---|---|
| Input data size | 428 M | 1.2 G | 2.4 G | 428 M | 1.2 G | 2.4 G | 428 M | 1.2 G | 2.4 G |
| Daily cost | $0.88 | $2.64 | $6.16 | $7.04 | $21.12 | NA | $21.12 | NA | NA |

## 2.7 DISCUSSIONS

Using AWS EMR, S3, CloudWatch, and the billing service, we are able to observe for both performance and cost metrics. Based on our experience, we discuss other features of the Hadoop platform related to the processing of incremental data sets.

*Programmability.* Hadoop is an open-source project, and it provides a simple programming interface for developers. Hadoop supports multiple programming languages through the Hadoop Streaming interface. Figure 2.4 is an example of a mapper programmed with Python.

*Deployment complexity.* Hadoop deployment is complicated and includes MapReduce modules, HDFS, and other complex configuration files. However, Hadoop is a widely used platform and has a mature community so that it is easy to find documents or choose an integrated Hadoop service like Amazon EMR. With Amazon EMR, developers no longer need to set up the Hadoop environment, and a simple configuration can get Hadoop jobs run as we present in "Deployment Architecture," Section 2.5.

*Integration with other tools.* Hadoop can output metrics to data files or real-time monitoring tools. Integration with other monitoring tools can be done by editing the configuration file. For example, Ganglia is a scalable distributed monitoring system for high-performance computing systems such as clusters and grids [14]. Hadoop can output metrics to Ganglia by editing *hadoop-metrics.properties* as follows:

```
hadoop-metrics.properties (@GANGLIA@ is the hostname or IP of the Ganglia endpoint):

dfs.class = org.apache.hadoop.metrics.ganglia.GangliaContext
dfs.period = 10
dfs.servers = @GANGLIA@:8649

mapred.class = org.apache.hadoop.metrics.ganglia.GangliaContext
mapred.period = 10
mapred.servers = @GANGLIA@:8649
```

*Autoscaling options.* Amazon EMR allows users to increase or decrease the number of task instances. For core instances that contain HDFS, users can increase but not decrease them. All these scaling operations can be made when the Hadoop cluster is running, which enables the autoscaling ability for EMR if the autoscaling rules are set up based on the Hadoop metrics.

## 2.8 RELATED WORK

Recent efforts have been dedicated to an incremental data processing system that would allow us to maintain a very large repository of data and update it efficiently as a small portion of the data mutates. Broadly speaking, there are two approaches. One approach is enhancing the Hadoop architecture and file systems to detect changes to the inputs and employ fine-grained result reuse [15]. Novel techniques of content-based chucking at both the programming languages and store system level were presented. The overall programming model still follows MapReduce. The other approach is proposing a new programming model and APIs that differ from MapReduce, such as Google's Percolator [4] and Yahoo!'s continuous bulk processing (CBP) [16]. Both approaches have mechanisms to identify where the updates occur, store intermediate results, and reuse these results.

Other performance evaluations of Hadoop include Hadoop with respect to virtual machine capacities [17], Hadoop running on high-performance clusters for scientific data [18], and Hadoop's integration with Not only Structured Query Language (NoSQL) databases (such as MongoDB) [19]. We focus on evaluating Hadoop's scalability and cost of processing incremental data in terms of data sizes and updating frequency. Zaharia et al. [20] identified that Hadoop's performance is closely tied to its task scheduler. The task scheduler implicitly assumes that cluster nodes are homogeneous and tasks make progress linearly, which is used to decide when to speculatively re-execute tasks that appear to be stragglers. Reference 20 shows that Hadoop's scheduler can cause severe performance degradation in heterogeneous environments.

## 2.9 CONCLUSION

In this chapter, we evaluate the scalability and cost of the Amazon Hadoop service through a movie recommendation scenario, which requires frequent updates on continuous input data streams. We implement this sample application with the MapReduce programming model and deploy it on the Amazon EMR service. We present the monitoring techniques to collect both system- and platform-level metrics to observe the performance and scalability status. Our evaluation experiments help to identify the minimal time interval or the highest frequency of data processing, which is an indicator on the autoscaling settings of optimal resource provisioning.

## ACKNOWLEDGMENT



This material is based upon work funded and supported by the Department of Defense under contract no. FA8721-05-C-0003 with Carnegie Mellon University for the operation of the Software Engineering Institute, a federally funded research and development center.



## REFERENCES

1. Paul Zikopoulos, and Chris Eaton. *Understanding Big Data: Analytics for Enterprise Class Hadoop and Streaming Data* (1st ed.). McGraw-Hill Osborne Media, Columbus, OH (2011).
2. Jeffrey Dean, and Sanjay Ghemawat. MapReduce: Simplified data processing on large clusters. *Communications of the ACM* 51, 1 (2008): 107–113.
3. What is apache hadoop? (2014). Available at http://hadoop.apache.org/.
4. Daniel Peng, and Frank Dabek. Large-scale incremental processing using distributed transactions and notifications. In *Proceedings of the 9th USENIX Symposium on Operating Systems Design and Implementation*. USENIX (2010).
5. Netflix Prize (2009). Available at http://www.netflixprize.com/.
6. Gediminas Adomavicius et al. Incorporating contextual information in recommender systems using a multidimensional approach. *ACM Transactions on Information Systems (TOIS)* 23, 1 (2005): 103–145.
7. John S. Breese, David Heckerman, and Carl Kadie. Empirical analysis of predictive algorithms for collaborative filtering. In *Proceedings of the Fourteenth Conference on Uncertainty in Artificial Intelligence*. Morgan Kaufmann Publishers Inc. (1998).
8. Jonathan L. Herlocker et al. An algorithmic framework for performing collaborative filtering. In *Proceedings of the 22nd Annual International ACM SIGIR Conference on Research and Development in Information Retrieval*. ACM (1999).

9. Paul Resnick et al. GroupLens: An open architecture for collaborative filtering of netnews. In *Proceedings of the 1994 ACM Conference on Computer Supported Cooperative Work*. ACM (1994).
10. Mukund Deshpande, and George Karypis. Item-based top-*n* recommendation algorithms. *ACM Transactions on Information Systems (TOIS)* 22, 1 (2004): 143–177.
11. Greg Linden, Brent Smith, and Jeremy York. Amazon.com recommendations: Item-to-item collaborative filtering. *Internet Computing, IEEE* 7, 1 (2003): 76–80.
12. Badrul Sarwar et al. Item-based collaborative filtering recommendation algorithms. In *Proceedings of the 10th International Conference on World Wide Web*. ACM (2001).
13. Amazon CloudWatch (2014). Available at http://aws.amazon.com/cloudwatch/.
14. Ganglia (2014). Available at http://ganglia.sourceforge.net/.
15. Pramod Bhatotia et al. Incoop: MapReduce for incremental computations. In *Proceedings of the 2nd ACM Symposium on Cloud Computing (SOCC '11)*. ACM, New York, Article 7, 14 pages (2011).
16. Dionysios Logothetis et al. Stateful bulk processing for incremental analytics. In *Proceedings of the 1st ACM symposium on Cloud computing (SoCC '10)*. ACM, New York, 51–62 (2010).
17. Shadi Ibrahim et al. Evaluating MapReduce on virtual machines: The hadoop case. *Cloud Computing, Lecture Notes in Computer Science* 5931 (2009): 519–528.
18. Zacharia Fadika et al. Evaluating hadoop for data-intensive scientific operations. In *Cloud Computing (CLOUD), 2012 IEEE 5th International Conference on*, 67, 74 (June 24–29, 2012).
19. Elif Dede et al. Performance evaluation of a MongoDB and hadoop platform for scientific data analysis. In *Proceedings of the 4th ACM Workshop on Scientific Cloud Computing (Science Cloud '13)*. ACM, New York, 13–20 (2013).
20. Matei Zaharia et al. Improving MapReduce performance in heterogeneous environments. In *Proceedings of the 8th USENIX Conference on Operating Systems Design and Implementation (OSDI '08)*. USENIX Association, Berkeley, CA, 29–42 (2008).

## APPENDIX 2.A: SOURCE CODE OF MAPPERS AND REDUCERS

**Source Code 2: Sample Code of Mapper in Round 2**

```
#!/usr/bin/env python

import sys

movie_ids = list()
user_id = 0

# output movies pairs that have mutual fan;
# key: movie pair, value: 1
defprint_id_pairs(id_list):
fori in range(0, len(id_list)-1):
for j in range(i+1, len(id_list)):
print '%s-%s\t1'% (id_list[i], id_list[j])

# input comes from standard input
for line in sys.stdin:
# remove leading and trailing whitespace
line = line.strip()
# split the line into key/value pairs
words = line.split('\t')
```

```
# write the results to standard output;
iflen(words) < 2:
continue
ifuser_id ! = words[0] and user_id ! = 0:
print_id_pairs(movie_ids)
delmovie_ids[:]
user_id = words[0]
movie_ids.append(words[1])

print_id_pairs(movie_ids)
```

**Source Code 3: Sample Code of Reducer in Round 2**

```
#!/usr/bin/env python

import sys

movie_pair_cnt = 0
movie_pair = 0

# input comes from standard input
for line in sys.stdin:
# remove leading and trailing whitespace
line = line.strip()

# split the line into key/value pairs
words = line.split('\t')

# accumulate the number of mutual fans and write the results to standard output;
iflen(words) < 2:
continue
ifmovie_pair ! = words[0] and movie_pair ! = 0:
print '%s\t%d'% (movie_pair, movie_pair_cnt)
movie_pair_cnt = 0
movie_pair = words[0]
movie_pair_cnt + = int(words[1])

print '%s\t%d'% (movie_pair, movie_pair_cnt)
```

**Source Code 4: Sample Code of Mapper in Round 3**

```
#!/usr/bin/env python

import sys

# input comes from standard input
for line in sys.stdin:
# remove leading and trailing whitespace
line = line.strip()
```

```
# split the line into key/value pairs
words = line.split()

# output for each movie in the movie pairs to standard output;
iflen(words) < 2:
continue
ids = words[0].split('-')
iflen(ids) = = 2:
print '%s\t%s\t%s'% (ids[0], ids[1], words[1])
print '%s\t%s\t%s'% (ids[1], ids[0], words[1])
```

### Source Code 5: Sample Code of Reducer in Round 3

```
#!/usr/bin/env python

import sys

top_lists = []
movie_id = 0

# output recommendation list
defprint_top_lists(m_id, top_list):
out_str = '%s\t'% m_id
fori, data in enumerate(sorted(top_list, key = lambda m:m[1], reverse = True)[:10]):
ifi ! = 0:
out_str + = ','
out_str + = '%s-%d'% (data[0], data[1])
printout_str

# input comes from standard input
for line in sys.stdin:
# remove leading and trailing whitespace
line = line.strip()
# split the line into key/value pairs
words = line.split('\t')
# write the recommendation lists to standard output;
iflen(words) < 3:
continue
ifmovie_id ! = words[0] and movie_id ! = 0:
print_top_lists(movie_id, top_lists)
deltop_lists[:]
movie_id = words[0]
top_lists.append((words[1], int(words[2])))

print_top_lists(movie_id, top_lists)
```

CHAPTER 3

# Singular Value Decomposition, Clustering, and Indexing for Similarity Search for Large Data Sets in High-Dimensional Spaces

Alexander Thomasian

CONTENTS

## ABSTRACT

Representing objects such as images by their feature vectors and searching for similarity according to the distances of the points representing them in high-dimensional space via *k*-nearest neighbors (*k*-NNs) to a target image is a popular paradigm. We discuss a combination of singular value decomposition (SVD), clustering, and indexing to reduce the cost of processing *k*-NN queries for large data sets with high-dimensional data. We first review dimensionality reduction methods with emphasis on SVD and related methods, followed by a survey of clustering and indexing methods for high-dimensional numerical data. We describe combining SVD and clustering as a framework and the main memory-resident ordered partition (OP)-tree index to speed up *k*-NN queries. We discuss techniques to save the OP-tree on disk and specify the stepwise dimensionality increasing (SDI) index suited for *k*-NN queries on dimensionally reduced data.

## 3.1 INTRODUCTION

IBM's *Query by Image Content* (*QBIC*) project in the 1990s [1] utilized *content-based image retrieval* (*CBIR*) or similarity search based on features extracted from images. We are concerned with the storage space and processing requirements for CBIR, which are important because of the high data volumes and the associated processing requirements, especially if the similarity search is conducted by a naive sequential scan of the whole data set. The *clustering with singular value decomposition* (*CSVD*) method and the *ordered partition* (*OP*)-tree index [2] are used in Reference 3 in applying CBIR to indexing satellite images [4].

CBIR has two steps [5]: (1) Characterize images by their features and define similarity measures. (2) Extract features such as color, texture, and shape [6]. Texture features include the fractal dimension, coarseness, entropy, circular Moran autocorrelation functions, and *spatial gray-level difference* (*SGLD*) statistics [7]. In addition to photographic, medical, and satellite images, content-based retrieval is used in conjunction with audio and video clips and so forth.

Given a target image represented by a point in high-dimensional space, images similar to it can be determined via a range query, which determines all points within a certain radius. Since a range query with a small radius may not yield any query points, *k-nearest neighbor* (*k-NN*) queries are used instead [8]. The distance may be based on the Euclidean or more complex distance functions, such as the Mahalanobis distance [8]. *k*-NN queries can be implemented by a sequential scan of the data set holding the feature vectors of all images, while keeping track of the nearest *k*. The naive algorithm has a running time of $O(MN)$, where $M$ is the cardinality of data set (number of images) and $N$ is the dimensionality (number of features). Amazingly, naive search may outperform multidimensional indexing methods in higher-dimensional data spaces [9].

Clustering is one method to reduce the cost of *k*-NN search from the viewpoint of CPU and disk processing. Given $H$ clusters with an equal number of points per cluster, the cost of *k*-NN queries is reduced $H$-fold, provided that only the data points in one cluster are to be searched. A survey of clustering methods with emphasis on disk-resident high-dimensional data is provided in this chapter. With the advent of *storage-class memory* (*SCM*) [10], more and more data are expected to migrate from magnetic disks [11] to such memory.

Clusters are usually envisioned as hyperspheres represented by their centroid and radius (the distance of the farthest point from the centroid). A query point may reside in the volume shared by two or more hyperspheres or hypercubes depending on the indexing structure, so that multiple clusters need to be searched to locate the $k$-NNs. $k$-NN search has to be trimmed as much as possible to reduce query processing cost [3]. Indexing is a form of clustering; however, most indexing structures, such as those in the R-tree family, perform poorly in higher-dimensional data spaces [8]. More details on this topic appear in Section 3.7.

*Principal component analysis* (*PCA*) [12], *singular value decomposition* (*SVD*), and *Karhunen–Loève transform* (*KLT*) are related methods to reduce the number of dimensions of high-dimensional data after rotating coordinates to attain minimal loss in information for the desired level of data compression [8,13]. Dimensionality reduction can be optionally applied to the original data set before applying clustering. Dimensionality reduction applied to clusters takes advantage of local correlations, resulting in a smaller normalized mean square error (NMSE) for the same reduction in data volume as when SVD is applied to the whole data set. The method is therefore called *recursive clustered SVD* (*RCSVD*), although SVD was applied to clusters in References 3 and 14.

As an example that combining clustering with SVD results in higher dimensionality reduction than applying SVD without clustering, consider points surrounding multiple nonintersecting lines in three-dimensional space, which can be specified by their single dominant dimension, that is, a threefold reduction in space requirements, but also computational cost for $k$-NN queries. Experimentation with several real-life data sets has shown lower information loss with the same compression ratio or vice versa when clustering is combined with SVD.

The *local dimensionality reduction* (*LDR*) method [15] finds local correlations and performs dimensionality reduction on these data. However, it is shown in Reference 3 that CSVD with an off-the-shelf clustering method, such as $k$-means, outperforms LDR.

A further reduction in cost can be attained by reducing the number of points to be considered by building an index for the data in each cluster. It turns out that popular indexes such as R-trees are inefficient in processing $k$-NN queries in high dimensions. Indexing methods are surveyed in References 16 and 17, but a more focused and up-to-date survey for high-dimensional data is provided in this chapter.

The chapter is organized as follows. Section 3.2 discusses the issue of dimensionality reduction. Section 3.3 presents clustering methods with emphasis on large, high-dimensional, disk-resident data sets. The method in Reference 3 to build the data structures for $k$-NN queries is described in Section 3.4. Section 3.5 specifies the method for determining similarity. Section 3.6 discusses the LDR method in Reference 15. Section 3.7 surveys indexing structures. Section 3.8 summarizes the discussion in the chapter. A more efficient method [18] to compute the Euclidean distance with dimensionally reduced data with SVD is given in the appendix.

## 3.2 DATA REDUCTION METHODS AND SVD

Data reduction methods are of interest in several domains, such as query cost estimation in relational databases [19], data mining, decision support, *online analytical processing* (*OLAP*) [20], data compression, information retrieval, and data visualization. We list the data reduction methods surveyed in Reference 21 as follows:

1. *SVD, PCA*, and *KLT* are related dimensionality reduction methods, which are discussed in this section.
2. *Discrete wavelet transform* (*DWT*) has many variations [13], such as Daubechies transform [22]. It has been utilized in Reference 23 for data compression in a relational database.
3. *Regression*, in its simplest form, is the linear regression of a single variable model, that is, variable $Y$ as a linear function of variable $X$, $Y = \alpha + \beta X$, but regression can be extended to more than one variable [13].
4. *Log-linear models* are a methodology for approximating discrete multidimensional probability distributions; for example, the multiway table of joint probabilities is approximated by a product of lower-order tables [24].
5. *Histograms* approximate the data in one or more attributes of a relation by their frequencies [25].
6. *Clustering techniques* are discussed in Section 3.3.
7. *Indexing structures* are discussed in Section 3.7.
8. *Sampling* is an attempt to represent a large set of data by a small random sample of its data elements. Sampling in databases is reviewed in Reference 26, and Reference 27 is a representative publication.

We only consider SVD and related methods in our discussion in the context of the discussion in References 3 and 14. Consider matrix $X$, whose $M$ rows are the feature vectors for images. Each image has $N$ features, which is the number of columns in the matrix. Subtracting the mean from the elements in each column yields columns with a zero mean. The columns may be furthermore studentized by dividing the elements by the standard deviation. The SVD of $X$ is given as follows [8,28]:

$$X = U \Sigma V^T, \tag{3.1}$$

where $U$ is an $M \times N$ column-orthonormal matrix, $V$ is an $N \times N$ unitary matrix of *eigenvectors*, and $\Sigma$ is a diagonal matrix of *singular values*, without loss of generality ($\sigma_1 \geq \sigma_2 \geq \ldots, \geq \sigma_N$). When the number of singular values that are not zero or close to zero is $n$, then the rank of $X$ is $n$ [8]. The cost of computing eigenvalues using SVD is $O(MN)^2$. The following example shows the decomposition of $X$ and that it has a rank of three, since two of its eigenvalues are zero.

$$X=\begin{pmatrix}1&0&0&0&2\\0&0&3&0&0\\0&0&0&0&0\\0&4&0&0&0\end{pmatrix}\quad U=\begin{pmatrix}0&0&1&0\\0&1&0&0\\0&0&0&-1\\1&0&0&0\end{pmatrix}$$

$$\Sigma=\begin{pmatrix}4&0&0&0&0\\0&3&0&0&0\\0&0&\sqrt{5}&0&0\\0&0&0&0&0\end{pmatrix}\quad V^T=\begin{pmatrix}0&1&0&0&0\\0&0&1&0&0\\\sqrt{0.2}&0&0&0&\sqrt{0.8}\\0&0&0&1&0\\-\sqrt{0.8}&0&0&0&\sqrt{0.2}\end{pmatrix}$$

*PCA* is based on eigenvalue decomposition of the *covariance matrix*:

$$C=\frac{1}{M}X^T X=V\Lambda V^T, \tag{3.2}$$

where $V$ is the matrix of eigenvectors (as in Equation 3.1) and $\Lambda$ is a diagonal matrix holding the *eigenvalues* of $C$. $C$ is positive-semidefinite, so that its $N$ eigenvectors are orthonormal and its eigenvalues are nonnegative. The trace (sum of eigenvalues) of $C$ is invariant under rotation. The computational cost in this case is higher than SVD and requires a matrix multiplication $O(MN^2)$ followed by eigenvalue decomposition $O(N^3)$.

We assume that the eigenvalues, similar to the singular values, are in nonincreasing order, so that $\lambda_1 \geq \lambda_2 \geq \ldots \geq \lambda_N$. To show the relation between singular values and eigenvalues, we substitute $X$ in Equation 3.2 with its decomposition in Equation 3.1.

$$C=\frac{1}{M}X^T X=\frac{1}{M}(V\Sigma U^T)(U\Sigma V^T)=\frac{1}{M}V\Sigma^2 V^T.$$

It follows that

$$\Lambda=\frac{1}{M}\Sigma^2. \tag{3.3}$$

In effect $\lambda_i=\sigma_i^2/M$ and, conversely, $\sigma_i=\sqrt{M\lambda_i}$ for $1 \leq i \leq N$.

The eigenvectors constitute the principal components of $X$; hence, the transformation below yields uncorrelated features:

$$Y=XV. \tag{3.4}$$

Retaining the first $n$ dimensions of $Y$ maximizes the fraction of preserved variance and minimizes the NMSE:

$$\text{NMSE} = \frac{\sum_{i=1}^{M} \sum_{j=n+1}^{N} y_{i,j}^2}{\sum_{i=1}^{M} \sum_{j=1}^{N} y_{i,j}^2} = \frac{\sum_{j=n+1}^{N} \lambda_j}{\sum_{j=1}^{N} \lambda_j}. \quad (3.5)$$

The computational method in References 13 and 29 were used in References 3 and 14, respectively. A more efficient method was used in References 3 and 14 than in Reference 18 in computing the Euclidean distance with dimensionally reduced data with SVD as specified in the appendix.

*Recall* and *precision* are two metrics that are used to quantify retrieval accuracy and efficiency for approximate $k$-NN search with dimensionally reduced data [8]. Since SVD is a lossy data compression method, to obtain $k$ points, we retrieve $k^* \geq k$ points. Given that $k'$ of the retrieved points are among the $k$ desired nearest neighbors, $k' \leq k \leq k^*$. Precision (efficiency) is defined as $\mathcal{P} = k'/k^*$, and recall (accuracy) is defined as $\mathcal{R} = k'/k$. Thus, precision is the percentage of relevant images among the retrieved ones, that is, the ratio of retrieved-and-relevant to retrieved. Recall is the percentage of relevant images that were retrieved over the total number of relevant images in the image collection, so that recall is the ratio of retrieved-and-relevant to relevant. Thus, high precision means that we have few false alarms; high recall means that we have few false dismissals. Since $k$-NN search on dimensionally reduced data is inherently approximate, so that $\mathcal{R} < 1$, postprocessing is required to achieve exact $k$-NN query processing on dimensionally reduced data [30].

## 3.3 CLUSTERING METHODS

Cluster analysis is the process of partitioning a set of objects specified by points in multiple dimensions into clusters, so that the objects in a cluster are similar to each other but dissimilar from objects in other clusters. The requirements for a function on pairs of points to be a distance measure are as follows: (1) Distances are nonnegative, and the distance of a point to itself is zero. (2) Distances are symmetric. (3) Distance measures obey the triangle inequality, that is, the distance from A to B plus B to C is larger than the distance going from A to C directly. Similarity is defined by Euclidean or some other form of distance, in which case the three conditions are satisfied, but in some cases, distance is given directly.

As noted in Section 3.1, the cost of processing $k$-NN queries using the naive method can be reduced by applying clustering. With $H$ clusters, the cost is reduced by a factor $H$ provided when only one cluster is to be searched and all clusters have the same number of points. An even higher reduction in cost is attained when the point belongs to a small cluster.

A large number of clustering algorithms, which are applicable to various domains, have been developed over the years and are discussed in numerous textbooks dedicated to this topic [31–34] but also in texts on data mining [20] and Big Data [28,35]. Clustering algorithms over numerical data are of interest in our discussion. Four problems need to be overcome for clustering high-dimensional data [36].

1. Distance defined as $(D_{max} - D_{min})/D_{min}$, where $D_{max}$ and $D_{min}$ are the points with the farthest and least distance from a target point, decreases with increasing dimensionality. Setting $0 \leq p \leq 1$ has been proposed in References 37 and 38 to provide more meaningful distance measures.

$$D(\mathbf{u},\mathbf{v}) = \left[ \sum_{i=1}^{n} |u_i - v_i|^p \right]^{1/p}. \quad (3.6)$$

   $p = 1$ yields the Manhattan distance and $p = 2$ the Euclidean distance.

2. Clusters are intended to group objects based on attribute values, but some of the large number of attributes may not be meaningful for a given cluster. This is known as the local feature relevance problem: Different clusters might be found in different subspaces.
3. Subsets of attributes might be correlated; hence, clusters might exist in arbitrarily oriented affine subspaces (in affine geometry, the coordinates are not necessarily orthonormal).
4. Problem 4 is similar to problem 3 in that there may be some correlations among subsets of attributes.

Given $n$ points and $k$ clusters, the number of possible clusters is a Stirling number of the second kind defined as follows:

$$S(n,k) = \left\{ \begin{matrix} n \\ k \end{matrix} \right\} = \frac{1}{k!} \sum_{t=0}^{k} (-1)^t \binom{k}{t} (k-t)^n.$$

$S(n, 1) = S(n, n) = 1$, $S(4, 2) = 7$ because there are seven ways to partition four objects, a, b, c, d, into two groups:

(a)(bcd), (b)(acd), (c)(abd), (d)(abc), (ab)(cd), (ac)(bd), (ad)(bc)

Recursion can be used to compute $S(n, k) = kS(n - 1, k) + S(n - 1, k - 1)$, $1 \leq k < n$, and to show that $S(n.k)$ increases rapidly, for example, $S(10, 5) = 42525$.

Cluster analysis can be classified into the following methods.

### 3.3.1 Partitioning Methods

Partition $n$ data points based on their distances from each other into $k$ clusters. Each cluster has at least one point, and each point belongs to one cluster, unless a fuzzy clustering algorithm is considered, in which case the allocation of data points to clusters is not all or nothing [39]. In fact, the *expectation maximization* (*EM*) method provides soft assignment of points to clusters, such that each point has a probability of belonging to each cluster.

$k$-means is a good example of partitioning-based clustering methods, which can be specified succinctly as follows [40]:

1. Randomly select $H$ points from the data set to serve as preliminary centroids for clusters. Centroid spacing is improved by using the method in Reference 41.
2. Assign points to the closest centroid based on the Euclidean distance to form $H$ clusters.
3. Recompute the centroid for each cluster.
4. Go back to step 2, unless there is no new reassignment.

The quality of the clusters varies significantly from run to run, so that the clustering experiment is repeated and the one yielding the smallest *sum of squares* (*SSQ*) distances to the centroids of all clusters is selected:

$$SSQ = \sum_{h=1}^{H} \sum_{i \in C^{(h)}} \sum_{j=1}^{N} \left( x_{i,j} - \mu_j^{(h)} \right)^2. \tag{3.7}$$

$C^{(h)}$ denotes the $h$th cluster, with centroid $C_h$ with coordinates given by $\vec{\mu}^{(h)}$ and radius $R^h$, which is the distance of the farthest point in the cluster from $C_h$. SSQ decreases with increasing $H$, but a point of diminishing returns is soon reached. The $k$-means clustering method does not scale well to large high-dimensional data sets [42].

*Clustering Large Applications Based on Randomized Search* (*CLARANS*) [43] extends *Clustering Large Applications* (*CLARA*), described in Reference 33. Clustering is carried out with respect to representative points referred to as *medoids*. This method, also known as $k$-medoids, is computationally expensive because (1) each iteration requires trying out new partitioning representatives through an exchange process and (2) a large number of passes over the data set may be required. CLARANS improves efficiency by performing iterative hill climbing over a smaller sample.

### 3.3.2 Hierarchical Clustering

This is also called connectivity-based clustering since it connects objects to form clusters based on their distance. Hierarchical clustering methods fall into two categories:

1. Agglomerative or bottom up, where at first, each point is its own cluster, and pairs of clusters are merged as one moves up the hierarchy
2. Divisive or top down, where all observations start in one cluster, and splits are performed recursively as one moves down the hierarchy

Agglomerative clustering is $O(N^3)$, although faster implementations are known, while divisive clustering is $O(2^N)$.

*Balanced Iterative Reducing and Clustering Using Hierarchies* (*BIRCH*) can find a good clustering with a single scan of data with a limited amount of main memory available [44]. It improves the quality further with a few additional scans. A hierarchical data structure called CF-tree, which is a height-balanced tree, is used to store the clustering features (CFs). It is an order-sensitive clustering algorithm, which may detect different clusters for different input orders of the same data. BIRCH can only generate spherically shaped clusters.

Consider a cluster of $n$ $d$-dimensional data objects or points. The *CF*, $(CF)=(m, \overrightarrow{LS}, \overrightarrow{SS})$, where $\overrightarrow{LS}$ and $\overrightarrow{SS}$ are the vectors for the $N$ dimensions of linear sums and square sums of the $m$ points, is as follows:

$$\overrightarrow{LS}=\sum_{i=1}^{m} x_{i,j} \quad \overrightarrow{SS}=\sum_{i=1}^{m} x_{i,j}^2 .$$

In the first phase, BIRCH scans the database to build an initial in-memory CF-tree, which can be viewed as a multilevel compression of the data that tries to preserve the inherent clustering structure of data. In the second phase, BIRCH applies a clustering algorithm to cluster the leaf nodes of the CF-tree, which removes sparse clusters as outliers and groups dense clusters into larger ones. Using CF, we can easily derive many useful statistics of a cluster: the cluster's centroid, radius, and diameter. To merge two clusters C1 and C2, we simply have the following:

$$CF=(CF_1+CF_2)=(m_1+m2, \overrightarrow{LS_1}+\overrightarrow{LS_2}, \overrightarrow{SS_1}+\overrightarrow{SS_2}).$$

For two dimensions $X$ and $Y$, we have the following:

$$\overrightarrow{LS}=(LS_X, LS_Y)=\left(\sum_{i=1}^{m} x_i, \sum_{i=1}^{m} y_i\right), \overrightarrow{SS}=(SS_X, SS_Y)=\left(\sum_{i=1}^{m} x_i^2, \sum_{i=1}^{m} y_i^2\right).$$

The centroid is $(x_0, y_0) = (SS_X/m, SS_Y/m)$, and the radius is as follows:

$$R=\frac{1}{m}\sqrt{SS_X+SS_Y+(1/m-2)\left(LS_X^2+LS_Y^2\right)}.$$

Generalizing to $N$ dimensions with subscript $1 \le n \le N$,

$$R=\frac{1}{m}\sqrt{\sum_{n=1}^{N} SS_n+(1/m-2)\sum_{n=1}^{N} LS_n^2}.$$

An algorithm similar to BIRCH is described in Reference 45.

*Chameleon* is a hierarchical clustering algorithm that uses dynamic modeling to determine the similarity between pairs of clusters [46]. Cluster similarity is assessed based on (1) how well connected objects are within a cluster and (2) the proximity of clusters. Two clusters are merged if their interconnectivity is high and they are close together. A $k$-NN graph approach constructs a sparse graph, where each vertex of the graph represents a data object, and there is an edge between two vertices if one object is among the $k$ most similar objects to the other. The edges are weighted to reflect the similarity between objects. A graph-partitioning algorithm partitions the $k$-NN graph into a large number of relatively small subclusters such that it minimizes the cost of cutting edges [47]. Chameleon is more capable of discovering arbitrarily shaped clusters of high quality than several well-known algorithms such as BIRCH and *density-based spatial clustering of applications with noise* (*DBSCAN*) (see also FastMap [FM] in Reference 8).

*Clustering Using Representatives* (*CURE*) uses multiple representations to denote each cluster, so that clusters with nonspherical shapes can be discovered and there is less sensitivity to outliers [48]. The representatives are generated by selecting well-scattered points in the cluster and then shrinking them toward the center of the cluster by a specified fraction. At each iteration, the process merges two clusters until the number of clusters reaches the specified value. CURE uses a minimum heap to keep track of distances between one cluster and its closest cluster and one $k$-d-tree [49] to keep track of the representatives in all clusters. A heap extracts the cluster with the minimum distance by searching the nearest neighbors for all representatives in a cluster. CURE tends to be slow due to its dependence on the $k$-d-tree when handling large high-dimensional data sets.

*BUBBLE/FM* is an extension of BIRCH to distance spaces. The only operation possible on data objects is the computation of distance between them, while all scalable algorithms in the literature assume a $k$-dimensional vector space, which allows vector operations on objects [50]. The distance function associated with a distance space can be computationally expensive. For example, the distance between two strings is the edit distance, which, for strings of lengths $N$, requires $O(N^2)$ comparisons. BUBBLE-FM improves upon BUBBLE by reducing the number of calls to the distance function, which tends to be computationally expensive.

### 3.3.3 Density-Based Methods

Most partitioning methods for cluster objects based on the distance between objects can find only spherical-shaped clusters and encounter difficulty in discovering clusters of arbitrary shapes. The general idea behind density-based clustering methods is to continue growing a given cluster as long as the neighborhood density exceeds some threshold, so that clusters are defined as areas of higher density than the remainder of the data set. Objects in the sparse areas are usually considered to be noise and border points. Density-based methods can divide a set of objects into multiple exclusive clusters or a hierarchy of clusters.

*DBSCAN* is so called because it finds a number of clusters starting from the estimated density distribution of corresponding nodes [51]. DBSCAN's definition of a cluster is based on the notion of density reachability, that is, a point $\mathbf{q}$ is directly density-reachable from a point $\mathbf{p}$ if it is not farther away than a given distance $\varepsilon$.

*Ordering Points to Identify the Clustering Structure* (*OPTICS*) [52] can be seen as a generalization of DBSCAN to multiple ranges, effectively replacing the ε parameter with a maximum search radius. To detect meaningful data clusters of varying densities, the points of the database are (linearly) ordered, and points that are spatially closest become neighbors in the ordering.

### 3.3.4 Grid-Based Methods

These methods are based on quantizing the object space into a finite number of cells that form a grid structure. All the clustering operations are performed on the grid structure in quantized space. The main advantage of this approach is fast processing time, which is linearly proportional to the number of data objects. Grid-based methods can be integrated with other clustering methods.

A *statistical information grid* (*STING*) is a grid-based multiresolution clustering technique in which the embedding spatial area of the input objects is divided into rectangular cells, which can be divided in a hierarchical and recursive way [53]. Higher-level cells are partitioned to form a number of cells at the next lower level. Statistical information regarding the attributes in each grid cell, such as the mean, maximum, and minimum values, is precomputed.

*Clustering in Quest* (*CLIQUE*) is a simple grid-based method and is also a subspace clustering method [54], which is discussed further in Section 3.3.5.

*WaveCluster* is a multiresolution clustering algorithm that first summarizes the data by imposing a multidimensional grid structure [55]. It then applies a wavelet transform to the original feature space, finding dense regions in the transformed space. In this approach, each grid cell summarizes the information of a group of points that map into the cell. This summary information typically fits into the main memory for use by the multiresolution wavelet transform and the subsequent cluster analysis.

*Optimal grid partitioning* (*OptiGrid*) constructs an optimal grid on data by calculating the best partitioning hyperplanes for each dimension [56]. The advantages of OptiGrid are as follows: (1) It has a firm mathematical basis. (2) It is by far more effective than existing clustering algorithms for high-dimensional data. (3) It is very efficient even for large data sets of high dimensionality.

### 3.3.5 Subspace Clustering Methods

Subspace clustering methods find clusters in different, lower-dimensional subspaces of a data set, taking advantage of the fact that many dimensions in high-dimensional data sets are redundant and hide clusters in noisy data. The curse of dimensionality also makes distance measures meaningless to an extent that data points are equidistant from each other.

A recent survey on subspace clustering clarifies the following issues [36]: (1) different problem definitions related to subspace clustering; (2) specific difficulties encountered in this field of research; (3) varying assumptions, heuristics, and intuitions forming the basis of different approaches; and (4) prominent solutions for tackling different problems.

The bottom-up search method takes advantage of the downward closure property of density to reduce the search space, using an a priori-style approach in data mining [20], which means that if there are dense units in $k$ dimensions, there are dense units in all $(k - 1)$ dimensional projections. Algorithms first create a histogram for each dimension and select those bins with densities above a given threshold. Candidate subspaces in two dimensions can then be formed using only those dimensions that contain dense units, dramatically reducing the search space. The algorithm proceeds until there are no more dense units found. Adjacent dense units are then combined to form clusters.

*CLIQUE* is the first subspace clustering algorithm combining density and grid-based clustering [54], which was developed in conjunction with the Quest data mining project at IBM [57]. An a priori-style search method was developed for *association rule mining* (*ARM*) to find dense subspaces [54]. Once the dense subspaces are found, they are sorted by coverage; only subspaces with the greatest coverage are kept, and the rest are pruned. The algorithm then looks for adjacent dense grid units in each of the selected subspaces using a depth-first search. Clusters are formed by combining these units using a greedy growth scheme. The algorithm starts with an arbitrary dense unit and greedily grows a maximal region in each dimension, until the union of all the regions covers the entire cluster. The weakness of this method is that the subspaces are aligned with the original dimensions.

The *Entropy-Based Clustering Method* (*ENCLUS*) is based heavily on the CLIQUE algorithm, but instead of measuring density or coverage, it measures entropy based on the observation that a subspace with clusters typically has lower entropy than a subspace without clusters, that is, entropy decreases as cell density increases [58]. ENCLUS uses the same a priori-style, bottom-up approach as CLIQUE to mine significant subspaces. Entropy can be used to measure all three clusterability criteria: coverage, density, and correlation. The search is accomplished using the downward closure property of entropy and the upward closure property of correlation to find minimally correlated subspaces. If a subspace is highly correlated, all of its superspaces must not be minimally correlated. Since non-minimally correlated subspaces might be of interest, ENCLUS searches for interesting subspaces by calculating interest gain and finding subspaces whose entropy exceeds a certain threshold. Once interesting subspaces are found, clusters can be identified using the same methodology as CLIQUE or any other existing clustering algorithms.

*Merging of Adaptive Finite Intervals (and more than a CLIQUE)* (*MAFIA*) extends CLIQUE by using an adaptive grid based on the distribution of data to improve efficiency and cluster quality [59]. MAFIA initially creates a histogram to determine the minimum number of bins for a dimension. The algorithm then combines adjacent cells of similar density to form larger cells. In this manner, the dimension is partitioned based on the data distribution, and the resulting boundaries of the cells capture the cluster perimeter more accurately than fixed-sized grid cells. Once the bins have been defined, MAFIA proceeds much like CLIQUE, using an a priori-style algorithm to generate the list of clusterable subspaces by building up from one dimension.

*Cell-based clustering* (*CBF*) addresses scalability issues associated with many bottom-up algorithms [60]. One problem for other bottom-up algorithms is that the number of bins created increases dramatically as the number of dimensions increases. CBF uses a cell creation algorithm that creates optimal partitions based on minimum and maximum values on a given dimension, which results in the generation of fewer bins. CBF also addresses scalability with respect to the number of instances in the data set, since other approaches often perform poorly when the data set is too large to fit in the main memory. CBF stores the bins in an efficient filtering-based index structure, which results in improved retrieval performance.

A *cluster tree* (*CLTree*) is built using a bottom-up strategy, evaluating each dimension separately and using dimensions with high density in further steps [61]. It uses a modified decision tree algorithm to adaptively partition each dimension into bins, separating areas of high density from areas of low density. The decision tree splits correspond to the boundaries of bins.

*Density-Based Optimal Projective Clustering* (*DOC*) is a Monte Carlo algorithm combining the grid-based approach used by the bottom-up approaches and the iterative improvement method from the top-down approaches [62].

*Projected clustering* (*PROCLUS*) is a top-down subspace clustering algorithm [63]. Similar to CLARANS, PROCLUS samples the data and then selects a set of $k$ medoids and iteratively improves the clustering. The three phases of PROCLUS are as follows. (1) *Initialization phase*: Select a set of potential medoids that are far apart using a greedy algorithm. (2) *Iteration phase*: Select a random set of $k$ medoids from this reduced data set to determine if clustering quality improves by replacing current medoids with randomly chosen new medoids. Cluster quality is based on the average distance between instances and the nearest medoid. For each medoid, a set of dimensions is chosen whose average distances are small compared to statistical expectation. Once the subspaces have been selected for each medoid, average Manhattan segmental distance is used to assign points to medoids, forming clusters. (3) *Refinement phase*: Compute a new list of relevant dimensions for each medoid based on the clusters formed and reassign points to medoids, removing outliers. The distance-based approach of PROCLUS is biased toward clusters that are hype-spherical in shape. PROCLUS is actually somewhat faster than CLIQUE due to the sampling of large data sets; however, using a small number of representative points can cause PROCLUS to miss some clusters entirely.

An *oriented projected cluster* (*ORCLUS*) looks for non–axis-parallel subspaces [64]. Similarly to References 3 and 14, this method is based on the observation that many data sets contain interattribute correlations. The algorithm can be divided into three steps. (1) *Assign phase:* Assign data points to the nearest cluster centers. (2) *Subspace determination:* Redefine the subspace associated with each cluster by calculating the covariance matrix for a cluster and selecting the orthonormal eigenvectors with the smallest eigenvalues. (3) *Merge phase:* Merge cluster that are near each other and have similar directions of least spread.

The *Fast and Intelligent Subspace Clustering Algorithm Using Dimension Voting* (*FINDIT*) uses *dimension-oriented distance* (*DOD*) to count the number of dimensions in which two points are within a threshold distance of each other, based on the assumption that in higher dimensions, it is more meaningful for two points to be close in several dimensions rather than in a few [65]. The algorithm has three phases. (1) *Sampling phase*: Select two small sets generated through random sampling to determine initial representative medoids of the clusters. (2) *Cluster forming phase*: Find correlated dimensions using the DOD measure for each medoid. (3) *Assignment phase*: Assign instances to medoids based on the subspaces found. FINDIT employs sampling techniques like the other top-down algorithms to improve performance with very large data sets.

The δ-*clusters* algorithm uses a distance measure to capture the coherence exhibited by a subset of instances on a subset of attributes [66]. Coherent instances may not be close, but instead, both follow a similar trend, offset from each other. One coherent instance can be derived from another by shifting by an offset. The algorithm starts with initial seeds and iteratively improves the overall quality of the clustering by randomly swapping attributes and data points to improve individual clusters. Residue measures the decrease in coherence that a particular entry (attribute or instance) brings to the cluster. The iterative process terminates when individual improvement levels off in each cluster.

*Clustering on Subsets of Attributes* (*COSA*) assigns weights to each dimension for each instance, not each cluster [67]. Starting with equally weighted dimensions, the algorithm examines the *k*-NNs of each instance. These neighborhoods are used to calculate the respective dimension weights for each instance, with higher weights assigned to those dimensions that have a smaller dispersion within the *k*-NN group. These weights are used to calculate dimension weights for pairs of instances, which are used to update the distances used in the *k*-NN calculation. The process is repeated using the new distances until the weights stabilize. The neighborhoods for each instance become increasingly enriched with instances belonging to its own cluster. The dimension weights are refined as the dimensions relevant to a cluster receive larger weights. The output is a distance matrix based on weighted inverse exponential distance and is suitable as input to any distance-based clustering method. After clustering, the weights for each dimension of cluster members are compared, and an overall importance value for each dimension is calculated for each cluster.

There are many others, which are reviewed in Reference 61. The *multilevel Mahalanobis-based dimensionality reduction* (*MMDR*) [68] clusters high-dimensional data sets using the low-dimensional subspace based on the Mahalanobis rather than the Euclidean distance, since it is argued that locally correlated clusters are elliptical shaped. The Mahalanobis distance uses different coefficients according to the dimension [8].

## 3.4 STEPS IN BUILDING AN INDEX FOR *k*-NN QUERIES

The following steps were followed combining SVD and clustering in References 3 and 14 for building the data structures to carry out *k*-NN queries.

1. Selecting an objective function in step 1
2. Selecting the number of clusters in step 2
3. Partitioning the data set of points in step 3
4. Rotating each partition into an uncorrelated frame of reference in step 4
5. Reducing the dimensionality of the partitions in step 5
6. Constructing the within-cluster index in step 6

A detailed description of the steps of building an index for $k$-NN queries is as follows.

1. CSVD supports three *alternative* objective functions: (1) the index space compression, (2) the NMSE (defined in Equation 3.5), and (3) desired recall.

   The compression of the index space is defined as the ratio of the original size of the database (equal to $N \cdot M$) to the size of its CSVD description, which is specified as follows:

$$V = N \cdot H + \sum_{h=1}^{H} \left( N \cdot p_h + m_h \cdot p_h \right), \tag{3.8}$$

   where $m_h$ is the number of points, $p_h$ is the number of dimensions retained in the $h$th cluster, and $H$ is the number of clusters. The term $N \cdot H$ accounts for the space required by the centroids, while $N \cdot p_h$ is the space occupied by the projection matrix and $m_h \cdot p_h$ is the space required by the projected points.

2. The system selects the number of clusters $H$ as a function of the database size and of the objective function. Alternatively, the desired number of clusters can be specified by the user. Some clustering algorithms determine the desired number of clusters.

3. The *partitioning step* divides the row of the database table $\mathbf{X}$ into $H$ clusters: $\mathbf{X}^{(h)}, h = 1, \ldots, H$, each containing vectors that are close to each other with $m_1, \ldots, m_H$ points. CSVD supports a wide range of classical clustering methods such as the *Linde–Buzo–Gray (LBG)* algorithm [69] (the default option in Reference 3) and the $k$-means methods [34]. For each cluster, this step also yields the centroid $\mu^{(h)}$ and the *cluster radius* $R^{(h)}$, defined as the distance between the centroid and the farthest point of the cluster.

4. The *rotation step* is carried out separately for each of the clusters $\mathbf{X}^{(h)}, h = 1, \ldots, H$. The corresponding centroid $\mu^{(h)}$ is subtracted from each group of vectors $\mathbf{X}^{(h)}$, and the eigenvector matrix $\mathbf{V}^{(h)}$ and the eigenvalues $\lambda_1^{(h)}, \ldots, \geq \lambda_N^{(h)}$ are computed as described in Section 3.2. Eigenvalues are ordered by decreasing magnitude. The vectors of $\mathbf{X}^{(h)}$ are rotated in the uncorrelated reference frame, having the eigenvectors as coordinate axes to produce $\tilde{\mathbf{X}}^{(h)}$.

5. The *dimensionality reduction step* is a global procedure, which depends on the selected objective function. The products of the eigenvalues $\lambda_i^{(h)}$ and the number of points of the corresponding clusters $m_h$ are computed and sorted in ascending order to produce an ordered list $\mathcal{L}$ of $H \cdot N$ elements; each element $j$ of the list contains the eigenvalue-cluster size product, the label $\kappa_j$ of the corresponding cluster, and the dimension $\partial_j$ associated with the eigenvalue. The list $\mathcal{L}$ is walked starting from its head. During step $j$, the $\partial_j^{th}$ dimension of the $\kappa_j^{th}$ cluster is discarded. The process ends when the target value of the objective function is reached.

The NMSE for a clustered data set is given as follows:

$$\text{NMSE} = \frac{\sum_{h=1}^{H}\sum_{i=1}^{m_h}\sum_{j=n_h+1}^{N}\left(y_{i,j}^{(h)}\right)^2}{\sum_{h=1}^{H}\sum_{i=1}^{m_h}\sum_{j=1}^{N}\left(y_{i,j}^{(h)}\right)^2} = \frac{\sum_{h=1}^{H} m_h \sum_{j=n_h+1}^{N} \lambda_j^{(h)}}{\sum_{h=1}^{H} m_h \sum_{j=1}^{N} \lambda_j^{(h)}}. \quad (3.9)$$

*Index space compression*: The index volume is computed as successive dimensions in the transposed space are removed.

*TNMSE*: The target NMSE (TNMSE) is recomputed as successive dimensions are omitted.

*Recall*: As dimensions are omitted, an experiment is run to determine the recall with a sufficiently large number of sample queries [3].

6. The *within-cluster index-construction step* operates separately on each cluster $\mathbf{Y}^{(h)}$. Following data reduction, each cluster is much more amenable to efficient indexing than the entire table. The OP-tree proposed in Reference 2 and utilized in Reference 3 is very efficient for $k$-NN search in medium- and even high-dimensionality spaces.

## 3.5 NEAREST NEIGHBORS QUERIES IN HIGH-DIMENSIONAL SPACE

Queries on high-dimensional data can be classified as follows [70]: (1) *point queries* to determine if an object with a specified feature vector is available in the database, (2) *range queries* to determine if objects within a certain distance are available in the database, (3) *k-NN queries*, and (4) *spatial join* an all closest pairs of points.

Similarity between two objects $U$ and $V$, $\mathbf{u} = [u_1, \ldots, u_N]^T$ and $\mathbf{v} = [v_1, \ldots, v_N]^T$, can be measured by their *Euclidean distance*:

$$D(\mathbf{u},\mathbf{v}) = \left[(\mathbf{u}-\mathbf{v})^T(\mathbf{u}-\mathbf{v})\right]^{1/2} = \left[\|\mathbf{u}\|^2 + \|\mathbf{v}\|^2 - 2\mathbf{u}^T\mathbf{v}\right]^{1/2} = \sqrt{\sum_{i=1}^{n}(u_i - v_i)^2}. \quad (3.10)$$

Given precomputed vector norms, we need to compute only the inner product of the two vectors.

Similarity search based on $k$-NN queries, which determine the $k$ images that are closest to a target image, is expensive when the number of dimensions per image ($N$) is in the hundreds and the number of images ($M$) is in the millions. The $k$-NNs can be determined by a sequential scan of an $M \times N$ matrix $X$. Retrieved points are inserted into a heap, as long as they are no further than the $k$ nearest points extracted so far. Determining $k$-NN queries in high dimensions is not a trivial task, due to "*the curse of dimensionality*" [71].

> As dimensionality increases, all points begin to appear almost equidistant from one another. They are effectively arranged in a d-dimensional sphere around the query, no matter where the query (point) is located. The radius of the sphere increases,

while the width of the sphere remains unchanged, and the space close to the query point remains unchanged [71].

Three oddities associated with high-dimensional spaces are as follows:

1. A $2^d$ partition of a unit hypercube has a limited number regions for small $d$, but with $d = 100$ there are $2^{100} \approx 10^{30}$ regions, so that even with billions of points almost all of the regions are empty.
2. A $0.95 \times 0.95$ square enclosed in a $1 \times 1$ square occupies 90.25% of its area, while for 100 dimensions, $0.95^{100} \approx 0.00592$.
3. A circle with radius 0.5 enclosed in a unit square occupies 3.14159/4 = 78.54% of its area. For $d = 40$, the volume of the hypersphere is $3.278 \times 10^{21}$. All the space is in the corners, and approximately $3 \times 10^{20}$ points are required to expect one point inside the sphere.

Nearest neighbor queries follow a branch-and-bound algorithm. When the search begins, the feasible region of the search problem is the entire space, and the partition is given by the clustering. The target function (the distance of the $k$th neighbor) is lower-bounded by the distance from the clusters; during the search, the target function is upper-bounded by the distance of the current $k$th neighbor, and the feasible regions are pruned by implicitly discarding the clusters having a distance larger than the running upper bound.

The following steps are used in carrying out the $k$-NN queries:

1. *Preprocessing and primary cluster identification.*
2. *Computation of distances from clusters*: The distance between the preprocessed query point $\mathbf{q}$ and cluster $i$ is defined as

$$\max\{[D(\mathbf{q}, \mu^{(i)} - R^{(i)}],0\}.$$

   The clusters are sorted in increasing order of distance, with the primary cluster in first position.
3. *Searching the primary cluster*: This step produces a list of $k$ results, in increasing order of distance; let $d_{\max}$ be the distance of the farthest point in the list.
4. *Searching the other clusters*: If the distance to the next cluster does not exceed $d_{\max}$, then the cluster is searched, otherwise, the search is terminated. While searching a cluster, if points closer to the query than $d_{\max}$ are found, they are added to the list of $k$ current best results, and $d_{\max}$ is updated.

Postprocessing takes advantage of *Lower-Bounding Property* (*LBP*) [8] (lemma 1 in Reference 72):

5. *Given that the distance of points in the subspace with reduced dimensions is less than the original distance, a range query guarantees no false dismissals.*

False alarms are discarded by referring to the original data set. Noting the relationship between range and $k$-NN queries, the latter can be processed as follows [72]: (1) Find the $k$-NNs of the query point $Q$ in the subspace. (2) Determine the farthest actual distance to $Q$ among these $k$ points ($d_{max}$). (3) Issue a range query centered on $Q$ on the subspace with radius $\varepsilon = d_{max}$. (4) For all points obtained in this manner, find their original distances to $Q$, by referring to the original data set and ranking the points, that is, select the $k$-NNs.

The exact $k$-NN processing method was extended to multiple clusters in Reference 30, where we compare the CPU cost of the two methods as the NMSE is varied using two data sets. An offline experiment was used to determine $k^*$, which yields a recall $R \approx 1$. The CPU time required by the exact method for a sequential scan is lower than the approximate method, even when $\mathcal{R} = 0.8$. This is attributable to the fact that the exact method issues a $k$-NN query only once, and this is followed by less costly range queries. Optimal data reduction is studied in the context of the exact query processing [73].

RCSVD was explored in Reference 14 in the context of sequential data set scans. The OP-tree index [2], which is a main memory index, which is suited for $k$-NN queries, is utilized in Reference 3. Two methods to make this index persistent are described in References 74 and 75. We also proposed a new index, the *stepwise dimensionally increasing* (*SDI*) index, which is shown to outperform other indices [74].

## 3.6 ALTERNATE METHOD COMBINING SVD AND CLUSTERING

The *LDR* method described in Reference 15 generates SVD-friendly clusters so that a higher-dimensionality reduction can be attained for the same NMSE. The description of the LDR method in what follows is based on the appendix in Reference 3.

1. *Initial selection of centroids. H* points with pairwise distances of at least *threshold* are selected.

2. *Clustering.* Each point is associated to the closest centroid, whose distance is less than $\varepsilon$ or ignored otherwise. The coordinates of the *centroid* are the average of the associated points.

3. Apply SVD to each cluster. Each cluster now has a reference frame.

4. *Assign points to clusters.* Each point in the data set is analyzed in the reference frame of each cluster. For each cluster, determine the minimum number of retained dimensions so that the squared error in representing points does not exceed $\text{MaxReconDist}^2$. The cluster requiring the minimum number of dimensions $N_{min}$ is determined. If $N_{min} \leq \text{MaxDim}$, the point is associated with that cluster, and the required number of dimensions is recorded; otherwise, the point becomes an outlier.

5. *Compute the subspace dimensionality of each cluster.* Find the minimum number of dimensions ensuring that at most, FracOutliers of the points assigned to the cluster in step 4 do not satisfy the $\text{MaxReconDist}^2$ constraints. Each cluster now has a *subspace.*

6. *Recluster points.* Each point is associated with the first cluster, whose subspace is at a distance less than MaxReconDist from the point.

7. *Remove small clusters.* Clusters with size smaller than MinSize are removed and the corresponding points reclustered in step 6.

8. *Recursively apply steps 1–7 to the outlier set.* Make sure that the centroids selected in step 1 have a distance at least *threshold* from the current valid clusters. The procedure terminates when no new cluster is identified.

9. *Create an outlier list.* As in Reference 18, this is searched using a linear scan. Varying the NMSE may yield different compression ratios. The smaller the NMSE, the higher the number of points that cannot be approximated accurately.

The values for $H$, threshold, $\varepsilon$, MaxReconDist, MaxDim, FracOutliers, and MinSize need to be provided by the user. This constitutes the main difficulty with applying the LDR method. The invariant parameters are MaxReconDist, MaxDim, and MinSize. The method can produce values of $H$, threshold, and overall FracOutliers, which are significantly different from the inputs. The fact that the LDR method determines the number of clusters is one of its advantages with respect to the $k$-means method.

The LDR method also takes advantage of *LBP* [8,72] to attain exact query processing. It produces fewer false alarms by incorporating the reconstruction distance (the squared distance of all dimensions that have been eliminated) as an additional dimension with each point. It is shown experimentally in Reference 3 that RCSVD outperforms LDR in over 90% of cases.

## 3.7 SURVEY OF HIGH-DIMENSIONAL INDICES

Multidimensional indexing has been an active research area in the last two decades [8,16,17,70,76]. Given the distinction between *point access methods* (*PAMs*) and *spatial access methods* (*SAMs*) [16], we are interested here in PAMs for high-dimensional data. Multidimensional indexing structures can be classified as *data partitioning* (*DP*), such as R-trees, similarity search (SS)-trees, and spherical–rectangular (SR)-trees, and *space partitioning* (*SP*), such as $k$-d-B-trees and holey Brick (hB) trees [70]. In DP, space is partitioned in a manner such that regions have an approximately equal number of points. SP divides space into nonoverlapping regions, so that any point in the space can then be identified to lie in exactly one of the regions. Another categorization is disk-resident versus main memory-resident indices. In the former (respectively latter) case, the number of disk pages accessed and CPU time are the major performance measures.

Quad trees [77] and $k$-d-trees [49] are two early main memory-resident indexing structures. $k$-d-B-trees combine $k$-d-trees and B-trees [78]. R-trees [79] were designed to handle spatial data on secondary storage. $R^+$-trees [80] and $R^*$-trees [81] are two improved versions of R-trees. SS-trees [5] use hyperspheres to partition the space, while SR-trees [82] use the intersection of hyperspheres and hyperrectangles to represent regions. SR-trees outperform both $R^*$-trees and SS-trees.

It is known that the nearest neighbor search algorithms based on the Hjaltason and Samet (HS) method [70,83] outperform the Roussopoulos, Kelley, and Vincent (RKV) method [84]. This was also ascertained experimentally as part of carrying out experimental studies on indexing structures in Reference 85. The RKV algorithm provided with SR-tree code was replaced with the HS algorithm in this study. In computing distances, we used the method specified in Reference 73, which is more efficient than the method in Reference 18. This method is specified in Section 3.8.

With the increasing dimensionality of feature vectors, most multidimensional indices lose their effectiveness. The so-called dimensionality curse [8] is due to an increased overlap among the nodes of the index and a low fan-out, which results in increased index height. For a typical feature vector-based hierarchical index structure, each node corresponds to an 8-kilobyte (KB) page. Given a fixed page size $S$ (minus space dedicated to bookkeeping), number of dimensions $N$, and $s$ = sizeof(data type), the fan-out for different index structures is as follows:

*Hyperrectangles*: R*-trees consist of a hierarchy of hyperrectangles, with those at the higher levels embedding lower-level ones [81]. A hyperrectangle is specified uniquely by the lower left and upper right coordinates in $N$-dimensional space. Alternatively, the centroid and its distance from the $N$ "sides" of the hyperrectangle may be specified. The cost is the same in both cases, so that the fan-out is $F \approx S/(2 \times N \times s)$. Implementing an R*-tree as an SAM, rather than a PAM, represents points as hyperrectangles, doubling the cost of representing points.

*Hyperspheres*: SS-trees consist of a tree of hyperspheres [5], with each node embedding the nodes at the lower level. SS-trees have been shown to outperform R-trees. Each hypersphere is represented by its centroid in $N$-dimensional space and its radius, that is, the fan-out is $F \approx S/((N + 1) \times s)$.

*Hyperspheres and hyperrectangles*: SR-trees [82] combine SS-trees with R-trees, which encapsulate all of the points in the index. The region of the index is the intersection of the bounding hyperrectangle and the bounding hypersphere, which results in a significant reduction in the size of the region, since the radius of the hypersphere is determined by the distance of the farthest point from its centroid. The space requirement per node is the sum of the space requirements in SS-trees and R*-trees, so that the fan-out is $F = S/((3 \times N + 1)s)$.

*Hybrid trees*, proposed in Reference 86, in conjunction with the LDR method discussed in Section 3.6, combine the positive aspects of DP and SP indexing structures into one. Experiments on "real" high-dimensional large-size feature databases demonstrate that the hybrid tree scales well to high dimensionality and large database sizes. It significantly outperforms both purely DP-based and SP-based index mechanisms as well as linear scan at all dimensionalities for large-sized databases.

*X-trees* combine linear and hierarchical structures using supernodes to avoid overlaps to attain improved performance for $k$-NN search for higher dimensions [87].

*Pyramid trees* [88] are not affected by the curse of dimensionality. The iMinMax(θ) [89] method, which maps points from high- to single-dimensional space, outperforms pyramid tress in experiments for range queries.

*iDistance* is an efficient method for $k$-NN search in a high-dimensional space [90,91]. Data are partitioned into several clusters, and each partition has a reference point, for example, the centroids of each cluster obtained using the $k$-means clustering method [40]. The data in each cluster are transformed into a single-dimensional space according to the similarity with respect to a reference point. The one-dimensional values of different clusters are disjoint. The one-dimensional space is indexed by a $B^+$ index, and $k$-NN search is implemented using range searches. The search starts with a small radius, but it is increased step by step to form a bigger query sphere. iDistance is lossy since multiple data points in the high-dimensional space may be mapped to the same value in the single-dimensional space. The SDI method is compared with the iDistance method in Reference 74.

A *vector approximation file* (*VA-file*) represents each data object using the cell into which it falls [9]. Cells are defined by a multidimensional grid, where dimension $i$ is partitioned $2^{b_i}$ ways. Due to the sparsity of high-dimensional space, it is very unlikely that several points can share a cell. Nearest neighbor search sequentially scans the VA-file to determine the upper-bound and lower-bound distance from the query to each cell. This is followed by a refinement or filtering step; if the approximate lower bound is greater than the current upper bound, it is not considered further. Otherwise, it is a candidate. During the refine step, all the candidates are sorted according to their lower-bound distances.

A *metric tree* transforms the feature vector space into metric space and then indexes the metric space [92]. A metric space is a pair, $\mathcal{M} = (\mathcal{F}, d)$, where $\mathcal{F}$ is a domain of feature value, and $d$ is a distance function with the following properties:

1. Symmetry $d(F_x, F_y) = d(F_y, F_x)$
2. Nonnegativity $d(F_x, F_y) > 0,\ F_x \neq F_y$
3. Triangle inequality $d(F_x, F_y) \leq d(F_x, F_z) + d(F_z, F_y)$

The search space is organized based on relative distances of objects, rather than their absolute positions in a multidimensional space.

A *vantage point* (*VP*)-*tree* [93] partitions a data set according to distances between the objects and a reference or vantage point. The corner point is chosen as the vantage point, and the median value of the distances is chosen as separating radius to partition the data set into two balanced subsets. The same procedure is applied recursively on each subset. The *multiple vantage point* (*mvp*)-tree [94] uses precomputed distances in the leaf nodes to provide further filtering during search operations. Both trees are built in a top-down manner; balance cannot be guaranteed during insertion and deletion. Costly reorganization is required to prevent performance degradation.

An *M-tree* is a paged metric-tree index [95]. It is balanced and able to deal with dynamic data. Leaf nodes of an M-tree store the feature vectors of the indexed objects $O_j$ and distances to their parents, whereas internal nodes store routing objects $O_r$, distances to their parents $O_p$, covering radii $r(O_r)$, and corresponding covering tree pointers. The M-tree reduces the number of distance computations by storing distances.

The *OMNI-family* is a set of indexing methods based on the same underlying theory that all the points $S_i$ located between $\ell$ and upper $u$ radii are candidates for a spherical query with radius $r$ and given point $Q$ for a specific focus $F_i$, where $\ell = d(Q, F_i) - r$, $u = d(Q, F_i) + r$ [96]. For multiple foci, the candidates are the intersections of $S_i$. Given a data set, a set of foci is to be found. For each point in the data set, the distance to all of the foci is calculated and stored. The search process can be applied to sequential scan, B⁺-trees, and R-trees. For B⁺-trees, the distances for each focus $F_i$ are indexed, a range query is performed on each index, and finally, the intersection is obtained. For R-trees, the distances for all the foci, which form lower-dimensional data, are indexed, and a single range query is performed.

A *Δ-tree* is a main memory-resident index structure, which represents each level with a different number of dimensions [97]. Each level of the index represents the data space starting with a few dimensions and expanding to full dimensions, while keeping the fan-out fixed. The number of dimensions increases toward the leaf level, which contains full dimensions of the data. This is intended to minimize the cache miss ratio as the dimensionality of feature vectors increases. At level $\ell \geq 1$, the number of dimensions is selected as the smallest $n_\ell$ satisfying

$$\sum_{k=1}^{n_\ell} \lambda_k \Big/ \sum_{k=1}^{N} \lambda_k \geq \min(\ell p, 1).$$

The nodes of the index increase in size from the highest level to the lowest level, and the tree may not be height balanced, but this is not a problem since the index is main memory resident. The index with shorter feature vector lengths attains a lower cache miss ratio, reduced cycles per instruction (CPI), and reduced CPU time [98].

The *telescopic vector* (*TV*)*-tree* is a disk-resident index with nodes corresponding to disk pages [99]. TV-trees partition the data space using *telescopic minimum bounding regions* (*TMBRs*), which have telescopic vectors as their centers. These vectors can be contracted and extended dynamically by telescopic functions defined in Reference 99, only if they have the same number of active dimensions ($\alpha$). Features are ordered using the KLT applied to the whole data set, so that the first few dimensions provide the most discrimination. The discriminatory power of the index is heavily affected by the value of the parameter $\alpha$, which is difficult to determine. In case the number of levels is large, the tree will still suffer from the curse of dimensionality. The top levels of TV-trees have higher fan-outs, which results in reducing the input/output (I/O) cost for disk accesses. Experimental results on a data set consisting of dictionary words are reported in Reference 99.

The *OP-tree index* described in Reference 2 for efficient processing of $k$-NN queries recursively equipartitions data points one dimension at a time in a round-robin manner until the data points can fit into a leaf node. The two properties of OP-trees are ordering and partitioning. Ordering partitions the search space, and partitioning rejects unwanted space without actual distance computation. A fast $k$-NN search algorithm by reducing the distance based on the structure is described. Consider the following sample data set:

**1**: (1, 2, 5), **2**: (3, 8, 7), **3**: (9, 10, 8), **4**: (12, 9.2), **5**: (8, 7, 20), **6**: (6.6.23),

**7**: (0, 3, 27), **8**: (2, 13.9), **9**: (11, 11, 15), **10**: (14, 17, 13), **11**: (7, 14, 12), **12**: (10, 12, 3).

As far as the first partition is concerned, we partition the points into three regions, $R_1(-\infty, 3)$, $R_2(3, 6)$, $R_3(6, +\infty)$.

$R_1$ is partitioned into subregions $R_{1,1}(-\infty, 2)$, $R_{1,2}(2, 8)$, $R_{1,3}(8, \infty)$;

$R_2$ is partitioned into subregions $R_{2,1}(-\infty, 4)$, $R_{2,2}(4, 7)$, $R_{2,3}(7, +\infty)$; and

$R_3$ is partitioned into subregions $R_{3,1}(-\infty, 9)$, $R_{3,2}(9, 12)$, $R_{3,3}(12, +\infty)$.

We next specify the points held by the partitions

$R_{1,2}$(**1**); $R_{2,1}$(**2, 7**); $R_{3,1}$(**8**);

$R_{2,1}$(**11**); $R_{2,2}$(**5, 6**); $R_{2,3}$(**3**);

$R_{3,1}$(**4**); $R_{3,2}$(**12**), (**9**); and $R_{3,3}$(**10**).

The building of a persistent semi-dynamic OP-tree index is presented in References 74 and 75. We use serialization to compact the dynamically allocated nodes of the OP-tree in the main memory, which form linked lists, into a contiguous area. The index can then be saved onto disk as a single file and loaded into the main memory with a single data transfer. This is because disk positioning time to load the index one page at a time is high and is improving at a very slow rate, while the improvement in disk transfer rates is significant [11]. The prefetching effect of modern operating systems results in the loading of the whole index into the main memory in experimental studies with small indices. The original OP-tree is static, in which case the OP-tree has to be rebuilt in the main memory as new points are added. We propose and compare the performance and space efficiency of several methods, which support inserting points dynamically.

The *SDI index* uses a reduced number of dimensions at the higher levels of the tree to increase the fan-out, so that the tree height is reduced [100]. The number of dimensions increases level by level, until full dimensionality is attained at the lower levels. The SDI index differs from the Δ-tree in that it is a disk-resident index structure with fixed node sizes, while the Δ-tree is a main memory-resident index with variable node sizes and fixed fan-outs. The SDI-tree differs from the TV-tree in that it uses a single parameter, specifying the fraction of variance to be added to each level, without the risk of having a large number of active dimensions.

SDI trees are compared to variance approximate median (VAM) SR-trees and the VA-file indexes in Reference 100. The VAMSR-tree uses the same split algorithm as the VAMSplit R-tree [101], but it is based on an SR-tree structure, which is statically built in a bottom-up manner. The data set is recursively split top down using the dimension with the maximum variance and choosing a pivot, which is approximately the median.

## 3.8 CONCLUSIONS

Similarity search via $k$-NN queries applied to feature vectors associated to images is a well-known paradigm in CBIR. Applying $k$-NN queries to a large data set, where the number of images is in the millions and the number of features is in the hundreds, is quite costly, so we present a combination of dimensionality reduction via SVD and related methods, clustering, and indexing to achieve a higher efficiency in $k$-NN queries.

SVD eliminates dimensions with little variance, which has little effect on the outcome of $k$-NN queries, although this results in an approximate search, so that precision and recall are required to quantify this effect. Dimensionality reduction allows more efficiency in applying clustering and building index structures. Both clustering and indexing reduce the cost of $k$-NN processing by reducing the number of points to be considered. Applying clustering before indexing allows more efficient dimensionality reduction, but in fact, we have added an extra indexing level.

The memory-resident OP-tree index, which was used in our earlier studies, was made disk resident. Unlike the pages of R-tree-like indices, which can be loaded one page at a time, the OP-tree index can be accessed from disk with one sequential access. The SDI index was also shown to outperform well-known indexing methods.

There are several interesting problems that are not covered in this discussion. The issue of adding points to the data set without the need for repeating the application of SVD is addressed in References 75 and 102. PCA can be applied to very large $M \times N$ data sets by computing the $N \times N$ covariance matrix after a single data pass. The eigenvectors computed by applying PCA to an $N \times N$ matrix can be used to compute the rotated matrix with a reduced number of dimensions. There is also the issue of applying SVD to very large data sets, which may be sparse [103,104].

Figures and experimental results are not reported for the sake of brevity. The reader is referred to the referenced papers for figures, graphs, and tables.

## ACKNOWLEDGMENTS

The work reported here is based mainly on the author's collaboration with colleagues Dr. Vittorio Castelli and Dr. Chung-Sheng Li at the IBM T. J. Watson Research Center in Yorktown Heights, New York, and PhD students Dr. Lijuan (Catherine) Zhang and Dr. Yue Li at the New Jersey Institute of Technology (NJIT), Newark, New Jersey.

## REFERENCES

1. W. Niblack, R. Barber, W. Equitz, M. Flickner, E. H. Glasman, D. Petkovic, P. Yanker, C. Faloutsos, and G. Taubin. "The QBIC project: Querying images by content, using color, texture, and shape." In *Proc. SPIE Vol. 1908: Storage and Retrieval for Image and Video Databases*, San Jose, CA, January 1993, pp. 173–187.
2. B. Kim, and S. Park. "A fast $k$-nearest-neighbor finding algorithm based on the ordered partition." *IEEE Transactions on Pattern Analysis and Machine Intelligence (PAMI)* 8(6): 761–766 (1986).

3. V. Castelli, A. Thomasian, and C. S. Li. "CSVD: Clustering and singular value decomposition for approximate similarity search in high dimensional spaces." *IEEE Transactions on Knowledge and Data Engineering (TKDE) 14*(3): 671–685 (2003).
4. C.-S. Li, and V. Castelli. "Deriving texture feature set for content-based retrieval of satellite image database." In *Proc. Int'l. Conf. on Image Processing (ICIP '97)*, Santa Barbara, CA, October 1997, pp. 576–579.
5. D. A. White, and R. Jain. "Similarity indexing with the SS-tree." In *Proc. 12th IEEE Int'l. Conf. on Data Engineering (ICDE)*, New Orleans, LA, March 1996, pp. 516–523.
6. V. Castelli, and L. D. Bergman (editors). *Image Databases: Search and Retrieval of Digital Imagery*. John Wiley and Sons, New York, 2002.
7. B. S. Manjunath, and W.-Y. Ma. "Texture features for image retrieval." In *Image Databases: Search and Retrieval of Digital Imagery*, V. Castelli, and L. D. Bergman (editors). Wiley-Interscience, 2002, pp. 313–344.
8. C. Faloutsos. *Searching Multimedia Databases by Content (Advances in Database Systems)*. Kluwer Academic Publishers (KAP)/Elsevier, Burlingame, MA, 1996.
9. R. Weber, H.-J. Schek, and S. Blott. "A quantitative analysis and performance study for similarity-search methods in high-dimensional spaces." In *Proc. 24th Int'l. Conf. on Very Large Data Bases (PVLDB)*, New York, August 1998, pp. 194–205.
10. R. F. Freitas, and W. W. Wilcke. "Storage-class memory: The next storage system technology." *IBM Journal of Research and Development 52*(4–5): 439–448 (2008).
11. B. Jacob, S. W. Ng, and D. T. Wang. *Memory Systems: Cache, DRAM, Disk*. Morgan Kauffman Publishers (MKP)/Elsevier, Burlingame, MA. 2008.
12. I. T. Jolliffe. *Principal Component Analysis*. Springer, New York, 2002.
13. W. H. Press, S. A. Teukolsky, W. T. Vetterling, and B. P. Flannery. *Numerical Recipes, 3rd Edition: The Art of Scientific Computing*. Cambridge University Press (CUP), Cambridge, UK, 2007.
14. A. Thomasian, V. Castelli, and C. S. Li. "RCSVD: Recursive clustering and singular value decomposition for approximate high-dimensionality indexing." In *Proc. 7th ACM Int'l. Conf. on Information and Knowledge Management (CIKM)*, Baltimore, MD, November 1998, pp. 201–207.
15. K. Chakrabarti, and S. Mehrotra. "Local dimensionality reduction: A new approach to indexing high dimensional space." In *Proc. Int'l. Conf. on Very Large Data Bases (PVLDB)*, Cairo, Egypt, August 2000, pp. 89–100.
16. V. Gaede, and O. Günther. "Multidimensional access methods." *ACM Computing Surveys 30*(2): 170–231 (1998).
17. V. Castelli. "Multidimensional indexing structures for content-based retrieval." In *Image Databases: Search and Retrieval of Digital Imagery*, V. Castelli, and L. D. Bergman (editors). Wiley-Interscience, Hoboken, NJ, pp. 373–434.
18. F. Korn, H. V. Jagadish, and C. Faloutsos. "Efficiently supporting ad hoc queries in large datasets of time sequences." In *Proc. ACM SIGMOD Int'l. Conf. on Management of Data*, Tucson, AZ, May 1997, pp. 289–300.
19. R. Ramakrishnan, and J. Gehrke. *Database Management Systems*, 3rd edition. McGraw-Hill, New York, 2003.
20. J. Han, M. Kamber, and J. Pei. *Data Mining: Concepts and Techniques*, 3rd edition. Morgan Kaufmann Publishers (MKP)/Elsevier, Burlingame, MA, 2011.
21. W. Barbara, W. DuMouchel, C. Faloutsos, P. J. Haas, J. M. Hellerstein, Y. Ioannidis, H. V. Jagadish, T. Johnson, R. Ng, V. Poosala, K. A. Ross, and K. C. Sevcik. "The New Jersey data reduction report." *Data Engineering Bulletin 20*(4): 3–42 (1997).
22. S. Mallat. *A Wavelet Tour of Signal Processing: The Sparse Way*. MKP, 2008.
23. K. Chakrabarti, M. N. Garofalakis, R. Rastogi, and K. Shim. "Approximate query processing using wavelets." In *Proc. Int'l. Conf. on Very Large Data Bases (PVLDB)*, Cairo, Egypt, August 2000, pp. 111–122.

24. Y. Bishop, S. Fienberg, and P. Holland. *Discrete Multivariate Analysis: Theory and Practice*. MIT Press, Cambridge, MA, 1975.
25. V. Poosala. "Histogram-based estimation techniques in databases." PhD Thesis, Univ. of Wisconsin-Madison, Madison, WI, 1997.
26. F. Olken. "Random sampling from databases." PhD Dissertation, University of California, Berkeley, CA, 1993.
27. J. M. Hellerstein, P. J. Haas, and H. J. Wang. "Online aggregation." In *Proc. ACM SIGMOD Int'l. Conf. on Management of Data*, Tucson, AZ, May 1997, pp. 171–182.
28. A. Rajaraman, J. Leskovec, and J. Ullman. *Mining of Massive Datasets*, 1.3 edition. Cambridge University Press (CUP), Cambridge, UK, 2013. Available at http://infolab.stanford.edu/~ullman/mmds.html.
29. IBM Corp. *Engineering and Scientific Subroutine Library (ESSL) for AIX V5.1, Guide and Reference*. IBM Redbooks, Armonk, NY. SA23-2268-02, 07/2012.
30. A. Thomasian, Y. Li, and L. Zhang. "Exact k-NN queries on clustered SVD datasets." *Information Processing Letters (IPL) 94*(6): 247–252 (2005).
31. J. A. Hartigan. *Clustering Algorithms*. John Wiley and Sons, New York, 1975.
32. A. K. Jain, and R. C. Dubes. *Algorithms for Clustering Data*. Prentice Hall, Upper Saddle River, NJ, 1988.
33. L. Kaufman, and P. J. Rousseeuw. *Finding Groups in Data: An Introduction to Cluster Analysis*. John Wiley and Sons, New York, 1990.
34. B. S. Everitt, S. Landau, M. Leese, and D. Stahl. *Cluster Analysis*, 5th edition. John Wiley and Sons, New York, 2011.
35. M. J. Zaki, and W. Meira Jr. *Data Mining and Analysis: Fundamental Concepts and Algorithms*. Cambridge University Press (CUP), Cambridge, UK, 2013.
36. H. P. Kriegel, P. Krger, and A. Zimek. "Clustering high-dimensional data: A survey on subspace clustering, pattern-based clustering, and correlation clustering." *ACM Transactions on Knowledge Discovery from Data (TKDD) 3*(1): 158 (2009).
37. C. Aggarwal, A. Hinneburg, and D. A. Keim. "On the surprising behavior of distance metrics in high dimensional spaces." In *Proc. Int'l. Conf. on Database Theory (ICDT)*, London, January 2001, pp. 420–434.
38. C. C. Aggarwal. "Re-designing distance functions and distance-based applications for high dimensional data." *ACM SIGMOD Record 30*(1): 13–18 (2001).
39. J. C. Bezdek. *Pattern Recognition with Fuzzy Objective Function Algorithms*. Plenum Press, New York, 1981. Available at http://home.deib.polimi.it/matteucc/Clustering/tutorial_html/cmeans.html.
40. F. Farnstrom, J. Lewis, and C. Elkan. "Scalability for clustering algorithms revisited." *ACM SIGKDD Explorations Newsletter 2*(1): 51–57 (2000).
41. T. F. Gonzalez. "Clustering to minimize the maximum intercluster distance." *Theoretical Computer Science 38*: 293–306 (1985).
42. A. McCallum, K. Nigam, and L. H. Unger. "Efficient clustering of high-dimensional data sets with applications to reference matching." In *Proc. 6th ACM Int'l. Conf. on Knowledge Discovery and Data Mining (KDD)*, Boston, August 2000, pp. 169–178.
43. R. T. Ng, and J. Han. "CLARANS: A method for clustering objects for spatial data mining." *IEEE Transactions on Knowledge and Data Engineering (TKDE) 14*(5): 1003–1016 (2002).
44. T. Zhang, R. Ramakrishnan, and M. Livny. "BIRCH: An efficient data clustering method for very large databases." In *Proc. 25th ACM SIGMOD Int'l. Conf. on Management of Data*, Montreal, Quebec, Canada, June 1996, pp. 103–114.
45. P. S. Bradley, U. M. Fayyad, and C. Reina. "Scaling clustering algorithms to large databases." In *Proc. Int'l. Conf. on Knowledge Discovery and Data Mining*, New York, August 1998, p. 915.

46. G. Karypis, E.-H. Han, and V. Kumar. "Chameleon: Hierarchical clustering using dynamic modeling." *IEEE Computer 32*(8): 68–75 (1999).
47. B. W. Kernighan, and S. Lin. "An efficient heuristic procedure for partitioning graph." *Bell Systems Technical Journal (BSTJ) 49*: 291–308 (1970).
48. S. Guha, R. Rastogi, and K. Shim. "CURE: An efficient clustering algorithm for large databases." In *Proc. ACM SIGMOD Int'l. Conf. on Management of Data*, Seattle, WA, June 1998, pp. 73–84.
49. J. L. Bentley. "Multidimensional binary search trees used for associative searching." *Communications of the ACM 18*(9): 509–517 (1975).
50. V. Ganti, R. Ramakrishnan, J. Gehrke, A. L. Powell, and J. C. French. "Clustering large datasets in arbitrary metric spaces." In *Proc. Int'l. Conf. on Data Engineering* (*ICDE*), Sidney, Australia, March 1999, pp. 502–511.
51. M. Ester, H.-P. Kriegel, J. Sander, and X. Xu. "A density-based algorithm for discovering clusters in large spatial databases with noise." In *Proc. 2nd Int'l. Conf. Knowledge Discovery and Data Mining (KDD-96)*, Portland, OR, 1996, pp. 226–231.
52. M. Ankerst, M. M. Breunig, H.-P. Kriegel, and J. Sander. "OPTICS: Ordering points to identify the clustering structure." In *Proc. ACM SIGMOD Int'l. Conf. on Management of Data*, Philadelphia, PA, June 1999, pp. 49–60.
53. W. Wang, J. Yang, and R. Muntz. "STING: A statistical information grid approach to spatial data mining." In *Proc. 23rd Int'l. Conf. Very Large Data Bases (VLDB)*, Athens, Greece, August 1997, pp. 186–195.
54. R. Agrawal, J. Gehrke, D. Gunopulos, and P. Raghavan. "Automatic subspace clustering of high dimensional data for data mining applications." In *Proc. ACM SIGMOD Int'l. Conf. on Management of Data*, Seattle, WA, June 1998, pp. 94–105.
55. G. Sheikholeslami, S. Chatterjee, and A. Zhang. "WaveCluster: A multi-resolution clustering approach for very large spatial databases." In *Proc. 24th Int'l. Conf. Very Large Data Bases (VLDB)*, New York, August 1998, pp. 428–439.
56. A. Hinneburg, and D. A. Keim. "Optimal grid-clustering: Towards breaking the curse of dimensionality in high-dimensional clustering." In *Proc. 25th Int'l. Conf. on Very Large Data Bases (PVLDB)*, Edinburgh, Scotland, UK, September 1999, pp. 506–517.
57. R. Agrawal, M. Mehta, J. Shafer, R. Srikant, A. Arning, and T. Bollinger. "The quest data mining system." In *Proc. 2nd Int'l. Conference on Knowledge Discovery in Databases and Data Mining (KDD)*, Portland, OR, 1996, pp. 244–249.
58. C.-H. Cheng, A. W. Fu, and Y. Zhang. "Entropy-based subspace clustering for mining numerical data." In *Proc. 5th ACM SIGKDD Int'l. Conf. on Knowledge Discovery and Data Mining (KDD)*, San Diego, CA, August 1999, pp. 84–93.
59. H. S. Nagesh, S. Goil, and A. Choudhary. "A scalable parallel subspace clustering algorithm for massive data sets." In *Proc. IEEE Int'l. Conf. on Parallel Processing (ICPP)*, Toronto, Ontario, August 2000, pp. 477–484.
60. J.-W. Chang, and D.-S. Jin. "A new cell-based clustering method for large, high-dimensional data in data mining applications." In *Proc. ACM Symposium on Applied Computing (SAC)*, Madrid Spain, March 2002, pp. 503–507.
61. L. Parsons, E. Haque, and H. Liu. "Subspace clustering for high dimensional data: A review." *ACM SIGKDD Explorations Newsletter 6*(1): 90–105 (2004).
62. C. M. Procopiuc, M. Jones, P. K. Agarwal, and T. M. Murali. "A Monte Carlo algorithm for fast projective clustering." In *Proc. ACM SIGMOD Int'l. Conf. on Management of Data*, Madison, WI, June 2002, pp. 418–427.
63. C. C. Aggarwal, J. L. Wolf, P. S. Yu, C. Precopiuc, and J. S. Park. "Fast algorithms for projected clustering." In *Proc. ACM SIGMOD Int'l. Conf. on Management of Data*, Philadelphia, PA, June 1999, pp. 61–72.

64. C. C. Aggarwal, and P. S. Yu. "Finding generalized projected clusters in high dimensional spaces." In *Proc. ACM SIGMOD Int'l. Conf. on Management of Data*, Dallas, May 2000, pp. 70–81.
65. K. G. Woo, J. H. Lee, M. H. Kim, and Y. J. Lee. "FINDIT: A fast and intelligent subspace clustering algorithm using dimension voting." *Information Software Technology 46*(4): 255–271 (2004).
66. J. Yang, W. Wang, H. Wang, and P. S. Yu. "δ-clusters: Capturing subspace correlation in a large data set." In *Proc. 18th Int'l. Conf. on Data Engineering (ICDE)*, San Jose, CA, February–March 2002, pp. 517–528.
67. J. H. Friedman, and J. J. Meulman. "Clustering objects on subsets of attributes." 2002. Available at http://statweb.stanford.edu/~jhf/.
68. H. Jin, B. C. Ooi, H. T. Shen, C. Yu, and A. Y. Zhou. "An adaptive and efficient dimensionality reduction algorithm for high-dimensional indexing." In *Proc. 19th IEEE Int'l. Conf. on Data Engineering (ICDE)*, Bangalore, India, March 2003, pp. 87–98.
69. Y. Linde, A. Buzo, and R. Gray. "An algorithm for vector quantizer design." *IEEE Transactions on Communications 28*(1): 84–95 (1980).
70. H. Samet. *Foundations of Multidimensional and Metric Data Structure*. Elsevier, 2007.
71. V. S. Cherkassky, J. H. Friedman, and H. Wechsler. *From Statistics to Neural Networks: Theory and Pattern Recognition Applications*. Springer-Verlag, New York, 1994.
72. F. Korn, N. Sidiropoulos, C. Faloutsos, E. Siegel, and Z. Protopapas. "Fast and effective retrieval of medical tumor shapes: Nearest neighbor search in medical image databases." *IEEE Transactions on Knowledge and Data Engineering (TKDE) 10*(6): 889–904 (1998).
73. A. Thomasian, Y. Li, and L. Zhang. "Optimal subspace dimensionality for *k*-nearest-neighbor queries on clustered and dimensionality reduced datasets with SVD." *Multimedia Tools and Applications (MTAP) 40*(2): 241–259 (2008).
74. A. Thomasian, and L. Zhang. "Persistent semi-dynamic ordered partition index." *The Computer Journal 49*(6): 670–684 (2006).
75. A. Thomasian, and L. Zhang. "Persistent clustered main memory index for accelerating k-NN queries on high dimensional datasets." *Multimedia Tools Applications (MTAP) 38*(2): 253–270 (2008).
76. C. Yu. *High-Dimensional Indexing: Transformational Approaches to High-Dimensional Range and Similarity Searches. Lecture Notes in Computer Science (LNCS)*, Volume 2431. Springer-Verlag, New York, 2002.
77. R. A. Finkel, and J. L. Bentley. "Quad trees: A data structure for retrieval of composite keys." *Acta Informatica 4*(1): 1–9 (1974).
78. J. T. Robinson. "The *k*-d-b tree: A search structure for large multidimensional dynamic indexes." In *Proc. ACM SIGMOD Int'l. Conf. on Management of Data*, Ann Arbor, MI, April 1981, pp. 10–18.
79. A. Guttman. "R-trees: A dynamic index structure for spatial searching." In *Proc. ACM SIGMOD Int'l. Conf. on Management of Data*, Boston, June 1984, pp. 47–57.
80. T. Sellis, N. Roussopoulos, and C. Faloutsos. "The R+-tree: A dynamic index for multidimensional objects." In *Proc. 13th Int'l. Conf. on Very Large Data Bases (VLDB)*, Brighton, England, September 1987, pp. 507–518.
81. N. Beckmann, H.-P. Kriegel, R. Schneider, and B. Seeger. "The R*-tree: An efficient and robust access method for points and rectangles." In *Proc. ACM SIGMOD Int'l. Conf. on Management of Data*, Atlantic City, NJ, May 1990, pp. 322–331.
82. N. Katayama, and S. Satoh. "The SR-tree: An index structure for high-dimensional nearest neighbor queries." In *Proc. ACM SIGMOD Int'l. Conf. on Management of Data*, Tucson, AZ, May 1997, pp. 369–380.
83. G. R. Hjaltason, and H. Samet. "Distance browsing in spatial databases." *ACM Transactions on Database Systems (TODS) 24*(2): 265–318 (1999).

84. N. Roussopoulos, S. Kelley, and F. Vincent. "Nearest neighbor queries." In *Proc. ACM SIGMOD Int'l. Conf. on Management of Data*, San Jose, CA, June 1995, pp. 71–79.
85. L. Zhang. "High-dimensional indexing methods utilizing clustering and dimensionality reduction." PhD Dissertation, Computer Science Department, New Jersey Institute of Technology (NJIT), Newark, NJ, May 2005.
86. K. Chakrabarti, and S. Mehrotra. "The hybrid tree: An index structure for high dimensional feature spaces." In *Proc. IEEE Int'l. Conf. on Data Engineering (ICDE)*, 1999, pp. 440–447.
87. S. Berchtold, D. A. Keim, and H. P. Kriegel. "The X-tree: An index structure for high-dimensional data." In *Proc. 22nd Int'l. Conf. on Very Large Data Bases (PVLDB)*, San Jose, CA, August 1996, pp. 28–39.
88. S. Berchtold, C. Böhm, and H.-P. Kriegel. "The pyramid-technique: Towards breaking the curse of dimensionality." In *Proc. ACM SIGMOD Int'l. Conf. on Management of Data*, Seattle, WA, June 1998, pp. 142–153.
89. B. C. Ooi, K.-L. Tan, C. Yu, and S. Bressan. "Indexing the edges—A simple and yet efficient approach to high-dimensional indexing." In *Proc. 19th ACM Int'l. Symp. on Principles of Database Systems (PODS)*, Dallas, May 2000, pp. 166–174.
90. C. Yu, B. C. Ooi, K.-L. Tan, and H. Jagadish. "Indexing the distance: An efficient method to knn processing." In *Proc. 27th Int'l. Conf. on Very Large Data Bases (VLDB)*, Rome, Italy, September 2001, pp. 421–430.
91. H. V. Jagadish, B. C. Ooi, K. L. Tan, C. Yu, and R. Zhang. "iDistance: An adaptive $B^+$-tree based indexing method for nearest neighbor search." *ACM Transactions on Database Systems (TODS) 30*(2): 364–397 (2000).
92. J. K. Uhlmann. "Satisfying general proximity/similarity queries with metric trees." *Information Processing Letters (IPL) 40*(4): 175–179 (1991).
93. P. N. Yianilos. "Data structures and algorithms for nearest neighbor search in general metric spaces." In *Proc. 4th Annual ACM-SIAM Symposium on Discrete Algorithms (SODA)*, Austin, TX, January 1993, pp. 311–321.
94. T. Bozkaya, and M. Ozsoyoglu. "Distance-based indexing for high-dimensional metric spaces." In *Proc. ACM SIGMOD Int'l. Conf. on Management of Data*, Tucson, AZ, May 1997, pp. 357–368.
95. P. Ciaccia, M. Patella, and P. Zezula. "M-tree: An efficient access method for similarity search in metric spaces." In *Proc. 23rd Int'l. Conf. on Very Large Data Bases (PVLDB)*, Athens, Greece, August 1997, pp. 426–435.
96. R. F. S. Filho, A. Traina, C. Traina Jr., and C. Faloutsos. "Similarity search without tears: The OMNI family of all-purpose access methods." In *Proc. 17th IEEE Int'l. Conf. on Data Engineering (ICDE)*, Heidelberg, Germany, April 2001, pp. 623–630.
97. B. Cui, B. C. Ooi, J. Su, and K.-L. Tan. "Indexing high-dimensional data for efficient in-memory similarity search." *IEEE Transactions on Knowledge and Data Engineering (TKDE) 17*(3): 339–353 (2005).
98. J. L. Hennessey, and D. A. Patterson. *Computer Architecture: A Quantitative Approach*, 5th edition. Elsevier, Burlingame, MA, 2011.
99. K. I. Lin, H. V. Jagadish, and C. Faloutsos. "The TV-tree: An index structure for high-dimensional data." *VLDB Journal 3*(4): 517–542 (1994).
100. A. Thomasian, and L. Zhang. "The Stepwise Dimensionality-Increasing (SDI) index for high dimensional data." *The Computer Journal 49*(5): 609–618 (2006).
101. D. A. White, and R. Jain. "Similarity indexing: Algorithms and performance." In *Storage and Retrieval for Image and Video Databases (SPIE)*, Volume 2670, San Jose, CA, 1996, pp. 62–73.
102. K. V. Ravikanth, D. Agrawal, and A. Singh. "Dimensionality-reduction for similarity searching in dynamic databases." In *Proc. ACM SIGMOD Int'l. Conf. on Management of Data*, Seattle, WA, June 1998, pp. 166–176.

103. S. Rajamanickam. "Efficient algorithms for sparse singular value decomposition." PhD Thesis, Computer Science Dept., University of Florida, Gainesville, FL, 2009. Available at http://www.cise.ufl.edu/~srajaman/Rajamanickam_S.pdf.
104. A. A. Amini. "High-dimensional principal component analysis." PhD Thesis, Report Number UCB/EECS-2011-104, EECS Dept., University of California, Berkeley, CA, 2011. Available at http://www.eecs.berkeley.edu/Pubs/TechRpts/2011/EECS-2011-104.html.

## APPENDIX 3.A: COMPUTING APPROXIMATE DISTANCES WITH DIMENSIONALITY-REDUCED DATA

The dimensionality reduction method in References 3 and 14 retains $n < N$ columns of the $Y$ matrix, which correspond to the largest eigenvalues, while the method described in References 8 and 18 retains the $n$ columns of the $U$ matrix corresponding to the largest eigenvalues. In both cases, the eigenvalues or singular values are in nonincreasing order, so that columns with the smallest indices need to be preserved. We first show that retaining the $n$ columns of the $Y$ matrix results in the same distance error as retaining the same dimensions of the $U$ matrix. We also show that our method is computationally more efficient than the other method. Retaining the first $n$ dimensions (columns) of the transformed matrix $Y = XV$, we have

$$y_{i,j} = \sum_{k=1}^{N} x_{i,k} v_{k,j}, \; 1 \le j \le n, 1 \le i \le M.$$

The rest of the columns of the matrix are set to the column mean, which is zero in our case. (Some data sets are studentized by dividing the zero-mean column by the standard deviation [3].) The squared error in representing point $P_i$, $1 \le i \le M$ is then

$$e_i = \sum_{j=n+1}^{N} y_{i,j}^2 = \sum_{j=n+1}^{N} \left( \sum_{k=1}^{N} x_{i,k} v_{k,j} \right)^2, \; 1 \le i \le M.$$

The elements of $X$ are approximated as $X' \approx USV^T$ by utilizing the first $n$ dimensions of $U$ [8,18]:

$$x'_{i,j} = \sum_{k=1}^{n} s_k u_{i,k} v_{k,j}, 1 \le j \le N, 1 \le i \le M.$$

Noting that $Y = XV = US$, the squared error in representing $P_i$ is

$$e'_i = \sum_{j=1}^{N} \left( \sum_{k=n+1}^{N} s_k u_{i,k} v_{k,j} \right)^2 = \sum_{j=1}^{N} \left( \sum_{k=n+1}^{N} y_{i,k} v_{k,j} \right)^2.$$

To show that $e_i = e'_i$, $1 \le i \le M$, we use the definition $\delta_{k,k'} = 1$ for $k = k'$ and $\delta_{k,k'} = 0$ for $k \ne k'$.

$$e'_i = \sum_{j=1}^{N} \left( \sum_{k=n+1}^{N} y_{i,k} v_{k,j} \right) \left( \sum_{k'=n+1}^{N} y_{i,k'} v_{k',j} \right) = \sum_{k=n+1}^{N} \sum_{k'=n+1}^{N} y_{i,k} y_{i,k'} \sum_{j=1}^{N} v_{k,j} v_{k',j}$$

$$= \sum_{k=n+1}^{N} \sum_{k'=n+1}^{N} y_{i,k} y_{i,k'} \delta_{k,k'} = \sum_{j=n+1}^{N} y_{i,j}^2 = e_i.$$

Our method is more efficient from the viewpoint of computational cost for nearest neighbor, range, and window queries. In the case of nearest neighbor queries, once the input or query vector is transformed and the appropriate $n$ dimensions are selected, we need only be concerned with these dimensions. This is less costly than applying the inverse transformation.

CHAPTER 4

# Multiple Sequence Alignment and Clustering with Dot Matrices, Entropy, and Genetic Algorithms

John Tsiligaridis

CONTENTS



ABSTRACT

The purpose of this project is to present a set of algorithms and their efficiency for Multiple Sequence Alignment (MSA) and clustering problems, including also solutions in distributive environments with Hadoop. The strength, the adaptability, and the effectiveness of the genetic algorithms (GAs) for both problems are pointed out. MSA is among the most important tasks in computational biology. In biological sequence comparison, emphasis is given to the simultaneous alignment of several sequences. GAs are stochastic approaches for efficient and robust search that can play a significant role for MSA and clustering. The divide-and-conquer principle ensures undisturbed consistency during vertical sequences' segmentations.

A set of algorithms has been developed for MSA: (a) the consistency with dot matrices (CDM) algorithm that discovers consistency and creates MSA using dot matrices; (b) the pilot entropy algorithm (PEA), based on entropy of columns, which can provide a new solution for MSA with no consistency consideration; and (c) the GA following the divide-and-conquer method (DCGA), and using the CDM as an internal part, which can provide a solution for MSA utilizing appropriate cut points.

CDM, with its internal phase, can minimize the number of pairwise comparisons in MSA, thus giving a better performance. CDM is also effective in finding conserved motifs (short sequence segments important in protein structure and/or function) among very distantly related sequences.

A distributed solution for CDM using MapReduce is also presented.

PEA is based on entropy of columns, and a new methodology for line or column shifting and insertion of gaps according to entropy values is developed.

DCGA, with the divide-and-conquer technique, can effectively reduce the space complexity for MSA so that the consistency achieved from the CDM is not broken.

As far as the clustering is concerned, the aim is to divide the objects into clusters so that the validity inside clusters is minimized. As an internal measure for cluster validity, the sum of squared error (SSE) is used. A clustering genetic algorithm with the SSE criterion (CGA_SSE), a hybrid approach, using the most popular algorithm, the *K*-means, is presented. The CGA_SSE combines local and global search procedures.

The fitness function works using the SSE criterion as a pilot. The combination of GA and *K*-means incorporates additional mechanisms to achieve better results than the classical GA. This improvement is particularly useful to the search for a global optimum.

Comparison of the *K*-means and CGA_SSE is provided in terms of accuracy and quality of solution for clusters of different sizes and densities. The complexity of all proposed algorithms is examined. The Hadoop for the distributed environment provides an alternate solution for the CGA_SSE, following the MapReduce paradigm. Simulation results are provided.

## 4.1 INTRODUCTION

Sequence alignment is an arrangement of two or more sequences, highlighting their similarity. Sequence alignment is useful for discovering *structural*, *functional*, and *evolutional* information in biological sequences. The idea of assigning a score to an alignment and then minimizing over all alignments is at the heart of all biological sequence alignments. We can define a distance or a similarity function to an alignment. A distance function will define negative values for mismatches or gaps and then aim at minimizing this distance. A similarity function will give high values to matches and low values to gaps and then maximize the resulting score. A dot matrix analysis is a method for comparing two sequences when looking for possible alignment. We use a dot plot to depict the pairwise comparison and identify a region of close similarity between them. The biggest asset of a dot matrix is its ability to find direct repeats and visualize the entire comparison. The consistent regions comprising aligned spans from all the sequences compared may permit multiple selections of fragment sets. For Multiple Sequence Alignment (MSA), the divide-and-conquer method is used with a genetic algorithm (GA) [1–3].

The plots may either contain enough points to trace a consecutive path for each pair of sequences or display only isolated spans of similarity. This can be achieved by the CDM. In addition, when there are no consecutive paths, an optimal subset with the longest subsequence may be selected. The criterion of consistency can guarantee the creation of consistent regions. First, the lists are updated with the indices of the aligned items, and then the MSA is created. The case of having more than one index in the next sequence is also examined. Vectors that follow each other (not crossing over) and an array of lists can provide a solution. Finally, after finding the consistency regions, using CDM, the divide-and-conquer method can be applied by selecting cut points that do not discontinue the consistency.

Moreover, based on entropy of an alignment, the pilot entropy algorithm (PEA) can provide MSA by shifting lines left or columns right. The difference between CDM and PEA is that PEA is not interested in finding consistency, but it makes alignment according to column entropy scores. CDM or PEA can cooperate with GA under DCGA using the divide-and-conquer method. In addition, faster alignment can be achieved for some part of the sequences (partial MSA) by the cut points and the preferred method (CDM, PEA, GA).

DCGA with the divide-and-conquer technique can effectively reduce the space complexity for MSA so that the consistency achieved from the CDM is not broken. The advantages and disadvantages of each of the methods are presented.

Clustering is the process of grouping data into clusters, where objects within each cluster have high similarity but are dissimilar to the objects in other clusters. The goal of clustering is to find groups of similar objects based on a similarity metric [4]. Some of the problems associated with current clustering algorithms are that they do not address all the requirements adequately and that they need high time complexity when dealing with a large number of dimensions and large data sets.

Since a clustering problem can be defined as an optimization problem, evolutionary approaches may be appropriate here. The idea is to use evolutionary operators and a population of clustering structures to converge into a globally optimal clustering. The most frequently used evolutionary technique in clustering problems is GAs. Candidate clusterings are encoded as chromosomes. A fitness value is associated with each cluster's structure. A higher fitness value indicates a better cluster structure. GAs perform a globalized search for solutions, whereas most other clustering procedures perform a localized search. In a localized search, the solution obtained at the "next iteration" of the procedure is in the vicinity of the current solution. The *K*-means algorithm is used in the localized search technique. In the hybrid approach that is proposed, the GA is used just to find good initial cluster centers, and the *K*-means algorithm is applied to find the final partition. This hybrid approach performs better than the GAs.

The parallel nature of GAs makes them an optimal base for parallelization. Beyond the Message Passing Interface (MPI), the traditional solution for the GA and hybrid GA implementation, Hadoop can improve the parallel behavior of the GAs, becoming a much better solution. In addition, for MPI programming, a sophisticated knowledge of parallel programming (i.e., synchronization, etc.) is needed. MPI does not scale well on clusters where a failure is not unlikely (i.e., long processing programs). MapReduce [5,6] provides an opportunity to easily develop large-scale distributed applications. Several different implementations of MapReduce have been developed [7,8]. Reference 7 is based on another architecture such as Phoenix, while Reference 8 deals with the issue of reducing the number of iterations for finding the right set of initial centroids.

The contribution of this chapter is the development of a set of algorithms for both MSA and GA and also the discovery of a distributed solution for CDM and for hybrid GA.

The chapter is organized as follows: CDM and PEA are developed in Sections 4.2 and 4.3, respectively. The divide-and-conquer principle is presented in Section 4.4. Sections 4.5 through 4.7 include the GAs, *K*-means, and GA following the divide-and-conquer method (DCGA), which represent the part of work that is based on GA. Section 4.8 refers to DCGA. Section 4.9 includes the DCGA distributed solution with Hadoop. Finally, Section 4.10 includes a simulation.

## 4.2 CDM

CDM has three phases: (1) the dot matrices' preparation, (2) the scan phase, and (3) the link phase. In the first phase, the dot matrices are examined, in the second, the diagonal similarities, and in the third phase, the vertical continuity. The dot matrix running time is proportional to the product of the size of the sequences. Working with a set of dot matrices poses the problem of how to define *consistency*. Scanning of graphs for a series of 1s reveals similarity or a string of the same characters. The steps of the algorithm are as follows: (1) Prepare the dot plot matrix [9] and examine the horizontal continuity (preparation phase). (2) Find gaps (horizontal) and examine the diagonal location (preparation phase). (3) Scan phase. (4) Examine the vertical continuity (link phase). Steps 1 and 2 belong to the preparation phase. Create vectors that hold the locations of 1s (as developed in the example). In the link phase, the locations are combined. The advantage of CDM is that with the link phase only $(n - 1)$, comparisons are needed (instead of $n$), where $n$ is the number of sequences.

```
CDM //input: the strings into 2 dim array in[][]
 //output: MSA using pointers method, out[][]
 //variable: ma[][], hold 1's in similar positions
 //fin[][]: hold the similar locations of all the pairs (there are
 //k-1 pairs for k sequences)
 //hor[][]: for horizontal/vertical gaps '-' in locations
 //V arrays for similar items (k-1) arrays for k sequences
 //scan phase
for each pair of strings (# pairs<= k-1)
   find the similar items, seed (m : location)    extend it
          (m+1,...)
```

```
//update the positions of fin[][] with similar items
fin (i,j) =1;
//finding gaps
if  (a(j) = b(j) && (i==j)) { there is no gap }
if  (a(j) = b(j) && (i>j))
   { //there are i-j+1 horizontal gaps
    hor(i,j)="-"; }
if  (a(j) = b(j) && (i<j))
   { //there are j-i+1 vertical gaps
    ver(i,j)="-"; }
//link phase
 examine the vertical continuity using V arrays
 if (the two vectors are in sequence)
     {link the two vectors
      append the location in the final vector}
```

Vectors have the locations of the similar items, and the index of a vector is the location of the item of the first sequence of each pair.

**Example**

Lets consider the following sequences.

A: a b c d e, B: b c x, C: a b c

A. Preparation phase

The dot plot A-B:

```
  a b c d e
b - 1
c   1
x
```

There is a gap (horizontal) before the match b-b. Hold the horizontal gap into an array hor[i,j]. i is the number of the string; j is the location of the gap.

The dot plot B-C:

```
  b c x
a -
b 1
c 1
```

B. Scan phase

*Read plot A-B*: Find that positions 2(A) and 1(B) have a similar item: b. Find that positions 3(A) and 2(B) have similar item: c.

There is continuity (horizontal) since there are numbers, one next to the other, [2(A), 3(A)] and [1(B), 2(B)]. There is a gap horizontally.

Create the vectors with the vertical locations of the same item b: $V_2$: 2, 1. Create the vectors with the vertical locations of the same item c: $V_3$: 3, 2.

*Read plot B-C*: Find that positions 1(B) and 2(C) have a similar item: b. There is a vertical gap. Find that positions 2(B) and 3(C) have a similar item: c.

There is also continuity (horizontal), since there are numbers next to each other, [1(B), 2(B)] and [2(C), 3(C)].

Create the vectors with the vertical locations (a) of the same item b: $V_1$: 1, 2 and (b) of the same item c: $V_2$: 2, 3. The vectors $V_2$: 2, 1 and $V_1$: 1, 2 are in sequence since the $y$ coordinate of $V_2$ (= 1) is equal to the $x$ coordinate of $V_1$, and there is a *vertical continuity.* The B (= "bcx") is considered as the intermediate string.

C. Link phase

The vectors can be appended since they have common item: for $V_2$–$V_1$, 1 (last), and for $V_3 - V_2$, 2 (last).

Hence, the final vectors are $FV_2$: 2, 1, 2 for b and $FV_3$: 3, 2, 3 for c. After that, using the hor[] of the MSA, a vertical continuity for b and c is discovered as follows:

A: a b c d e
B: - b c x
C: a b c

Another solution could be obtained by using associative arrays.

The complexity is $O(n^2) * (k - 1)$ where $n$ is the size of the sequences and $k$ is the number of sequences. MapReduce is the heart of Hadoop [5,10,11]. Google proposed the MapReduce [6] abstraction that enables users to easily develop large-scale distributed applications.

The Hadoop and MapReduce programming paradigm already has a substantial base in the next-generation sequencing analysis [12,13]. The design of CDM with the three phases offers a good example for the MapReduce implementation. The mapper consists of the preparation and the scan phase. In the map phase, the file name is the key, and all file contents are the value. Vectors with scores and positions of similar items will be created by the dot plots. The reducer will contain the link phase combining all (key/values) pairs.

## 4.3 PEA

PEA develops a methodology for line or column shifting and inserts gaps according to the entropy values. The pilot is the line (which will be moved left) or the column (which will move right). PEA is a dynamic solution for finding the alignment of $k$ columns ($k < n$ = number of columns). It starts with the entropy of column 1, and if the value of the entropy (entr) is less than a predefined limit (lm), then the line with the smaller contribution value (different item) is searching (from left to right) until it finds the item with the higher contribution in the entropy for the examined column.

PEA looks ahead for $m$ steps. It searches the lists (one for each sequence) in order to find if there are similar items in $k$ columns and try to figure out if the new alignment will have a better (lower) column score.

There are two moves, the left move of the line (LML) where pilot is the line with the lower participation item (case 1) and the right move of a column (RMC) where pilot is the column with the lower score (case 2). The LML and RMC are visible only when, after the moves, the total score of the column is greater than the previous total score. In the LML for a column i, the left move happens when a line is moving one step left, and then "left gaps" are put on all positions of column i – 1 except the line with the lower score item. For a column I, when the LML is not visible, then the RMC is examined. If RMC happens, gaps are put before the tested column.

```
PEA: input: set of sequences
output: alignment sequences
//d: is the final # of columns to be aligned
//p: the pointer that show a col
//lm_i: left move of a col (boolean)
//entr_i: entropy of column i
//m: is the number of col for "looking ahead"
//p_entr_i: is the entropy of col i before any alignment change
//n_entr_i: is the entropy of col i after the temporal
//alignment of the sequences
//LML_F: LML flag (Y/N), RLM_F:RLM flag (Y/N)
step 1:
prepare the lists with each sequence (list_i for seq_i)
find the entropy from the initial alignment
LML_F="N", RLM_F="N"
step 2:
for all the columns i of range (1: d)
start with testing the col: i
if (entr_i >0) //entropy of col i
   //search for left shifting
   find the line (k) of item with the lower participation
   search the list j for similar item to the high frequency
     items of col: i
    //    for step m (look ahead)
    while (n=1; n<=m; n++)
     { //create the alignment for col n
         //search lists of the sequences for similar items
          //test for LML (case 1)
          read next item of line k, recompute entropy
          if (the new entropy for col i is <
                 previous entropy of col i or
                 (n_entr)_i+1 <(p_entr)_i)
                 {//accept the new alignment, LML visible
                  //left shift of the items in line i, put "left gaps"
                  //   for the items ≠ i col
                    d=n; LML_F='Y'    }
           //test for RMC (case 2)
```

```
          if (LML_F='N')
             {read next items of lines except line k,
              recompute entropy }
             if ((the new entropy of col i < previous
                  entropy of col i, or
              (n_entr)i+1 <(p_entr)i) && RMC test is valid))
              {//accept the new alignment, RLM visible
               //right shift of the items in lines (\k line), put
               //"right gaps" for the items ≠ k line
               d=n;    }  }//end while
if (entri =0)    go to the next col (i+1)   }
```

**Example**

A set of three sequences is given.

1 2 3 4
A B C (1)
C A K B (2)
A B K (3)

The column 1 in line 2 has to be tested. Line 2 is the line with the lower participation, since it has "C." The LML is examined. From list 2, the item A appears in the second location, and the $entr_2 = 0$. Case 2 is visible $(n_entr)_{i+1} < (p_entr)_i$. The new left scored gaps are put for all the lines except line 2.

1 2 3 4
- A B C (1)
C A K B (2)
- A B K (3)

In testing column 3, we see that there is, in line 2, the item K. If we try to use LML with a new left move of line 2, then it is not valid, since it will destroy the previous alignments. So the LML test is invalid, and the RMC is examined. The pilot is the column (B, K, B), and line 2 provides the lower participation.

Case 2 is visible since $(n_entr)_{i+1} < (p_entr)_i$, and in column 4, the $entr_4 = 0$.

1 2 3 4 5
- A - B C (1)
C A K B (2)
- A - B K (3)

The complexity of PEA is $k * h^k * n^k$, where $k$ is the number of sequences, $h$ is the size of "look-ahead" items that have to be examined for alignment, and $n$ is the size of the sequences.

## 4.4 DIVIDE AND CONQUER

MSA belongs to a class of optimization problems with exponential time complexity. It exhibits $O(L^N)$ time complexity where $L$ is the mean length of the sequences to be aligned and $N$ is the number of sequences.

If the size of the sequences is greater than bound 1 (predefined size), it is better to work with the divide-and-conquer method with GA. The divide-and-conquer principle is as follows: Each of the given $k$ sequences is being cut at an appropriate chosen site somewhere near its midpoint. This may reduce the original alignment problem to the two subproblems of aligning the two resulting groups of prefix and suffix subsequences, respectively. The main problem is to find a tuple of slicing sites $(c_1, c_2, c_3,\ldots,c_k)$ so that the simple concatenation of the two optimal alignments of the prefixes $s_1(\leq c_1), s_2(\leq c_2),\ldots,s_k(\leq c_k)$ and the suffixes $s_1(>c_1), s_2(>c_2),\ldots,s_k(>c_k)$ forms an optimal alignment of the original sequences. The midpoint solution is not always the best approach for alignments since it does not consider the conserved regions. This ends up in additional waste of time in the compose phase of the substrings. On the other hand, if the divide-and-conquer principle is based on the results of the CDM or PEA, then it can guarantee the maintenance of the conserved regions, which provides better performance.

## 4.5 GAs

By imitating the evolution process of organisms in the complicated natural environment, an evolutionary system contains the operators of selection, crossover, and mutation. In GAs, chromosomes form a population, and a set of genes forms a chromosome [14–16].

Selection: The system selects chromosomes to perform crossover and mutation operations according to the fitness values of the chromosomes. The larger the fitness value of a chromosome, the higher the chance of the chromosome to be chosen. The roulette wheel is used as a method of selection. The selection of the survivors for the next generation is achieved through the fitness function. The fitness score for each alignment is calculated by summing up the individual scores for each column of the sequences. The columns are scored by looking at matches (+1), mismatches (0), and gaps (−1) in the sequences.

Crossover: The system randomly chooses a pair of chromosomes from a population, and a crossover point is randomly generated. The single-point crossover method is used.

Mutation: a selected chromosome will mutate or not.

```
GA: Generate the initial population P;
  for i = 1 to genSize do
            for j = 1 to (popSize/2) do
            Select two chromosomes X and Y
                from P by Selection operation;
            Crossover(X,Y) and let X' and Y'   be the offspring;
            Mutate(X');            Mutate(Y');
            Put X'and Y' into the mating pool     M
  end
Let M be the new population P
```

GA is used after the pairwise alignment for all the sequences using dynamic programming (DP). The time complexity of DP is $O(n^2)$, where $n$ is the longest length of sequences for pairwise alignment. To compare different alignments, for MSA, a fitness function is defined based on the number of matching symbols and number of sizes of gaps. For the clustering problem, the sum of squared error (SSE) criterion is used.

## 4.6 DCGA

DCGA, with the divide-and-conquer technique, can effectively reduce the space complexity for MSA. The cut points are defined, and one part is covered by PEA, while the rest, by the GA. DCGA is a combination of PEA or CDM and GA. DCGA diminishes the search space into a part of sequences using PEA or CDM for finding the "conserved areas" (left part), and GA will be applied for the remaining sequences (right part). DCGA with the "conserved areas" has superiority over the GA solution where the chromosome is used from DP without any cut.

```
DCGA: input: the alignment of the sequences
output: new alignment
step 1: prepare the vertical segmentation
step 2: prepare the new alignment using for the left part the PEA
and for the rest part the GA
step 3: merging the two set of subsequences
```

**Example**

A1 = A A T T C C T
A2 = A T C T
A3 = A A T T C G A

The sequences need to be cut according to the cut points (the locations with no consistency) using PEA or CDM so that they do not disturb the consistency. Starting with PEA from left to right, the subsequences' sets are as follows.

SA11 = A A T SA12 = T C C T
SA21 = - A T SA22 = C T
SA31 = A A T SA32 = T C G A

The set of cut points is $C_1 = \{3,2,3\}$.

Scores for each column (pair work) are as follows: match pair (2), mismatch pair (−1), space (−2). Then we calculate the score of multiple DNA by taking the sum of the score of each column of the first segment. The advantage of DCGA is that the search space after the vertical segmentation (right segment or second segment) is now diminished to the set of the subsequences: SA12, SA22, SA32. PEA or CDM can be applied into the left segment. It is extended until similar subsequences are discovered. The GA, within DCGA, is applied only to a right set of subsequences that requires a search space smaller than the entire sequences' space. On the contrary, the GA solution requires all the sequence spaces. For GA, the pairwise DP is used in order to find the alignment of the subsequences. Finally, the total alignment can be obtained by merging the two subsets.

## 4.7 *K*-MEANS

The centroid-based *K*-means algorithm generates initially a random set of *K* patterns from the data set, known as the centroids. They represent the problem's initial solution. Each point is then assigned to the closest centroid, and each collection of points assigned to a centroid is a cluster. The quality of a cluster $C_i$ can be measured by the within-cluster variation, which is the SSE between all objects in $C_i$ and the centroid $c_i$, defined as $SSE = \sum_{i=1}^{K} \sum_{p \in Ci} dist(p, c_i)^2$, where SSE is the sum of the squared error for all objects in the data set; p is the point in space representing a given object; and $c_i$ is the centroid of cluster $C_i$ (both p and $c_i$ are multidimensional) [4]. The SSE can measure the quality of a clustering by calculating the error of each data point. For two different sets of clusters that are produced by two different runs of *K*-means, the cluster with the smallest squared error is preferred. This means that the centroids of the preferred cluster provide better representation of the points in that cluster.

A major problem with this algorithm is that it is sensitive to the selection of the initial partition and may converge to a local minimum of variation if the original partition is not properly chosen. The *K*-means fails to recognize some clusters. The time complexity is $O(n * k * d * i)$, where $n$ = number of data points, $k$ = number of clusters, $d$ = dimension of data, and $i$ = number of iterations.

## 4.8 CLUSTERING GENETIC ALGORITHM WITH THE SSE CRITERION

The clustering genetic algorithm with the SSE criterion (CGA_SSE) is used for clustering. For the encoding, a straightforward scheme is used [15,17]. The individual is a vector of the length N; each position refers to one object from the training set and contains an identification of the cluster to which the object belongs. The initial population is randomly selected. Each allele in the population can be initialized to a cluster number selected from uniform distribution over set 1,...,$k$ where $k$ is the number of clusters. So given a selection that is encoded by $a_1, a_2, a_3, \ldots, a_n$, we replace $a_n$ with $a_n'$ for $n = 1, \ldots,$ simultaneously where $a_n'$ is the number of clusters where the centroid is closest to $x_n$ in Euclidean distance. Another encoding technique is based on the idea of that the solution can be represented by the centers of clusters [18]. The basic single-point crossover and random mutation has been applied. The *K*-means is used in order to provide better results. This is achieved by applying the *K*-means to each population and performing a local search. The parents are selected using the roulette-wheel method. After the selection, crossover, and mutation, the *K*-means calculates the cluster centers and reassigns each data point (or individual) to the nearest cluster center, providing the best solution. The current population is evaluated, and each individual has a new fitness value, updated using the SSE criterion. The fitness function is defined as the inverse of the SSE metric. A new population is selected and replaces the previous one.

The pseudocode of CGA_SSE is as follows:

```
CGA_SSE Input: X, set of N patterns with d dimensions;
K: number of centroids; Tp: population size;
Pc: crossover rate; Pm: mutation rate;
Ng: maximum number of generations;

Output: Best solution found S*
//the initialization
 t=0, Set the initial population P(t) with size Tp, Evaluate P(t)
S = select_best_individual(P(t)), S* = S
SSSE = SSE computed with the solution S*
Pp = Ø
//the main part
while (t < Ng ) {
      t= t +1 //for next population
      select Pp (t) from P(t-1) with crossover rate Pc
      P(t) = crossover Pp(t)
      apply mutation P(t) with mutation rate Pm
      //Local Search
      for all Ii ∈P(t)
        { apply K-Means, - LocalSearch(Ii),
           calculate cluster centers.
           reassign each data point to the nearest cluster center}
      //GA steps
     evaluate P(t)
      S' = select_best_individual (P(t))
      S'SSE = SSE computed with the solution S', -fitness function
      find the minimum S'SSE (S*)
      from minimum S'SSE define the set of clusters (the best
      solution, S*).
    }//end while
    return S*
```

The process is repeated until there is no change in $k$ cluster centers. This can be achieved by minimizing the objective function of SSE. Hence, the minimum SSE defines an optimum set of clusters. The time complexity of GCA_SSE is $O(t_m * p * n * k * d)$ where $t_m$ = maximum number of iterations, $p$ = population size, $n$ = number of points, $k$ = number of clusters, and $d$ = dimension of data. The GA with the selection and mutation operations may take time to converge because the initial assignments are arbitrary. For improvement purposes, a $K$-means algorithm is used. The $K$-means calculates cluster centers and reassigns each data point to the cluster with the nearest cluster center according to the SSE value.

## 4.9 MapReduce SECTION

The map stage for the distributed CGA_SSE involves execution of a function on data of the form key/values and generates an intermediate data set. The reduce function using the particular key compiles all the data sets for all the various key values.

In this work, our main focus is on the MapReduce solution of the CGA_SSE algorithm. For the distributed implementation, two MapReduce models are used: the MR_GA, which is the model related to the generations, and the GA, which is the model relevant to the *K*-means (MR_Kmeans).

For the MR_GA, there is a set of key/values pairs, and the output with the format key/values is created.

The mapper of the MR_GA focuses on the initial population generation and the evaluation of the fitness values of each individual. It also discovers the best individual and transfers it as a file into the file system. Instead of the basic format of key/values, there is now gene ID/gene value. For the local search, the mapper computes the distance between given data values and each cluster center, and after that, the given data are assigned to the closest calculated cluster. The initial set of centers is stored as input to the map routine, and they form the "key" field of the key/values pair.

The assignment of the individuals, emitted from mappers to the reducers, can be achieved with a partitioner by using a hash function (h) based on the module function [h(i)%*r*] where i is the ID of the individual and *r* is the number of reducers.

Generally, the reducer for each intermediate key finds assets of values. The reducer will concentrate on the select and crossover operations. It also calculates the $S'_{SSE}$ and tests the exit condition for the predefined number of generations. It also continues with the recalculation of the centroids of any particular cluster.

After the creation of the new population, there is a new MapReduce job for the new generation.

## 4.10 SIMULATION

1. Three scenarios have been developed for the MSA problem.

   a. *PEA versus GA*: PEA and GA have been developed with five sequences of size 10. According to the divide-and-conquer method, two cut points have been defined: (a) 3 and (b) 7. For test 1, PEA is applied to three columns and GA to the rest (seven). For test 2, similarly, PEA is applied to seven columns and GA to the rest (three). The fact is that the vertical segmentation of the sequences plays a role. In Figure 4.1, in test 1, PEA is faster than GA since it uses only a small size of sequences. GA is more time consuming since it uses the DP for the alignment. In test 2, GA is faster than PEA due to the small size of the sequence alignment. After that, from test 1, there is faster MSA for the area of 1–3 considering the cut

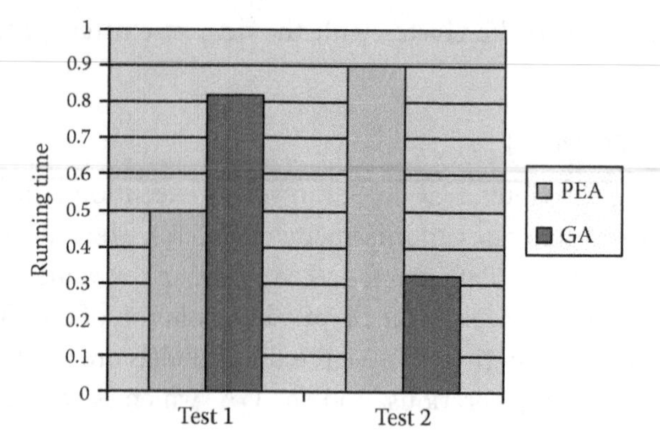


FIGURE 4.1 PEA versus GA for various cut points.

point of 3. Similarly, for test 2, there is faster MSA for the area of 8–10 considering the cut point of 7.

b. *CDM and GA for various sequence sizes*: Different sizes of sequences are compared for their running time using CDM and GA. In Figure 4.2, there is the case for two sets (5, 8) of sequences where the smaller-size sequences have better performance.

c. *DCGA versus GA*: A set of sequences with sizes 10, 20, and 30 has been examined. The CDM finds the "conserved areas" starting from left to right. In the next step, DCGA applies the GA only for the right segment. DCGA outperforms GA, since GA has to use the operations (selection, crossover, and mutate) for all the sizes of sequences. The greater the size of sequences, the better the performance of DCGA is (Figure 4.3).

2. Two scenarios have been developed for clustering.

a. *2-D, 3-D data*: 2-D and 3-D data for 300 points have been considered for CGA_SSE with the measure of the running time. The running time is greater for the 3-D data (Figure 4.4).

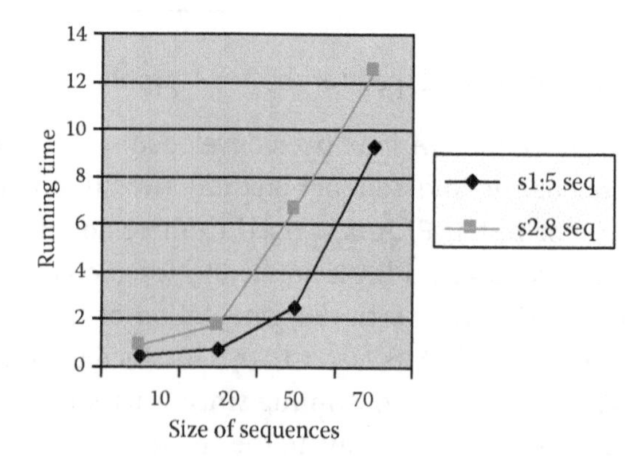


FIGURE 4.2 MSA for two sets (5, 8) of sequences.

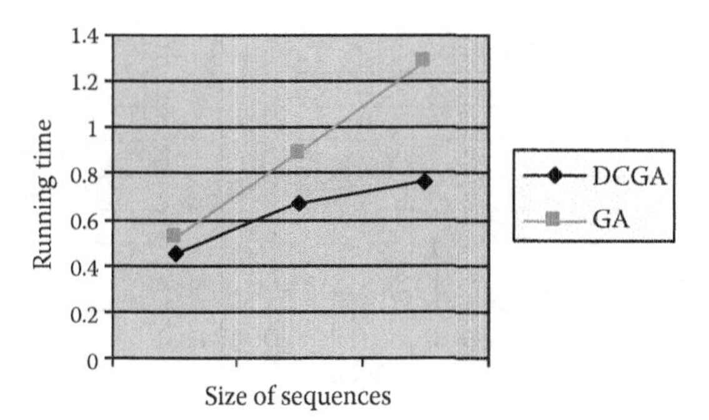


FIGURE 4.3 DCGA versus GA.

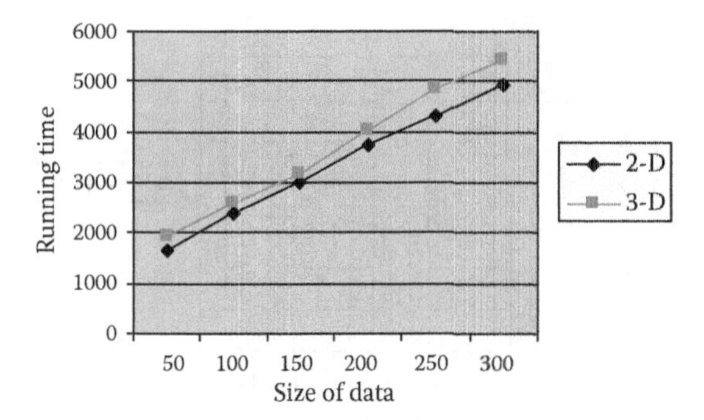


FIGURE 4.4 GCA_SSE for 2-D and 3-D data.

b. *CGA_SSE, K-means accuracy*: CGA_SSE has greater accuracy than *K*-means. *K*-means is capable of recognizing many of the clusters, while CGA_SSE can discover more clusters but needs more time. Accuracy is the percentage of correctly assigned examples. For a set of 3-D data and the wine data, the accuracy is shown below (Figure 4.5).

3. Four scenarios have been developed for the distributed environment.

a. *CGA_SSE (distributed)*: The CGA_SSE is examined for the sequential and the distributed (one node, Hadoop) cases with the "wine" file (178 instances). This is evident (Figure 4.6) due to the small file size and because the Hadoop operations, map, reading, copying, and so forth, have overhead.

b. *CGA_SSE (distributed with two nodes)*: The CGA_SSE is examined with the "wine quality" file (4898 instances) for the sequential, distributed (one node), and two-nodes cases. In Figure 4.7, the two nodes give the opportunity to Hadoop to

| | 3-D | wine |
|---|---|---|
| *K*-means | 95.2% | 96.3% |
| CGA_SSE | 97.7% | 98.2% |

FIGURE 4.5 Accuracy *K*-means, CGA_SSE.

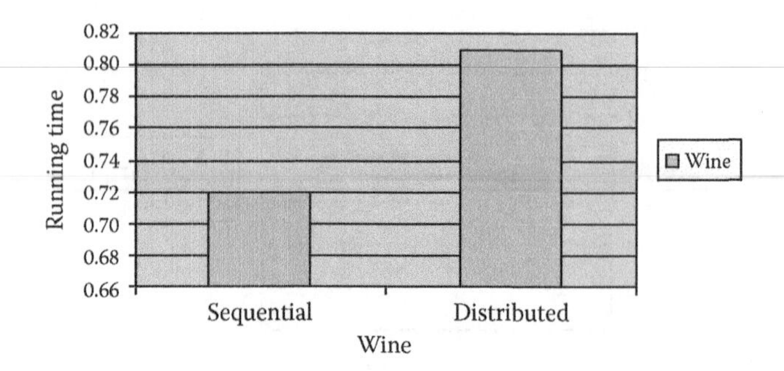


FIGURE 4.6 CGA_SSE distributed.

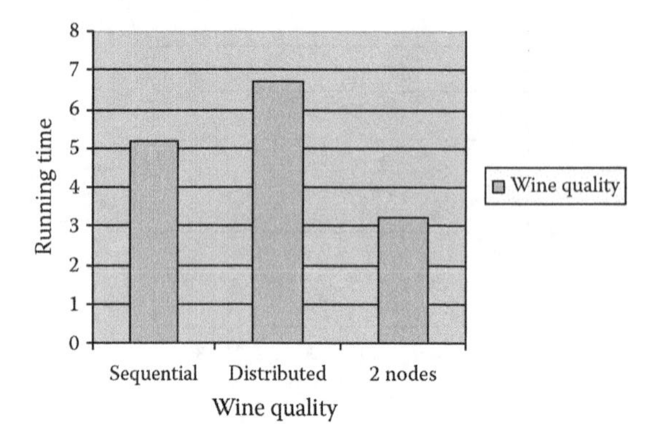


FIGURE 4.7 CGA_SSE with two nodes.

increase the performance of the processes due to the parallel instead of sequential execution of the code. For the distributed (one node) case, more processing time is needed due to the Hadoop overheads. The use of a large number of nodes decreases the required processing time.

c. *The CDM (distributed) with various size sequences*: The time for four sequences with varying size and the number of nodes is examined. From Figure 4.8, the

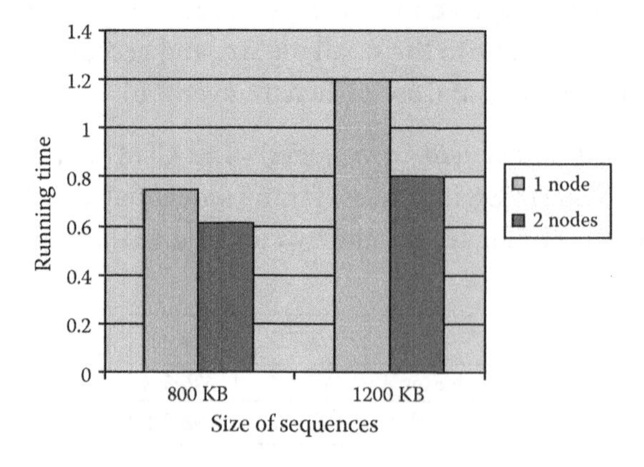


FIGURE 4.8 CDM with various sizes of sequences.

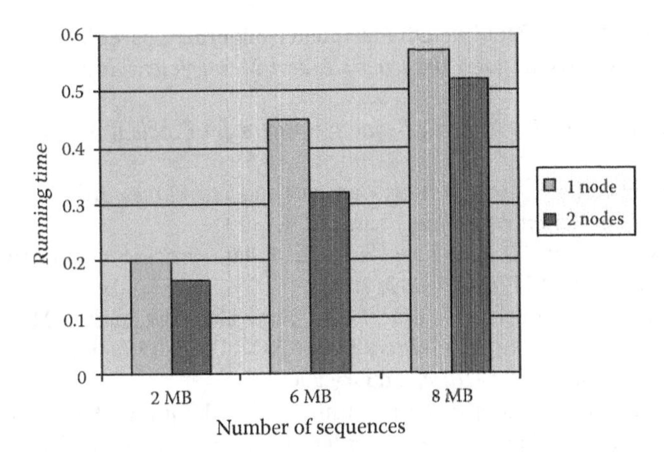


FIGURE 4.9 CDM with various numbers of sequences.

larger the sequence size, the longer the alignment time. When more nodes are involved, there is significant reduction of time.

d. *CDM (distributed) with various numbers of sequences*: For all the sizes of sequences, the Hadoop with two nodes gives better time performance than with just one node (Figure 4.9).

## 4.11 CONCLUSION

A set of algorithms has been developed for finding MSA using dot matrices and the divide-and-conquer method with GA. CDM discovers the consistency, while PEA provides the MSA. In DCGA, there are different results in favor of PEA or GA depending on the values of the cut points or the size of the subsequences that are under processing. This provides the opportunity for partial MSA. Moreover, DCGA with the divide-and-conquer technique, can effectively reduce the space complexity for MSA so that the consistency achieved from the CDM is not broken. For long sequences with "conserved areas" on their left side, DCGA outperforms GA. This is due to the fact that GA uses the whole size of the sequences. Finally, CGA_SSE, compared with *K*-means, can discover an optimal clustering using the SSE criterion.

Hadoop with the MapReduce framework can provide easy solutions by dividing the applications into many small blocks of work that can be executed in parallel. CGA_SSE and CDM with Hadoop are examples of MapReduce's ability to solve large-scale processing problems. Generally, for problems with massive populations, Hadoop provides a new direction for solving hard problems using MapReduce with a number of nodes. Future work would deal with MSA, neural networks, and Hadoop.

## REFERENCES

1. J. Stoye, A. Dress, S. Perrey, "Improving the divide and conquer approach to sum-of-pairs multiple sequence alignment," *Applied Mathematical Literature* 10(2):67–73, 1997. Mathematical Reviews, 1980.

2. S.-M. Chen, C.-H. Lin, "Multiple DNA sequence alignment based on genetic algorithms and divide and conquer techniques," *Information and Management Science Journal* 18(2):76–111, 2007.
3. H. Garillo, D. Lipman, *The Multiple Sequence Alignment Problem in Biology*, Addison-Wesley, Boston, 1989.
4. J. Han, M. Kamber, J. Pei, *Data Mining: Concepts and Techniques*, 3rd ed., MK, 2012.
5. C. Lam, *Hadoop in Action*, Manning, Stamford, CT, 2011.
6. J. Dean, S. Ghemawai, "MapReduce: Simplified data processing on large clusters," *Communications of the ACM* 51(1):107–113, 2008.
7. R. Raghuraman, A. Penmetsa, G. Bradski, C. Kozyrakis, "Evaluating MapReduce for multi-core and multiprocessor systems," *Proceedings 2007 IEEE, 13th Intern. Symposium on High Performance Computer Architecture*, January 2007.
8. J. Ekanayake, H. Li, B. Zhang, T. Gunarathne, S. Bae, J. Qiu, G. Fox, "Twister: Runtime for iterative MapReduce," *Proc. 19th ACM Inter. Symposium on High Performance Distributed Computing*, 2010, pp. 810–818.
9. D. Mount, *Bioinformatics: Sequence and Genome Analysis*, 2nd ed., CSHL Press, 2004.
10. Apache Hadoop. Available at http://hadoop.apache.org.
11. T. White, *Hadoop: The Definite Guide*, O'Reilly Media, Yahoo Press, Cambridge, MA, June 5, 2009.
12. R. Taylor, "An overview of the Hadoop/MapReduce/HBase framework and its current applications in bioinformatics," *BMC Bioinformatics* 11(Suppl 12):S1, 2010. Available at http://www.biomedicalcentral.com/1471-2105/11/512/51.
13. A. Hughes, Y. Ruan, S. Ekanayake, S. Bae, Q. Dong, M. Rho, J. Qiu, G. Fox, "Interpolative multidimensional scaling techniques for the identification of clusters in very large sequence sets," *BMC Bioinformatics* 12(Suppl 2):59, 2012. Available at http://www.biomedcentral.com/1471-2105/13/52/59.
14. C. Zhang, A. Wong, "Towards efficient multiple molecular sequence alignment: A system of genetic algorithm and dynamic programming," *IEEE Transaction System Man and Cybernetics B* 27(6):918–932, 1997.
15. D. Colberg, *Genetic Algorithms in Search, Optimization, and Machine Learning*, Addison-Wesley, Boston, 1989.
16. K. Jong, "Learning with genetic algorithms: An overview," *Machine Learning* 3:121–138, 1988.
17. E. Hruschka, R. Campello, L. Castro, "Improving the efficiency of a clustering genetic algorithm," *Advances in Artificial Intelligence, IBERAMIA 2004*, vol. 3315 of LNCS, 2004, pp. 861–870.
18. K. Maulik, S. Bandyopabhyay, "Genetic algorithm-based clustering technique," *Pattern Recognition* 33:1455–1465, 2000.

# II

## Big Data Processing

CHAPTER 5

# Approaches for High-Performance Big Data Processing

## *Applications and Challenges*

Ouidad Achahbar, Mohamed Riduan Abid, Mohamed Bakhouya, Chaker El Amrani, Jaafar Gaber, Mohammed Essaaidi, and Tarek A. El Ghazawi

CONTENTS



ABSTRACT

Social media websites, such as Facebook, Twitter, and YouTube, and job posting websites like LinkedIn and CareerBuilder involve a huge amount of data that are very useful to economy assessment and society development. These sites provide sentiments and interests of people connected to web communities and a lot of other information. The Big Data collected from the web is considered as an unprecedented source to fuel data processing and business intelligence. However, collecting, storing, analyzing, and processing these Big Data as quickly as possible create new challenges for both scientists and analytics. For example, analyzing Big Data from social media is now widely accepted by many companies as a way of testing the acceptance of their products and services based on customers' opinions. Opinion mining or sentiment analysis methods have been recently proposed for extracting positive/negative words from Big Data. However, highly accurate and timely processing and analysis of the huge amount of data to extract their meaning require new processing techniques. More precisely, a technology is needed to deal with the massive amounts of unstructured and semistructured information, in order to understand hidden user behavior. Existing solutions are time consuming given the increase in data volume and complexity. It is possible to use high-performance computing technology to accelerate data processing, through MapReduce ported to cloud computing. This will allow companies to deliver more business value to their end customers in the dynamic and changing business environment. This chapter discusses approaches proposed in literature and their use in the cloud for Big Data analysis and processing.

## 5.1 INTRODUCTION

Societal and technological progress have brought a widespread diffusion of computer services, information, and data characterized by an ever-increasing complexity, pervasiveness, and social meaning. Ever more often, people make use of and are surrounded by devices enabling the rapid establishment of communication means on top of which people may socialize and collaborate, supply or demand services, and query and provide knowledge as it has never been possible before [1]. These novel forms of social services and social organization are a promise of new wealth and quality for all. More precisely, social media, such as Facebook, Twitter, and YouTube, and job posting websites like LinkedIn and CareerBuilder are providing a communication medium to share and distribute information among users.

These media involve a huge amount of data that are very useful to economy assessment and society development. In other words, the Big Data collected from social media is considered as an unprecedented source to fuel data processing and business intelligence. A study done in 2012 by the American Multinational Corporation (AMC) has estimated that from 2005 to 2020, data will grow by a factor of 300 (from 130 exabytes to 40,000 exabytes), and therefore, digital data will be doubled every 2 years [2]. IBM estimates that every day, 2.5 quintillion bytes of data are created, and 90% of the data in the world today has been created in the last 2 years [3]. In addition, Oracle estimated that 2.5 zettabytes of data were generated in 2012, and this will grow significantly every year [4].

The growth of data constitutes the "Big Data" phenomenon. Big Data can be then defined as a large amounts of data, which requires new technologies and architectures so that it becomes possible to extract value from them by a capturing and analysis process. Due to such a large size of data, it becomes very difficult to perform effective analysis using existing traditional techniques. More precisely, as Big Data grows in terms of volume, velocity, and value, the current technologies for storing, processing, and analyzing data become inefficient and insufficient. Furthermore, unlike data that are structured in a known format (e.g., rows and columns), data extracted, for example, from social media, have unstructured formats and are very complex to process directly using standards tools. Therefore, technologies that enable a scalable and accurate analysis of Big Data are required in order to extract value from it.

A Gartner survey stated that data growth is considered as the largest challenge for organizations [5]. With this issue, analyzing and high-performance processing of these

Big Data as quickly as possible create new challenges for both scientists and analytics. However, high-performance computing (HPC) technologies still lack the tool sets that fit the current growth of data. In this case, new paradigms and storage tools were integrated with HPC to deal with the current challenges related to data management. Some of these technologies include providing computing as a utility (cloud computing) and introducing new parallel and distributed paradigms. Recently, cloud computing has played an important role as it provides organizations with the ability to analyze and store data economically and efficiently. For example, performing HPC in the cloud was introduced as data has started to be migrated and managed in the cloud. For example, Digital Communications Inc. (DCI) stated that by 2020, a significant portion of digital data will be managed in the cloud, and even if a byte in the digital universe is not stored in the cloud, it will pass, at some point, through the cloud [6].

Performing HPC in the cloud is known as high-performance computing as a service (HPCaaS). Therefore, a scalable HPC environment that can handle the complexity and challenges related to Big Data is required [7]. Many solutions have been proposed and developed to improve computation performance of Big Data. Some of them tend to improve *algorithm efficiency*, provide new *distributed paradigms*, or develop powerful clustering environments, though few of those solutions have addressed a whole picture of integrating HPC with the current emerging technologies in terms of storage and processing.

This chapter introduces the Big Data concepts along with their importance in the modern world and existing projects that are effective and important in changing the concept of science into big science and society too. In this work, we will be focusing on efficient approaches for high-performance mining of Big Data, particularly existing solutions that are proposed in the literature and their efficiency in HPC for Big Data analysis and processing. The remainder of this chapter is structured as follows. The definition of Big Data and related concepts are introduced in Section 5.2. In Section 5.3, existing solutions for data processing and analysis are presented. Some remaining issues and further research directions are presented in Section 5.4. Conclusions and perspectives are presented in Section 5.5.

## 5.2 BIG DATA DEFINITION AND CONCEPTS

Social media websites involve a huge amount of data that are very useful to economy assessment and society development. Big Data is then defined as large and complex data sets that are generated from different sources including social media, online transactions, sensors, smart meters, and administrative services [8]. With all these sources, the size of Big Data goes beyond the ability of typical tools of storing, analyzing, and processing data. Literature reviews on Big Data divide the concept into four dimensions: *volume*, *velocity*, *variety*, and *value* [8]. In other words, (1) the size of data generated is very large, and it goes from terabytes to petabytes; (2) data grows continuously at an exponential rate; (3) data are generated in different forms: structured, semistructured, and unstructured data, which require new techniques that can handle data heterogeneity; and (4) the challenge in Big Data is to identify what is valuable so as to be able to capture, transform, and extract data for analysis. Data incorporates information of great benefit and insight for users. For example, many social media websites provide sentiments and opinions of people connected to web communities and a lot of other information. Sentiments or opinions are clients' personal judgments about something, such as whether the product is good or bad [9]. Sentiment analysis or opinion mining is now widely accepted by companies as an important core element for automatically detecting sentiments. For example, this research field has been active since 2002 as a part of online market research tool kits for processing of large numbers of texts in order to identify clients' opinions about products.

In past years, many machine learning algorithms for text analysis have been proposed. They can be classified into two main families: *linguistic* and *nonlinguistic*. Linguistic-based algorithms use knowledge about language (e.g., syntactic, semantic) in order to understand and analyze texts. Nonlinguistic-based algorithms use a learning process from training data in order to guess correct sentiment values. However, these algorithms are time consuming, and plenty of efforts have been dedicated to parallelize them to get better speed. Therefore, new technologies that can support the volume, velocity, variety, and value of data were recently introduced. Some of the new technologies are Not Only Structured Query Language (NoSQL), parallel and distributed paradigms, and new cloud computing trends that can support the four dimensions of Big Data. For example, NoSQL is the transition from relational databases to nonrelational databases [10]. It is characterized by the ability to scale horizontally, the ability to replicate and to partition data over many servers, and the ability to provide high-performance operations. However, moving from relational to NoSQL systems has eliminated some of the atomicity, consistency, isolation, and durability (ACID) transactional properties [11]. In this context, NoSQL properties are defined by consistency, availability, and partition (CAP) tolerance [12] theory, which states that developers must make trade-off decisions between consistency, availability, and partitioning. Some examples of NoSQL tools are Cassandra [13], HBase [14], MongoDB [15], and CouchDB [16].

In parallel to these technologies, cloud computing becomes the current innovative and emerging trend in delivering information technology (IT) services that attracts the interest of both academic and industrial fields. Using advanced technologies, cloud computing provides end users with a variety of services, starting from hardware-level services to the application level. Cloud computing is understood as utility computing over the Internet. This means that computing services have moved from local data centers to hosted services that are offered over the Internet and paid for based on a pay-per-use model [17]. As stated in References 18 and 19, cloud deployment models are classified as follows: *software as a service* (SaaS), *platform as a service* (PaaS), and *infrastructure as a service* (IaaS). SaaS represents application software, operating systems (OSs), and computing resources. End users can view the SaaS model as a web-based application interface where services and complete software applications are delivered over the Internet. Some examples of SaaS applications are Google Docs, Microsoft Office Live, Salesforce Customer Relationship Management (CRM), and so forth. PaaS allows end users to create and deploy applications on a provider's cloud infrastructure. In this case, end users do not manage or control the underlying cloud infrastructure like the network, servers, OSs, or storage. However, they do have control over the deployed applications by being allowed to design, model, develop, and test them. Examples of PaaS are Google App Engine, Microsoft Azure, Salesforce, and so forth. IaaS consists of a set of virtualized computing resources such as network bandwidth, storage capacity, memory, and processing power. These resources can be used to deploy and run arbitrary software, which can include OSs and applications. Examples of IaaS providers are Dropbox, Amazon Web Services (AWS), and so forth.

As illustrated in Table 5.1, there are many other providers who offer cloud services with different features and pricing [20]. For example, Amazon (AWS) [21] offers a number of cloud services for all business sizes. Some AWS services are Elastic Compute Cloud; Simple Storage Service; SimpleDB (relational data storage service that stores, processes, and queries data sets in the cloud); and so forth. Google [22] offers high accessibility and usability in its cloud services. Some of Google's services include Google App Engine, Gmail, Google Docs, Google Analytics, Picasa (a tool used to exhibit products and upload their images in the cloud), and so forth.

Recently, HPC has been adopted to provide high computation capabilities, high bandwidth, and low-latency networks in order to handle the complexity of Big Data. HPC fits these requirements by implementing large physical clusters. However, traditional HPC faces a set of challenges, which consist of peak demand, high capital, and high expertise, to acquiring and operating the physical environment [23]. To deal with these issues, HPC experts have leveraged the benefits of new technology trends including cloud technologies and large storage infrastructures. Merging HPC with these new technologies has led to a new HPC model, HPCaaS. HPCaaS is an emerging computing model where end users have on-demand access to pre-existing needed technologies that provide high performance and a scalable HPC computing environment [24]. HPCaaS provides unlimited benefits because of the better quality of services provided by the cloud technologies and the better parallel processing and storage provided by, for example, the Hadoop Distributed File System (HDFS) and MapReduce paradigm. Some HPCaaS benefits are stated in Reference 23 as follows: (1) high scalability, in which resources are scaling up as to ensure essential resources that fit users' demand in terms of processing large and complex data sets; (2) low cost, in which end users can eliminate the initial capital outlay, time, and complexity to procure HPC; and (3) low latency, by implementing the placement group concept that ensures the execution and processing of data in the same server.

TABLE 5.1 Examples of Services Provided in Cloud Computing Environment

| Deployment Models | Types | Examples |
|---|---|---|
| Software as a service | Gmail, Google Docs, finance, collaboration, communication, business, CRM | Zaho, Salesforce, Google Apps |
| Platform as a service | Web 2 application runtime, developer tools, middleware | Windows Azure, Aptana, Google App Engine |
| Infrastructure as a service | Servers, storage, processing power, networking resources | Amazon Web Services, Dropbox, Akamai |

*Source*: Youssef, A. E., *J. Emerg. Trends Comput. Inf. Sci.*, 3, 2012.

There are many HPCaaS providers in the market. An example of an HPCaaS provider is Penguin Computing [25], which has been a leader in designing and implementing high-performance environments for over a decade. Nowadays, it provides HPCaaS with different options: on demand, HPCaaS as private services, and hybrid HPCaaS services. AWS [26] is also an active HPCaaS in the market; it provides simplified tools to perform HPC over the cloud. AWS allows end users to benefit from HPCaaS features with different pricing models: On-Demand, Reserved [27], or Spot Instances [28]. HPCaaS on AWS is currently used for computer-aided engineering, molecular modeling, genome analysis, and numerical modeling across many industries, including oil and gas, financial services, and manufacturing [21]. Other leaders of HPCaaS in the market are Microsoft (Windows Azure HPC) [29] and Google (Google Compute Engine) [30].

For Big Data processing, two main research directions can be identified: (1) deploying popular tools and libraries in the cloud and providing the service as SaaS and (2) providing computing and storage resources as IaaS or HPCaaS to allow users to create virtual clusters and run jobs.

## 5.3 CLOUD COMPUTING FOR BIG DATA ANALYSIS

Recent efforts have investigated cloud computing platforms either to create parallel versions in the form of libraries and statistic tools or to allow users to create clusters on these platforms and run them in the cloud [31]. This section highlights existing work in both directions, data analytics tools as SaaS and computing as IaaS.

### 5.3.1 Data Analytics Tools as SaaS

Data analytics solutions provide optimized statistical toolboxes that include parallel algorithms (e.g., Message Passing Interface [MPI], Hadoop) for data analysis. Furthermore, users also can write their algorithms and run them in parallel over computer clusters. For example, the *Apache Mahout* (https://mahout.apache.org/) is a project of the Fondation Apache aiming to build scalable machine learning libraries using Apache Hadoop and the MapReduce paradigm. These libraries are self-contained and highly optimized in order to get better performance while making them easy to use. In other words, these libraries include optimized algorithms (e.g., clustering, classification, collaborative filtering), making them among the most popular libraries for machine learning projects.

*GraphLab* (http://graphlab.org/projects/index.html) is another project from Carnegie Mellon University with the main aim to develop new parallel machine learning algorithms for graph programming Application Programming Interface (API). It is a graph-based high-performance framework that includes different machine learning algorithms for data analysis. It includes several libraries and algorithms, for example, feature extraction (e.g., linear discriminant analysis); graph analytics (e.g., PageRank, triangle counting); and clustering (e.g., *K*-means). *Jubatus* (http://jubat.us/en/) is another open-source machine learning and distributed computing framework that provides several features, such as classification, recommendation, regression, and graph mining.

The *IBM Parallel Machine Learning Toolbox* (PML, https://www.research.ibm.com/haifa/projects/verification/ml_toolbox/index.html) was developed, similar to the MapReduce programming model and Hadoop system, to allow users with little knowledge in parallel and distributed systems to easily implement parallel algorithms and run them on multiple processor environments or on multithreading machines. It also provides preprogrammed parallel algorithms, such as support vector machine classifiers, linear and transform regression, nearest neighbors, *K*-means, and principal component analysis (PCA). *NIMBLE* is another tool kit for implementing parallel data mining and machine learning algorithms on MapReduce [32]. Its main aim is to allow users to develop parallel machine learning algorithms and run them on distributed- and shared-memory machines.

Another category of machine learning systems has been developed and uses Hadoop as a processing environment for data analysis. For example, the *Kitenga Analytics* (http://software.dell.com/products/kitenga-analytics-suite/) platform provides analytical tools with an easy-to-use interface for Big Data sophisticated processing and analysis. It combines Hadoop, Mahout machine learning, and advanced natural language processing in a fully integrated platform for performing fast content mining and Big Data analysis. Furthermore, it can be considered as the first Big Data search and analytics platform to integrate and process diverse, unstructured, semistructured, and structured information. *Pentaho Business Analytics* (http://www.pentaho.fr/explore/pentaho-business-analytics/) is another platform for data integration and analysis. It offers comprehensive tools that support data preprocessing, data exploration, and data extraction together with tools for visualization and for distributed execution on the Hadoop platform. Other systems, such as *BigML* (https://bigml.com/), *Google Prediction API* (https://cloud.google.com/products/prediction-api/), and *Eigendog*, recently have been developed by offering services for data processing and analysis. For example, Google Prediction API is Google's cloud-based machine learning tool used for analyzing data. However, these solutions cannot be used for texts that are extracted, for example, from social media and social networks (e.g., Twitter, Facebook).

Recently, an increased interest has been devoted to text mining and natural language processing approaches. These solutions are delivered to users as cloud-based services. It is worth noting that the main aim of text mining approaches is to extract features, for example, concept extraction and sentiment or opinion extraction. However, the size and number of documents that need to be processed require the development of new solutions. Several solutions are provided via web services. For example, *AlchemyAPI* (http://www.alchemyapi.com/) provides natural language processing web services for processing and analyzing vast amounts of unstructured data. It can be used for performing key word/entity extraction and sentiment analysis on large amounts of documents and tweets. In other words, it uses linguistic parsing, natural language processing, and machine learning to analyze and extract semantic meaning (i.e., valuable information) from web content.

### 5.3.2 Computing as IaaS

Providing high-performance data analysis allows users to avoid installing and managing their own clusters. This can be done by offering computing clusters using public providers. In fact, a high-performance platform is easy to create and effortless to maintain. These solutions provide users the possibility to experiment with complex algorithms on their customized clusters on the cloud.

It is worth noting that performing HPC in the cloud was introduced as data has started to be migrated and managed in the cloud. As stated in Section 5.2, performing HPC in the cloud is known as HPCaaS.

In short, HPCaaS offers high-performance, on-demand, and scalable HPC environments that can handle the complexity and challenges related to Big Data [7]. One of the most known and adopted parallel and distributed systems is the MapReduce model that was developed by Google to meet the growth of their web search indexing process [33]. MapReduce computations are performed with the support of a data storage system known as Google File System (GFS). The success of both GFS and MapReduce inspired the development of Hadoop, which is a distributed and parallel system that implements MapReduce and HDFS [34]. Nowadays, Hadoop is widely adopted by big players in the market because of its scalability, reliability, and low cost of implementation. Hadoop is also proposed to be integrated with HPC as an underlying technology that distributes the work across an HPC cluster [35,36]. With these solutions, users no longer need high skills in HPC-related fields in order to access computing resources.

Recently, several cloud-based HPC clusters have been provided to users together with software packages (e.g., Octave, R system) for data analysis [31]. The main objective is to provide users with suitable environments that are equipped with scalable high-performance resources and statistical software for their day-to-day data analysis and processing. For example, *Cloudnumbers.com* (http://www.scientific-computing.com/products/product_details.php?product_id=1086) is a cloud-based HPC platform that can be used for time-intensive processing of Big Data from different domains, such as finance and social science. Furthermore, a web interface can be used by users to create, monitor, and easily maintain their work environments. Another similar environment to Cloudnumbers.com is *Opani* (http://opani.com), which provides additional functionalities that allow users to adapt resources according to data size. While these solutions are scalable, high-level expertise statistics are required, and thus, there are a limited number of providers in this category of solutions. To overcome this drawback, some solutions have been proposed to allow users to build their cloud Hadoop clusters and run their applications. For example, *RHIPE* (http://www.datadr.org/) provides a framework that allows users to access Hadoop clusters and launch map/reduce analysis of complex Big Data. It is an environment composed of R, an interactive language for data analysis; HDFS; and MapReduce. Other environments, such as *Anaconda* (https://store.continuum.io/cshop/anaconda/) and *Segue* (http://code.google.com/p/segue/), were developed for performing map/reduce jobs on top of clusters for Big Data analysis.

## 5.4 CHALLENGES AND CURRENT RESEARCH DIRECTIONS

Big Data are ubiquitous in many domains such as social media, online transactions, sensors, smart meters, and administrative services. In the past few years, many solutions have been proposed and developed to improve computation performance of Big Data. Some of them tend to improve algorithm efficiency, provide new distributed paradigms, or develop powerful clustering environments. For example, machine learning provides a wide range of techniques, such as association rule mining, decision trees, regression, support vector machines, and other data mining techniques [37].

However, few of those solutions have addressed the whole picture of integrating HPC with the current emerging technologies in terms of storage and processing of Big Data, mainly those extracted from social media. Some of the most popular technologies currently used in hosting and processing Big Data are cloud computing, HDFS, and Hadoop MapReduce [38]. At present, the use of HPC in cloud computing is still limited. The first step towards this research was done by the Department of Energy National Laboratories (DOE), which started exploring the use of cloud services for scientific computing [24]. Besides, in 2009, Yahoo Inc. launched partnerships with major top universities in the United States to conduct more research about cloud computing, distributed systems, and high computing applications.

Recently, there have been several studies that evaluated the performance of high computing in the cloud. Most of these studies used Amazon Elastic Compute Cloud (Amazon EC2) as a cloud environment [26,39–42]. Besides, only few studies have evaluated the performance of high computing using the combination of both new emerging distributed paradigms and the cloud environment [43]. For example, in Reference 39, the authors have evaluated HPC on three different cloud providers: Amazon EC2, GoGrid cloud, and IBM cloud. For each cloud platform, they ran HPC on Linux virtual machines (VMs), and they came to the conclusion that the tested public clouds do not seem to be optimized for running HPC applications. This was explained by the fact that public cloud platforms have slow network connections between VMs. Furthermore, the authors in Reference 26 evaluated the performance of HPC applications in today's cloud environments (Amazon EC2) to understand the trade-offs in migrating to the cloud. Overall results indicated that running HPC on the EC2 cloud platform limits performance and causes significant variability. Besides Amazon EC2, research [44] has evaluated the performance–cost trade-offs of running HPC applications on three different platforms. The first and second platforms consisted of two physical clusters (Taub and Open Cirrus cluster), and the third platform consisted of Eucalyptus with kernel-based virtualization machine (KVM) virtualization. Running HPC on these platforms led authors to conclude that the cloud is more cost effective for low-communication-intensive applications.

Evaluation of HPC without relating it to new cloud technologies was also performed using different virtualization technologies [45–48]. For example, in Reference 45, the authors performed an analysis of virtualization techniques including VMware, Xen, and OpenVZ. Their findings showed that none of the techniques matches the performance of the base system perfectly; however, OpenVZ demonstrates high performance in both file system performance and industry-standard benchmarks. In Reference 46, the authors compared the performance of KVM and VMware. Overall findings showed that VMware performs better than KVM. Still, in a few cases, KVM gave better results than VMware. In Reference 47, the authors conducted quantitative analysis of two leading open-source hypervisors, Xen and KVM. Their study evaluated the performance isolation, overall performance, and scalability of VMs for each virtualization technology. In short, their findings showed that KVM has substantial problems with guests crashing (when increasing the number of guests); however, KVM still has better performance isolation than Xen. Finally, in Reference 48, the authors have extensively compared four hypervisors: Hyper-V, KVM, VMware, and Xen. Their results demonstrated that there is no perfect hypervisor. However, despite the evaluation of different technologies, HPCaaS still needs more investigation to decide on appropriate environments for Big Data analysis.

In our recent work, different experiments have been conducted on three different clusters, Hadoop physical cluster (HPhC), Hadoop virtualized cluster using KVM (HVC-KVM), and Hadoop virtualized cluster using VMware Elastic sky X Integrated (HVC-VMware ESXi), as illustrated in Figure 5.1 [49].

Two main benchmarks, TeraSort and HDFS I/O saturation (TestDFSIO) benchmarks [50], were used to study the impact of machine virtualization on HPCaaS. TeraSort does considerable computation, networking, and storage input/output (I/O) and is often

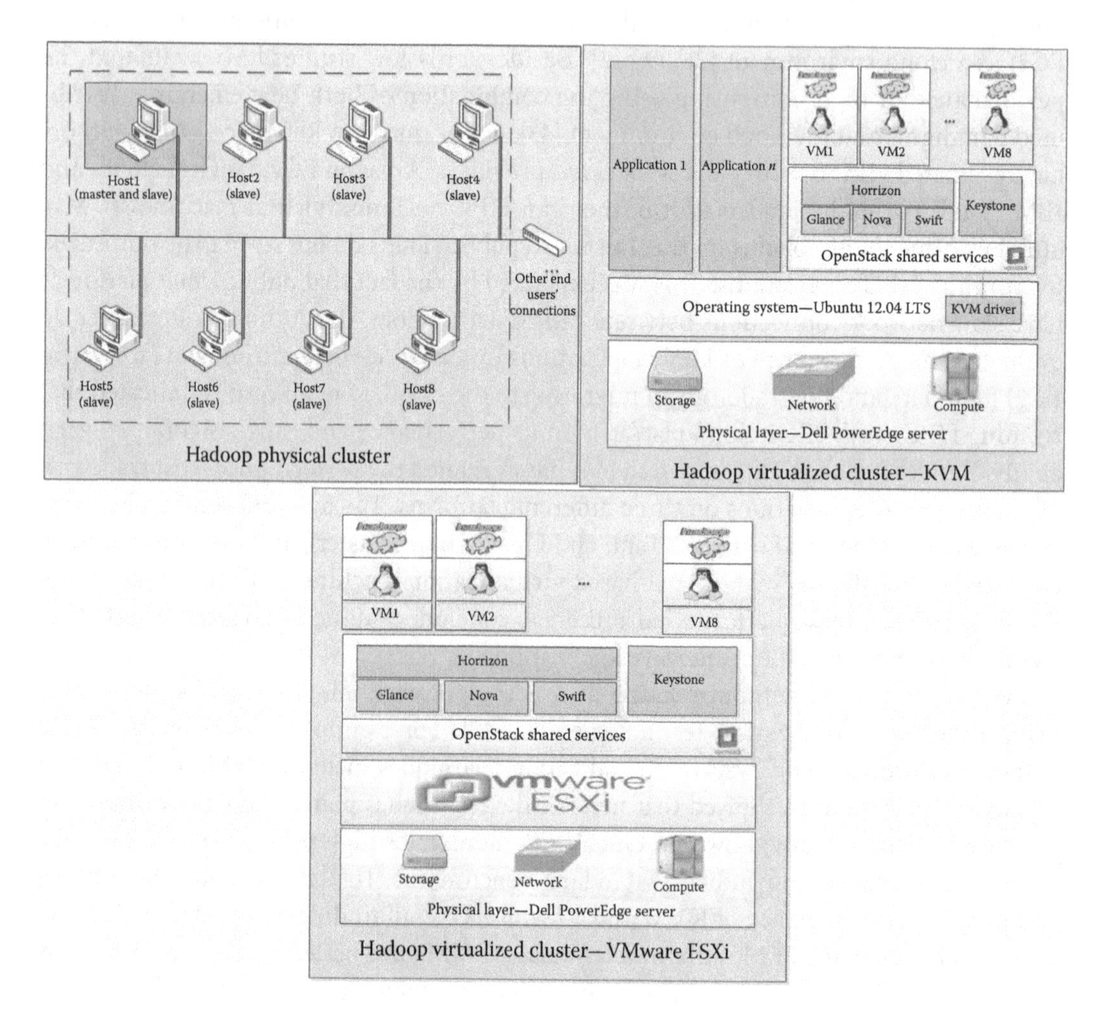


FIGURE 5.1 Hadoop physical and virtualized clusters. LTS, long term support. (From Achahbar, O., The Impact of Virtualization on High Performance Computing Clustering in the Cloud, Master's thesis report, Al Akhawayn University in Ifrane, Morocco, 2014.)

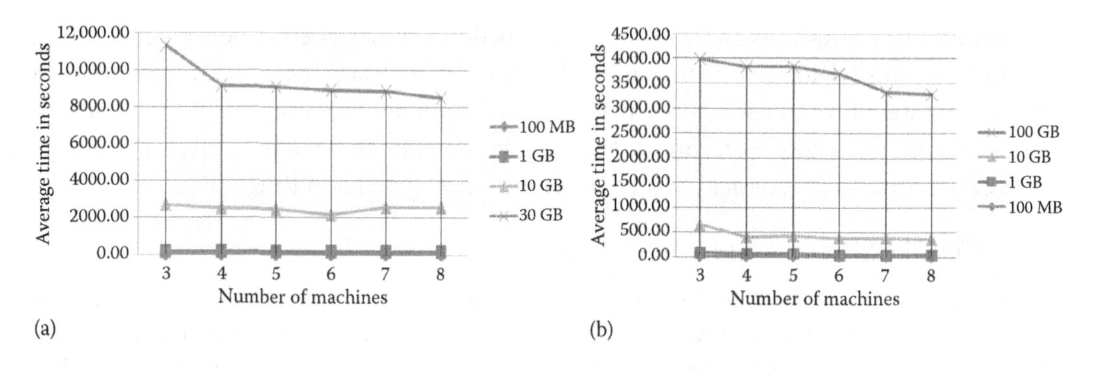


FIGURE 5.2 Performance on Hadoop physical cluster of (a) TeraSort and (b) TestDFSIO-Write. (From Achahbar, O., The Impact of Virtualization on High Performance Computing Clustering in the Cloud, Master's thesis report, Al Akhawayn University in Ifrane, Morocco, 2014.)

considered to be representative of real Hadoop workloads. The TestDFSIO benchmark is used to check the I/O rate of a Hadoop cluster with write and read operations. Such a benchmark can be helpful for testing HDFS by checking network performance and testing hardware, OS, and Hadoop setup. For example, Figure 5.2 shows some experimental results obtained when running TeraSort benchmark and TestDFSIO-Write on HPhC. Figure 5.2a shows that it needs much time to sort large data sets like 10 and 30 GB. However, scaling the cluster to more nodes led to significant time reduction in sorting data sets. Figure 5.2b shows that as the number of VMs increases, the average time decreases when writing different data set sizes. In fact, the Hadoop VMware ESXi cluster performs better at sorting Big Data sets (more computations), and the Hadoop KVM cluster performs better at I/O operations.

Furthermore, while several parallel machine algorithms have been developed and integrated into the cloud environments using traditional HPC architectures, still other techniques are not fully exploited in emerging technologies such as graphics processing units (GPUs) [51]. It is worth noting that GPU-based computing is a huge shift of paradigm in parallel computing that promises a dramatic increase in performance. GPUs are designed for computing-intensive applications; they allow for the execution of threads on a larger number of processing elements. Having a large number of threads makes it possible to surpass the performance of current multicore CPUs.

GPU programming is now a much richer environment than it used to be a few years ago. On top of the two major programming languages, Compute Unified Device Architecture (CUDA) [52] and Open Computing Language (OpenCL), libraries have been developed that allow fast access to the computing power of GPUs without detailed knowledge or programming of GPU hardware. GPUs [53] could be used as accelerators for Big Data analytics such as in social media, statistical analysis in sensor data streams, predictive models, and so forth. GPUs could also be integrated in cloud computing and provide GPU-accelerated cloud services for Big Data. For example, since GPUs are not expensive and have performed better than multicore machines, they can serve as a practical infrastructure for cloud environments. Also, most efforts have been dedicated to machine learning techniques (e.g., association rule mining), and little work has been done for accelerating other algorithms related to text mining and sentiment analysis. These techniques are data and computationally intensive, making them good candidates for implementation in cloud environments. Generally, experimental results demonstrate that vitalized clusters can perform much better than physical clusters when processing and handling HPC.

## 5.5 CONCLUSIONS AND PERSPECTIVES

In this chapter, approaches proposed in literature and their use in the cloud for Big Data analysis and processing are presented. More precisely, we have first introduced the Big Data technology and its importance in different fields. We focused mainly on approaches for high-performance mining of Big Data and their efficiency in HPC for Big Data analysis and processing. We highlighted existing work on data analytics tools such as SaaS and computing such as IaaS. Some remaining research issues and directions are introduced together with some results from our recent work, especially the introduction of HPC in cloud environments. Future work will focus on studying the impact of cloud open sources on improving HPCaaS and conducting experiments using real machine learning and text mining applications on HPCaaS.

## REFERENCES

1. V. De Florio, M. Bakhouya, A. Coronato and G. Di Marzo, "Models and Concepts for Socio-Technical Complex Systems: Towards Fractal Social Organizations," *Systems Research and Behavioral Science*, vol. 30, no. 6, pp. 750–772, 2013.
2. J. Gantz and D. Reinsel, "The Digital Universe in 2020: Big Data, Bigger Digital Shadows, and Biggest Growth in the Far East," IDC IVIEW, 2012, pp. 1–16.
3. M. K. Kakhani, S. Kakhani, and S. R. Biradar, "Research Issues in Big Data Analytics," *International Journal of Application or Innovation in Engineering and Management*, vol. 2, no. 8, pp. 228–232, 2013.
4. C. Hagen, "Big Data and the Creative Destruction of Today's," ATKearney, 2012.
5. Gartner, Inc., "Hunting and Harvesting in a Digital World," Gartner CIO Agenda Report, 2013, pp. 1–8.
6. J. Gantz and D. Reinsel, "The Digital Universe Decade—Are You Ready?" IDC IVIEW, 2010, pp. 1–15.
7. Ch. Vecchiola, S. Pandey and R. Buyya, "High-Performance Cloud Computing: A View of Scientific Applications," in the 10th International Symposium on Pervasive Systems, Algorithms and Networks I-SPAN 2009, IEEE Computer Society, 2009, pp. 4–16.
8. J.-P. Dijcks, "Oracle: Big Data for the Enterprise," white paper, Oracle Corp., 2013, pp. 1–16.
9. B. Pang and L. Lee, "Opinion Mining and Sentiment Analysis," *Foundations and Trends in Information Retrieval*, vol. 2, no. 1–2, pp. 1–135, 2008.
10. Oracle Corporation, "Oracle NoSQL Database," white paper, Oracle Corp., 2011, pp. 1–12.
11. S. Yu, "ACID Properties in Distributed Databases," Advanced eBusiness Transactions for B2B-Collaborations, 2009.
12. S. Gilbert and N. Lynch, "Brewer's Conjecture and the Feasibility of Consistent, Available, Partition-Tolerant Web Services," *ACM SIGACT News*, vol. 33, no. 2, p. 51, 2002.
13. A. Lakshman and P. Malik, "Cassandra—A Decentralized Structured Storage System," *ACM SIGOPS Operating Systems Review*, vol. 44, no. 2, pp. 35–40, 2010.
14. G. Lars, *HBase: The Definitive Guide*, 1st edition, O'Reilly Media, Sebastopol, CA, 556 pp., 2011.

15. MongoDB. Available at http://www.mongodb.org/.
16. Apache CouchDBTM. Available at http://couchdb.apache.org/.
17. D. Boulter, "Simplify Your Journey to the Cloud," Capgemini and SOGETI, 2010, pp. 1–8.
18. P. Mell and T. Grance, "The NIST Definition of Cloud Computing," National Institute of Standards and Technology, 2011, pp. 1–3.
19. A. E. Youssef, "Exploring Cloud Computing Services and Applications," *Journal of Emerging Trends in Computing and Information Sciences*, vol. 3, no. 6, pp. 838–847, 2012.
20. A. T. Velte, T. J. Velte, and R. Elsenpeter, *Cloud Computing, A Practical Approach*, 1st edition, McGraw-Hill Osborne Media, New York, 352 pp., 2009.
21. Amazon Web Services. Available at http://aws.amazon.com/.
22. Google Cloud Platform. Available at https://cloud.google.com/.
23. J. Bernstein and K. McMahon, "Computing on Demand—HPC as a Service," pp. 1–12, Penguin On Demand. Available at http://www.penguincomputing.com/files/whitepapers/PODWhitePaper.pdf.
24. Y. Xiaotao, L. Aili and Z. Lin, "Research of High Performance Computing with Clouds," in the Third International Symposium on Computer Science and Computational Technology (ISCSCT), 2010, pp. 289–293.
25. Self-service POD Portal. Available at http://www.penguincomputing.com/services/hpc-cloud/pod.
26. K. R. Jackson, "Performance Analysis of High Performance Computing Applications on the Amazon Web Services Cloud," in Cloud Computing Technology and Science (CloudCom), 2010 IEEE Second International Conference on, 2010, pp. 159–168.
27. Amazon Cloud Storage. Available at http://aws.amazon.com/ec2/reserved-instances/.
28. Amazon Cloud Drive. Available at http://aws.amazon.com/ec2/spot-instances/.
29. Microsoft High Performance Computing for Developers. Available at http://msdn.microsoft.com/en-us/library/ff976568.aspx.
30. Google Cloud Storage. Available at https://cloud.google.com/products/compute-engine.
31. D. Pop, "Machine Learning and Cloud Computing: Survey of Distributed and SaaS Solutions," Technical Report Institute e-Austria Timisoara, 2012.
32. A. Ghoting, P. Kambadur, E. Pednault and R. Kannan, "NIMBLE: A Toolkit for the Implementation of Parallel Data Mining and Machine Learning Algorithms on MapReduce," in KDD, 2011.
33. J. Dean and S. Ghemawat, "MapReduce: Simple Data Processing on Large Clusters," in OSDI, 2004, pp. 1–12.
34. Hadoop. Available at http://hadoop.apache.org/.
35. S. Krishman, M. Tatineni and C. Baru, "myHaddop—Hadoop-on-Demand on Traditional HPC Resources," in the National Science Foundation's Cluster Exploratory Program, 2011, pp. 1–7.
36. E. Molina-Estolano, M. Gokhale, C. Maltzahn, J. May, J. Bent and S. Brandt, "Mixing Hadoop and HPC Workloads on Parallel Filesystems," in the 4th Annual Workshop on Petascale Data Storage, 2009, pp. 1–5.
37. S. R. Upadhyaya, "Parallel Approaches to Machine Learning—A Comprehensive Survey," *Journal of Parallel and Distributed Computing*, vol. 73, pp. 284–292, 2013.
38. C. Cranor, M. Polte and G. Gibson, "HPC Computation on Hadoop Storage with PLFS," Parallel Data Laboratory at Carnegie Mellon University, 2012, pp. 1–9.
39. S. Zhou, B. Kobler, D. Duffy and T. McGlynn, "Case Study for Running HPC Applications in Public Clouds," in Science Cloud '10, 2012.
40. E. Walker, "Benchmarking Amazon EC2 for High-Performance Scientific Computing," Texas Advanced Computing Center at the University of Texas, 2008, pp. 18–23.
41. J. Ekanayake and G. Fox, "High Performance Parallel Computing with Clouds and Cloud Technologies," School of Informatics and Computing at Indiana University, 2009, pp. 1–20.

42. Y. Gu and R. L. Grossman, “Sector and Sphere: The Design and Implementation of a High Performance Data Cloud,” National Science Foundation, 2008, pp. 1–11.
43. C. Evangelinos and C. N. Hill, “Cloud Computing for Parallel Scientific HPC Applications: Feasibility of Running Coupled Atmosphere-Ocean Climate Models on Amazon's EC2,” Department of Earth, Atmospheric and Planetary Sciences at Massachusetts Institute of Technology, 2009, pp. 1–6.
44. A. Gupta and D. Milojicic, “Evaluation of HPC Applications on Cloud,” Hewlett-Packard Development Company, 2011, pp. 1–6.
45. C. Fragni, M. Moreira, D. Mattos, L. Costa, and O. Duarte, “Evaluating Xen, VMware, and OpenVZ Virtualization Platforms for Network Virtualization,” Federal University of Rio de Janeiro, 2010, 1 p. Available at http://www.gta.ufrj.br/ftp/gta/TechReports/FMM10b.pdf.
46. N. Yaqub, “Comparison of Virtualization Performance: VMWare and KVM,” Master Thesis, DUO Digitale utgivelser ved UiO, Universitetet i Oslo, Norway, 2012, pp. 30–44. Available at http://urn.nb.no/URN:NBN:no-33642.
47. T. Deshane, M. Ben-Yehuda, A. Shah and B. Rao, “Quantitative Comparison of Xen and KVM,” in Xen Summit, 2008, pp. 1–3.
48. J. Hwang, S. Wu and T. Wood, “A Component-Based Performance Comparison of Four Hypervisors,” George Washington University and IBM T.J. Watson Research Center, 2012, pp. 1–8.
49. O. Achahbar, “The Impact of Virtualization on High Performance Computing Clustering in the Cloud,” Master Thesis Report, Al Akhawayn University in Ifrane, Morocco, 2014.
50. M. G. Noll, “Benchmarking and Stress Testing and Hadoop Cluster with TeraSort, Test DFSIO & Co.,” 2011.
51. E. Wu and Y. Liu, “Emerging Technology about GPGPU,” APCCAS. IEEE Asia Pacific Conference on Circuits and Systems, 2008.
52. NVIDIA, “NVIDIA CUDA Compute Unified Device Architecture: Programming Guide,” Version 2.3, July 2009. Available at http://www.cs.ucla.edu/~palsberg/course/cs239/papers/CudaReferenceManual_2.0.pdf.
53. W. Fang, “Parallel Data Mining on Graphics Processors,” Technical Report HKUST-CS08-07, October 2008.

CHAPTER 6

# The Art of Scheduling for Big Data Science

Florin Pop and Valentin Cristea

CONTENTS



ABSTRACT

Many applications generate Big Data, like social networking and social influence programs, cloud applications, public websites, scientific experiments and simulations, data warehouses, monitoring platforms, and e-government services. Data grow rapidly, since applications produce continuously increasing volumes of unstructured and structured data. The impact on data processing, transfer, and storage is the need to reevaluate the approaches and solutions to better answer user needs. In this context, scheduling models and algorithms have an important role. A large variety of solutions for specific applications and platforms exist, so a thorough and systematic analysis of existing solutions for scheduling models, methods, and algorithms used in Big Data processing and storage environments has high importance. This chapter presents the best of existing solutions and creates an overview of current and near-future trends. It will highlight, from a research perspective, the performance and limitations of existing solutions and will offer the scientists from academia and designers from industry an overview of the current situation in the area of scheduling and resource management related to Big Data processing.

## 6.1 INTRODUCTION

The rapid growth of data volume requires processing of petabytes of data per day. Cisco estimates that mobile data traffic alone will reach 11.16 exabytes of data per month in 2017. The produced data is subject to different kinds of processing, from real-time processing with impact for context-aware applications to data mining analysis for valuable information extraction. The multi-V (volume, velocity, variety, veracity, and value) model is frequently used to characterize Big Data processing needs. *Volume* defines the amount of data, *velocity* means the rate of data production and processing, *variety* refers to data types, *veracity* describes how data can be a trusted function of its source, and *value* refers to the importance of data relative to a particular context [1].

Scheduling plays an important role in Big Data optimization, especially in reducing the time for processing. The main goal of scheduling in Big Data platforms is to plan the processing and completion of as many tasks as possible by handling and changing data in an efficient way with a minimum number of migrations. Various mechanisms are used for resource allocation in cloud, high performance computing (HPC), grid, and peer-to-peer systems, which have different architectural characteristics. For example, in HPC, the cluster used for data processing is homogeneous and can handle many tasks in parallel by applying predefined rules. On the other side, cloud systems are heterogeneous and widely distributed; task management and execution are aware of communication rules and offer the possibility to create particular rules for the scheduling mechanism. The actual scheduling methods used in Big Data processing frameworks are as follows: first in first out, fair scheduling, capacity scheduling, *Longest Approximate Time to End* (LATE) scheduling, deadline constraint scheduling, and adaptive scheduling [2,3]. Finding the best method for a particular processing request remains a significant challenge. We can see the Big Data processing as a big "batch" process that runs on an HPC cluster by splitting a job into smaller tasks and distributing the work to the cluster nodes. The new types of applications, like social networking, graph analytics, and complex business work flows, require data transfer and data storage. The processing models must be aware of data locality when deciding to move data to the computing nodes or to create new computing nodes near data locations. The workload optimization strategies are the key to guarantee profit to resource providers by using resources to their maximum capacity. For applications that are both computationally and data intensive, the processing models combine different techniques like in-memory, CPU, and/or graphics processing unit (GPU) Big Data processing.

Moreover, Big Data platforms face the problem of environments' heterogeneity due to the variety of distributed systems types like cluster, grid, cloud, and peer-to-peer, which actually offer support for advanced processing. At the confluence of Big Data with widely distributed platforms, scheduling solutions consider solutions designed for efficient problem solving and parallel data transfers (that hide transfer latency) together with techniques for failure management in highly heterogeneous computing systems. In addition, handling heterogeneous data sets becomes a challenge for interoperability among various software systems.

This chapter highlights the specific requirements of scheduling in Big Data platforms, scheduling models and algorithms, data transfer scheduling procedures, policies used in different computing models, and optimization techniques. The chapter concludes with a case study on Hadoop and Big Data, and the description of the new fashion to integrate New Structured Query Language (NewSQL) databases with distributed file systems and computing environments.

## 6.2 REQUIREMENTS FOR SCHEDULING IN BIG DATA PLATFORMS

The requirements of traditional scheduling models came from applications, databases, and storage resources, which did exponentially grow over the years. As a result, the cost and complexity of adapting traditional scheduling models to Big Data platforms have increased, prompting changes in the way data is stored, analyzed, and accessed. The traditional model is being expanded to incorporate new building blocks. They address the challenges of Big Data with new information processing frameworks built to meet Big Data's requirements.

The integration of applications and services in Big Data platforms requires a cluster resource management layer as a middleware component but also particular scheduling and execution engines specific to different models: batch tasks, data flow, NewSQL tasks, and so forth (Figure 6.1). So, we have scheduling methods and algorithms in both cluster and execution engine layers.

The general requirements for scheduling in Big Data platforms define the functional and nonfunctional specifications for a scheduling service. The general requirements are as follows:

- *Scalability and elasticity*: A scheduling algorithm must take into consideration the peta-scale data volumes and hundred thousands of processors that can be involved in processing tasks. The scheduler must be aware of execution environment changes and be able to adapt to workload changes by provisioning or deprovisioning resources.
- *General purpose*: A scheduling approach should make assumptions about and have few restrictions to various types of applications that can be executed. Interactive jobs,

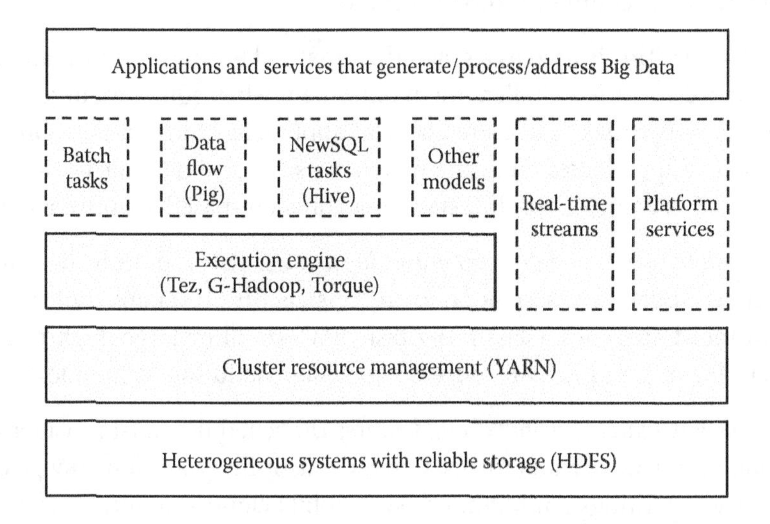


FIGURE 6.1 Integration of application and service integration in Big Data platforms.

distributed and parallel applications, as well as noninteractive batch jobs should all be supported with high performance. For example, a noninteractive batch job requiring high throughput may prefer time-sharing scheduling; similarly, a real-time job requiring short-time response prefers space-sharing scheduling.

- *Dynamicity*: The scheduling algorithm should exploit the full extent of available resources and may change its behavior to cope, for example, with many computing tasks. The scheduler needs to continuously adapt to resource availability changes, paying special attention to cloud systems and HPC clusters (data centers) as reliable solutions for Big Data [4].
- *Transparency*: The host(s) on which the execution is performed should not affect tasks' behavior and results. From the user perspective, there should be no difference between local and remote execution, and the user should not be aware about system changes or data movements for Big Data processing.
- *Fairness*: Sharing resources among users is a fair way to guarantee that each user obtains resources on demand. In a pay-per-use model in the cloud, a cluster of resources can be allocated dynamically or can be reserved in advance.
- *Time efficiency*: The scheduler should improve the performance of scheduled jobs as much as possible using different heuristics and state estimation suitable for specific task models. Multitasking systems can process multiple data sets for multiple users at the same time by mapping the tasks to resources in a way that optimizes their use.
- *Cost (budget) efficiency*: The scheduler should lower the total cost of execution by minimizing the total number of resources used and respect the total money budget. This aspect requires efficient resource usage. This can be done by optimizing the execution for mixed tasks using a high-performance queuing system and by reducing the computation and communication overhead.
- *Load balancing*: This is used as a scheduling method to share the load among all available resources. This is a challenging requirement when some resources do not match with tasks' properties. There are classical approaches like round-robin scheduling, but also, new approaches that cope with large scale and heterogeneous systems were proposed: least connection, slow start time, or agent-based adaptive balancing.
- *Support of data variety and different processing models*: This is done by handling multiple concurrent input streams, structured and unstructured content, multimedia content, and advanced analytics. Classifying tasks into small or large, high or low priority, and periodic or sporadic will address a specific scheduling technique.
- *Integration with shared distributed middleware*: The scheduler must consider various systems and middleware frameworks, like the sensor integration in any place following the Internet of Things paradigm or even mobile cloud solutions that use offloading techniques to save energy. The integration considers the data access and consumption and supports various sets of workloads produced by services and applications.

The scheduling models and algorithms are designed and implemented by enterprise tools and integration support for applications, databases, and Big Data processing environments. The workload management tools must manage the input and output data at every data acquisition point, and any data set must be handled by specific solutions. The tools used for resource management must solve the following requirements:

- *Capacity awareness*: Estimate the percentage of resource allocation for a workload and understand the volume and the velocity of data that are produced and processed.
- *Real-time, latency-free delivery and error-free analytics*: Support the service-level agreements, with continuous business process flow and with an integrated labor-intensive and fault-tolerant Big Data environment.
- *API integration*: With different operating systems, support various-execution virtual machines (VMs) and wide visibility (end-to-end access through standard protocols like hypertext transfer protocol [HTTP], file transfer protocol [FTP], and remote login, and also using a single and simple management console) [5].

## 6.3 SCHEDULING MODELS AND ALGORITHMS

A *scheduling model* consists of a scheduling policy, an algorithm, a programing model, and a performance analysis model. The design of a scheduler that follows a model should specify the architecture, the communication model between entities involved in scheduling, the process type (static or dynamic), the objective function, and the state estimation [6]. It is important that all applications are completed as quickly as possible, all applications receive the necessary proportion of the resources, and those with a close deadline have priority over other applications that could be finished later. The scheduling model can be seen from another point of view, namely, the price. It is important for cloud service providers to have a higher profit but also for user applications to be executed so that the cost remains under the available budget [7].

Several approaches to the scheduling problem have been considered over time. These approaches consider different scenarios, which take into account the applications' types, the execution platform, the types of algorithms used, and the various constraints that might be imposed. The existing schedulers suitable for large environments and also for Big Data platforms are as follows:

- *First In First Out (FIFO) (oldest job first)*—jobs are ordered according to the arrival time. The order can be affected by job priority.
- *Fair scheduling*—each job gets an equal share of the available resources.
- *Capacity scheduling*—provides a minimum capacity guarantee and shares excess capacity; it also considers the priorities of jobs in a queue.
- *Adaptive scheduling*—balances between resource utilization and jobs' parallelism in the cluster and adapts to specific dynamic changes of the context.

- *Data locality-aware scheduling*—minimizes data movement.
- *Provisioning-aware scheduling*—provisions VMs from larger physical clusters; virtual resource migration is considered in this model.
- *On-demand scheduling*—uses a batch scheduler as resource manager for node allocation based on the needs of the virtual cluster.

The use of these models takes into account different specific situations: For example, FIFO is recommended when the number of tasks is less than the cluster size, while fairness is the best one when the cluster size is less than the number of tasks. Capacity scheduling is used for multiple tasks and priorities specified as response times. In Reference 8, a solution of a *scheduling bag of tasks* is presented. Users receive guidance regarding the plausible response time and are able to choose the way the application is executed: with more money and faster or with less money but slower. An important ingredient in this method is the phase of profiling the tasks in the actual bag. The basic scheduling is realized with a bounded knapsack algorithm. Reference 9 presents the idea of scheduling based on scaling up and down the number of the machines in a cloud system. This solution allows users to choose among several types of VM instances while scheduling each instance's start-up and shutdown to respect the deadlines and ensure a reduced cost.

A scheduling solution based on *genetic algorithms* is described in Reference 10. The scheduling is done on grid systems. Grids are different from cloud systems, but the principle used by the authors in assigning tasks to resources is the same. The scheduling solution works with applications that can be modeled as directed acyclic graphs. The idea is minimizing the duration of the application execution while the budget is respected. This approach takes into account the system's heterogeneity.

Reference 11 presents a scheduling model for instance-intensive work flows in the cloud, which takes into consideration both *budget* and *deadline constraints*. The level of user interaction is very high, the user being able to change dynamically the cost and deadline requirements and provide input to the scheduler during the work flow execution. The interventions can be made every scheduled round. This is an interesting model because the user can choose to pay more or less depending on the scenario. The main characteristic is that the user has more decision power on work flow execution. In addition, the cloud estimates the time and cost during the work flow execution to provide hints to users and dynamically reschedule the workload.

The *Apache Hadoop framework* is a software library that allows the processing of large data sets distributed across clusters of computers using a simple programming model [3]. The framework facilitates the execution of MapReduce applications. Usually, a cluster on which the Hadoop system is installed has two masters and several slave components (Figure 6.2) [12]. One of the masters is the *JobTracker*, which deals with processing projects coming from users and sends them to the scheduler used in that moment. The other master is *NameNode*, which manages the file system namespace and the user access control. The other machines act as slaves. A *TaskTracker* represents a machine in Hadoop, while a *DataNode* handles the operations with the Hadoop Distributed File System (HDFS), which

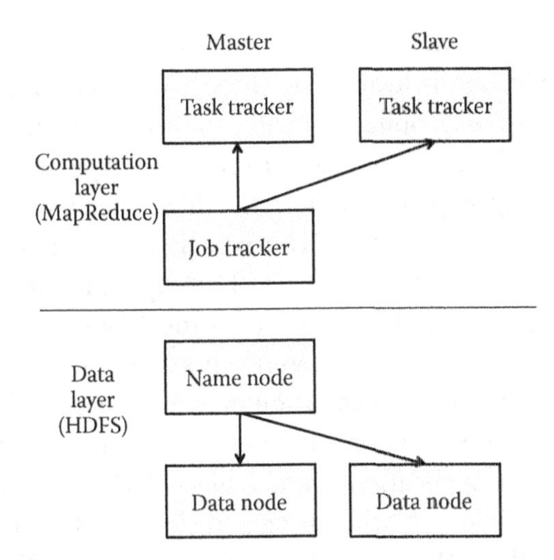


FIGURE 6.2 Hadoop general architecture.

deals with data replication on all the slaves in the system. This is the way input data gets to map and to reduce tasks. Every time an operation occurs on one of the slaves, the results of the operation are immediately propagated into the system [13].

The Hadoop framework considers the capacity and fair scheduling algorithms. The *Capacity Scheduler* has a number of queues. Each of these queues is assigned a part of the system resources and has specific numbers of map and reduce slots, which are set through the configuration files. The queues receive users' requests and order them by the associated priorities. There is also a limitation for each queue per user. This prevents the user from seizing the resources for a queue.

The *Fair Scheduler* has *pools* in which job requests are placed for selection. Each user is assigned to a pool. Also, each pool is assigned a set of shares and uses them to manage the resources allocated to jobs, so that each user receives an equal share, no matter the number of jobs he/she submits. Anyway, if the system is not loaded, the remininig shares are distributed to existing jobs. The Fair Scheduler has been proposed in Reference 14. The authors demonstrated its special qualities regarding the reduced response time and the high throughput.

There are several extensions to scheduling models for Hadoop. In Reference 15, a new scheduling algorithm is presented, LATE, which is highly robust to heterogeneity without using specific information about nodes. The solution solves the problems posed by heterogeneity in virtualized data centers and ensures good performance and flexibility for speculative tasks. In Reference 16, a scheduler is presented that meets *deadlines*. This scheduler has a preliminary phase for estimating the possibility to achieve the deadline claimed by the user, as a function of several parameters: the runtimes of map and reduce tasks, the input data size, data distribution, and so forth. Jobs are scheduled only if the deadlines can be met. In comparison with the schedulers mentioned in this section, the genetic scheduler proposed in Reference 7 approaches the deadline constraints but also takes into account the environment heterogeneity. In addition, it uses speculative techniques in order to increase the scheduler's power. The genetic scheduler has an estimation phase, where the processing data speed for each application is measured. The scheduler ensures that, once an application's execution has started, that application will end successfully in normal conditions. The Hadoop On Demand (HOD) [3] virtual cluster uses the Torque resource manager for node allocation and automatically prepares configuration files. Then it initializes the system based on the nodes within the virtual cluster. HOD can be used in a relatively independent way.

To support multiuser situations [14,17], the Hadoop framework incorporates several components that are suitable for Big Data processing (Figure 6.3) since they offer high scalability through a large volume of data and support access to widely distributed data. Here is a very short description of these components: *Hadoop Common* consists of common utilities that support any Hadoop modules and any new extension. *HDFS* provides high-throughput access to application data. *Hadoop YARN* (Apache Hadoop NextGen MapReduce) is a framework for job scheduling and cluster resource management that can be extended across multiple platforms. *Hadoop MapReduce* is a YARN-based system for parallel processing of large data sets.

Facebook solution *Corona* [18] extends and improves the Hadoop model, offering better scalability and cluster utilization, lower latency for small jobs, the ability to upgrade without disruption, and scheduling based on actual task resource requirements rather than a count of map and reduce tasks. Corona was designed to answer the most important Facebook challenges: unique scalability (the largest cluster has more than 100 PB of data) and processing needs (crunch more than 60,000 Hive queries a day). The data warehouse inside Facebook has grown by 2500 times between 2008 and 2012, and it is expected to grow by the same factor until 2016.

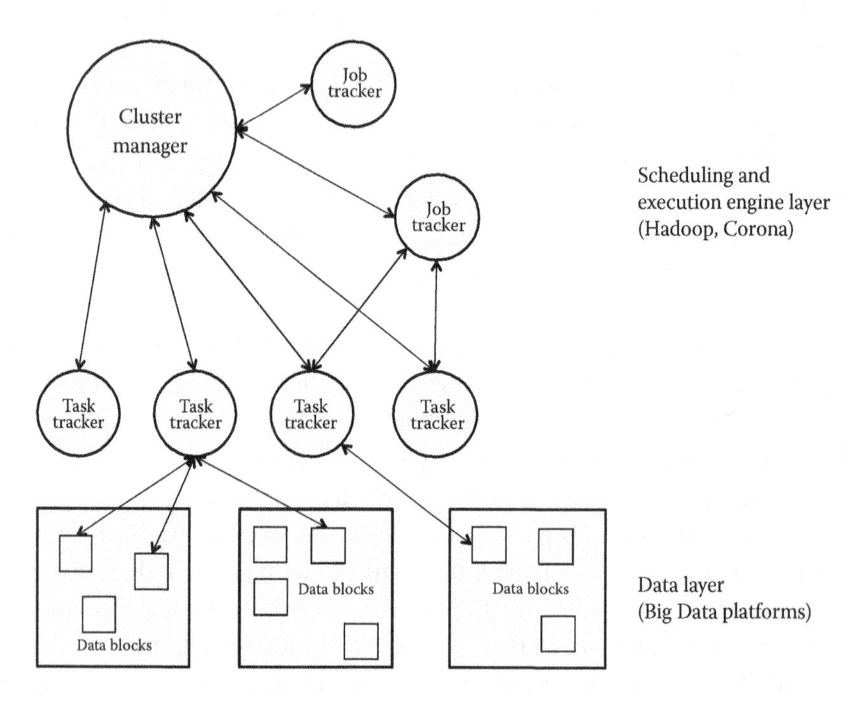


FIGURE 6.3 Hadoop processing and Big Data platforms.

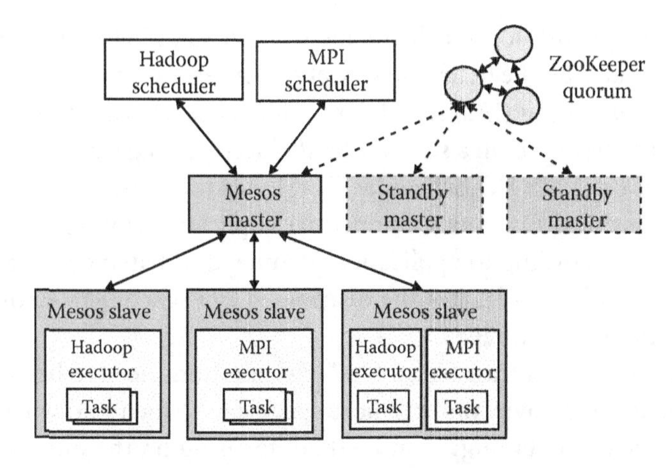


FIGURE 6.4 Mesos architecture.

*Mesos* [19] uses a model of resource isolation and sharing across distributed applications using Hadoop, message passing interface (MPI), Spark, and Aurora in a dynamic way (Figure 6.4). The ZooKeeper quorum is used for master replication to ensure fault tolerance. By integration of multiple slave executors, Mesos offers support for multiresource scheduling (memory and CPU aware) using isolation between tasks with Linux Containers. The expected scalability goes over 10,000 nodes.

*YARN* [20] splits the two major functionalities of the Hadoop JobTracker in two separate components: resource management and job scheduling/monitoring (application management). The *Resource Manager* is integrated in the data-computation framework and coordinates the resource for all jobs processing alive in a Big Data platform. The Resource Manager has a pure scheduler, that is, it does not monitor or track the application status and does not offer guarantees about restarting failed tasks due to either application failure or hardware failures. It offers only matching of applications' jobs on resources.

The new processing models based on bioinspired techniques are used for fault-tolerant and self-adaptable handling of data-intensive and computation-intensive applications. These evolutionary techniques approach the learning based on history with the main aim to find a near-optimal solution for problems with multiconstraint and multicriteria optimizations. For example, adaptive scheduling is needed for dynamic heterogeneous systems where we can change the scheduling strategy according to available resources and their capacity.

## 6.4 DATA TRANSFER SCHEDULING

In many cases, depending on applications' architecture, data must be transported to the place where tasks will be executed [21]. Consequently, scheduling schemes should consider not only the task execution time but also the data transfer time for finding a more convenient mapping of tasks. Only a handful of current research efforts consider the simultaneous optimization of computation and data transfer scheduling. The general data scheduling techniques are the Least Frequently Used (LFU), Least Recently Used (LRU), and economical models. Handling multiple file transfer requests is made in parallel with maximizing the bandwidth for which the file transfer rates between two end points are calculated and considering the heterogeneity of server resources.

The Big Data input/output (I/O) scheduler in Reference 22 is a solution for applications that compete for I/O resources in a shared MapReduce-type Big Data system. The solution, named Interposed Big Data I/O Scheduler (IBIS), has the main aims to solve the problem of differentiating the I/Os among competitive applications on separate data nodes and perform scheduling according to applications' bandwidth demands. IBIS acts as a metascheduler and efficiently coordinates the distributed I/O schedulers across data nodes in order to allocate the global storage.

In the context of Big Data transfers, a few "big" questions need to be answered in order to have an efficient cloud environment, more specifically, when and where to migrate. In this context, an efficient data migration method, focusing on the minimum global time, is presented in Reference 23. The method, however, does not try to minimize individual migrations' duration. In Reference 24, two migration models are described: offline and online. The *offline* scheduling model has as a main target the minimization of the maximum bandwidth usage on all links for all time slots of a planning period. In the *online* scheduling model, the scheduler has to make fast decisions, and the migrations are revealed to the migration scheduler in an a priori undefined sequence. Jung et al. [25] treat the data mining parallelization by considering the data transfer delay between two computing nodes. The delay is estimated by using the autoregressive moving average filter. In Reference 26, the impact of two resource reservation methods is tested: reservation in source machines and reservation in target machines. Experimental results proved that resource reservation in target machines is needed, in order to avoid migration failure. The performance overheads of live migration are affected by memory size, CPU, and workload types.

The model proposed in Reference 27 uses the greedy scheduling algorithm for data transfers through different cloud data centers. This algorithm gets the transfer requests on a first-come-first-serve order and sets a time interval in which they can be sent. This interval is reserved on all the connections the packet has to go through (in this case, there is a maximum of three hops to destination, because of the full mesh infrastructure). This is done until there are no more transfers to schedule, taking into account the previously reserved time frames for each individual connection. The connections are treated individually to avoid the bottleneck. For instance, the connections between individual clouds need to transfer more messages than connections inside the cloud. This way, even if the connection from a physical machine to a router is unused, the connection between the routers can be oversaturated. There is no point in scheduling the migration until the transfers that are currently running between the routers end, even if the connection to the router is unused.

## 6.5 SCHEDULING POLICIES

The scheduling policies are used to determine the relative ordering of requests. Large distributed systems with different administrative domains will most likely have different resource utilization policies. A policy can take into consideration the priority, the deadlines, the budgets, and also the dynamic behavior [28]. For Big Data platforms, dynamic scheduling with soft deadlines and hard budget constraints on hybrid clouds are an open issue.

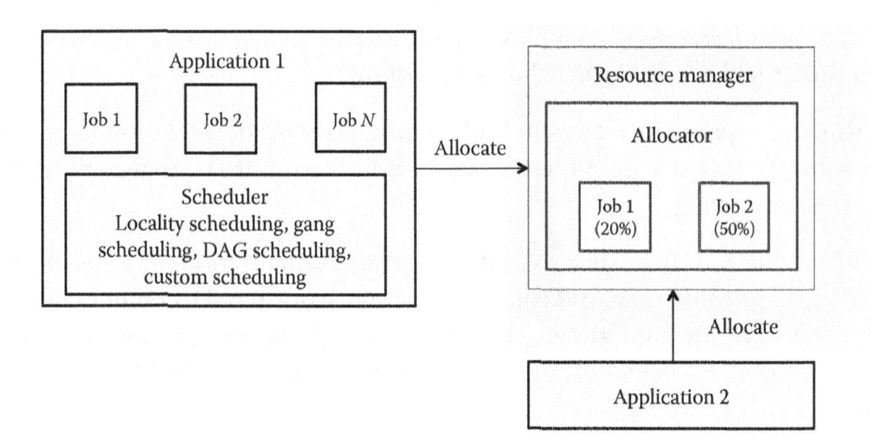


FIGURE 6.5 Resource allocation process.

A general-purpose resource management approach in a cluster used for Big Data processing should make some assumptions about policies that are incorporated in service-level agreements. For example, interactive tasks, distributed and parallel applications, as well as noninteractive batch tasks should all be supported with high performance. This property is a straightforward one but, to some extent, difficult to achieve. Because tasks have different attributes, their requirements to the scheduler may contradict in a shared environment. For example, a real-time task requiring short-time response prefers space-sharing scheduling; a noninteractive batch task requiring high throughput may prefer time-sharing scheduling [6,29]. To be general purpose, a trade-off may have to be made. The scheduling method focuses on parallel tasks, while providing an acceptable performance to other kinds of tasks.

YARN has a pluggable solution for dynamic policy loading, considering two steps for the resource allocation process (Figure 6.5). Resource allocation is done by YARN, and task scheduling is done by the application, which permits the YARN platform to be a generic one while still allowing flexibility of scheduling strategies. The specific policies in YARN are oriented on resource splitting according with schedules provided by applications. In this way, the YARN's scheduler determines how much and in which cluster to allocate resources based on their availability and on the configured sharing policy.

## 6.6 OPTIMIZATION TECHNIQUES FOR SCHEDULING

The scheduling in Big Data platforms is the main building block for making data centers more available to applications and user communities. An example of optimization is multiobjective and multiconstrained scheduling of many tasks in Hadoop [30] or optimizing short jobs with Hadoop [31]. The optimization strategies for scheduling are specific to each model and for each type of application. The most used techniques for scheduling optimization are as follows:

- *Linear programming* allows the scheduler to find the suitable resources or cluster of resources, based on defined constraints.

- *Dynamic partitioning* splits complex applications in a cluster of tasks and schedules each cluster with a specific scheduling algorithm.
- *Combinatorial optimization* aims to find an optimal allocation solution for a finite set of resources. This is a time-consuming technique, and it is not recommended for real-time processing.
- *Evolutionary optimization* aims to find an optimal configuration for a specific system within specific constraints and consists of specific bioinspired techniques like genetic and immune algorithms, ant and bee computing, particle swarm optimization, and so forth. These methods usually find a near-optimal solution for the scheduling problem.
- *Stochastic optimization* uses random variables that appear in the formulation of the optimization problem itself and are used especially for applications that have a deterministic behavior, that is, a normal distribution (for periodic tasks) or Poisson distribution (for sporadic tasks), and so forth.
- *Task-oriented optimization* considers the task's properties, arrival time (slack optimization), and frequency (for periodic tasks).
- *Resource-oriented optimization* considers completion time constraints while making the decision to maximize resource utilization.

## 6.7 CASE STUDY ON HADOOP AND BIG DATA APPLICATIONS

We can consider that, on top of Apache Hadoop or a Hadoop distribution, you can use a Big Data suite and scheduling application that generate Big Data. The Hadoop framework is suitable for specific high-value and high-volume workloads [32]. With optimized resource and data management by putting the right Big Data workloads in the right systems, a solution is offered by the integration of YARN with specific schedulers: G-Hadoop [33], MPI Scheduler, Torque, and so forth. The cost-effectiveness, scalability, and streamlined architectures of Hadoop have been developed by IBM, Oracle, and Hortonworks Data Platform (HDP) with the main scope to construct a Hadoop operating system. Hadoop can be used in public/private clouds through related projects integrated in the Hadoop framework, trying to offer the best answer to the following question: What type of data/tasks should move to a public cloud, in order to achieve a cost-aware cloud scheduler [34]?

Different platforms that use and integrate Hadoop in application development and implementation consider different aspects. *Modeling* is the first one. Even in Apache Hadoop, which offers the infrastructure for Hadoop clusters, it is still very complicated to build a specific MapReduce application by writing complex code. Pig Latin and Hive query language (HQL), which generate MapReduce code, are optimized languages for application development. The integration of different scheduling algorithms and policies is also supported. The application can be modeled in a graphical way, and all required code is generated. Then, the *tooling* of Big Data services within a dedicated development environment (Eclipse) uses various dedicated plugins. *Code generation* of all the code is automatically generated for a MapReduce application. Last, but not least, *execution* of Big Data jobs has to be *scheduled* and *monitored*. Instead of writing jobs or other code for scheduling, the Big Data suite offers the possibility to define and manage execution plans in an easy way.

In a Big Data platform, Hadoop needs to integrate data of all different kinds of technologies and products. Besides files and SQL databases, Hadoop needs to integrate the NoSQL databases, used in applications like social media such as Twitter or Facebook; the messages from middleware or data from business-to-business (B2B) products such as Salesforce or SAP products; the multimedia streams; and so forth. A Big Data suite integrated with Hadoop offers connectors from all these different interfaces to Hadoop and back.

The main Hadoop-related projects, developed under the Apache license and supporting Big Data application execution, include the following [35]:

- *HBase* (https://hbase.apache.org) is a scalable, distributed database that supports structured data storage for large tables. HBase offers random, real-time read/write access to Big Data platforms.
- *Cassandra* (http://cassandra.apache.org) is a multimaster database with no single points of failure, offering scalability and high availability without compromising performance.
- *Pig* (https://pig.apache.org) is a high-level data-flow language and execution framework for parallel computation. It also offers a compiler that produces sequences of MapReduce programs, for which large-scale parallel implementations already exist in Hadoop.
- *Hive* (https://hive.apache.org) is a data warehouse infrastructure that provides data summarization and ad hoc querying. It facilitates querying and managing large data sets residing in distributed storage. It provides the SQL-like language called HiveQL.
- *ZooKeeper* (https://zookeeper.apache.org) is a high-performance coordination service for distributed applications. It is a centralized service for maintaining configuration information, naming, providing distributed synchronization, and providing group services.
- *Ambari* (https://ambari.apache.org) is a web-based tool for provisioning, managing, and monitoring Apache Hadoop clusters, which includes support for Hadoop HDFS, Hadoop MapReduce, Hive, HCatalog, HBase, ZooKeeper, Oozie, Pig, and Sqoop.
- *Tez* (http://tez.incubator.apache.org) is a generalized data-flow managing framework, built on Hadoop YARN, which provides an engine to execute an arbitrary directed acyclic graph (DAG) of tasks (an application work flow) to process data for both batch and interactive use cases. Tez is being adopted by other projects like Hive and Pig.
- *Spark* (http://spark.apache.org) is a fast and general compute engine for Hadoop data in large-scale environments. It combines SQL, streaming, and complex analytics and support application implementation in Java, Scala, or Python. Spark also has an advanced DAG execution engine that supports cyclic data flow and in-memory computing.

## 6.8 CONCLUSIONS

In the past years, scheduling models and tools have been faced with fundamental rearchitecting to fit as well as possible with large many-task computing environments. The way that existing solutions were redesigned consisted of splitting the tools into multiple components and adapting each component according to its new role and place. Workloads and work flow sequences are scheduled at the application side, and then, a resource manager allocates a pool of resources for the execution phase. So, the scheduling became important at the same time for users and providers, being the most important key for any optimal processing in Big Data science.

## REFERENCES


1. M. D. Assuncao, R. N. Calheiros, S. Bianchi, M. A. Netto, and R. Buyya. Big Data computing and clouds: Challenges, solutions, and future directions. arXiv preprint arXiv:1312.4722, 2013. Available at http://arxiv.org/pdf/1312.4722v2.pdf.
2. A. Rasooli, and D. G. Down. COSHH: A classification and optimization based scheduler for heterogeneous Hadoop systems. *Future Generation Computer Systems*, vol. 36, pp. 1–15, 2014.
3. M. Tim Jones. Scheduling in Hadoop. An introduction to the pluggable scheduler framework. IBM Developer Works, Technical Report, December 2011. Available at https://www.ibm.com/developerworks/opensource/library/os-hadoop-scheduling.
4. L. Zhang, C. Wu, Z. Li, C. Guo, M. Chen, and F. C. M. Lau. Moving Big Data to the cloud: An online cost-minimizing approach. *Selected Areas in Communications, IEEE Journal on*, vol. 31, no. 12, pp. 2710–2721, 2013.
5. CISCO Report. Big Data Solutions on Cisco UCS Common Platform Architecture (CPA). Available at http://www.cisco.com/c/en/us/solutions/data-center-virtualization/big-data/index.html, May 2014.
6. V. Cristea, C. Dobre, C. Stratan, F. Pop, and A. Costan. *Large-Scale Distributed Computing and Applications: Models and Trends*. IGI Global, Hershey, PA, pp. 1–276, 2010. doi:10.4018/978-1-61520-703-9.
7. D. Pletea, F. Pop, and V. Cristea. Speculative genetic scheduling method for Hadoop environments. In *2012 14th International Symposium on Symbolic and Numeric Algorithms for Scientific Computing*. SYNASC, pp. 281–286, 2012.
8. A.-M. Oprescu, T. Kielmann, and H. Leahu. Budget estimation and control for bag-of-tasks scheduling in clouds. *Parallel Processing Letters*, vol. 21, no. 2, pp. 219–243, 2011.
9. M. Mao, J. Li, and M. Humphrey. Cloud auto-scaling with deadline and budget constraints. In *The 11th ACM/IEEE International Conference on Grid Computing (Grid 2010)*. Brussels, Belgium, 2010.
10. J. Yu, and R. Buyya. Scheduling scientific workflow applications with deadline and budget constraints using genetic algorithms. *Scientific Programming Journal*, vol. 14, nos. 3–4, pp. 217–230, 2006.
11. K. Liu, H. Jin, J. Chen, X. Liu, D. Yuan, and Y. Yang. A compromised-time-cost scheduling algorithm in swindew-c for instance-intensive cost-constrained workflows on a cloud computing platform. *International Journal of High Performance Computing Applications*, vol. 24, no. 4, pp. 445–456, 2010.
12. T. White. *Hadoop: The Definitive Guide*. O'Reilly Media, Inc., Yahoo Press, Sebastopol, CA, 2012.
13. X. Hua, H. Wu, Z. Li, and S. Ren. Enhancing throughput of the Hadoop Distributed File System for interaction-intensive tasks. *Journal of Parallel and Distributed Computing*, vol. 74, no. 8, pp. 2770–2779, 2014. ISSN 0743-7315.

14. M. Zaharia, D. Borthakur, J. S. Sarma, K. Elmeleegy, S. Shenker, and I. Stoica. Job scheduling for multi-user MapReduce clusters. Technical Report, EECS Department, University of California, Berkeley, CA, April 2009. Available at http://www.eecs.berkeley.edu/Pubs/TechRpts/2009/EECS-2009-55.html.
15. M. Zaharia, A. Konwinski, A. D. Joseph, R. Katz, and I. Stoica. Improving MapReduce performance in heterogeneous environments. In *Proceeding OSDI '08 Proceedings of the 8th USENIX Conference on Operating Systems Design and Implementation*. USENIX Association, Berkeley, CA, pp. 29–42, 2008.
16. K. Kc, and K. Anyanwu. Scheduling Hadoop jobs to meet deadlines. In *Proceeding CLOUDCOM '10 Proceedings of the 2010 IEEE Second International Conference on Cloud Computing Technology and Science*. IEEE Computer Society, Washington, DC, pp. 388–392, 2010.
17. Y. Tao, Q. Zhang, L. Shi, and P. Chen. Job scheduling optimization for multi-user MapReduce clusters. In *Parallel Architectures, Algorithms and Programming (PAAP), 2011 Fourth International Symposium on*, Tianjin, China, pp. 213, 217, December 9–11, 2011.
18. Corona. Under the Hood: Scheduling MapReduce jobs more efficiently with Corona, 2012. Available at https://www.facebook.com/notes/facebook-engineering/under-the-hood-scheduling-mapreduce-jobs-more-efficiently-with-corona/10151142560538920.
19. B. Hindman, A. Konwinski, M. Zaharia, A. Ghodsi, A. D. Joseph, R. Katz, S. Shenker, and I. Stoica. Mesos: A platform for fine-grained resource sharing in the data center. In *Proceedings of the 8th USENIX Conference on Networked Systems Design and Implementation (NSDI '11)*. USENIX Association, Berkeley, CA, 22 p., 2011.
20. Hortonworks. Hadoop YARN: A next-generation framework for Hadoop data processing, 2013. Available at http://hortonworks.com/hadoop/yarn/.
21. J. Celaya, and U. Arronategui. A task routing approach to large-scale scheduling. *Future Generation Computer Systems*, vol. 29, no. 5, pp. 1097–1111, 2013.
22. Y. Xu, A. Suarez, and M. Zhao. IBIS: Interposed big-data I/O scheduler. In *Proceedings of the 22nd International Symposium on High-Performance Parallel and Distributed Computing (HPDC '13)*. ACM, New York, pp. 109–110, 2013.
23. J. Hall, J. Hartline, A. R. Karlin, J. Saia, and J. Wilkes. On algorithms for efficient data migration. In *Proceedings of the Twelfth Annual ACM-SIAM Symposium on Discrete Algorithms (SODA '01)*. Society for Industrial and Applied Mathematics, Philadelphia, PA, pp. 620–629, 2001.
24. A. Stage, and T. Setzer. Network-aware migration control and scheduling of differentiated virtual machine workloads. In *Proceedings of the 2009 ICSE Workshop on Software Engineering Challenges of Cloud Computing (CLOUD '09)*. IEEE Computer Society, Washington, DC, pp. 9–14, 2009.
25. G. Jung, N. Gnanasambandam, and T. Mukherjee. Synchronous parallel processing of big-data analytics services to optimize performance in federated clouds. In *Proceedings of the 2012 IEEE Fifth International Conference on Cloud Computing (CLOUD '12)*. IEEE Computer Society, Washington, DC, pp. 811–818, 2012.
26. K. Ye, X. Jiang, D. Huang, J. Chen, and B. Wang. Live migration of multiple virtual machines with resource reservation in cloud computing environments. In *Proceedings of the 2011 IEEE 4th International Conference on Cloud Computing (CLOUD '11)*. IEEE Computer Society, Washington, DC, pp. 267–274, 2011.
27. M.-C. Nita, C. Chilipirea, C. Dobre, and F. Pop. A SLA-based method for big-data transfers with multi-criteria optimization constraints for IaaS. In *Roedunet International Conference (RoEduNet), 2013 11th*, Sinaia, Romania, pp. 1–6, January 17–19, 2013.
28. R. Van den Bossche, K. Vanmechelen, and J. Broeckhove. Online cost-efficient scheduling of deadline-constrained workloads on hybrid clouds. *Future Generation Computer Systems*, vol. 29, no. 4, pp. 973–985, 2013.

29. H. Karatza. Scheduling in distributed systems. In *Performance Tools and Applications to Networked Systems*, M. C. Calzarossa and E. Gelenbe (Eds.), Springer Berlin, Heidelberg, pp. 336–356, 2004. ISBN: 978-3-540-21945-3.
30. F. Zhang, J. Cao, K. Li, S. U. Khan, and K. Hwang. Multi-objective scheduling of many tasks in cloud platforms. *Future Generation Computer Systems*, vol. 37, pp. 309–320.
31. K. Elmeleegy. Piranha: Optimizing short jobs in Hadoop. *Proceedings of the VLDB Endowment*, vol. 6, no. 11, pp. 985–996, 2013.
32. R. Ramakrishnan, and Team Members CISL. Scale-out beyond MapReduce. In *Proceedings of the 19th ACM SIGKDD International Conference on Knowledge Discovery and Data Mining (KDD '13)*, I. S. Dhillon, Y. Koren, R. Ghani, T. E. Senator, P. Bradley, R. Parekh, J. He, R. L. Grossman, and R. Uthurusamy (Eds.). ACM, New York, 1 p., 2013.
33. L. Wang, J. Tao, R. Ranjan, H. Marten, A. Streit, J. Chen, and D. Chen. G-Hadoop: MapReduce across distributed data centers for data-intensive computing. *Future Generation Computer Systems*, vol. 29, no. 3, pp. 739–750, 2013.
34. B. C. Tak, B. Urgaonkar, and A. Sivasubramaniam. To move or not to move: The economics of cloud computing. In *Proceedings of the 3rd USENIX Conference on Hot Topics in Cloud Computing (HotCloud '11)*. USENIX Association, Berkeley, CA, 5 p., 2011.
35. D. Loshin. Chapter 7—Big Data tools and techniques. In *Big Data Analytics*, D. Loshin, and M. Kaufmann (Eds.). Boston, pp. 61–72, 2013.

CHAPTER 7

# Time–Space Scheduling in the MapReduce Framework

Zhuo Tang, Ling Qi, Lingang Jiang, Kenli Li, and Keqin Li

CONTENTS



ABSTRACT

As data are the basis of information systems, using Hadoop to rapidly extract useful information from massive data of an enterprise has become an efficient method for programmers in the process of application development. This chapter introduces the MapReduce framework, an excellent distributed and parallel computing model. For the increasing data and cluster scales, to avoid scheduling delays, scheduling skews, poor system utilization, and low degrees of parallelism, some improved methods that focus on the time and space scheduling of reduce tasks in MapReduce are proposed in this chapter. Through analyzing the MapReduce scheduling mechanism, this chapter first illustrates the reasons for system slot resource wasting, which results in reduce tasks waiting around, and proposes the development of a method detailing the start times of reduce tasks dynamically according to each job context. And then, in order to mitigate network traffic and improve the performance of Hadoop, this chapter addresses several optimizing algorithms to solve the problems of reduce placement. It makes a Hadoop reduce task scheduler aware of partitions' network locations and sizes. Finally, as the implementation, a parallel biomedical data processing model using the MapReduce framework is presented as an application of the proposed methods.

## 7.1 INTRODUCTION

Data are representations of information, the information content of data is generally believed to be valuable, and data form the basis of information systems. Using computers to process data, extracting information is a basic function of information systems. In today's highly information-oriented society, the web can be said to be currently the largest information system, of which the data are massive, diverse, heterogeneous, and dynamically changing. Using Hadoop to rapidly extract useful information from the massive data of an enterprise has become an efficient method for programmers in the process of application development.

The significance of Big Data is to analyze people's behavior, intentions, and preferences in the growing and popular social networks. It is also to process data with nontraditional structures and to explore their meanings. Big Data is often used to describe a company's large amount of unstructured and semistructured data. Using analysis to create these data in a relational database for downloading will require too much time and money. Big Data analysis and cloud computing are often linked together, because real-time analysis of large data requires a framework similar to MapReduce to assign work to hundreds or even thousands of computers. After several years of criticism, questioning, discussion, and speculation, Big Data finally ushered in the era belonging to it.

Hadoop presents MapReduce as an analytics engine, and under the hood, it uses a distributed storage layer referred to as the *Hadoop distributed file system* (HDFS). As an open-source implementation of MapReduce, Hadoop is, so far, one of the most successful realizations of large-scale data-intensive cloud computing platforms. It has been realized that when and where to start the reduce tasks are the key problems to enhance MapReduce performance.

For *time scheduling* in MapReduce, the existing work may result in a block of reduce tasks. Especially when the map tasks' output is large, the performance of a MapReduce task scheduling algorithm will be influenced seriously. Through analysis for the current MapReduce scheduling mechanism, Section 7.3 illustrates the reasons for system slot resource wasting, which results in reduce tasks waiting around. Then, the section proposes a self-adaptive reduce task scheduling policy for reduce tasks' start times in the Hadoop platform. It can decide the start time point of each reduce task dynamically according to each job context, including the task completion time and the size of the map output.

Meanwhile, another main performance bottleneck is caused by all-to-all communications between mappers and reducers, which may saturate the top-of-rack switch and inflate job execution time. The bandwidth between two nodes is dependent on their relative locations in the network topology. Thus, moving data repeatedly to remote nodes becomes the bottleneck. For this bottleneck, reducing cross-rack communication will improve job performance. Current research proves that moving a task is more efficient than moving data [1], especially in the Hadoop distributed environment, where data skews are widespread.

Data skew is an actual problem to be resolved for MapReduce. The existing Hadoop reduce task scheduler is not only locality unaware but also partitioning skew unaware. The parallel and distributed computation features may cause some unforeseen problems. Data skew is a typical such problem, and the high runtime complexity amplifies the skew and leads to highly varying execution times of the reducers. Partitioning skew causes shuffle skew, where some reduce tasks receive more data than others. The shuffle skew problem can degrade performance, because a job might get delayed by a reduce task fetching large input data. In the presence of data skew, we can use a reducer placement method to minimize all-to-all communications between mappers and reducers, whose basic idea is to place related map and reduce tasks on the same node, cluster, or rack.

Section 7.4 addresses *space scheduling* in MapReduce. We analyze the source of data skew and conclude that partitioning skew exists within certain Hadoop applications. The node at which a reduce task is scheduled can effectively mitigate the shuffle skew problem. In these cases, reducer placement can decrease the traffic between mappers and reducers and upgrade system performance. Some algorithms are released, which synthesize the network locations and sizes of reducers' partitions in their scheduling decisions in order to mitigate network traffic and improve MapReduce performance. Overall, Section 7.4 introduces several ways to avoid scheduling delay, scheduling skew, poor system utilization, and low degree of parallelism.

Some typical applications are discussed in this chapter. At present, there is an enormous quantity of biomedical literature, and it continues to increase at high speed. People urgently need some automatic tools to process and analyze the biomedical literature. In the current methods, the model training time increases sharply when dealing with large-scale training samples. How to increase the efficiency of named entity recognition (NER) in biomedical Big Data becomes one of the key problems in biomedical text mining. For the purposes of improving the recognition performance and reducing the training time, through implementing the model training process based on MapReduce, Section 7.5 proposes an optimization method for two-phase recognition using conditional random fields (CRFs) with some new feature sets.

## 7.2 OVERVIEW OF BIG DATA PROCESSING ARCHITECTURE

MapReduce is an excellent model for distributed computing, introduced by Google in 2004 [2]. It has emerged as an important and widely used programming model for distributed and parallel computing, due to its ease of use, generality, and scalability. Among its open-source implementation versions, Hadoop has been widely used in industry around the whole world [3]. It has been applied to production environments, such as Google, Yahoo,

Amazon, Facebook, and so on. Because of the short development time, Hadoop can be improved in many aspects, such as the problems of intermediate data management and reduce task scheduling [4].

As shown in Figure 7.1, map and reduce are two sections in a MapReduce scheduling algorithm. In Hadoop, each task contains three function phases, that is, copy, sort, and reduce [5]. The goal of the copy phase is to read the map tasks' output. The sort phase is to sort the intermediate data, which are the output from map tasks and will be the input to the reduce phase. Finally, the eventual results are produced through the reduce phase, where the copy and sort phases are to preprocess the input data of the reduce phase. In real applications, copying and sorting may cost a considerable amount of time, especially in the copy phase. In the theoretical model, the reduce functions start only if all map tasks are finished [6]. However, in the Hadoop implementation, all copy actions of reduce tasks will start when the first map action is finished [7]. But in slot duration, if there is any map task still running, the copy actions will wait around. This will lead to the waste of reduce slot resources.

In traditional MapReduce scheduling, reduce tasks should start when all the map tasks are completed. In this way, the output of map tasks should be read and written to the reduce tasks in the copy process [8]. However, through the analysis of the slot resource usage in the reduce process, this chapter focuses on the slot idling and delay. In particular, when the map tasks' output becomes large, the performance of MapReduce scheduling algorithms will be influenced seriously [9]. When multiple tasks are running, inappropriate scheduling of the reduce tasks will lead to the situation where other jobs in the system cannot be scheduled in a timely manner. These are the stumbling blocks of Hadoop popularization.

A user needs to serve two functions in the Hadoop framework, that is, mapper and reducer, to process data. Mappers produce a set of files and send it to all the reducers. Reducers will receive files from all the mappers, which is an all-to-all communication model. Hadoop runs in a data center environment in which machines are organized in

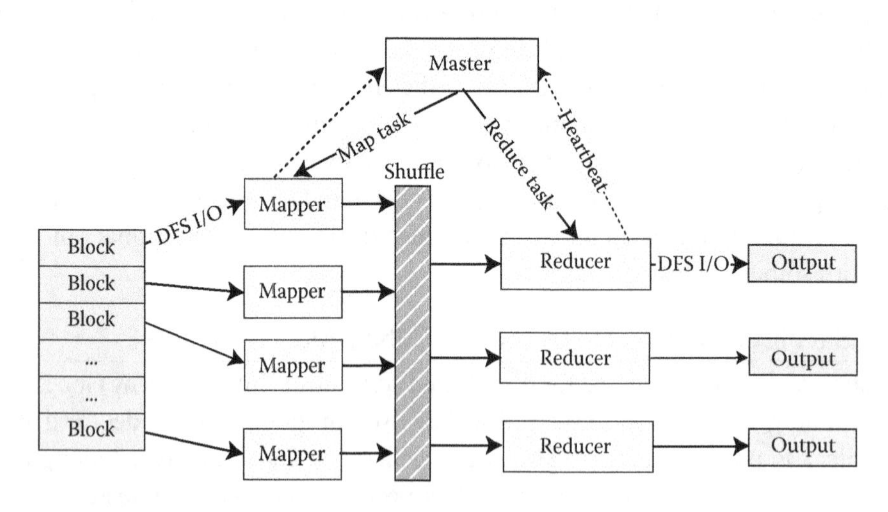


FIGURE 7.1 The typical process of MapReduce.

racks. Cross-rack communication happens if a mapper and a reducer reside in different racks. Every cross-rack communication needs to travel through the root switch, and hence, the all-to-all communication model becomes a bottleneck.

This chapter points out the main affecting factors for the system performance in the MapReduce framework. The solutions to these problems constitute the content of the proposed time–space scheduling algorithms. In Section 7.3, we present a self-adaptive reduce task scheduling algorithm to resolve the problem of slot idling and waste. In Section 7.4, we analyze the source of data skew in MapReduce and introduce some methods to minimize cross-rack communication and MapReduce traffic. To show the application of this advanced MapReduce framework, in Section 7.5, we describe a method to provide the parallelization of model training in NER in biomedical Big Data mining.

## 7.3 SELF-ADAPTIVE REDUCE TASK SCHEDULING

### 7.3.1 Problem Analysis

Through studying reduce task scheduling in the Hadoop platform, this chapter proposes an optimizing policy called *self-adaptive reduce scheduling* (SARS) [10]. This method can decrease the waiting around of copy actions and enhance the performance of the whole system. Through analyzing the details of the map and reduce two-phase scheduling process at the runtime of the MapReduce tasks [8], SARS can determine the start time point of each reduce task dynamically according to each job's context, such as the task completion time, the size of the map output [11], and so forth. This section makes the following contributions: (1) analysis of the current MapReduce scheduling mechanism and illustration of the reasons for system slot resource wasting, which results in reduce tasks waiting around; (2) development of a method to determine the start times of reduce tasks dynamically according to each job context, including the task completion time and the size of the map output; and (3) description of an optimizing reduce scheduling algorithm, which decreases the reduce completion time and system average response time in a Hadoop platform.

Hadoop allows the user to configure the job, submit it, control its execution, and query the state. Every job consists of independent tasks, and each task needs to have a system slot to run. Figure 7.2 shows the time delay and slot resource waste problem in reduce task scheduling. Through Figure 7.2a, we can know that $Job_1$ and $Job_2$ are the current running jobs, and at the initial time, each job is allocated two map slots to run respective tasks. Since the execution time of each task is not the same, as shown in Figure 7.2a, $Job_2$ finishes its map tasks at time t2. Because the reduce tasks will begin once any map task finishes, from the duration t1 to t2, there are two reduce tasks from $Job_1$ and $Job_2$, which are running respectively. As indicated in Figure 7.2b, at time t3, when all the reduce tasks of $Job_2$ are finished, two new reduce tasks from $Job_1$ are started. Now all the reduce slot resources are taken up by $Job_1$. As shown in Figure 7.2c, at the moment t4, when $Job_3$ starts, two idle map slots can be assigned to it, and the reduce tasks from this job will then start. However, we can find that all reduce slots are already occupied by $Job_1$, and the reduce tasks from $Job_3$ have to wait for slot release.

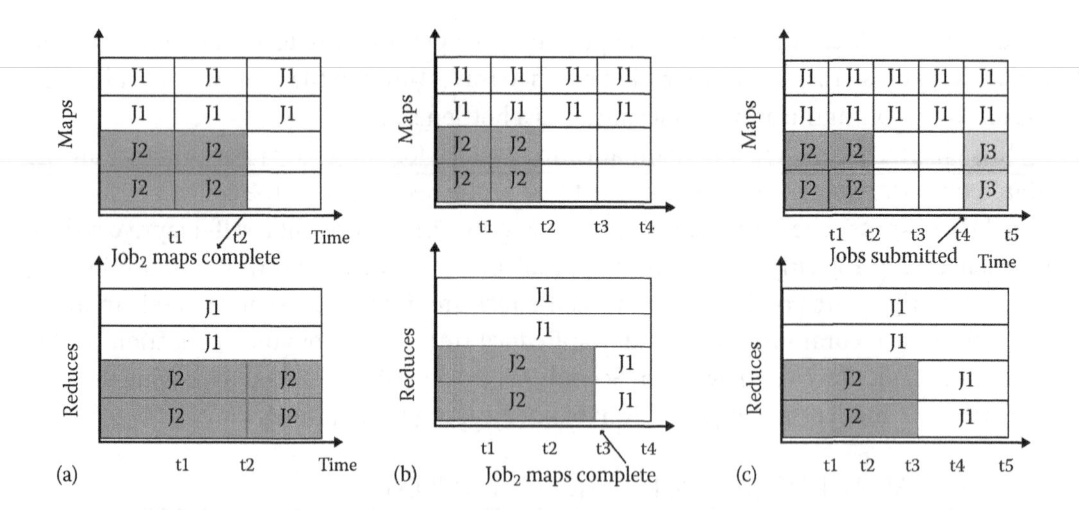


FIGURE 7.2 Performance of the policies with respect to various graph sizes. (a) $Job_2$ map tasks finished. (b) $Job_2$ reduce tasks finished. (c) $Job_3$ submitted.

The root cause of this problem is that the reduce task of $Job_3$ must wait for all the reduce tasks of $Job_1$ to be completed, as $Job_1$ takes up all the reduce slots and the Hadoop system does not support preemptive action acquiescently. In early algorithm design, a reduce task can be scheduled once any map tasks are finished [12]. One of the benefits is that the reduce tasks can copy the output of the map tasks as soon as possible. But reduce tasks will have to wait before all map tasks are finished, and the pending tasks will always occupy the slot resources, so that other jobs that finish the map tasks cannot start the reduce tasks. All in all, this will result in long waiting of reduce tasks and will greatly increase the delay of Hadoop jobs.

In practical applications, there are often different jobs running in a shared cluster environment, which are from multiple users at the same time. If the situation described in Figure 7.2 appears among the different users at the same time, and the reduce slot resources are occupied for a long time, the submitted jobs from other users will not be pushed ahead until the slots are released. Such inefficiency will extend the average response time of a Hadoop system, lower the resource utilization rate, and affect the throughput of a Hadoop cluster.

### 7.3.2 Runtime Analysis of MapReduce Jobs

Through the analysis in Section 7.3.1, one method to optimize the MapReduce tasks is to select an adaptive time to schedule the reduce tasks. By this means, we can avoid the reduce tasks' waiting around and enhance the resource utilization rate. This section proposes a self-adaptive reduce task scheduling policy, which gives a method to estimate the start time of a task, instead of the traditional mechanism where reduce tasks are started once any map task is completed.

The reduce process can be divided into the following several phases. First, reduce tasks request to read the output data of map tasks in the copy phase, this will bring data transmissions from map tasks to reduce tasks. Next, in the sort process, these intermediate data are ordered by merging, and the data are distributed in different storage locations. One type is the data in memory. When the data are read from the various maps at the same time, the data set should be merged as the same keys. The other type is data as the circle buffer. Because the memory belonging to the reduce task is limited, the data in the buffer should be written to disks regularly in advance.

In this way, subsequent data which are written into the disk earlier need to be merged, so-called external sorting. The external sorting needs to be executed several times if the number of map tasks is large in the practical works. The copy and sort processes are customarily called the shuffle phase. Finally, after finishing the copy and sort processes, the subsequent functions start, and the reduce tasks can be scheduled to the compute nodes.

### 7.3.3 A Method of Reduce Task Start-Time Scheduling

Because Hadoop employs a greedy strategy to schedule the reduce tasks, to schedule the reduce tasks fastest, as described in Section 7.3.1, some reduce tasks will always take up the system resources without actually performing operations for a long time. Reduce task start time is determined by the advanced algorithm SARS. In this method, the start times of the reduce tasks are delayed for a certain duration to lessen the utilization of system resources. The SARS algorithm schedules the reduce tasks at a special moment, when some map tasks are finished but not all. By this means, how to select an optimal time point to start the reduce scheduling is the key problem of the algorithm. Distinctly, the optimum point can minimize the system delay and maximize the resource utilization.

As shown in Figure 7.3, assuming that $Job_1$ has 16 map tasks and 1 reduce task, and there are 4 map slots and only 1 reduce slot in this cluster system. Figures 7.3 and 7.4 describe the time constitution of the life cycle for a special job:

$$(FT_{lm} - STf_m) + (FT_{cp} - FT_{lm}) + (FT_{lr} + ST_{sr}). \tag{7.1}$$

The denotations in Equation 7.1 are defined as follows. $FT_{lm}$ is the completion time of the last map task; $STf_m$ is the start time of the first map task; $FT_{cp}$ is the finish time of the copy phase; $FT_{lr}$ is the finish time of reduce; and $ST_{sr}$ is the start time of reduce sort.

In Figure 7.3, t1 is the start time of Map1, Map2, and the reduce task. During t1 to t3, the main work of the reduce task is to copy the output from Map1 to Map14. The output of

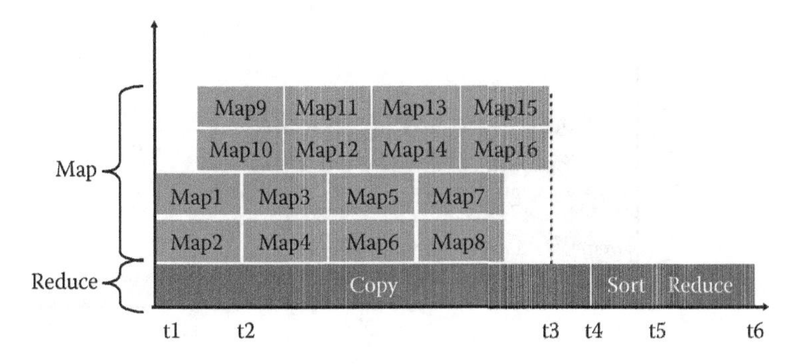


FIGURE 7.3 Default scheduling of reduce tasks.

Map15 and Map16 will be copied by the reduce task from t3 to t4. The duration from t4 to t5 is called the sort stage, which ranks the intermediate results according to the key values. The reduce function is called at time t5, which continues from t5 to t6. Because during t1 to t3, in the copy phase, the reduce task only copies the output data intermittently, once any map task is completed, for the most part, it is always waiting around. We hope to make the copy operations completed at a concentrated duration, which can decrease the waiting time of the reduce tasks.

As Figure 7.4 shows, if we can start the reduce tasks at t2′, which can be calculated using the following equations, and make sure these tasks can be finished before t6, then during t1 to t2′, the slots can be used by any other reduce tasks. But if we let the copy operation start at t3, because the output of all map tasks should be copied from t3, delay will be produced in this case. As shown in Figure 7.3, the copy phase starts at t2, which just collects the output of the map tasks intermittently. By contrast, the reduce task's waiting time is decreased obviously in Figure 7.4, in which case the copy operations are started at t2′.

The SARS algorithm works by delaying the reduce processes. The reduce tasks are scheduled when part but not all of the map tasks are finished. For a special key value, if we assume that there are $s$ map slots and $m$ map tasks in the current system, and the completion time and the size of output data of each map task are denoted as $t_map_i$ and $m_out_j$, respectively, where $i, j \in [1,m]$. Then, we can know that the data size of the map tasks can be calculated as

$$N_m = \sum_{j=1}^{m} m_\text{out}_j, \quad j \in [1,m]. \tag{7.2}$$

In order to predict the time required to transmit the data, we define the speed of the data transmission from the map tasks to the reduce tasks as transSpeed in the cluster environment, and the number of concurrent copy threads with reduce tasks is denoted as copyThread. We denote the start time of the first map task and the first reduce task as $\text{start}_{\text{map}}$ and $\text{start}_{\text{reduce}}$, respectively. Therefore, the optimal start time of reduce tasks can be determined by the following equation:

$$\text{start}_{\text{reduce}} = \text{start}_{\text{map}} + \frac{\sum_{i=1}^{m} t_\text{map}_i}{s} - \frac{N_m}{transSpeed \times copyThread}. \tag{7.3}$$

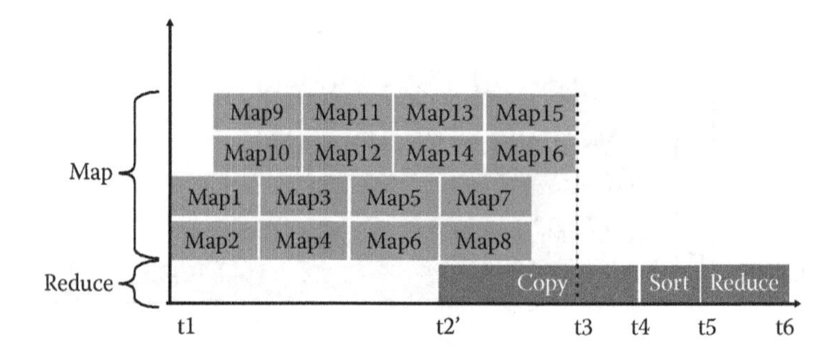


FIGURE 7.4 Scheduling method for reduce tasks in SARS.

As shown by time t2′ in Figure 7.4, the most appropriate start time of a reduce task is when all the map tasks about the same key are finished, which is between the times when the first map is started and when the last map is finished. The second item in Equation 7.3 denotes the required time of the map tasks, and the third item is the time for data transmission. Because the reduce tasks will be started before the copy processes, the time cost should be cut from the map tasks' completion time. The waiting around of the reduce tasks may make the jobs in need of the slot resources not able to work normally. Through adjusting the reduce scheduling time, this method can decrease the time waste for the data replication process and advance the utilization of the reduce slot resources effectively. Through adjusting the reduce scheduling time, this method can decrease the time waste for data replication process and advance the utilization of the reduce slot resources effectively. The improvement of these policies is especially important for the CPU-type jobs. For these jobs that need more CPU computing, the data input/output (I/O) of the tasks is less, so more slot resources will be wasted in the default schedule algorithm.

## 7.4 REDUCE PLACEMENT

As the mapper and reducer functions use an all-to-all communication model, this section presents some exciting and popular solutions in Sections 7.4.1–7.4.3, where we introduce several algorithms to optimize the communication traffic, which could increase the performance of data processing. In Sections 7.4.4–7.4.5, we mention the existence of data skew and propose some methods based on space scheduling, that is, reduce placement, to solve the problem of data skew.

### 7.4.1 Optimal Algorithms for Cross-Rack Communication Optimization

In the Hadoop framework, a user needs to provide two functions, that is, mapper and reducer, to process data. Mappers produce a set of files and send to all the reducers, and a reducer will receive files from all the mappers, which is an all-to-all communication model. Cross-rack communication [13] happens if a mapper and a reducer reside in different racks, which happens very often in today's data center environments. Typically, Hadoop runs in a data center environment in which machines are organized in racks. Each rack has a top-of-rack switch, and each top-of-rack switch is connected to a root switch. Every cross-rack communication needs to travel through the root switch, and hence, the root switch becomes a bottleneck [14]. MapReduce employs an all-to-all communication model between mappers and reducers. This results in saturation of network bandwidth of the top-of-rack switch in the shuffle phase, straggles some reducers, and increases job execution time.

There are two optimal algorithms to solve the reducer placement problem (RPP) and an analytical method to find the minimum (may not be feasible) solution of RPP, which considers the placement of reducers to minimize cross-rack traffic. One algorithm is a *greedy algorithm* [15], which assigns one reduce task to a rack at a time. When assigning a reduce task to a rack, it chooses the rack that incurs the minimum total traffic (up and down) if the reduce task is assigned to that rack. The second algorithm, called *binary search* [16], uses binary search to find the minimum bound of the traffic function for each rack and then uses that minimum bound to find the number of reducers on each rack.

### 7.4.2 Locality-Aware Reduce Task Scheduling

MapReduce assumes the master–slave architecture and a tree-style network topology [17]. Nodes are spread over different racks encompassed in one or many data centers. A salient point is that the bandwidth between two nodes is dependent on their relative locations in the network topology. For example, nodes that are in the same rack have higher bandwidth between them than nodes that are off rack. As such, it pays to minimize data shuffling across racks. The master in MapReduce is responsible for scheduling map tasks and reduce tasks on slave nodes after receiving requests from slaves for that regard. Hadoop attempts to schedule map tasks in proximity to input splits in order to avoid transferring them over the network. In contrast, Hadoop schedules reduce tasks at requesting slaves without any data locality consideration. As a result, unnecessary data might get shuffled in the network, causing performance degradation.

Moving data repeatedly to distant nodes becomes the bottleneck [18]. We rethink reduce task scheduling in Hadoop and suggest making Hadoop's reduce task scheduler aware of partitions' network locations and sizes in order to mitigate network traffic. There is a practical strategy that leverages network locations and sizes of partitions to exploit data locality, named *locality-aware reduce task scheduler* (LARTS) [17]. In particular, LARTS attempts to schedule reducers as close as possible to their maximum amount of input data and conservatively switches to a relaxation strategy seeking a balance between scheduling delay, scheduling skew, system utilization, and parallelism. LARTS attempts to collocate reduce tasks with the maximum required data computed after recognizing input data network locations and sizes. LARTS adopts a cooperative paradigm seeking good data locality while circumventing scheduling delay, scheduling skew, poor system utilization, and low degree of parallelism. We implemented LARTS in Hadoop-0.20.2. Evaluation results show that LARTS outperforms the native Hadoop reduce task scheduler by an average of 7% and up to 11.6%.

### 7.4.3 MapReduce Network Traffic Reduction

Informed by the success and the increasing prevalence of MapReduce, we investigate the problems of data locality and partitioning skew present in the current Hadoop implementation and propose the *center-of-gravity reduce scheduler* (CoGRS) algorithm [19], a locality-aware and skew-aware reduce task scheduler for saving MapReduce network traffic. CoGRS attempts to schedule every reduce task $R$ at its center-of-gravity node determined by the network locations of $R$'s feeding nodes and the skew in the sizes of $R$'s partitions. Notice that the center-of-gravity node is computed after considering partitioning skew as well.

The network is typically a bottleneck in MapReduce-based systems. By scheduling reducers at their center-of-gravity nodes, we argue for reduced network traffic, which can possibly allow more MapReduce jobs to coexist in the same system. CoGRS controllably avoids scheduling skew, a situation where some nodes receive more reduce tasks than others, and promotes pseudoasynchronous map and reduce phases. Evaluations show that CoGRS is superior to native Hadoop. When Hadoop schedules reduce tasks, it neither exploits data locality nor addresses partitioning skew present in some MapReduce applications. This might lead to increased cluster network traffic.

We implemented CoGRS in Hadoop-0.20.2 and tested it on a private cloud as well as on Amazon Elastic Compute Cloud (EC2). As compared to native Hadoop, our results show that CoGRS minimizes off-rack network traffic by an average of 9.6% and 38.6% on our private cloud and on an Amazon EC2 cluster, respectively. This reflects on job execution times and provides an improvement of up to 23.8%.

Partitioning skew refers to the significant variance in intermediate keys' frequencies and their distribution across different data nodes. In essence, a reduce task scheduler can determine the pattern of the communication traffic in the network, affect the quantity of shuffled data, and influence the runtime of MapReduce jobs.

### 7.4.4 The Source of MapReduce Skews

Over the last few years, MapReduce has become popular for processing massive data sets. Most research in this area considers simple application scenarios like log file analysis, word count, and sorting, and current systems adopt a simple hashing approach to distribute the load to the reducers. However, processing massive amounts of data exhibits imperfections to which current MapReduce systems are not geared. The distribution of scientific data is typically skewed [20]. The high runtime complexity amplifies the skew and leads to highly varying execution times of the reducers.

There are three typical skews in MapReduce. (1) *Skewed key frequencies*—If some keys appear more frequently in the intermediate data tuples, the number of tuples per cluster owned will be different. Even if every reducer receives the same number of clusters, the overall number of tuples per reducer received will be different. (2) *Skewed tuple sizes*—In applications that hold complex objects within the tuples, unbalanced cluster sizes can arise from skewed tuple sizes. (3) *Skewed execution times*—If the execution time of the reducer is worse than linear, processing a single large cluster may take much longer than processing a higher number of small clusters. Even if the overall number of tuples per reducer is the same, the execution times of the reducers may differ.

According to those skew types, we propose several processes to improve the performance of MapReduce.

### 7.4.5 Reduce Placement in Hadoop

In Hadoop, map and reduce tasks typically consume a large amount of data, and the total intermediate output (or total reduce input) size is sometimes equal to the total input size of all map tasks (e.g., sort) or even larger (e.g., 44.2% for *K*-means). For this reason, optimizing the placement of reduce tasks to save network traffic becomes very essential for optimizing the placement of map tasks, which is already well understood and implemented in Hadoop systems.

This section explores scheduling to ensure that the data that a reduce task handles the most are localized, so that it can save traffic cost and diminish data skew [21].

*Sampling*—Input data are loaded into a file or files in a distributed file system (DFS), where each file is partitioned into smaller chunks, called input splits. Each split is assigned to a map task. Map tasks process splits [22] and produce intermediate outputs, which are usually partitioned or hashed to one or many reduce tasks. Before a MapReduce computation

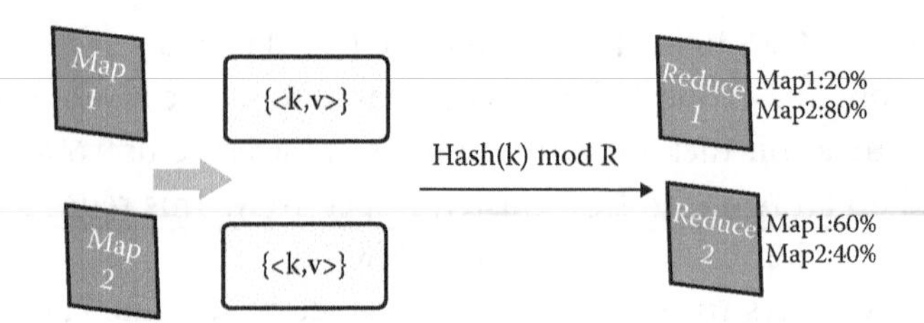


FIGURE 7.5 Intermediate result distribution in reduce tasks.

begins with a map phase, where each input split is processed in parallel, a random sample of the required size will be produced. The split of samples is submitted to the auditor group (AG); meanwhile, the master and map tasks will wait for the results of the auditor.

*AG*—The AG carries out a statistical and predicted test to calculate the distribution of reduce tasks and then start the reduce VM [23] at the appropriate place in the PM. The AG will receive several samples and then will assign its members that contain map and reduce tasks to them. The distribution of intermediate key/value pairs that adopt a hashing approach to distribute the load to the reducers will be computed in reduces.

*Placement of reduce virtual machine* (*VM*)—The results of AG will decide the placement of reduce VMs. For example, in Figure 7.5, if 80% of key/value pairs of Reduce1 come from Map2 and the remaining intermediate results are from Map1, the VM of Reduce1 will be started in the physical machine (PM) that contains the VM of Map2. Similarly, the VM of Reduce2 will be started in the PM that includes the VM of Map1.

## 7.5 NER IN BIOMEDICAL BIG DATA MINING: A CASE STUDY

Based on the above study of time–space Hadoop MapReduce scheduling algorithms in Sections 7.3 and 7.4, we present a case study in the field of biomedical Big Data mining. Compared to traditional methods and general MapReduce for data mining, our project makes an originally inefficient algorithm become time-bearable in the case of integrating the algorithms in Sections 7.3 and 7.4.

### 7.5.1 Biomedical Big Data

In the past several years, massive data have been accumulated and stored in different forms, whether in business enterprises, scientific research institutions, or government agencies. But when faced with more and more rapid expansion of the databases, people cannot set out to obtain and understand valuable knowledge within Big Data.

The same situation has happened in the biomedical field. As one of the most concerned areas, especially after the human genome project (HGP), literature in biomedicine has appeared in large numbers, reaching an average of 600,000 or more per year [24]. Meanwhile, the completion of the HGP has produced large human gene sequence data. In addition, with the fast development of science and technology in recent years, more and more large-scale biomedical experiment techniques, which can reveal the law of life activities on the molecular level, must use the Big Data from the entire genome or the entire proteome, which results in a huge amount of biological data. These mass biological data contain a wealth of biological information, including significant gene expression situations and protein–protein interactions. What is more, a disease network, which contains hidden information associated with the disease and gives biomedical scientists the basis of hypothesis generation, is constructed based on disease relationship mining in these biomedical data.

However, the most basic requirements for biomedical Big Data processing are difficult to meet efficiently. For example, key word searching in biomedical Big Data or the Internet can only find lots of relevant file lists, and the accuracy is not high, so that a lot of valuable information contained in the text cannot be directly shown to the people.

### 7.5.2 Biomedical Text Mining and NER

In order to explore the information and knowledge in biomedical Big Data, people integrate mathematics, computer science, and biology tools, which promote the rapid development of large-scale biomedical text mining. This refers to the biomedical Big Data analysis process of deriving high-quality information that is implicit, previously unknown, and potentially useful from massive biomedical data.

Current research emphasis on large-scale biomedical text mining is mainly composed of two aspects, that is, information extraction and data mining. Specifically, it includes biomedical named entity recognition (Bio-NER), relation extraction, text classification, and an integration framework of the above work.

Bio-NER is the first and an important and critical step in biomedical Big Data mining. It aims to help molecular biologists recognize and classify professional instances and terms, such as protein, DNA, RNA, cell_line, and cell_type. It is meant to locate and classify atomic elements with some special significance in biomedical text into predefined categories. The process of Bio-NER systems is structured as taking an unannotated block of text and then producing an annotated block of text that highlights where the biomedical named entities are [25].

However, because of lots of unique properties in the biomedical area, such as unstable quantity, nonunified naming rules, complex form, the existence of ambiguity, and so on, Bio-NER is not mature enough, especially since it takes much time. Most current Bio-NER systems are based on machine learning, which needs multiple iterative calculations from corpus data. Therefore, it is computationally intensive and seriously increases recognition time, including model training and inference. For example, it takes almost 5 hours for the CRF model training process using the Genia4ER training corpus, which is only about 14 MB [26]. How do we deal with huge amounts of biomedical text data? How do we cope with the unbearable wait for recognition for a very long time? It is natural to seek distributed computing and parallel computing to solve the problem.

### 7.5.3 MapReduce for CRFs

CRFs are an important milestone in the field of machine learning, put forward in 2001 by John Lafferty et al. [27]. CRFs, a kind of discriminant model and an undirected graph model at the same time, define a single logarithmic linear distribution for a joint probability of an entire label sequence based on a given particular observation sequence. The model is widely used in natural language processing (NLP), including NER, part-of-speech tagging, and so on.

Figure 7.6 shows the CRF model, which computes the conditional probability $p(\vec{y}|\vec{x})$ of an output sequence $\vec{y}=(y_1,y_2,...,y_n)$ under the condition of a given input sequence $\vec{x}=(x_1,x_2,\ldots,x_n)$.

A linear CRF, which is used in Bio-NER, is as follows:

$$P(\vec{y}|\vec{x})=\frac{1}{Z(\vec{x})}\cdot\exp\left(\sum_{i=1}^{n}\sum_{k=1}^{K}\lambda_k f_k(\vec{x},i,y_{i-1},y_i)\right), \tag{7.4}$$

where

$$Z(\vec{x})=\sum_{y}\exp\left(\sum_{i=1}^{n}\sum_{k=1}^{K}\lambda_k f_k(\vec{x},i,y_{i-1},y_i)\right), \tag{7.5}$$

$i$ is the position in the input sequence $\vec{x}=(x_1,x_2,\ldots,x_n)$, $\lambda_k$ is a weight of a feature that does not depend on location $i$, and $\left\{f_k(\vec{x},i,y_{i-1},y_i)\right\}_{k=1}^{K}$ are feature functions.

For the training process of the CRF model, the main purpose is to seek for the parameter $\vec{\lambda}=(\lambda_1,\lambda_2,\ldots,\lambda_K)$ that is most in accordance with the training data $T=\left\{(\vec{x}_i,\vec{y}_i)\right\}_{i=1}^{N}$. Presume every $(\vec{x},\vec{y})$ is independently and identically distributed. The parameter is obtained generally in this way:

$$L(\lambda)=\sum_{T}\log P(y|x). \tag{7.6}$$

When the log-likelihood function $L(\lambda)$ reaches the maximum value, the parameter is almost the best. However, to find the parameter to maximize the training data likelihood, there is no closed-form solution. Hence, we adopt parameter estimation, that is, the limited-memory Broyden–Fletcher–Goldfarb–Shanno (L-BFGS) algorithm [28], to find the optimum solution.

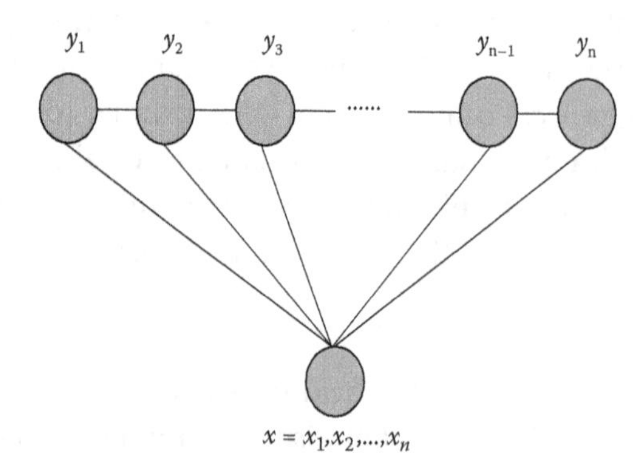


FIGURE 7.6 Linear CRFs.

To find the parameter $\vec{\lambda} = (\lambda_1, \lambda_2, \ldots, \lambda_K)$ to make convex function $L(\lambda)$ reach the maximum, algorithm L-BFGS makes its gradient vector $\nabla L = \left(\frac{\partial L}{\partial \lambda_1}, \frac{\partial L}{\partial \lambda_2}, \ldots, \frac{\partial L}{\partial \lambda_K}\right)\vec{0}$ by iterative computations with initial value $\lambda_0 = 0$ at first. Research shows that the first step, that is, to calculate $\nabla L_i$, which is on behalf of the gradient vector in iteration $i$, calls for much time. Therefore, we focus on the optimized improvement for it.

Every component in $\nabla L_i$ is computed as follows:

$$\frac{\partial L'(\lambda)}{\partial \lambda_k} = \sum_T \left[ \sum_{i=1}^{n} f_k(\vec{x}, i, y_{i-1}, y_i) - \sum_y P(y'|\vec{x}) \sum_{i=1}^{n} f_k(\vec{x}, i, y'_{i-1}, y'_i) \right] - \frac{\lambda_k}{\sigma^2}. \tag{7.7}$$

It can be linked with every ordered pair $(\vec{x}, \vec{y})$ within $\sum_T$ that is mutually independent. So we can calculate the difference between $\sum_{i=1}^{n} f_k(\vec{x}, i, y_{i-1}, y_i)$ and $\sum_{y'} P(y'|\vec{x}) \sum_{i=1}^{n} f_k(\vec{x}, i, y'_{i-1}, y'_i)$ on each of the input sequences in the training set $T$ and then put the results of all the sequences together. As a result, they can be computed in parallel as shown in Figure 7.7.

We split the calculation process in-house $\sum_T$ into several map tasks and summarize the results by a reduce task. And the difference between penalty term $\frac{\lambda_k}{\sigma^2}$ is designed to be the postprocessing.

In the actual situation, it is impossible to schedule one map task for one ordered pair $(\vec{x}, \vec{y})$ because the number of ordered pairs in the large scale of training samples is too huge to estimate effectively. We must syncopate the training data $T$ into several small parts and then start the MapReduce plan as shown in the two above paragraphs at this section.

For a MapReduce Bio-NER application, the data skew leads to uneven load in the whole system. Any specific corpus has its own uneven distribution of the entity (as shown in Table 7.1), resulting in the serious problem of data skew. And protean, artificial defined feature sets exacerbate the problem in both training and inference processes.

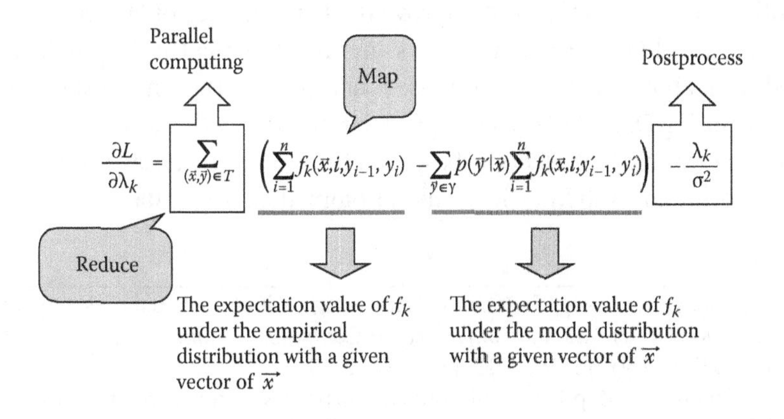


FIGURE 7.7 MapReduce plan for computing component.

TABLE 7.1 Proportion of Each Type of Entity in the Corpus Joint Workshop on Natural Language Processing in Biomedicine and Its Application (JNLPBA2004)

| | Protein | DNA | RNA | Cell Line | Cell Type |
|---|---|---|---|---|---|
| Training Set | 59.00% | 18.58% | 1.85% | 7.47% | 13.10% |
| Test Set | 58.50% | 12.19% | 1.36% | 5.77% | 22.18% |

Combined with schemes given in this chapter, the uneven load can be solved based on the modified Hadoop MapReduce. The implementation will further improve system performance on MapReduce with time–space scheduling.

## 7.6 CONCLUDING REMARKS

As data are the basis of information systems, how to process data and extract information becomes one of the hottest topics in today's information society. This chapter introduces the MapReduce framework, an excellent distributed and parallel computing model. As its implementation, Hadoop plays a more and more important role in a lot of distributed application systems for massive data processing.

For the increasing data and cluster scales, to avoid scheduling delay, scheduling skew, poor system utilization, and low degree of parallelism, this chapter proposes some improved methods that focus on the time and space scheduling of reduce tasks in MapReduce.

Through analyzing the MapReduce scheduling mechanism, this chapter illustrates the reasons for system slot resource wasting that result in reduce tasks waiting around, and it proposes the development of a method detailing the start times of reduce tasks dynamically according to each job context, including the task completion time and the size of the map output. There is no doubt that the use of this method will decrease the reduce completion time and system average response time in Hadoop platforms.

Current Hadoop schedulers often lack data locality consideration. As a result, unnecessary data might get shuffled in the network, causing performance degradation. This chapter addresses several optimizing algorithms to solve the problem of reduce placement. We make a Hadoop reduce task scheduler aware of partitions' network locations and sizes in order to mitigate network traffic and improve the performance of Hadoop.

Finally, a parallel biomedical data processing model using the MapReduce framework is presented as an application of the proposed methods. As the United States proposed the HGP, biomedical Big Data shows its unique position among the academics. A widely used CRF model and an efficient Hadoop-based method, Bio-NER, have been introduced to explore the information and knowledge under biomedical Big Data.

## REFERENCES

1. J. Tan, S. Meng, X. Meng, L. Zhang. Improving reduce task data locality for sequential MapReduce jobs. *International Conference on Computer Communications (INFOCOM), 2013 Proceedings IEEE*, April 14–19, 2013, pp. 1627–1635.
2. J. Dean, S. Ghemawat. MapReduce: Simplified data processing on large clusters, *Communications of the ACM—50th Anniversary Issue: 1958–2008*, 2008, Volume 51, Issue 1, pp. 137–150.

3. X. Gao, Q. Chen, Y. Chen, Q. Sun, Y. Liu, M. Li. A dispatching-rule-based task scheduling policy for MapReduce with multi-type jobs in heterogeneous environments. *2012 7th ChinaGrid Annual Conference (ChinaGrid)*, September 20–23, 2012, pp. 17–24.
4. J. Xie, F. Meng, H. Wang, H. Pan, J. Cheng, X. Qin. Research on scheduling scheme for Hadoop clusters. *Procedia Computer Science*, 2013, Volume 18, pp. 2468–2471.
5. Z. Tang, M. Liu, K. Q. Li, Y. Xu. A MapReduce-enabled scientific workflow framework with optimization scheduling algorithm. *2012 13th International Conference on Parallel and Distributed Computing, Applications and Technologies (PDCAT)*, December 14–16, 2012, pp. 599–604.
6. F. Ahmad, S. Lee, M. Thottethodi, T. N. Vijaykumar. MapReduce with communication overlap (MaRCO). *Journal of Parallel and Distributed Computing*, 2013, Volume 73, Issue 5, pp. 608–620.
7. M. Lin, L. Zhang, A. Wierman, J. Tan. Joint optimization of overlapping phases in MapReduce. *Performance Evaluation*, 2013, Volume 70, Issue 10, pp. 720–735.
8. Y. Luo, B. Plale. Hierarchical MapReduce programming model and scheduling algorithms. *12th IEEE/ACM International Symposium on Cluster, Cloud and Grid Computing (CCGRID)*, May 13–16, 2012, pp. 769–774.
9. H. Mohamed, S. Marchand-Maillet. MRO-MPI: MapReduce overlapping using MPI and an optimized data exchange policy. *Parallel Computing*, 2013, Volume 39, Issue 12, pp. 851–866.
10. Z. Tang, L. G. Jiang, J. Q. Zhou, K. L. Li, K. Q. Li. A self-adaptive scheduling algorithm for reduce start time. *Future Generation Computer Systems*, 2014. Available at http://dx.doi.org/10.1016/j.future.2014.08.011.
11. D. Linderman, D. Collins, H. Wang, H. Meng. Merge: A programming model for heterogeneous multi-core systems. *ASPLOSXIII Proceedings of the 13th International Conference on Architectural Support for Programming Languages and Operating Systems*, March 2008, Volume 36, Issue 1, pp. 287–296.
12. B. Palanisamy, A. Singh, L. Liu, B. Langston. Cura: A cost-optimized model for MapReduce in a cloud. *IEEE International Symposium on Parallel and Distributed Processing (IPDPS)*, IEEE Computer Society, May 20–24, 2013, pp. 1275–1286.
13. L. Ho, J. Wu, P. Liu. Optimal algorithms for cross-rack communication optimization in MapReduce framework. *2011 IEEE International Conference on Cloud Computing (CLOUD)*, July 4–9, 2011, pp. 420–427.
14. M. Isard, V. Prabhakaran, J. Currey, U. Wieder, K. Talwar, A. Goldberg. Quincy: Fair scheduling for distributed computing clusters. *Proceedings of the ACM SIGOPS 22nd Symposium on Operating Systems Principles (SOSP)*, October 11–14, 2009, pp. 261–276.
15. Wikipedia, Greedy algorithm [EB/OL]. Available at http://en.wikipedia.org/wiki/Greedy_algorithm, September 14, 2013.
16. D. D. Sleator, R. E. Tarjan. Self-adjusting binary search trees. *Journal of the ACM (JACM)*, 1985, Volume 32, Issue 3, pp. 652–686.
17. M. Hammoud, M. F. Sakr. Locality-aware reduce task scheduling for MapReduce. *Cloud Computing Technology and Science (CloudCom), 2011 IEEE Third International Conference on*, November 29–December 1, 2011, pp. 570–576.
18. S. Huang, J. Huang, J. Dai, T. Xie, B. Huang. The HiBench benchmark suite: Characterization of the MapReduce-based data analysis. *Date Engineering Workshops (ICDEW), IEEE 26th International Conference on*, March 1–6, 2010, pp. 41–51.
19. M. Hammoud, M. S. Rehman, M. F. Sakr. Center-of-gravity reduce task scheduling to lower MapReduce network traffic. *Cloud Computing (CLOUD), IEEE 5th International Conference on*, June 24–29, 2012, pp. 49–58.
20. Y. C. Kwon, M. Balazinska, B. Howe, J. Rolia. Skew-resistant parallel processing of feature-extracting scientific user-defined functions. *Proceedings of the 1st ACM Symposium on Cloud Computing (SoCC)*, June 2010, pp. 75–86.

21. P. Dhawalia, S. Kailasam, D. Janakiram. Chisel: A resource savvy approach for handling skew in MapReduce applications. *Cloud Computing (CLOUD), IEEE Sixth International Conference on*, June 28–July 3, 2013, pp. 652–660.
22. R. Grover, M. J. Carey. Extending Map-Reduce for efficient predicate-based sampling. *Data Engineering (ICDE), 2012 IEEE 28th International Conference on*, April 1–5, 2012, pp. 486–497.
23. S. Ibrahim, H. Jin, L. Lu, L. Qi, S. Wu, X. Shi. Evaluating MapReduce on virtual machines: The Hadoop case cloud computing. *Lecture Notes in Computer Science*, 2009, Volume 5931, pp. 519–528.
24. Wikipedia, MEDLINE [EB/OL]. Available at http://en.wikipedia.org/wiki/MEDLINE, September 14, 2013.
25. J. Kim, T. Ohta, Y. Tsuruoka, Y. Tateisi, N. Collier. Introduction to the bio-entity recognition task at JNLPBA. *Proceedings of the International Joint Workshop on Natural Language Processing in Biomedicine and Its Applications (JNLPBA)*, August 2004, pp. 70–75.
26. L. Li, R. Zhou, D. Huang. Two-phase biomedical named entity recognition using CRFs. *Computational Biology and Chemistry*, 2009, Volume 33, Issue 4, pp. 334–338.
27. J. Lafferty, A. McCallum, F. Pereira. Conditional random fields: Probabilistic models for segmenting and labeling sequence data. *Proceedings of the 18th International Conference on Machine Learning (ICML)*, June 28–July 1, 2001, pp. 282–289.
28. D. Liu, J. Nocedal. On the limited memory BFGS method for large scale optimization. *Mathematical Programming*, 1989, Volume 45, Issue 1–3, pp. 503–528.

CHAPTER 8

# GEMS: Graph Database Engine for Multithreaded Systems

Alessandro Morari, Vito Giovanni Castellana, Oreste Villa, Jesse Weaver, Greg Williams, David Haglin, Antonino Tumeo, and John Feo

CONTENTS



ABSTRACT

Many fields require organizing, managing, and analyzing massive amounts of data. Among them, we can find social network analysis, financial risk management, threat detection in complex network systems, and medical and biomedical databases. For these areas, there is a problem not only in terms of size but also in terms of performance, because the processing should happen sufficiently fast to be useful. Graph databases appear to be a good candidate to manage these data: They provide an efficient data structure for heterogeneous data or data that are not themselves rigidly structured. However, exploring large-scale graphs on modern high-performance machines is challenging. These systems include processors and networks optimized for regular, floating-point intensive computations and large, batched data transfers. At the opposite, exploring graphs generates fine-grained, unpredictable memory and network accesses, is mostly memory bound, and is synchronization intensive. Furthermore, graphs often are difficult to partition, making their processing prone to load unbalance.

In this book chapter, we describe Graph Engine for Multithreaded Systems (GEMS), a full software stack that implements a graph database on a commodity cluster and enables scaling in data set size while maintaining a constant query throughput when adding more cluster nodes. GEMS employs the SPARQL (SPARQL Protocol and RDF Query Language) language for querying the graph database. The GEMS software stack comprises a SPARQL-to-data parallel C++ compiler; a library of distributed data structures; and a custom, multithreaded, runtime system.

We provide an overview of the stack, describe its advantages compared with other solutions, and focus on how we solved the challenges posed by irregular behaviors.

We finally propose an evaluation of GEMS on a typical SPARQL benchmark and on a Resource Description Format (RDF) data set currently curated at Pacific Northwest National Laboratory.

## 8.1 INTRODUCTION

Many very diverse application areas are experiencing an explosive increase in the availability of data to process. They include finance; science fields such as astronomy, biology, genomics, climate and weather, and material sciences; the web; geographical systems; transportation; telephones; social networks; and security. The data sets of these applications have quickly surpassed petabytes in size and keep exponentially growing. Enabling computing systems to process these large amounts of data, without compromising performance, is becoming of paramount importance. Graph databases appear a most natural and convenient way to organize, represent, and store the data of these applications, as they usually contain a large number of relationships among their elements, providing significant benefits with respect to relational databases.

Graph databases organize the data in the form of subject–predicate–object triples following the Resource Description Framework (RDF) data model (Klyne et al. 2004). A set of triples naturally represents a labeled, directed multigraph. An analyst can query semantic graph databases through languages such as SPARQL (W3C SPARQL Working Group 2013), in which the fundamental query operation is graph matching. This is different from conventional relational databases that employ schema-specific tables to store data and perform select and conventional join operations when executing queries. With relational approaches, graph-oriented queries on large data sets can quickly become unmanageable in both space and time due to the large sizes of immediate results created when performing conventional joins.

Expressing a query as pattern-matching operations provides both advantages and disadvantages for modern high-performance systems. Exploring a graph is inherently parallel, because the system can potentially spawn a thread for each pattern, vertex, or edge, but also inherently irregular, because it generates unpredictable fine-grained data accesses with poor spatial and temporal locality. Furthermore, graph exploration is mainly network and memory bandwidth bound.

These aspects make clusters promising platforms for scaling both size and performance graph databases at the same time: when adding a node, both the available memory (for in-memory processing) and the number of available cores in the system increase (Weaver 2012). On the other hand, modern clusters are optimized for regular computation, batched data transfers, and high flop ratings, and they introduce several challenges for adequately supporting the required graph processing techniques.

This chapter presents the Graph Engine for Multithreaded Systems (GEMS) database stack. GEMS implements a full software system for graph databases on a commodity cluster. GEMS comprises a SPARQL-to-C++ data parallel compiler; a library of distributed data structures; and custom, multithreaded runtime systems. Differently from other database systems, GEMS executes queries mostly by employing graph exploration methods. The compiler optimizes and converts the SPARQL queries directly to parallel graph-crawling algorithms expressed in C++. The C++ code exploits the distributed data structures and the graph primitives to perform the exploration. The C++ code runs on top of the Global Memory and Threading (GMT) library, a runtime library that provides a global address space across cluster nodes and implements fine-grained software multithreading to tolerate data access latencies, and message aggregation to maximize network bandwidth utilization.

## 8.2 RELATED INFRASTRUCTURES

Currently, many commercial and open-source SPARQL engines are available. We can distinguish between purpose-built databases for the storage and retrieval of triples (triplestores) and solutions that try to map triplestores on top of existing commercial databases, usually relational structured query language (SQL)-based systems. However, obtaining feature-complete SPARQL-to-SQL translation is difficult and may introduce performance penalties. Translating SPARQL to SQL implies the use of relational algebra to perform optimizations and the use of classical relational operators (e.g., conventional joins and selects) to execute the query.

By translating SPARQL to graph pattern-matching operations, GEMS reduces the overhead for intermediate data structures and can exploit optimizations that look at the execution plan (i.e., order of execution) from a graph perspective.

SPARQL engines can further be distinguished between solutions that process queries in memory and solutions that store data on disks and perform swapping. Jena (Apache Jena, n.d.), with the ARQ SPARQL engine (ARQ—A SPARQL Processor for Jena, n.d.); Sesame (openRDF.org, n.d.); and Redland (Redland RDF Libraries, n.d.), also known as librdf, are all examples of RDF libraries that natively implement in-memory RDF storage and support integration with some disk-based, SQL back ends. OpenLink Virtuoso (Virtuoso Universal Server, n.d.) implements an RDF/SPARQL layer on top of their SQL-based column store for which multinode cluster support is available. GEMS adopts in-memory processing: It stores all data structures in RAM. In-memory processing potentially allows increasing the data set size while maintaining constant query throughput by adding more cluster nodes.

Some approaches leverage MapReduce infrastructures for RDF-encoded databases. SHARD (Rohloff and Schantz 2010) is a triplestore built on top of Hadoop, while YARS2 (Harth et al. 2007) is a bulk-synchronous, distributed query-answering system. Both exploit hash partitioning to distribute triples across nodes. These approaches work well for simple index lookups, but they also present high communication overheads for moving data through the network with more complex queries as well as introduce load-balancing issues in the presence of data skew.

More general graph libraries, such as Pregel (Malewicz et al. 2010), Giraph (Apache Giraph, n.d.), and GraphLab (GraphLab, n.d.), may also be exploited to explore semantic databases, once the source data have been converted into a graph. However, they require significant additions to work in a database environment, and they still rely on bulk-synchronous, parallel models that do not perform well for large and complex queries. Our system relies on a custom runtime that provides specific features to support exploration of a semantic database through graph-based methods.

Urika is a commercial shared-memory system from YarcData (YarcData, Inc., n.d.) targeted at Big Data analytics. Urika exploits custom nodes with purpose-built multithreaded processors (barrel processors with up to 128 threads and a very simple cache) derived from the Cray XMT. Besides multithreading, which allows tolerating latencies for accessing data on remote nodes, the system has hardware support for a scrambled global address space and fine-grained synchronization. These features allow more efficient execution of irregular applications, such as graph exploration. On top of this hardware, YarcData interfaces with the Jena framework to provide a front-end application programming interface (API). GEMS, instead, exploits clusters built with commodity hardware that are cheaper to acquire and maintain and that are able to evolve more rapidly than custom hardware.

## 8.3 GEMS OVERVIEW

Figure 8.1 presents a high-level overview of the architecture of GEMS. The figure shows the main component of GEMS: the SPARQL-to-C++ compiler, the Semantic Graph Library (SGLib), and the GMT runtime. An analyst can load one or more graph databases, write the SPARQL queries, and execute them through a client web interface that connects to a server on the cluster.

The top layer of GEMS consists of the compilation phases: The compiler transforms the input SPARQL queries into intermediate representations that are analyzed for optimization opportunities. Potential optimization opportunities are discovered at multiple levels. Depending on statistics of the data sets, certain query clauses can be moved, enabling early pruning of the search. Then, the optimized intermediate representation is converted into C++ code that contains calls to the SGLib API. SGLib APIs completely hide the low-level APIs of GMT, exposing to the compiler a lean, simple, pseudosequential shared-memory programming model. SGLib manages the graph database and query execution. SGLib

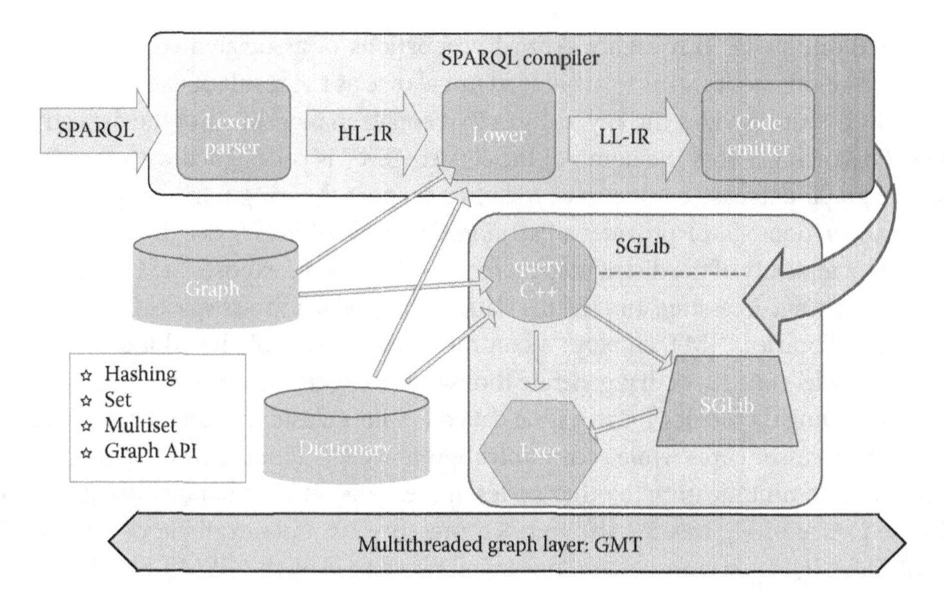


FIGURE 8.1 High-level overview of Graph Engine for Multithreaded Systems (GEMS). HL-IR, high level-intermediate representation; LL-IR, low level-intermediate representation.

generates the graph database and the related dictionary by ingesting the triples. Triples can, for example, be RDF triples stored in N-Triples format.

The approach implemented by our system to extract information from the semantic graph database is to solve a structural graph pattern-matching problem. GEMS employs a variation of Ullmann's (1976) subgraph isomorphism algorithm. The approach basically enumerates all the possible mappings of the graph pattern vertices to those in the graph data set through a depth-first tree search. A path from root to leaves in the resulting tree denotes a complete mapping. If all the vertices that are neighbors in a path are also neighbors both in the graph pattern and in the graph data set (i.e., adjacency is preserved), then the path represents a match. Even if the resulting solution space has exponential complexity, the algorithm can prune early subtrees that do not lead to feasible mappings. Moreover, the compiler can perform further optimizations by reasoning on the query structure and the data set statistics.

The data structures implemented within the SGLib layer support the operation of a query search using the features provided by the GMT layer. When loaded, the size of the SGLib data structures is expected to be larger than what can fit into a single node's memory, and these are therefore implemented using the global memory of GMT. The two most fundamental data structures are the graph and dictionary. Supplemental data structures are the array and table.

The ingest phase of the GEMS workflow initializes a graph and dictionary. The dictionary is used to encode string labels with unique integer identifiers (UIDs). Therefore, each RDF triple is encoded as a sequence of three UIDs. The graph indexes RDF triples in subject–predicate–object and object–predicate–subject orders. Each index is range-partitioned so that each node in the cluster gets an almost equal number of triples. Subject–predicate and object–predicate pairs with highly skewed proportions of associated triples are specially distributed among nodes so as to avoid load imbalance as the result of data skew.

The GMT layer provides the key features that enable management of the data structures and load balancing across the nodes of the cluster. GMT is codesigned with the upper layers of the graph database engine so as to better support the irregularity of graph pattern-matching operations. GMT provides a Partitioned Global Address Space (PGAS) data model, hiding the complexity of the distributed memory system. GMT exposes to SGLib an API that permits allocating, accessing, and freeing data in the global address space. Differently from other PGAS libraries, GMT employs a control model typical of shared-memory systems: fork-join parallel constructs that generate thousands of lightweight tasks. These lightweight tasks allow hiding the latency for accessing data on remote cluster nodes; they are switched in and out of processor cores while communication proceeds. Finally, GMT aggregates operations before communicating to other nodes, increasing network bandwidth utilization.

The sequence of Figures 8.2 through 8.6 presents a simple example of a graph database, of a SPARQL query, and of the conversion performed by the GEMS' SPARQL-to-C++ compiler.

Figure 8.2 shows the data set in N-Triples format (a common serialization format for RDF) (Carothers and Seaborne 2014). Figure 8.3 shows the corresponding graph representation

```
PERSON1 has_name    JOHN .
PERSON1 has_address ADDR1 .
PERSON1 owns        CAR1 .
CAR1 of_type SEDAN .
CAR1 year           1997 .
PERSON2 has_name BOB .
PERSON2 has_address ADDR2 .
PERSON2 owns        CAR2 .
CAR2 of_type SEDAN .
CAR2  year          2012 .
PERSON2 owns        CAR3 .
CAR3 of_type SUV .
CAR3 year           2001 .
PERSON3 has_name MARY .
PERSON3 has_address ADDR2 .
PERSON3 owns        CAR3 .
```

FIGURE 8.2 Data set in simplified N-Triples format.

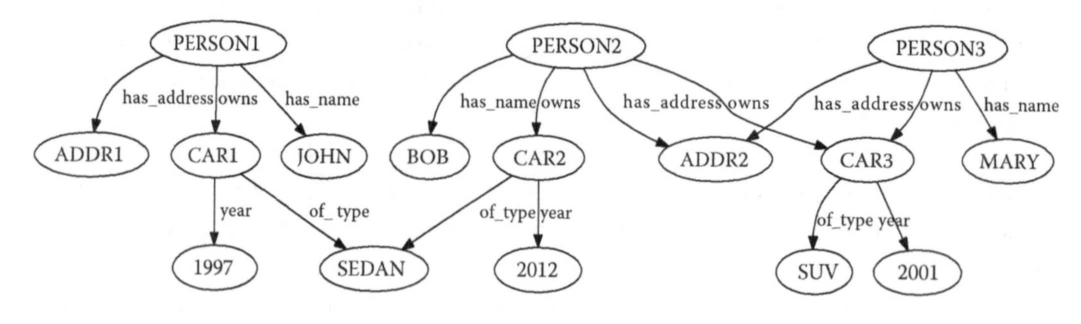


FIGURE 8.3 RDF graph.

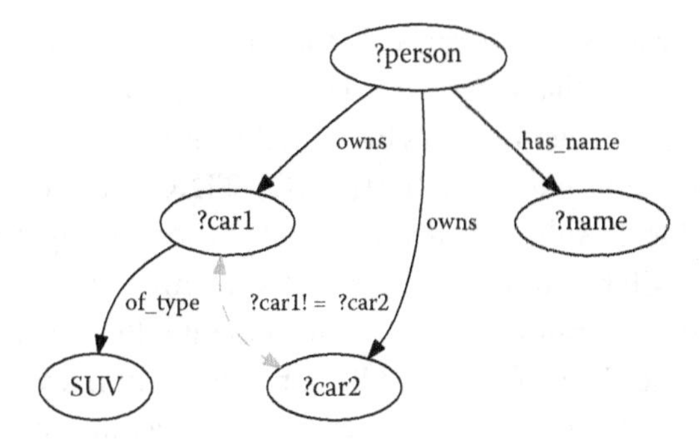


FIGURE 8.4 SPARQL query.

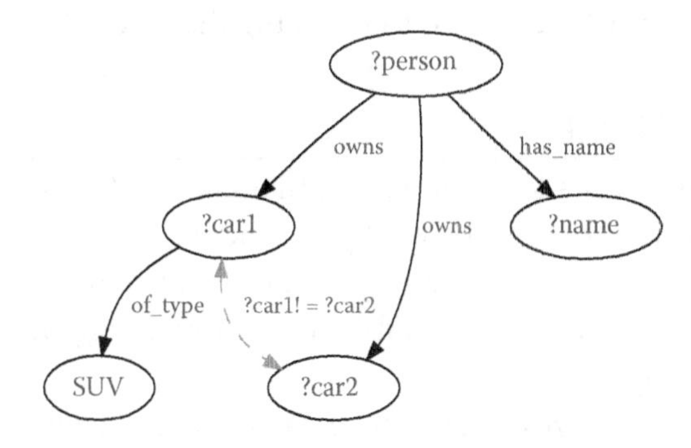


FIGURE 8.5 Pattern graph of the SPARQL query.

```
has_name = get_label("has_name")
of_type = get_label("of_type")
owns = get_label("owns")
suv = get_label("SUV")
forall e1 in edges(*, of_type, suv)
        ?car1 = source_node(e1)
        forall e2 in edges(*, owns, ?car1)
                ?person = source_node(e2)
                forall e3 in edges(?person, owns, *)
                        ?car2 = target_node(e3)
                        if (?car1 != ?car2)
                                forall e4 in edges(?person,has_name,*)
                                        ?name = target_node(e4)
                                        tuples.add(<?name>)
```

FIGURE 8.6 Translated pseudocode of the query.

of the data set. Figure 8.4 shows the SPARQL description of the query. Figure 8.5 illustrates the corresponding graph of the query. Finally, Figure 8.6 shows a pseudocode conceptually similar to the C/C++ generated by the GEMS' compiler and executed, through SGLib, by the GMT runtime library.

As illustrated in Figure 8.6, the bulk of the query is executed as a parallel graph walk. The forAll method is used to call a function for all matching edges, in parallel. Thus, the graph walk can be conceptualized as nested loops. At the end (or "bottom") of a graph walk, results are buffered in a loader object that is associated with a table object. When all the parallel edge traversals are complete, the loader finalizes by actually inserting the results into its associated table. At this point, operations like deduplication (DISTINCT) or grouping (GROUP BY) are performed using the table. Results are effectively structs containing different variables of the results. Many variables will be bound to UIDs, but some may contain primitive values.

GEMS has minimal system-level library requirements: Besides Pthreads, it only needs message passing interface (MPI) for the GMT communication layer and Python for some compiler phases and for glue scripts. The current implementation of GEMS also requires x86-compatible processors because GMT employs optimized context switching routines. However, because of the limited requirements in terms of libraries, the porting to other architectures should be mostly confined to replacing these context switching routines.

## 8.4 GMT ARCHITECTURE

The GMT runtime system is a key component that enables GEMS to operate on a larger data set while maintaining constant throughput as nodes are added to the cluster.

GMT relies on three main elements to provide its scalability: a *global address space*, latency tolerance through fine-grained software *multithreading*, and message *aggregation* (also known as coalescing).

GMT provides a global address space across the various cluster nodes by implementing a custom PGAS data substrate. The PGAS model relieves the other layers of GEMS (and the developers) from partitioning the data structures and from orchestrating communication. PGAS does not neglect completely the concept of locality but instead acknowledges that there are overheads in accessing data stored on other cluster nodes. Indeed, PGAS libraries normally enable accessing data through put and get operations that move the data from/to a globally shared array to/from a node-local memory location.

Message aggregation maximizes network bandwidth utilization, despite the small data accesses typical of graph methods. Aggregation is completely transparent to the developer.

Fine-grained multithreading allows hiding the latency for remote data transfers, and the additional latency for aggregation, by exploiting the inherent parallelism of graph algorithms.

Figure 8.7 shows the high-level design of GMT. Each node in the cluster executes an instance of GMT. Different instances communicate through *commands*, which describe data, synchronization, and thread management operations. GMT is a *parallel* runtime, in the sense that it exploits threads also to perform its operations, with three types of specialized threads. The main idea is to exploit the cores of modern processors to support the functionalities of the runtime. The specialized threads are as follows:

1. *Worker*: executes application code, in the form of lightweight user tasks, and generates commands directed to other nodes
2. *Helper*: manages global address space and synchronization and handles incoming commands from other nodes

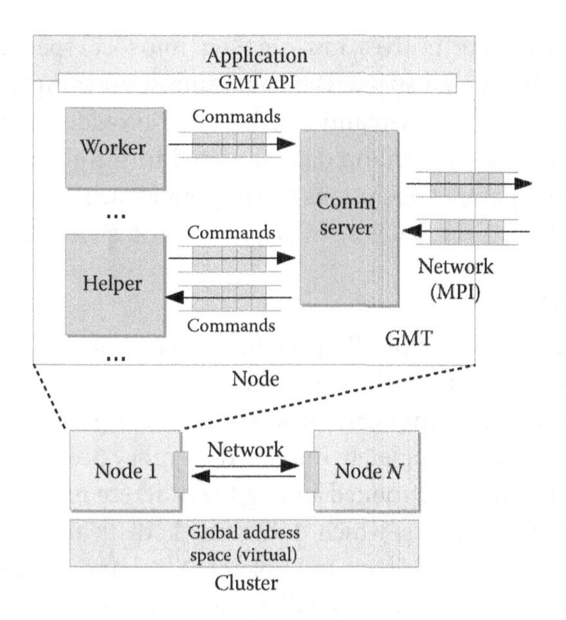


FIGURE 8.7 High-level GMT architecture.

3. *Communication server*: end point for the network, which manages all incoming/outgoing communication at the node level in the form of network messages, which contain the commands

The specialized threads are implemented as *POSIX* threads, each one pinned to a core. The communication server employs MPI to send and receive messages to and from other nodes. There are multiple workers and helpers per node. Usually, we use an equal number of workers and helpers, but this is one of the parameters that can be adjusted, depending on empirical testing on the target machine. There is, instead, a single communication server because while building the runtime, we verified that a single communication server is already able to provide the peak MPI bandwidth of the network with reasonably sized packets, removing the need to manage multiple network end points (which, in turn, may bring challenges and further overheads for synchronization).

SGLib data structures are implemented using shared arrays in GMT's global address space. Among them, there are the graph data structure and the terms dictionary. The SPARQL-to-C++ compiler assumes operation on a shared-memory system and does not need to reason about the physical partitioning of the database. However, as is common in PGAS libraries, GMT also exposes locality information, allowing reduction in data movements whenever possible. Because graph exploration algorithms mostly have loops that run through edge or vertex lists (as the pseudocode in Figure 8.6 shows), GMT provides a parallel loop construct that maps loop iterations to lightweight tasks. GMT supports task generation from nested loops and allows specifying the number of iterations of a loop mapped to a task. GMT also allows controlling code locality, enabling spawning (or moving) of tasks on preselected nodes, instead of moving data. SGLib routines exploit these features to better manage its internal data structures. SGLib routines access data through *put* and *get* communication primitives, moving them into local space for manipulation and writing them back to the global space. The communication primitives are available with both blocking and nonblocking semantics. GMT also provides atomic operations, such as atomic addition and test-and-set, on data allocated in the global address space. SGLib exploits them to protect parallel operations on the graph data sets and to implement global synchronization constructs for database management and querying.

### 8.4.1 GMT: Aggregation

Graph exploration algorithms typically present *fine-grained* data accesses. They mainly operate through for-loops that run through edges and/or vertices. Edges usually are pointers that connect one vertex to another. Depending on the graph structure, each pointer may point to a location in a completely unrelated memory area. On distributed memory systems, the data structure is partitioned among the different memories of each node. This is true also with PGAS substrates, which only provide distributed memory abstractions. Expert programmers thus usually have to implement by hand optimizations to aggregate requests and reduce the overhead due to small network transactions. GMT hides these complexities from the other layers of GEMS by implementing automatic message aggregation.

GMT collects commands directed towards the same destination nodes in aggregation queues. GMT copies commands and their related data (e.g., values requested from the global address space with a get) into aggregation buffers and sends them in bulk. Commands are then unpacked and executed at the destination node. At the node level, GMT employs high-throughput, nonblocking aggregation queues, which support concurrent access from multiple workers and helpers. Accessing these queues for every generated command would have a very high cost. Thus, GMT employs a two-level aggregation mechanism: Workers (or helpers) initially collect commands in local command blocks, and then they insert command blocks into the aggregation queues. Figure 8.8 shows how the aggregation mechanism operates. When aggregation starts, workers (or helpers) request a preallocated command block from the command block pool (1). Command blocks are preallocated and reused for performance reasons. Commands generated during program execution are collected into the local command block (2). A command block is pushed

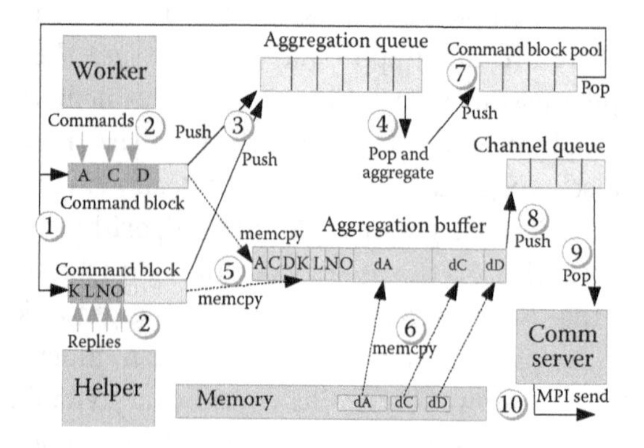


FIGURE 8.8 Message aggregation.

into aggregation queues when (a) it is full or (b) it has been waiting longer than a predetermined time interval. Condition (a) is true when all the available entries are occupied with commands or when the equivalent size in bytes of the commands (including any attached data) reaches the size of the aggregation buffer. Condition (b) allows setting a configurable upper bound for the latency added by aggregation. After pushing a command block, when a worker or a helper finds that the aggregation queue has sufficient data to fill an aggregation buffer, it starts popping command blocks from the aggregation queue and copying them with the related data into an aggregation buffer (4, 5, and 6). Aggregation buffers also are preallocated and recycled to save memory space and eliminate allocation overhead. After the copy, command blocks are returned to the command block pool (7). When the aggregation buffer is full, the worker (or helper) pushes it into a channel queue (8). Channel queues are high-throughput, single-producer, single-consumer queues that workers and helpers use to exchange data with the communication server. If the communication server finds a new aggregation buffer in one of the channel queues, it pops it (9) and performs a nonblocking MPI send (10). The aggregation buffer is then returned into the pool of free aggregation buffers.

The size of aggregation buffers and the time intervals for pushing out aggregated data are configurable parameters that depend on the interconnection of the cluster on which GEMS runs. Thumb rules to set these parameters are as follows: Buffers should be sufficiently large to maximize network throughput, while time intervals should not increase the latency over the values maskable through multithreading. The right values may be derived through empirical testing on the target cluster with toy examples.

### 8.4.2 GMT: Fine-Grained Multithreading

Concurrency, through fine-grained software multithreading, allows tolerance for both the latency for accessing data on remote nodes and the added latency for aggregating communication operations. Each worker executes a set of GMT tasks. The worker switches among tasks' contexts every time it generates a blocking command that requires a remote memory operation. The task that generated the command executes again only when the command itself completes (i.e., it gets a reply back from the remote node). In case of nonblocking commands, the task continues executing until it encounters a wait primitive.

GMT implements custom context switching primitives that avoid some of the lengthy operations (e.g., saving and restoring the signal mask) performed by the standard *libc* context switching routines.

Figure 8.9 schematically shows how GMT executes a task. A node receives a message containing a *spawn* command (1) that is generated by a worker on a remote node when encountering a parallel construct. The communication server passes the buffer containing the command to a helper that parses the buffer and executes the command (2). The helper then creates an *iteration block* (itb). The itb is a data structure that contains the function to execute, the arguments of the function itself, and number of tasks that will execute the function. This way of representing a set of tasks avoids the cost of creating a large number of function arguments and sending them over the network. In the following step, the helper pushes the itb into the itb queue (3). Then, an idle worker pops an itb from the itb

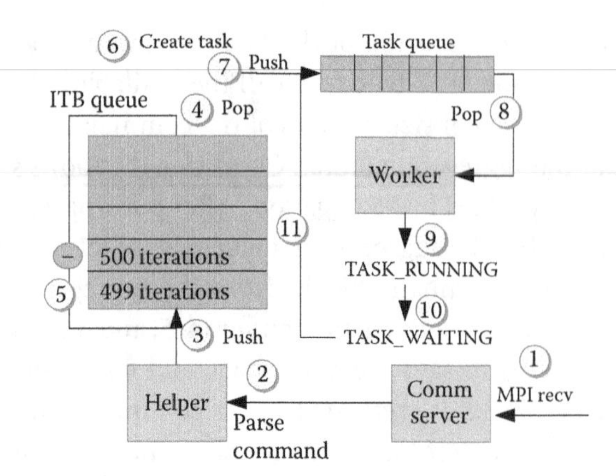


FIGURE 8.9 Fine-grained task management in GMT. ITB, iteration block.

queue (5), and it decreases the counter of *t* and pushes it back into the queue (6). The worker creates *t* tasks (6) and pushes them into its private task queue (7).

At this point, the idle worker can pop a task from its task queue (8). If the task is executable (i.e., all the remote operations completed), the worker restores the task's context and executes it (9). Otherwise, it pushes the task back into the task queue. If the task contains a blocking remote request, the task enters a waiting state (10) and is reinserted into the task queue for future execution (11).

This mechanism provides load balancing at the node level because each worker gets new tasks from the itb queue as soon as its task queue is empty. At the cluster level, GMT evenly splits tasks across nodes when it encounters a parallel for-loop construct.

## 8.5 EXPERIMENTAL RESULTS

We evaluated GEMS on the Olympus supercomputer at Pacific Northwest National Laboratory's Institutional Computing center, listed in Top500.org (Top500.org, n.d.). Olympus is a cluster of 604 nodes interconnected through a quad data rate (QDR) InfiniBand switch with 648 ports (theoretical bandwidth peak of 4 GB/s). Each of Olympus' nodes features two AMD Opteron 6272 processors at 2.1 GHz and 64 GB of double data rate 3 (DDR3) memory clocked at 1600 MHz. Each socket hosts eight processor modules (two integer cores, one floating-point core per module) on two different dies, for a total of 32 integer cores per node. Dies and sockets are interconnected through HyperTransport.

We configured a GEMS stack with 15 workers, 15 helpers, and 1 communication server per node. Each worker hosts up to 1024 lightweight tasks. We measured the MPI bandwidth of Olympus with the OSU Micro-Benchmarks 3.9 (Ohio State University, n.d.), reaching the peak of 2.8 GB/s with messages of at least 64 KB. Therefore, we set the aggregation buffer size at 64 KB. Each communication channel hosts up to four buffers. There are two channel queues per helper and one channel queue per worker.

We initially present some synthetic benchmarks of the runtime, highlighting the combined effects of multithreading and aggregation to maximize network bandwidth utilization. We then show experimental results of the whole GEMS system on a well-established benchmark, the Berlin SPARQL Benchmark (BSBM) (Bizer and Schultz 2009), and on a data set currently curated at Pacific Northwest National Laboratory (PNNL) for the Resource Discovery for Extreme Scale Collaboration (RDESC) project.

### 8.5.1 Synthetic Benchmarks

Figures 8.10 and 8.11 show the transfer rates reached by GMT with small messages (from 8 to 128 bytes) when increasing the number of tasks. Every task executes 4096 blocking put operations. Figure 8.10 shows the bandwidth between two nodes, and Figure 8.11 shows the

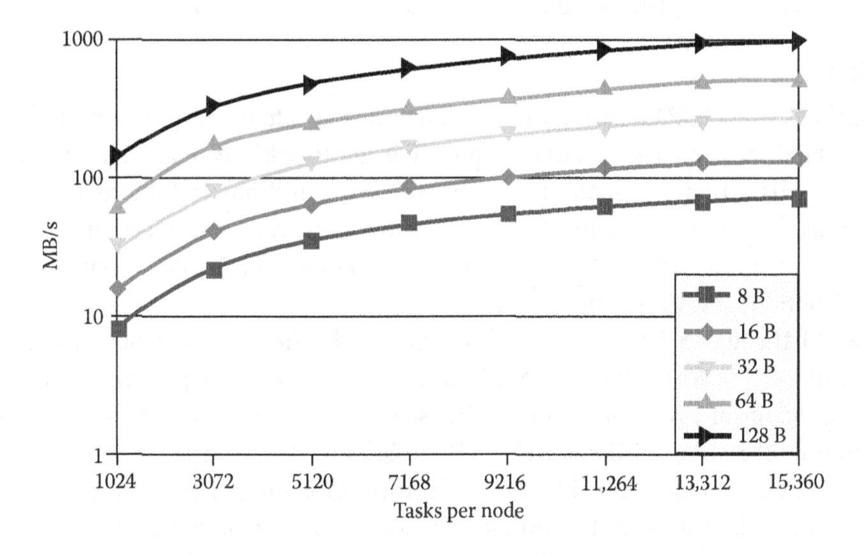


FIGURE 8.10 Transfer rates of put operations between two nodes while increasing concurrency.

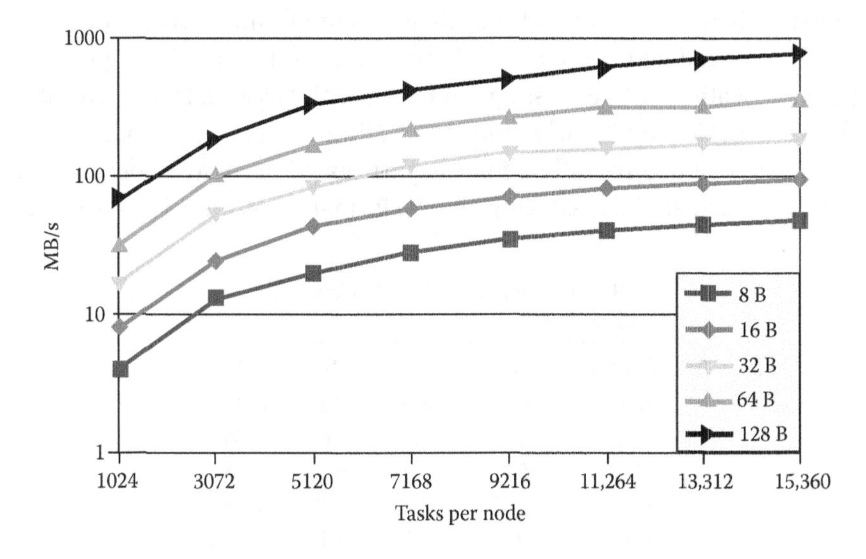


FIGURE 8.11 Transfer rates of put operations among 128 nodes (one to all) while increasing concurrency.

bandwidth among 128 nodes. The figures show how increasing the concurrency increases the transfer rates because there is a higher number of messages that GMT can aggregate. For example, across two nodes (Figure 8.10) with 1024 tasks each, puts of 8 bytes reach a bandwidth of 8.55 MB/s. With 15,360 tasks, instead, GMT reaches 72.48 MB/s. When increasing message sizes to 128 bytes, 15,360 tasks provide almost 1 GB/s. For reference, 32 MPI processes with 128 bytes messages only reach 72.26 MB/s. With more destination nodes, the probability of aggregating enough data to fill a buffer for a specific remote node decreases. Although there is a slight degradation, Figure 8.11 shows that GMT is still very effective. For example, 15,360 tasks with 16 bytes messages reach 139.78 MB/s, while 32 MPI processes only provide up to 9.63 MB/s.

### 8.5.2 BSBM

BSBM defines a set of SPARQL queries and data sets to evaluate the performance of semantic graph databases and systems that map RDF into other kinds of storage systems. Berlin data sets are based on an e-commerce use case with millions to billions of commercial transactions, involving many product types, producers, vendors, offers, and reviews. We run queries 1 through 6 (Q1–Q6) of the business intelligence use case on data sets with 100 million, 1 billion, and 10 billion triples.

Tables 8.1 through 8.3 respectively show the build time of the database and the execution time of the queries on 100 million, 1 billion, and 10 billion triples, while progressively increasing the number of cluster nodes. The sizes of the input files respectively are 21 GB (100 million), 206 GB (1 billion), and 2 TB (10 billion). In all cases, the build time scales with the number of nodes. Considering all the three tables together, we can appreciate how GEMS scales in data set sizes by adding new nodes and how it can exploit the additional parallelism available. With 100 million triples, Q1 and Q3 scale for all the experiments up to 16 nodes. Increasing the number of nodes for the other queries, instead, provides constant or slightly worse execution time. Their execution time is very short (under 0.5 s), and the small data set does not provide sufficient data parallelism. These queries only have two graph walks with two-level nesting, and even with larger data sets, GEMS is able to exploit all the available parallelism already with a limited number of nodes. Furthermore, the database has the same overall size but is partitioned on more nodes; thus, the communication increases, slightly reducing the performance. With 1 billion triples, we see

TABLE 8.1 BSBM 100 Million Triples, 2 to 16 Cluster Nodes

| Nodes | 2 | 4 | 8 | 16 |
|---|---|---|---|---|
| Build | 199.00 | 106.99 | 59.85 | 33.42 |
| Q1 | 1.83 | 1.12 | 0.67 | 0.40 |
| Q2 | 0.07 | 0.07 | 0.07 | 0.05 |
| Q3 | 4.07 | 2.73 | 1.17 | 0.65 |
| Q4 | 0.13 | 0.13 | 0.14 | 0.15 |
| Q5 | 0.07 | 0.07 | 0.07 | 0.11 |
| Q6 | 0.01 | 0.02 | 0.02 | 0.03 |

TABLE 8.2 BSBM 1 Billion Triples, 8 to 64 Cluster Nodes

| Nodes | 8 | 16 | 32 | 64 |
|---|---|---|---|---|
| Build | 628.87 | 350.47 | 200.54 | 136.69 |
| Q1 | 5.65 | 3.09 | 1.93 | 2.32 |
| Q2 | 0.30 | 0.34 | 0.23 | 0.35 |
| Q3 | 12.79 | 6.88 | 4.50 | 2.76 |
| Q4 | 0.31 | 0.25 | 0.22 | 0.27 |
| Q5 | 0.11 | 0.12 | 0.14 | 0.18 |
| Q6 | 0.02 | 0.03 | 0.04 | 0.05 |

TABLE 8.3 BSBM 10 Billion Triples, 64 and 128 Cluster Nodes

| Nodes | 64 | 128 |
|---|---|---|
| Build | 1066.27 | 806.55 |
| Q1 | 27.14 | 39.78 |
| Q2 | 1.48 | 1.91 |
| Q3 | 24.27 | 18.32 |
| Q4 | 2.33 | 2.91 |
| Q5 | 2.13 | 2.82 |
| Q6 | 0.40 | 0.54 |

similar behavior. In this case, however, Q1 stops scaling at 32 nodes. With 64 nodes, GEMS can execute queries on 10 billion triples. Q3 still scales in performance up to 128 nodes, while the other queries, except Q1, approximately maintain stable performance. Q1 experiences the highest decrease in performance when using 128 nodes because its tasks present higher communication intensity than the other queries, and GEMS already exploited all the available parallelism with 64 nodes. These data confirm that GEMS can maintain constant throughput when running sets of mixed queries in parallel, that is, in typical database usage.

### 8.5.3 RDESC

In this section, we describe some results in testing the use of GEMS to answer queries for the RDESC project (Table 8.4). The RDESC project is a 3-year effort funded by the

TABLE 8.4 Specifications of the Query Executed on the RDESC Data Set

| | |
|---|---|
| Q1 | Find all instruments related to data resources containing measurements related to soil moisture |
| Q2 | Find all locations having measurements related to soil moisture that are taken with at least ten instruments |
| Q3 | Find locations (spatial locations) that have daily soil moisture profile data since 1990 with at least 10 points |
| Q4 | Find locations that have soil moistures profile data ranked by how many resources are available for that location |

Department of Energy (DOE) Advanced Scientific Computing Research (ASCR) program. RDESC aims to provide a prototypical "collaborator" system to facilitate the discovery of science resources. The project involves curating diverse metadata about soil, atmospheric, and climate data sets. The quantity and complexity of the metadata are ultimately so great that they present data scaling challenges. For example, curating metadata about data sets and data streams from the Atmospheric Radiation Measurement (ARM) climate research facility generates billions of relationships between abstract entities (e.g., data sets, data streams, sites, facilities, variables/measured properties). At present, the RDESC metadata consists of nearly 1.4 billion RDF triples, mostly metadata about ARM data streams and GeoNames locations, although it includes, to a lesser proportion, descriptions of Global Change Master Directory (GCMD) (National Aeronautics and Space Administration 2013) locations, key words, and data sets, as well as metadata from the International Soil Moisture Network (ISMN) (International Soil Moisture Network, n.d.).

| Nodes | 8 | | 16 | | 32 | |
|---|---|---|---|---|---|---|
| 1 run | Exe [s] | Throughput [q/s] | Exe [s] | Throughput [q/s] | Exe [s] | Throughput [q/s] |
| Load | 2577.45909 | | 1388.4894 | | 1039.5300 | |
| Build | 607.482135 | | 301.1487 | | 161.8761 | |
| Q1 | 0.0059 | 169.2873 | 0.0076 | 131.6247 | 0.0066 | 151.9955 |
| Q2 | 0.0078 | 127.729 | 0.0088 | 113.3613 | 0.0093 | 107.9400 |
| Q3 | 0.0017 | 592.6100 | 0.0020 | 477.9567 | 0.0017 | 592.2836 |
| Q4 | 0.0154 | 64.9224 | 0.0119 | 83.7866 | 0.0119 | 84.2356 |
| **Nodes** | **8** | | **16** | | **32** | |
| **Avg. 100 runs** | **Exe [s]** | **Throughput [q/s]** | **Exe [s]** | **Throughput [q/s]** | **Exe [s]** | **Throughput [q/s]** |
| Load | 2593.6120 | | 1367.2713 | | 1062.4514 | |
| Build | 583.0361 | | 303.8100 | | 153.7527 | |
| Q1 | 0.0057 | 174.1774 | 0.0059 | 168.5413 | 0.0070 | 142.5891 |
| Q2 | 0.0074 | 135.0213 | 0.0080 | 124.4752 | 0.0091 | 109.6614 |
| Q3 | 0.0017 | 582.8798 | 0.0044 | 229.1323 | 0.0017 | 640.1768 |
| Q4 | 0.0124 | 80.3632 | 0.0147 | 67.8223 | 0.0101 | 98.6041 |

Using 8, 16, and 32 nodes of Olympus (equating to 120, 240, and 480 GMT workers, respectively), the Q3 correctly produces no results since there are currently no data that match the query, in the range of 0.0017 to 0.0044 s. Although there are no results, the query still needs to explore several levels of the graph. However, our approach allows early pruning of the search, significantly limiting the execution time.

Q1, Q2, and Q4, instead, under the same circumstances, run in the ranges of 0.0059–0.0076, 0.0074–0.0093, and 0.0101–0.0154 s with 76, 1, and 68 results, respectively.

For the query in Q2, the one location with at least 10 "points" (instruments) for soil moisture data is the ARM facility sgp.X1 (that is, the first experimental facility in ARM's

Southern Great Plains site). The reason these queries take approximately the same amount of time regardless of the number of nodes is that the amount of parallelism in executing the query is limited so that eight nodes is sufficient for fully exploiting the parallelism in the query (in the way that we perform the query). We could not run the query on four nodes or less due to the need for memory of the graph and dictionary. This emphasizes our point that the need for more memory exceeds the need for more parallel computation. We needed the eight nodes not necessarily to make the query faster but, rather, to keep the data and query in memory to make the query *feasible*.

## 8.6 CONCLUSIONS

In this chapter, we presented GEMS, a full software stack for semantic graph databases on commodity clusters. Different from other solutions, GEMS proposes an integrated approach that primarily utilizes graph-based methods across all the layers of its stack. GEMS includes a SPARQL-to-C++ compiler, a library of algorithms and data structures, and a custom runtime. The custom runtime system (GMT) provides to all the other layers several features that simplify the implementation of the exploration methods and makes more efficient their execution on commodity clusters. GMT provides a global address space, fine-grained multithreading (to tolerate latencies for accessing data on remote nodes), remote message aggregation (to maximize network bandwidth utilization), and load balancing. We have demonstrated how this integrated approach provides scaling in size and performance as more nodes are added to the cluster on two example data sets (the BSBM and the RDESC data set).

## REFERENCES

*Apache Giraph*. (n.d.). Retrieved May 1, 2014, from http://incubator.apache.org/giraph/.

*Apache Jena—Home*. (n.d.). Retrieved March 19, 2014, from https://jena.apache.org/.

*ARQ—A SPARQL Processor for Jena*. (n.d.). Retrieved May 1, 2014, from http://jena.sourceforge.net/ARQ/.

Bizer, C., and Schultz, A. (2009). The Berlin SPARQL benchmark. *Int. J. Semantic Web Inf. Syst., 5* (2), 1–24.

Carothers, G., and Seaborne, A. (2014). *RDF 1.1 N-Triples*. Retrieved March 19, 2014, from http://www.w3.org/TR/2014/REC-n-triples-20140225/.

*GraphLab*. (n.d.). Retrieved May 1, 2014, from http://graphlab.org.

Harth, A., Umbrich, J., Hogan, A., and Decker, S. (2007). YARS2: A federated repository for querying graph structured data from the web. *ISWC '07/ASWC '07: 6th International Semantic Web and 2nd Asian Semantic Web Conference* (pp. 211–224).

*International Soil Moisture Network*. (n.d.). Retrieved February 17, 2014, from http://ismn.geo.tuwien.ac.at.

Klyne, G., Carroll, J. J., and McBride, B. (2004). *Resource Description Framework (RDF): Concepts and Abstract Syntax*. Retrieved December 2, 2013, from http://www.w3.org/TR/2004/REC-rdf-concepts-20040210/.

Malewicz, G., Austern, M. H., Bik, A. J., Dehnert, J. C., Horn, I., Leiser, N. et al. (2010). Pregel: A system for large-scale graph processing. *SIGMOD '10: ACM International Conference on Management of Data* (pp. 135–146).

National Aeronautics and Space Administration. (2013). *Global Change Master Directory, Online, Version 9.9*. Retrieved May 1, 2014, from http://gcmd.nasa.gov.

Ohio State University. (n.d.). *OSU Micro-Benchmarks.* Retrieved May 1, 2014, from http://mvapich.cse.ohio-state.edu/benchmarks/.
*openRDF.org.* (n.d.). Retrieved May 1, 2014, from http://www.openrdf.org.
*Redland RDF Libraries.* (n.d.). Retrieved May 1, 2014, from http://librdf.org.
Rohloff, K., and Schantz, R. E. (2010). High-performance, massively scalable distributed systems using the MapReduce software framework: The SHARD triple-store. *PSI EtA '10: Programming Support Innovations for Emerging Distributed Applications, 4,* Reno (pp. 1–5).
Top500.org. (n.d.). *PNNL's Olympus Entry.* Retrieved May 1, 2014, from http://www.top500.org/system/177790.
Ullmann, J. (1976). An algorithm for subgraph isomorphism. *J. ACM, 23* (1), 31–42.
*Virtuoso Universal Server.* (n.d.). Retrieved May 1, 2014, from http://virtuoso.openlinksw.com.
W3C SPARQL Working Group. (2013). *SPARQL 1.1 Overview.* Retrieved February 18, 2014, from http://www.w3.org/TR/2013/REC-sparql11-overview-20130321/.
Weaver, J. (2012). A scalability metric for parallel computations on large, growing datasets (like the web). *Proceedings of the Joint Workshop on Scalable and High-Performance Semantic Web Systems,* Boston.
YarcData, Inc. (n.d.) *Urika Big Data Graph Appliance.* Retrieved May 1, 2014, from http://www.cray.com/Products/BigData/uRiKA.aspx.

CHAPTER 9

# KSC-net

## *Community Detection for Big Data Networks*

Raghvendra Mall and Johan A.K. Suykens

CONTENTS



ABSTRACT

In this chapter, we demonstrate the applicability of the kernel spectral clustering (KSC) method for community detection in Big Data networks. We give a practical exposition of the KSC method [1] on large-scale synthetic and real-world networks with up to $10^6$ nodes and $10^7$ edges. The KSC method uses a primal–dual framework to construct a model on a smaller subset of the Big Data network. The original large-scale kernel matrix cannot fit in memory. So we select smaller subgraphs using a fast and unique representative subset (FURS) selection technique as proposed in Reference 2. These subsets are used for training and validation, respectively, to build the model and obtain the model parameters. It results in a powerful out-of-sample extensions property, which allows inferring of the community affiliation for unseen nodes. The KSC model requires a kernel function, which can have kernel parameters and what is needed to identify the number of clusters $k$ in the network. A memory-efficient and computationally efficient model selection technique named balanced angular fitting (BAF) based on angular similarity in the eigenspace is proposed in Reference 1. Another parameter-free KSC model is proposed in Reference 3. In Reference 3, the model selection technique exploits the structure of projections in eigenspace to automatically identify the number of clusters and suggests that a normalized linear kernel is sufficient for networks with millions of nodes. This model selection technique uses the concept of entropy and balanced clusters for identifying the number of clusters $k$.

We then describe our software called KSC-net, which obtains the representative subset by FURS, builds the KSC model, performs one of the two (BAF and parameter-free) model selection techniques, and uses out-of-sample extensions for community affiliation for the Big Data network.

## 9.1 INTRODUCTION

In the modern era, with the proliferation of data and ease of its availability with the advancement of technology, the concept of Big Data has emerged. Big Data refers to a massive amount of information, which can be collected by means of cheap sensors and wide usage of the Internet. In this chapter, we focus on Big Data networks that are ubiquitous in current life. Their omnipresence is confirmed by the modern massive online social networks, communication networks, collaboration networks, financial networks, and so forth.

Figure 9.1 represents a synthetic network with 5000 nodes and a real-world YouTube network with over a million nodes. The networks are visualized using the Gephi software (https://gephi.org/).

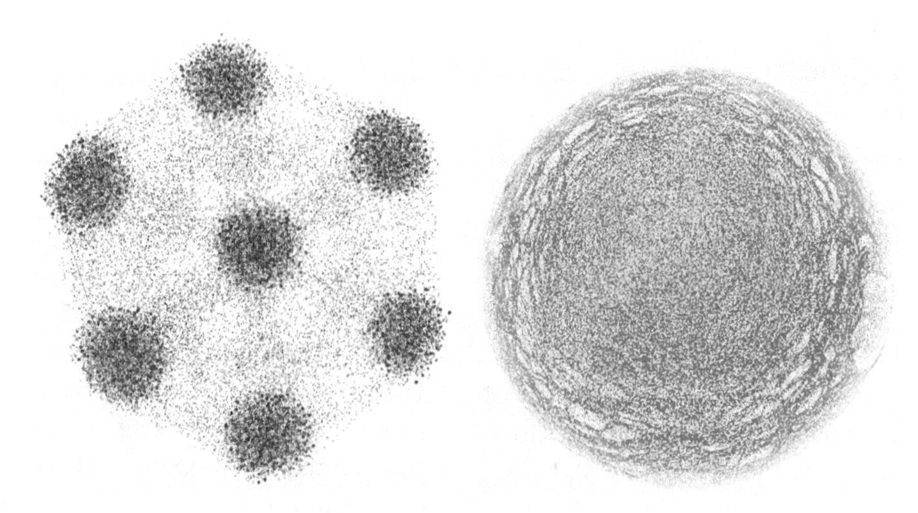

FIGURE 9.1 A synthetic network and a real-world YouTube network.

Real-world complex networks are represented as graphs $G(V, E)$ where the vertices $V$ represent the entities and the edges $E$ represent the relationship between these entities. For example, in a scientific collaboration network, the entities are the researchers, and the presence or absence of an edge between two researchers depicts whether they have collaborated or not in the given network. Real-life networks exhibit community-like structure. This means nodes that are part of one community are densely connected to each other and sparsely connected to nodes belonging to other communities. This problem of community detection can also be framed as graph partitioning and graph clustering [4] and, of late, has received wide attention [5–16]. Among these techniques, one class of methods used for community detection is referred to as spectral clustering [13–16].

In spectral clustering, an eigen-decomposition of the graph Laplacian matrix ($L$) is performed. $L$ is derived from the similarity or affinity matrix of the nodes in the network. Once the eigenvectors are obtained, the communities can be detected using the $k$-means clustering technique. The major disadvantage of the spectral clustering technique is that we need to create a large $N \times N$ kernel matrix for the entire network to perform unsupervised community detection. Here, $N$ represents the number of nodes in the large-scale network. However, as the size of network increases, calculating and storing the $N \times N$ matrix becomes computationally infeasible.

A spectral clustering method based on weighted kernel principal component analysis (PCA) with a primal–dual framework is proposed in Reference 17. The formulation resulted in a model built on a representative subset of the data with a powerful out-of-sample extensions property. This representative subset captures the inherent cluster structure present in the data set. This property allows community affiliation for previously unseen data points in the data set. The kernel spectral clustering (KSC) method was extended for community detection on moderate-size graphs in Reference 11 and for large-scale graphs in Reference 1.

We use the fast and unique representative subset (FURS) selection technique introduced in Reference 2 for selection of the subgraphs on which we build our training and validation models (Figure 9.2). An important step in the KSC method is to estimate the model parameters, that is, kernel parameters, if any, and the number of communities $k$ in the large-scale network. In the case of large-scale networks, the normalized linear kernel is an effective kernel

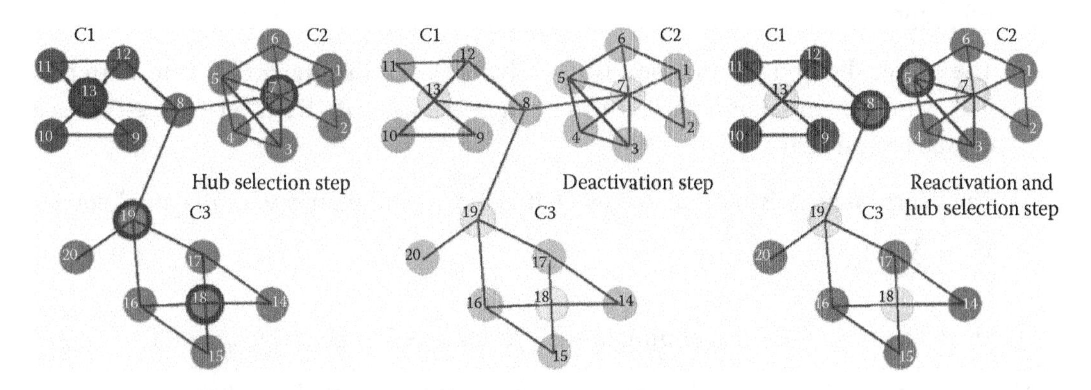


FIGURE 9.2 Steps of FURS for a sample of 20 nodes.

as shown in Reference 18. Thus, we use the normalized linear kernel, which is parameter-less in our experiments. This saves us from tuning for an additional kernel parameter. In order to estimate the number of communities $k$, we exploit the eigen-projections in eigenspace to come up with a balanced angular fitting (BAF) criterion [1] and another self-tuned approach [3] where we use the concept of entropy and balance to automatically get $k$.

In this chapter, we give a practical exposition of the KSC method by first briefly explaining the steps involved and then showcasing its usage with the KSC-net software. We demonstrate the options available during the usage of KSC-net by means of two demos and explain the internal structure and functionality in terms of the KSC methodology.

## 9.2 KSC FOR BIG DATA NETWORKS

We first provide a brief description of the KSC methodology along with the FURS selection technique. We also explain the BAF and self-tuned model selection techniques. The notations used in this chapter are explained in Section 9.2.1.

### 9.2.1 Notations

1. A graph is represented as $G = (V, E)$ where $V$ represents the set of nodes and $E \subseteq V \times V$ represents the set of edges in the network. The nodes represent the entities in the network, and the edges represent the relationship between them.
2. The cardinality of the set $V$ is given as $N$.
3. The cardinality of the set $E$ is given as $E_{\text{total}}$.
4. The adjacency matrix $A$ is an $N \times N$ sparse matrix.
5. The affinity/kernel matrix of nodes is given by $\Omega$, and the affinity matrix of projections is depicted by $S$.
6. For unweighted graphs, $A_{ij} = 1$ if $(v_i, v_j) \in E$, else $A_{ij} = 0$.
7. The subgraph generated by the subset of nodes $B$ is represented as $G(B)$. Mathematically, $G(B) = (B, Q)$ where $B \subset V$ and $Q = (S \times S) \cap E$ represents the set of edges in the subgraph.
8. The degree distribution function is given by $f(V)$. For the graph $G$, it can written as $f(V)$, while for the subgraph $B$, it can be presented as $f(B)$. Each vertex $v_i \in V$ has a degree represented as $f(v_i)$.
9. The degree matrix is represented as $D$. It is a diagonal matrix with diagonal entries $d_{i,i} = \sum_j A_{ij}$.
10. The adjacency list corresponding to each vertex $v_i \in V$ is given by $x_i = A(:,i)$.
11. The set of neighboring nodes of a given node $v_i$ is represented by $Nbr(v_i)$.
12. The median degree of the graph is represented as $M$.

The KSC methodology comprises three major steps:

1. Building a model on the training data
2. Identifying the model parameters using the validation data
3. The out-of-sample extension to determine community affiliation for the unseen test set of nodes

### 9.2.2 FURS Selection

Nodes that have higher-degree centrality or more connections represent the presence of more influence and have the tendency to be located in the center of the network. The goal of the FURS selection technique is to select several such nodes of high-degree centrality from different dense regions in the graph to capture the inherent community structure present in the network. However, this problem of selection of such a subset ($B$) takes non-deterministic polynomial time (NP-hard) and is formulated as

$$\begin{aligned} \max_{B} \quad & J(B) = \sum_{j=1}^{s} D(v_j) \\ \text{s.t.} \quad & v_j \in c_i, \\ & c_i \in \{c_1, \ldots, c_k\} \end{aligned} \tag{9.1}$$

where $D(v_j)$ represents the degree centrality of the node $v_j$, $s$ is the size of the subset, $c_i$ represents the $i$th community, and $k$ represents the number of communities in the network that cannot be determined explicitly beforehand.

The FURS selection technique is a greedy solution to the aforementioned problem. It is formulated as an optimization problem where we maximize the sum of the degree centrality of the nodes in the selected subset $B$, such that neighbors of the selected node are deactivated or cannot be selected in that iteration. By deactivating its neighborhood, we move from one dense region in the graph to another, thereby approximately covering all the communities in the network. If all the nodes are deactivated in one iteration and we have not yet reached the required subset size $s$, then these deactivated nodes are reactivated, and the procedure is repeated till we reach the required size $s$. The optimization problem is formulated in Equation 9.2:

$$\begin{aligned} & J(B) = 0 \\ & \text{While } |B| < s \\ & \max_{B} \quad J(B) := J(B) + \sum_{j=1}^{s^t} D(v_j) \\ & \text{s.t.} \quad Nbr(v_j) \rightarrow \text{deactivated, iteration } t, \\ & \qquad\; Nbr(v_j) \rightarrow \text{activated, iteration } t+1, \end{aligned} \tag{9.2}$$

where $s^t$ is the size of the set of nodes selected by FURS during iteration $t$.

Several approaches have been proposed for sampling a graph, including References 2, 19 through 21. The FURS selection approach was proposed in Reference 2 and used in Reference 1. A comprehensive comparison of various sampling techniques has been explained in detail in Reference 2. We use the FURS selection technique in KSC-net software for training and validation set selection.

### 9.2.3 KSC Framework

For large-scale networks, the training data comprise the adjacency list of all the nodes $v_i$, $i = 1,\ldots,N_{\text{tr}}$. Let the training set of nodes be represented by $V_{\text{tr}}$ and the training set cardinality be $N_{\text{tr}}$. The validation and test sets of nodes are represented by $V_{\text{valid}}$ and $V_{\text{test}}$, respectively. The cardinalities of these sets are $N_{\text{valid}}$ and $N_{\text{test}}$, respectively. These sets of adjacency lists can efficiently be stored in the memory as real-world networks are highly sparse and there are limited connections for each node $v_i \in V_{\text{tr}}$. The maximum length of the adjacency list can be equal to $N$. This is the case when a node is connected to all the other nodes in a network.

#### 9.2.3.1 *Training Model*

For $V_{\text{tr}}$ training nodes selected by the FURS selection technique, $\mathcal{D}=\{x_i\}_{i=1}^{N_{tr}}$, such that $x_i \in \mathbb{R}^N$. Here, $x_i$ represents the adjacency list of the $i$th training node. Given $\mathcal{D}$ and a user-defined max$k$ (maximum number of clusters in the network), the primal formulation of the weighted kernel PCA [17] is given by

$$\min_{w^{(l)},e^{(l)},b_l} \quad \frac{1}{2}\sum_{l=1}^{\max k-1} w^{(l)\intercal}w^{(l)} - \frac{1}{2N_{tr}}\sum_{l=1}^{\max k-1} \gamma_l e^{(l)\intercal} D_\Omega^{-1} e^{(l)} \tag{9.3}$$

$$\text{such that} \quad e^{(l)} = \Phi w^{(l)} + b_l 1_{N_{tr}}, l = 1,\ldots,\max k-1$$

where $e^{(l)} = \left[e^{(l)\intercal};\ldots;e_{N_{tr}}^{(l)\intercal}\right]$ are the projections onto the eigenspace; $l = 1;\ldots;\max k - 1$ indicates the number of score variables required to encode the max$k$ communities; $D_\Omega^{-1} \in \mathbb{R}^{N_{tr}\times N_{tr}}$ is the inverse of the degree matrix associated to the kernel matrix $\Omega$; $\Phi$ is the $N_{\text{tr}} \times d_h$ feature matrix; and $\Phi = \left[\phi(x_1)^\intercal;\ldots;\phi(x_{N_{tr}})^\intercal\right]$ and $\gamma_l \in \mathbb{R}^+$ are the regularization constants. We note that $N_{tr} \ll N$, that is, the number of nodes in the training set is much less than the total number of nodes in the large-scale network. The kernel matrix $\Omega$ is obtained by calculating the similarity between the adjacency list of each pair of nodes in the training set. Each element of $\Omega$, denoted as $\Omega_{ij} = K(x_i, x_j) = \phi(x_i)^\intercal\phi(x_j)$, is obtained by calculating the cosine similarity between the adjacency lists $x_i$ and $x_j$. Thus, $\Omega_{ij} = \frac{x_i^\intercal x_j}{\|x_i\|\|x_j\|}$ can be calculated efficiently using notions of set unions and intersections. This corresponds to using a normalized linear kernel function $K(x,z) = \frac{x^\intercal z}{\|x\|\|z\|}$ [22]. The clustering model is then represented by

$$e_i^{(l)} = w^{(l)\intercal}\phi(x_i) + b_l, i = 1,\ldots,N_{tr} \tag{9.4}$$

where $\phi:\mathbb{R}^N \rightarrow \mathbb{R}^{d_h}$ is the mapping to a high-dimensional feature space of dimension $d_h$; $b_l$ are the bias terms; and $l = 1;\ldots;\max k - 1$. However, for large-scale networks, we can utilize the explicit expression of the underlying feature map and $d_h = N$. Since the KSC formulation is valid for any positive definite kernel, we use a normalized linear kernel function to avoid any kernel parameter. The projections $e_i^{(l)}$ represent the latent variables of a set of max$k$ – 1 binary cluster indicators given by $\text{sign}\left(e_i^{(l)}\right)$, which can be combined with the final groups using an encoding/decoding scheme. The dual problem corresponding to this primal formulation is

$$D_\Omega^{-1} M_D \Omega \alpha^{(l)} = \lambda_l \alpha^{(l)} \tag{9.5}$$

where $M_D = I_{N_{tr}} - \left(\frac{\left(1_{N_{tr}} 1_{N_{tr}}^\mathsf{T} D_\Omega^{-1}\right)}{1_{N_{tr}}^\mathsf{T} D_\Omega^{-1} 1_{N_{tr}}}\right)$. The $\alpha^{(l)}$ are the dual variables, and the kernel function $K:\mathbb{R}^N \times \mathbb{R}^N \rightarrow \mathbb{R}$ plays the role of similarity function.

#### 9.2.3.2 *Model Selection*

KSC-net provides the user with the option to select one of the two model selection techniques, that is, the BAF criterion [1] or the self-tuned procedure [3].

**Algorithm 1: BAF Model Selection Criterion**

**Data**: Given codebook $\mathcal{CB}$, max$k$, and $P = \left[e_{valid_1},\ldots,e_{valid_N}\right]$.
**Result**: Maximum BAF value and optimal $k$.

1 **foreach** $k \in (2, \max k]$ **do**
2 Calculate the cluster memberships of training nodes using codebook $\mathcal{CB}$.
3 Get the clustering $\Delta = \{C_1,\ldots,C_k\}$ and calculate cluster mean $u_i$ for each $C_i$.
4 For each validation node $valid_i$, obtain $\max_j \cos(\theta_{j,valid_i})$, where $\cos(\theta_{j,valid_i}) = \frac{\mu_j^\mathsf{T} e_{valid_i}}{\|\mu_j\| \|e_{valid_i}\|}, j = 1,\ldots,k.$
5 Maintain dictionary $MaxSim(valid_i) = \max_j \cos(\theta_{j,valid_i}), j = 1,\ldots,k.$
6 Obtain clustering $\Delta_{valid} = \{C_{valid_1},\ldots,C_{valid_k}\}$ for the validation nodes.
7 Define $BAF$ as $BAF(k) = \sum_{i=1}^{k} \sum_{valid_j \in C_i} \frac{1}{k} \cdot \frac{MaxSim(valid_j)}{|C_i|}.$
8 **end**
9 Save the maximum BAF value along with the corresponding $k$.

#### 9.2.3.3 *Out-of-Sample Extension*

Ideally, when the communities are nonoverlapping, we will obtain $k$ well-separated communities, and the normalized Laplacian has $k$ piecewise constant eigenvectors. This is because the multiplicity of the largest eigenvalue, that is, 1, is $k$ as depicted in Reference 23.

In the case of KSC due to the centering matrix $M_D$, the eigenvectors have zero mean, and the optimal threshold for binarizing the eigenvectors is self-determined (equal to 0). So we need $k-1$ eigenvectors. However, in real-world networks, the communities exhibit overlap and do not have piecewise constant eigenvectors.

**Algorithm 2: Algorithm to Automatically Identify *k* Communities**

**Data**: $P=[e_{valid_1}, e_{valid_2}, \ldots, e_{valid_N}]$.
**Result**: The number of clusters $k$ in the given large-scale network.
1 Construct $S$ using the projection vectors $e_{valid_i} \in P$ and their CosDist().
2 Set $td = [0.1, 0.2, \ldots, 1]$.
3 **foreach** $t \in td$ **do**
4     Save $S$ in a temporary variable $R$, that is, $R := S$ and $SizeC_t = []$.
5     **while** *S is not an empty matrix* **do**
6         Find $e_{valid_i}$ with maximum number of nodes whose cosine distance is $< t$.
7         Determine the number of these nodes and append it to $SizeC_t$.
8         Remove rows and columns corresponding to the indices of these nodes.
9     **end**
10     Calculate the entropy and expected balance from $SizeC_t$ as in Reference 3.
11     Calculate the F-measure $F(t)$ using entropy and balance.
12 **end**
13 Obtain the threshold corresponding to which F-measure is maximum as max$t$.
14 Estimate $k$ as the number of terms in the vector $SizeC_{\max t}$.

The decoding scheme consists of comparing the cluster indicators obtained in the validation/test stage with the codebook $\mathcal{CB}$ and selecting the nearest codebook based on Hamming distance. This scheme corresponds to the error correcting output codes (ECOC) decoding procedure [24] and is used in out-of-sample extensions as well.

The out-of-sample extension is based on the score variables that correspond to the projections of the mapped out-of-sample nodes onto the eigenvectors found in the training stage. The cluster indicators can be obtained by binarizing the score variables for the out-of-sample nodes as

$$\text{sign}\left(e_{test}^{(l)}\right) = \text{sign}(\Omega_{test}\alpha^{(l)} + b_l 1_{N_{test}}) \tag{9.6}$$

where $l = 1;\ldots;k-1$; $\Omega_{\text{test}}$ is the $N_{\text{test}} \times N_{\text{tr}}$ kernel matrix evaluated using the test nodes with entries $\Omega_{r,i} = K(x_r, x_i)$; $r = 1;\ldots;N_{\text{test}}$; and $i = 1;\ldots;N_{\text{tr}}$. This natural extension to out-of-sample nodes is the main advantage of KSC. In this way, the clustering model can be trained, validated, and tested in an unsupervised learning procedure. For the test set, we use the entire network. If the network cannot fit in memory, we divide it into blocks to calculate the cluster memberships for each test block in parallel.

### 9.2.4 Practical Issues

In this chapter, all the graphs are undirected and unweighted unless otherwise specified. We conducted all our experiments on a machine with a 12 GB RAM, 2.4 GHz Intel Xeon processor. The maximum cardinality allowed for the training ($N_{tr}$) and the validation set ($N_{valid}$) is 5000 nodes. This is because the maximum-size kernel matrix that we can store in the memory of our PC is 5000 × 5000. We use the entire network as test set $V_{test}$. We perform the community affiliation for unseen nodes in an iterative fashion by dividing $V_{test}$ into blocks of 5000 nodes. This is because the maximum-size $\Omega_{test}$ that we can store in the memory of our PC is 5000 × 5000. However, this operation can easily be extended to a distributed environment and performed in parallel.

## 9.3 KSC-net SOFTWARE

The KSC-net software is available for Windows and Linux platforms at ftp://ftp.esat.kuleuven.be/SISTA/rmall/KSCnet_Windows.tar.gz and ftp://ftp.esat.kuleuven.be/SISTA/rmall/KSCnet_Linux.tar.gz, respectively. In our demos, we use a benchmark synthetic network obtained from Reference 9 and an anonymous network available freely from the Stanford SNAP library, http://snap.stanford.edu/data/index.html.

### 9.3.1 KSC Demo on Synthetic Network

We first provide a demo script to showcase the KSC methodology on a synthetic network that has 5000 nodes and 124,756 edges. The network comprises communities that are known beforehand and act as ground truth. We provide the demo script below. Figure 9.3 is also obtained as an output of the demo script to show the effectiveness of the BAF model selection criterion. From Figure 9.3, we can observe that the maximum BAF value is 0.9205, and it occurs corresponding to $k$ = 7, which is equal to the actual number of communities in the

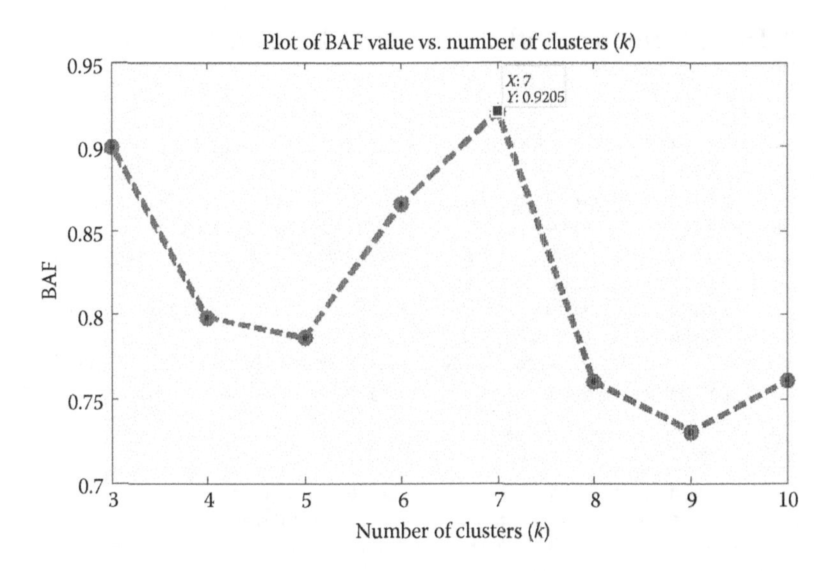


FIGURE 9.3 Selection of optimal $k$ by BAF criterion for synthetic network.

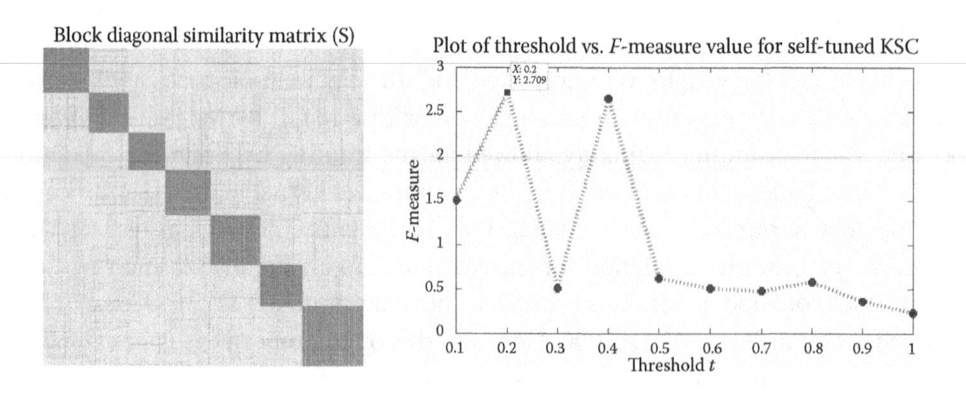


FIGURE 9.4 Selection of optimal $k$ by self-tuned criterion for synthetic network.

network. The adjusted Rand index (ARI) [25] value is 0.999, which means that the community membership provided by the KSC methodology is as good as the original ground truth.

If we put `mod_sel = 'Self'` in the demo script of Algorithm 3, then we perform the self-tuned model selection. It uses the concept of entropy and balance clusters to identify the optimal number of clusters $k$ in the network. Figure 9.4 showcases the block diagonal affinity matrix $S$ generated from the eigen-projections of the validation set using the self-tuned model selection procedure. From Figure 9.4, we can observe that there are seven blocks in the affinity matrix $S$. In order to calculate the number of these blocks, the self-tuned procedure uses the concept of entropy and balance together as an F-measure. Figure 9.4 shows that the F-measure is maximum for threshold $t = 0.2$, for which it takes the value 2.709. The number of clusters $k$ identified corresponding to this threshold value is 7, which is equivalent to the number of ground truth communities.

### Algorithm 3: Demo Script for Synthetic Network Using BAF Criterion

**Data:**
```
netname = 'network';//Name of the file and extension should be '.txt'
baseinfo = 1;                                    //Ground truth exists
cominfo = 'community'; //Name of ground truth file and extension should be '.txt'
weighted = 0;  //Whether a network is weighted or unweighted
frac_network = 15;     //Percentage of network to be used as training and
                       //validation set
maxk = 10;       //maxk is maximum value of k to be used in eigen-decomposition, use
                 //maxk = 10 for k < = 10 else use maxk = 100
mod_sel = 'BAF';       //Method for model selection (options are 'BAF' or 'Self')
output = KSCnet(netname,baseinfo,cominfo,weighted,frac_network,maxk,mod_sel);
```

**Result**
```
output = 1;      //When community detection operation completes
netname_numclu_mod_sel.csv;      //Number of clusters detected
netname_outputlabel.csv;     //Cluster labels assigned to the nodes
netname_ARI.csv;         //Valid only when ground truth is known, provides information
                         //about ARI and number of communities covered
```

### 9.3.2 KSC Subfunctions

We briefly describe some of the subfunctions that are called within the KSCnet() function. First, the large-scale network is provided as an edge list with or without the ground truth information; we convert the edge list into a sparse adjacency matrix $A$. The frac network provides information about what percentage of the total number of nodes in the network is to be used as a training and validation set. We put a condition to check that the $N_{\text{tr}}$ or $N_{\text{valid}}$ cannot exceed a maximum value of 5000. We then use the "system" command to run the FURS selection technique as it is implemented in the scripting language Python. The Python prerequisites are provided in the "ReadMe.txt." Figure 9.5 demonstrates the output obtained when the FURS selection technique is used for selecting the training and validation set.

From Figure 9.5, we can observe that the statement "First Step of the Sampling Process Completed" appears twice (once for training and once for validation).

- *KSCmod.* The KSCmod function takes as input a Boolean state variable to determine whether it is the training or test phase. If it is 0, then we provide as input the training and validation set; otherwise, the input is the training and test sets. It also takes as input max$k$ and the eigenvalue decomposition algorithm (options are "eig" or "eigs"). It calculates the kernel matrix using a Python code and then solves the eigen-decomposition problem to obtain the model. Output is the model along with its parameters, training/validation/test set eigen-projections, and codebook $\mathcal{CB}$. For the validation set, we also obtain the community affiliation corresponding to max$k$, but for the test set, we obtain cluster membership corresponding to optimal $k$.

- *KSCcodebook.* KSCcodebook takes as input the eigen-projections and eigenvectors. It takes the sign of these eigen-projections, calculates the top $k$ most frequent code words, and outputs them as a codebook. It also estimates the cluster memberships for the training nodes using the KSCmembership function.

- *KSCmembership.* KSCmembership takes as input the eigen-projections of the train/validation/test set and codebook. It takes a sign of these eigen-projections and calculates the Hamming distance between the codebook and the sign $(e_i)$. It assigns the corresponding node to that cluster with which it has minimum Hamming distance.

In the case of the BAF model selection criterion, we utilize the KSCcodebook and KSCmembership functionality to estimate the optimal number of communities $k$ in the

```
>> demoscript
First Step of the Sampling Process Completed
Degree Matrix Empty for the 1 time
Degree Matrix Empty for the 2 time
0.960400104523 seconds
Elapsed time is 1.072561 seconds.
First Step of the Sampling Process Completed
Degree Matrix Empty for the 1 time
Degree Matrix Empty for the 2 time
0.637423992157 seconds
Elapsed time is 0.743813 seconds.
```

FIGURE 9.5 Output of FURS for selecting training and validation set.

```
Tuning of the number of clusters
/usr/lib64/python2.6/site-packages/scipy/maxentropy/__init__.py:19: DeprecationWarning:
The scipy.maxentropy module is deprecated in scipy 0.10, and scheduled to be
removed in 0.11.

If you are using some of the functionality in this module and are of the
opinion that it should be kept or moved somewhere - or you are even interested
to maintain/improve this whole module - please ask on the scipy-dev mailing
list.

The logsumexp function has already been moved to scipy.misc.
  DeprecationWarning)
Elapsed time is 17.605555 seconds.
/usr/lib64/python2.6/site-packages/scipy/maxentropy/__init__.py:19: DeprecationWarning:
The scipy.maxentropy module is deprecated in scipy 0.10, and scheduled to be
removed in 0.11.

If you are using some of the functionality in this module and are of the
opinion that it should be kept or moved somewhere - or you are even interested
to maintain/improve this whole module - please ask on the scipy-dev mailing
list.

The logsumexp function has already been moved to scipy.misc.
  DeprecationWarning)
Elapsed time is 17.627895 seconds.
```

FIGURE 9.6 Output snippet obtained while identifying $k$.

large-scale network. However, the "Self" model selection technique works independent of these functions and uses the concept of entropy and balance to calculate the number of block diagonals in the affinity matrix $S$ generated by the validation set eigen-projections after reordering by means of the ground truth information. Figure 9.6 represents a snippet that we obtain when trying to identify the optimal number of communities $k$ in the large-scale network.

In Figure 9.6, we observe a "DeprecationWarning," which we ignore as it is related to some functionality in the "scipy" library, which we are not using. We also observe that we obtained "Elapsed time..." at two places. Once, it occurs after building the training model, and the second time, it appears after we have performed the model selection using the validation set. Figure 9.7 represents the time required to obtain the test set community affiliation. Since there are only 5000 nodes in the entire network and we divide the test set into blocks of 5000 nodes due to memory constraints, we observe "I am in block 1" in Figure 9.7. Figure 9.8 showcases the community-based structure for the synthetic network. We get the same result in either case when we use the BAF selection criterion or the "Self" selection technique.

```
I am in block 1
/usr/lib64/python2.6/site-packages/scipy/maxentropy/__init__.py:19: DeprecationWarning:
The scipy.maxentropy module is deprecated in scipy 0.10, and scheduled to be
removed in 0.11.

If you are using some of the functionality in this module and are of the
opinion that it should be kept or moved somewhere - or you are even interested
to maintain/improve this whole module - please ask on the scipy-dev mailing
list.

The logsumexp function has already been moved to scipy.misc.
  DeprecationWarning)
Elapsed time is 110.644138 seconds.
```

FIGURE 9.7 Output snippet obtained while estimating the test cluster membership.

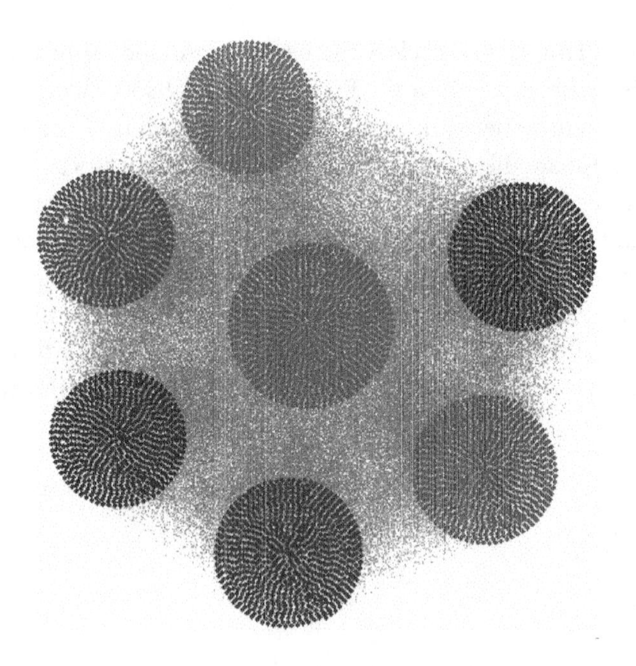

FIGURE 9.8 Communities detected by KSC methodology for the synthetic network.

### 9.3.3 KSC Demo on Real-Life Network

We provide a demo script to showcase the KSC methodology on a real-life large-scale YouTube network that has 1,134,890 nodes with 2,987,624 edges. The network is highly sparse. It is a social network where users form friendship with each other and can create groups that other users can join. We provide the demo script below.

Figure 9.9 shows the effectiveness of the BAF model selection criterion on the YouTube social network. From Figure 9.9, we observe that there are multiple local peaks on the

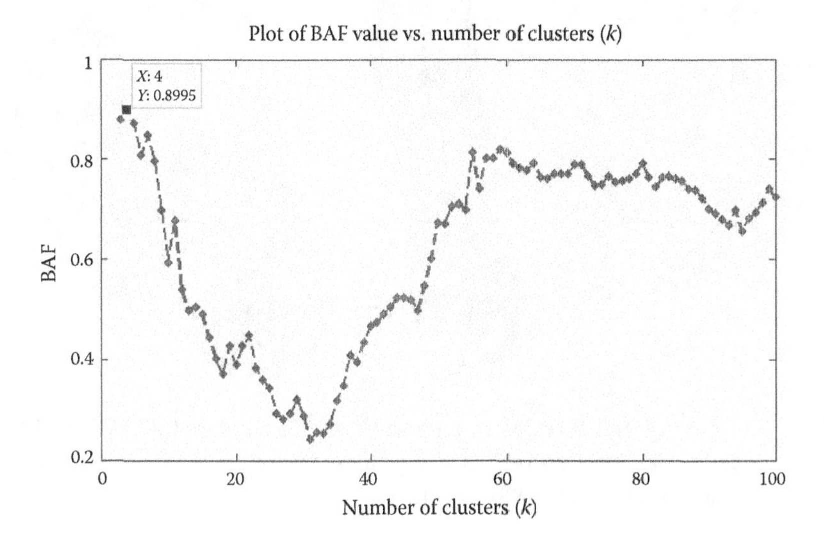


FIGURE 9.9 Selection of optimal $k$ by BAF criterion for YouTube network.

BAF-versus-$k$ curve. However, we select the peak corresponding to which the BAF value is maximum. This occurs at $k = 4$ for the BAF value of 0.8995. Since we provide the cluster memberships for the entire network to the end user, the end user can use various internal quality criteria to estimate the quality of the resulting communities.

If we put `mod_sel = 'Self'` in the demo script of Algorithm 4, then we would be using the parameter-free model selection technique. Since the ground truth communities are not known beforehand, we try to estimate the block diagonal structure using the greedy selection technique mentioned in Algorithm 2 to determine the $SizeC_t$ vector. Figure 9.10 shows that the F-measure is maximum for $t = 0.1$, for which it takes the value 0.224. The number of communities $k$ identified corresponding to this threshold value is 36 (Figure 9.11).

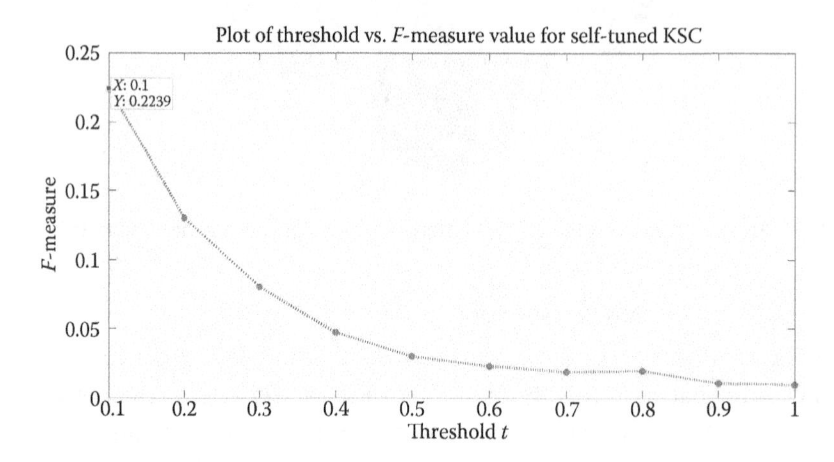


FIGURE 9.10 Selection of optimal $k$ by self-tuned criterion for YouTube network.

FIGURE 9.11 Community structure detected for the YouTube network by "BAF" and "Self" model selection techniques respectively visualized using the software provided in Reference 26. (From Lancichinetti, A. et al., *Plos One*, 6(e18961), 2011.)

**Algorithm 4: Demo Script for YouTube Network Using BAF Criterion**

**Data:**

```
netname = 'youtube_mod';        //Name of the file and extension should be '.txt'
baseinfo = 0;   //Ground truth does not exists
cominfo = [];   //No ground truth file so empty set as input
weighted = 0;   //Whether a network is weighted or unweighted
frac_network = 10;      //Percentage of network to be used as training and
                        //validation set
maxk = 100;     //maxk is maximum value of k to be used in eigen-decomposition,
                //use maxk = 10 for k < = 10 else use maxk = 100
mod_sel = 'BAF';        //Method for model selection (options are 'BAF' or 'Self')
output = KSCnet(netname,baseinfo,cominfo,weighted,frac_network,maxk,mod_sel);
```

**Result:**

```
output = 1;     //When community detection operation completes
netname_numclu_mod_sel.csv;     //Number of clusters detected
netname_outputlabel.csv;        //Cluster labels assigned to the nodes
```

## 9.4 CONCLUSION

In this chapter, we gave a practical exposition of a methodology to perform community detection for Big Data networks. The technique was built on the concept of spectral clustering and referred to as KSC. The core concept was to build a model on a small representative subgraph that captured the inherent community structure of the large-scale network. The subgraph was obtained by the FURS [2] selection technique. Then, the KSC model was built on this subgraph. The model parameters were obtained by one of the two techniques: (1) BAF criterion or (2) self-tuned technique. The KSC model has a powerful out-of-sample extensions property. This property was used for community affiliation of previously unseen nodes in the Big Data network. We also explained and demonstrated the usage of the KSC-net software, which uses the KSC methodology as the underlying core concept.

## ACKNOWLEDGMENTS

EU: The research leading to these results has received funding from the European Research Council under the European Union's Seventh Framework Programme (FP7/2007-2013)/ ERC AdG A-DATADRIVE-B (290923). This chapter reflects only the authors' views; the Union is not liable for any use that may be made of the contained information. Research Council KUL: GOA/10/09 MaNet, CoE PFV/10/002 (OPTEC), BIL12/11T; PhD/postdoc grants. Flemish government—FWO: projects G.0377.12 (structured systems), G.088114N (tensor-based data similarity); PhD/postdoc grants. IWT: project SBO POM (100031); PhD/ postdoc grants. iMinds Medical Information Technologies: SBO 2014. Belgian Federal Science Policy Office: IUAP P7/19 (DYSCO, dynamical systems, control and optimization, 2012–2017).

## REFERENCES

1. R. Mall, R. Langone and J.A.K. Suykens; Kernel spectral clustering for Big Data networks. *Entropy (Special Issue: Big Data)*, 15(5):1567–1586, 2013.
2. R. Mall, R. Langone and J.A.K. Suykens; FURS: Fast and Unique Representative Subset selection retaining large scale community structure. *Social Network Analysis and Mining*, 3(4):1–21, 2013.
3. R. Mall, R. Langone and J.A.K. Suykens; Self-tuned kernel spectral clustering for large scale networks. In *Proceedings of the IEEE International Conference on Big Data* (IEEE BigData 2013). Santa Clara, CA, October 6–9, 2013.
4. S. Schaeffer; Algorithms for nonuniform networks. PhD thesis. Espoo, Finland: Helsinki University of Technology, 2006.
5. L. Danaon, A. Diáz-Guilera, J. Duch and A. Arenas; Comparing community structure identification. *Journal of Statistical Mechanics: Theory and Experiment*, 9:P09008, 2005.
6. S. Fortunato; Community detection in graphs. *Physics Reports*, 486:75–174, 2009.
7. A. Clauset, M. Newman and C. Moore; Finding community structure in very large scale networks. *Physical Review E*, 70:066111, 2004.
8. M. Girvan and M. Newman; Community structure in social and biological networks. *Proceedings of the National Academy of Sciences of the United States of America*, 99(12):7821–7826, 2002.
9. A. Lancichinetti and S. Fortunato; Community detection algorithms: A comparative analysis. *Physical Review E*, 80:056117, 2009.
10. M. Rosvall and C. Bergstrom; Maps of random walks on complex networks reveal community structure. *Proceedings of the National Academy of Sciences of the United States of America*, 105:1118–1123, 2008.
11. R. Langone, C. Alzate and J.A.K. Suykens; Kernel spectral clustering for community detection in complex networks. In *IEEE WCCI/IJCNN*, pp. 2596–2603, 2002.
12. V. Blondel, J. Guillaume, R. Lambiotte and L. Lefebvre; Fast unfolding of communities in large networks. *Journal of Statistical Mechanics: Theory and Experiment*, 10:P10008, 2008.
13. A.Y. Ng, M.I. Jordan and Y. Weiss; On spectral clustering: Analysis and an algorithm. In *Proceedings of the Advances in Neural Information Processing Systems*, Dietterich, T.G., Becker, S., Ghahramani, Z., editors. Cambridge, MA: MIT Press, pp. 849–856, 2002.
14. U. von Luxburg; A tutorial on spectral clustering. *Statistics and Computing*, 17:395–416, 2007.
15. L. Zelnik-Manor and P. Perona; Self-tuning spectral clustering. In *Advances in Neural Information Processing Systems*, Saul, L.K., Weiss, Y., Bottou, L., editors. Cambridge, MA: MIT Press, pp. 1601–1608, 2005.
16. J. Shi and J. Malik; Normalized cuts and image segmentation. *IEEE Transactions on Pattern Analysis and Intelligence*, 22(8):888–905, 2000.
17. C. Alzate and J.A.K. Suykens; Multiway spectral clustering with out-of-sample extensions through weighted kernel PCA. *IEEE Transactions on Pattern Analysis and Machine Intelligence*, 32(2):335–347, 2010.
18. L. Muflikhah; Document clustering using concept space and concept similarity measurement. In *ICCTD*, pp. 58–62, 2009.
19. A. Maiya and T. Berger-Wolf; Sampling community structure. In *WWW*, pp. 701–710, 2010.
20. U. Kang and C. Faloutsos; Beyond 'caveman communities': Hubs and spokes for graph compression and mining. In *Proceedings of ICDM*, pp. 300–309, 2011.
21. N. Metropolis, A. Rosenbluth, M. Rosenbluth, A. Teller and E. Teller; Equation of state calculations by fast computing machines. *Journal of Chemical Physics*, 21(6):1087–1092, 1953.
22. J.A.K. Suykens, T. Van Gestel, J. De Brabanter, B. De Moor and J. Vandewalle; *Least Squares Support Vector Machines*. Singapore: World Scientific, 2002.
23. F.R.K. Chung; *Spectral Graph Theory*. Providence, RI: American Mathematical Society, 1997.

24. J. Baylis; *Error Correcting Codes: A Mathematical Introduction*. Boca Raton, FL: CRC Press, 1988.
25. R. Rabbany, M. Takaffoli, J. Fagnan, O.R. Zaiane and R.J.G.B. Campello; Relative validity criteria for community mining algorithms. In *International Conference on Advances in Social Networks Analysis and Mining (ASONAM)*, pp. 258–265, 2012.
26. A. Lancichinetti, F. Radicchi, J.J. Ramasco and S. Fortunato; Finding statistically significant communities in networks. *PLoS One*, 6:e18961, 2011.

CHAPTER 10

# Making Big Data Transparent to the Software Developers' Community

Yu Wu, Jessica Kropczynski, and John M. Carroll

## CONTENTS



## ABSTRACT

Software developers in the digital age participate in a set of social media services, such as GitHub, Twitter, and Stack Overflow, to gain access to different resources. Recent research indicates that these social media services form an interconnected ecosystem for software developers to make connections, stay aware of the latest news, and coordinate work. Due to the complexity of the online ecosystem and the large volume of information generated from it, software developers encounter information overload wherein abundant information makes it hard to be aware of the most relevant resources to meet information needs. Software developers' participation traces in the ecosystem of social media services generate Big Data, which is publicly available for retrieval and analysis. Applying software analytics and related techniques to Big Data can reduce information overload, satisfy information needs, and make collaboration more effective. This chapter reviews the literature on a software developer's information needs and participation in social media services, describes the information overload issue, and illustrates the available Big Data sources in the ecosystem and their potential applications.

## 10.1 INTRODUCTION

The open-source software (OSS) development community has allowed technology to progress at a rapid pace around the globe through shared knowledge, expertise, and collaborations. The broad-reaching open-source movement bases itself on a share-alike principle that allows anybody to use or modify software, and upon completion of a project, its source code is made publicly available. Programmers who are a part of this community contribute by voluntarily writing and exchanging code through a collaborative development process in order to produce high-quality software. This method has led to the creation of popular software products including Mozilla Firefox, Linux, and Android. Most OSS development activities are carried out online through formalized platforms (such as GitHub), incidentally creating a vast amount of interaction data across an ecosystem of platforms that can be used not only to characterize open-source development work activity more broadly but also to create Big Data awareness resources for OSS developers. The intention of these awareness resources is to enhance the ability to seek out much-needed information necessary to produce high-quality software in this unique environment that is not conducive to ownership or profits. Currently, it is problematic that interconnected resources are archived across stand-alone websites.

This chapter describes the process through which these resources can be more conspicuous through Big Data, in three interrelated sections about the context and issues of the collaborating process in online space (Figure 10.1) and a fourth section on how Big Data can be obtained and utilized. In Section 10.2, we will describe the information needs that developers draw upon to perform basic tasks. Section 10.3 highlights the multiplatform suite of resources utilized by this community and the ways in which these platforms disperse development activities. Section 10.4 discusses the information overload and

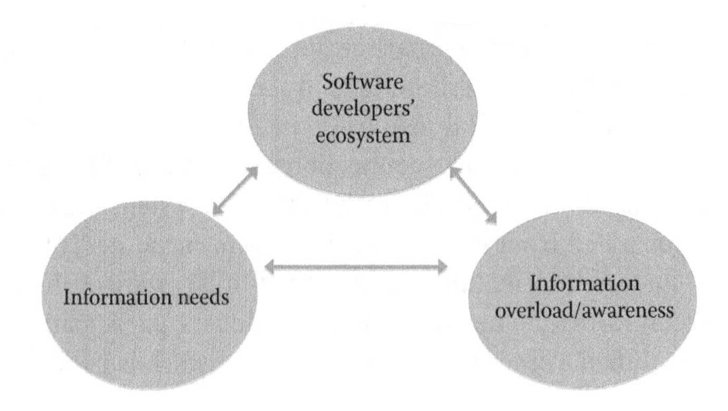


FIGURE 10.1 Interrelated components within the software community.

awareness issue that is generated from the information needs and the ecosystem. Section 10.5 explores ways that the seemingly boundless data generated by this community might be synthesized to increase awareness and lead to a deeper sense of community among developers and strengthen cooperation.

Software developers have pervasive needs for data, such as activity traces, issue reporting, and coding resources necessary to transform existing code, due to the revision control methods (or version control) necessary for collaboration through web-based interfaces. Developers spend a significant amount of time processing data just to maintain awareness of the activities of fellow developers (Biehl et al. 2007). With the rise of social media and its application in software engineering, software developers need to be aware of an overwhelming amount of information from various sources, such as coworkers, interesting developers, trends, repositories, and conversations. A paradox exists wherein there is simply too much information and too little awareness of the knowledge and resources that this information contains. Community awareness is not only important for day-to-day activities, but it is also necessary in order for developers to stay current with the development landscape and industry change (Singer et al. 2014), to make decisions about projects (Buse and Zimmerman 2012), and to make connections with others based on the data revealed in public (Singer et al. 2013). Millions of software developers all over the world continuously generate new social and productive activities across multiple platforms. It is difficult for developers to maintain awareness of the breadth of the information generated by the larger community while also maintaining awareness of individual projects of which they are a part. The overwhelming amount of activity awareness can reduce the transparency of the underlying community of software developers. In an effort to increase the transparency of a thriving community and to improve awareness of available resources, this chapter will describe how developers' activity log data may be collected and analyzed and will provide potential applications of how results may be presented to the public in order to make resources more available and searchable.

Recent research on OSS developers' collaboration suggests that their interactions are not limited to revision control programming platforms (Singer et al. 2013; Vasilescu et al. 2013, 2014; Wu et al. 2014). Instead, developers make use of social features, such as profile, activity traces, and bookmarking, to coordinate work across several platforms. All these sites and platforms form an "ecosystem" wherein software developers achieve productive goals, such as code production, making social connections, and professional development. Later in this chapter, we will illustrate the ecosystem of resources that software developers draw upon and describe the purposes that they fulfill, including social media functions that facilitate peer-to-peer interaction and the software practices that are associated with productivity. As one can imagine, the Big Data possibilities with which to explore software developers' interactions become exponentially greater when considering the multiplatform dimensions with which to explore their interactions.

Section 10.5 will provide the greatest emphasis on the kinds of Big Data we can be collecting about the software developers' community, as well as the quantities of data that exist. Online work production platforms for software development, such as GitHub (Dabbish et al. 2012; Marlow et al. 2013), CodePlex, and Bitbucket, among others, are common platforms where software productive activities take place. These activities are publicly available through application programming interfaces (APIs) (Feeds 2014). The synthesis of the information from this Big Data set is likely to increase the awareness of software developers towards code artifacts, latest news, and other peer developers, which in turn increases a sense of community among individual developers and facilitates further collaboration. In Sections 10.2 through 10.5, we will provide further details about the software developers' information needs, characterize the ecosystem of resources available to software developers, enumerate types of interaction data for OSS that are available to researchers, and review recent efforts to utilize Big Data analytic techniques to analyze OSS interaction data and provide summaries of information that developers will find useful to their work.

## 10.2 SOFTWARE DEVELOPERS' INFORMATION NEEDS

Software development is a complicated process, which requires developers to acquire related information in order to correctly and efficiently carry out their work. Software developers have three broad categories of information needs: (1) information needs that directly relate to software development, that is, the "core" practice; (2) information needs that are helpful for them to build and maintain social connections; and (3) information needs for professional development, such as learning or following activity from experts in their field. The amount of time devoted to locating appropriate information to meet needs reduces cost in terms of time and energy spent on productive activities (Perlow 1999). In Section 10.5, we will describe the importance of using Big Data tools to aggregate and summarize extensive data resources in order to increase OSS performance. Here, we will first list the extensive resources that developers use to satisfy these needs.

### 10.2.1 Information Needs: Core Work Practice

Software developers perform various types of activities on a daily basis, and the "core" of these contributes to the software development process directly. These activities include code submission, bug fixes, and documentation. Developers usually resort to information generated from software code repositories to revise code or to solve a bug. Besides, software development is an extremely complicated process, which involves frequent coordination of work among many developers. Therefore, information needs for software core practice can be broadly categorized as information about software artifacts and information about collaborators. Ko and colleagues (2007) identify the following information needs for the development process: (1) writing code, (2) submitting a change, (3) triaging bugs, (4) reproducing a failure, (5) understanding execution behavior, (6) reasoning about design, and (7) maintaining awareness.

For source code collaboration, software repositories usually apply revision control or a version control system (VCS) so that multideveloper projects undertaking numerous revisions can be compared, restored, and merged as necessary. This system allows a logical way to organize and control revisions. VCS can become increasingly complicated when a team of people must maintain information to track changes to code, maintain a bug-tracking system, and raise/fix issues in the maintenance process.

Bug reporting and fixing is another central software core practice, where differing information needs must be satisfied in order to solve bugs and issues efficiently. Bug tracking plays a central role in supporting collaboration between the developers and users of the software. Breu et al. (2010) found that information needs change over a bug's life cycle; for example, in the beginning, a bug report may include missing information or details for debugging. Later, questions develop, focusing on the correction of a bug and on status inquiries. A bug-tracking system should account for such evolving information needs.

### 10.2.2 Information Needs: Constructing and Maintaining Relationships

The newest approach to OSS development combines version controlled code repositories with social media features, such as GitHub, CodePlex, and so forth (Dabbish et al. 2012). The interfaces of these platforms often include user profiles, code artifacts, and activity traces performed by a certain user. All these are publicly available through an online interface, which is referred to as the "transparency" of software development in recent literature (Dabbish et al. 2012; Marlow et al. 2013). Dabbish et al. (2012) report that developers make intensive use of these social cues in order to effectively coordinate work. While attempting collaborative work, developers usually resort to various cues to make social inferences on such platforms (Marlow et al. 2013).

In addition to those activities that directly relate to code production and the development of professional relationships, developers also perform various social activities, such as communicating new ideas with others, asking and answering questions, and following one another's projects. Marlow et al. (2013) report that developers rely heavily on the social features of social media in order to form impressions of others in order to perform certain types of activities, such as staying up to date with repositories and other peers, discovering a peer's interaction style to better communicate, and assessing expertise of others in order to respond to code contribution. The work led by Dabbish et al. (2012) and Marlow et al. (2013) illustrates how software developers seek information to interact with others within a single platform. However, recent work has uncovered that all of the information necessary to complete OSS projects and for software developers to make social connections is not available on a single platform. Software developers actively engage in the creation of their social and professional profiles on various sites, tailoring each to the functional advantages of that site and assessing other peers for the purpose of connection, collaboration, and coordination. Thus, software developers' interactions are coordinated over a variety of platforms with varying purposes and resources. While their profile information may remain similar, these coordinated interactions among platforms are not explicit to the development community (Singer et al. 2013; Wu et al. 2014).

### 10.2.3 Information Needs: Professional/Career Development

Members of the OSS community find opportunities to build their personal resume and seek out talents among others in the community to build their workforce for ongoing projects. Working on OSS projects demonstrates coding skills, collaboration abilities, work ethic, and technology interests. Programmers with an interest in a particular field or type of programming but lacking appropriate qualifications may strategically participate in projects that will help them develop the skills and experience necessary to break into a hot area of information technology. In such a role, programmers do not have to start out by taking key development roles but can instead follow a project, try to take on a small role, and ask questions of experts in the field. From the perspective of those hiring programmers, a list of projects and contributions that a programmer has made in an open-source community may offer a better perspective of technical and team skills than a reference. Potential employers seek out developers who are taking full advantage of resources available and are on the cutting edge of the software industry.

While experiences on such projects might help land a job, this is not the only reason that programmers might utilize OSS in their career. Corporations have begun relying on enterprise editions of OSS platforms. Working on OSS is no longer a domain of hobbyists coding in their spare time. Instead, many developers are now working on open-source programs as part of their daily jobs. Many large organizations now rely on having their employees engage with the open-source community, who may contribute to major bug reports and fixes (Arora 2012). In addition to acquiring a workforce to perform these smaller tasks, building in the open-source environment tends to allow for a broader perspective than in teams typically working on single-use cases in the context of one individual company. This environment naturally lends itself to new and innovative practices that encourage adaptation and learning from additional data resources.

## 10.3 SOFTWARE DEVELOPERS' ECOSYSTEM

In the digital economy, software developers find it increasingly necessary to publicize their work and collaborate with others over online space through social media sites. Recent work (Singer et al. 2013; Wu et al. 2014) suggests important interrelations among a suite of platforms and sites that software developers use to coordinate work. In this section, we will address how social media is used in software practice to coordinate social interactions and profile evidence of the interconnectivity of the software developers' ecosystem.

### 10.3.1 Social Media Use

Software development is becoming more and more social, where the social context of software development and the social interactions among developers becomes increasingly important. In recent literature, software developers are even characterized as "social programmers" because of their intensive use of social media sites (Singer et al. 2013). According to Lietsala and Sirkkunen (2008), social media can be categorized into the following categories: (1) collaborative content creation tools, (2) content sharing tools, (3) virtual worlds tools, and (4) social networking (microblogging) tools. Past literature reports that software developers make intensive use of collaborative content creation tools and social networking tools to carry out their work. A survey study of software developers reveals that developers access these tools by visiting between one and seven social media sites on a regular basis (Black et al. 2010).

Research has also found that software developers leverage social media sites/platforms for a wide range of purposes. Bougie and colleagues (2011) reported that software developers leverage Twitter to make social conversation and share information, which is largely different from software engineer groups. Zhao and Rosson (2009) established that the use of Twitter reduces the cost of information sharing. A study led by Black et al. (2010) found that developers also share source code, specification, and design information over social media. Social media use brings other benefits to software developers as well, such as increasing the quality of communication and the visibility of activities across all levels of organizations (Black and Jacobs 2010).

Beyond the functional purposes of the tools embedded in social media, recent research shows that the context of that use can vary widely. Even within one social media platform, software developers discuss a variety of issues relevant to their work. By analyzing 300 sample Tweets sent by software developers, Tian et al. (2012) discovered that software developers tweet on the following topics:

1. Software engineering related (work related), such as projects being worked on or seeking or providing technical help
2. Gadgets and technological topics (e.g., iPhone 4)
3. Events outside of technical topics (e.g., things of interest)
4. Daily chatter (e.g., family, weekend activities)

Moreover, social media also allows a new way for software developers to form teams and work collaboratively (Begel et al. 2010). Begel and colleagues (2010) report several typical small group communication functions that social media sites serve for online software development groups to work together: (1) storming, when social media allows rapid feedback that can be used to quickly identify and respond to changing user needs; (2) norming, which emphasizes the knowledge-sharing function of social media among developers; (3) performing, which considers that social media helps various development processes; and (4) adjourning, which highlights the recording mechanism of social media from interactions left behind. Activity traces of all of these interactions can be aggregated from social media to form institutional memory that is accessible to all users. Researchers have also made efforts to incorporate social media in software development tools. Reinhardt (2009) embedded a Twitter client in Eclipse to support Global Software Development.

In short, social media has infiltrated many aspects of software development, from software developers' daily life to team formation, and is not limited to source code sharing and collaboration. On one hand, it greatly connects developers with one another across multiple platforms that serve different purposes; on the other hand, it also introduces complexities, where developers have to manage their profiles and activities in many platforms, and potential issues, such as information overload from various social media sites.

### 10.3.2 The Ecosystem

In Section 10.3.1, software developers frequent social media sites to coordinate work, to share information, and to make social connections. The users of these sites are not isolated from each other. The most recent work suggests that online communities create and organize complex information space among community members through several categories of social tools—such as wikis, blogs, forums, social bookmarks, social file repositories, and task-management tools (Matthews et al. 2014). The "core" functions of each type of tool are leveraged in combination to coordinate work effectively. The software community as a whole acts similarly, where many online social media sites serve distinct functions for developers, and their coordination and collaborations are spread over the whole ecosystem space.

Among those types of social media adopted by software developers, GitHub is one of the most prominent platforms since it has successfully attracted millions of developers to host their own projects and collaborate with other peers (Marlow et al. 2013). Its growing success has become a popular research focus in the past several years. However, previous research on GitHub has not tapped into the expansiveness of the data resource, because it mainly addresses how software developers utilize GitHub features, like user profiles and activity traces, to coordinate work (Begel et al. 2010; Dabbish et al. 2012; Marlow et al. 2013). GitHub is more than a collaborative tool. It also contains an extensive network of relations among millions of software developers. Thung and colleagues (2013) present an initial attempt to analyze the social network structures on GitHub. The sample presented in their study, however, is not well articulated and seems to lack whole network or partial network data necessary to generate a meaningful correlation. The rationale for the formation of GitHub networks is not discussed in previous literature. The gap missing here is how developers' social connections are correlated with their collaborations.

Researchers have already observed that, with the advancement of social media, individuals are all connected (through several degrees or with the potential to connect) in online space (Hanna et al. 2011). Followship, which is a common feature for most social media sites, allows users to constantly pay attention to others they are interested in to receive the latest news and updates. For example, Twitter users can follow one another to receive messages or tweets from interesting people (Kwak et al. 2010). Besides followship, researchers also have found increasing correlations and reciprocity among microblogging services through follow relationships (Java et al. 2007). GitHub provides a similar function: By following others, one will receive activity traces of people he/she is following, which is similar to following on other social media, with the exception that it is broadcasting activities rather than short messages. As GitHub is similar to Twitter in terms of the follow function, activities that demonstrate paying attention to others, such as following, are also possibly used to coordinate collaborative work on software projects.

Social media also largely extends users' ability to create, modify, share, and discuss Internet content (Kietzmann et al. 2011). Seven fundamental building blocks of social media were identified by researchers, including identity, conversations, sharing, presence, relationships, reputation, and groups, which form a rich and diverse ecology of social media sites (Kietzmann et al. 2011). The authors argue that each social media site emphasizes these seven perspectives differently. In the software industry, although GitHub is a crucial platform that supports software core practices, many sites' other platforms, like Stack Overflow and Hacker News, are adopted by software developers in order to support the fundamental purposes (Kietzmann et al. 2011). Sites/platforms emphasize the functions differently, as suggested by Kietzmann et al. (2011); however, little is known about the ways these sites/platforms are organized and form an integral system for software developers.

The software ecosystem has been discussed before but more from an industry's perspective (Bosch 2009; Campbell and Ahmed 2010; Jansen et al. 2009; Messerschmitt and Szyperski 2005), which describes the software industry in the context of users, developers, buyers, sellers, government, and economics, leaving the software developers' perspective missing from the analysis. The recent work of Singer et al. (2013) reported the existence of the ecosystem for software developers, but the missing parts are the exact components that make up the system surrounding software code repositories and software practices and what effort can be made to make this valuable ecosystem of social and productive resources more transparent to all software developers. The large data set generated from software developers' activity traces in the online environment is a valuable resource to address the issue.

## 10.4 INFORMATION OVERLOAD AND AWARENESS ISSUE

In Section 10.3, software developers adopt various online services to coordinate work, to make social connections, and to build their careers. However, these activities are complicated processes and widely spread over numerous platforms. The number and types of social media and their interrelations are likely to increase complexity of collaborations and cause problems as well, such as information overload.

Information overload is the phenomenon wherein the high rate and volume of information will become noise when they reach overload (Klapp 1986). Klapp (1986) deems that if high-rate information cannot be processed without distraction, stress, and increasing errors, the quality of information actually becomes poorer. By reviewing the literature on information overload, Edmunds and Morris (2000) identify the problem of information overload as a situation wherein there is an excessive amount of information but it is difficult for acquire relevant and useful information when in need.

The situation in software development is likely to be severe. In Section 10.2, software developers have intensive information needs over a number of different activities; the activities are sometimes very complicated, like submitting pull requests to a not-very-familiar code repository, and they need to work with each other in an intricate ecosystem, which encompasses many sites and platforms in online space. Researchers in software development sometimes refer to this problem as an awareness issue.

Awareness, knowing what is going on, is the precursor of software practices. Although direct communication, like face-to-face communication, is the most effective way to maintain awareness of other peers, a software development team, especially distributed software development teams, lacks for discursive communication, where collaboration is happening through implicit activities (Bolici et al. 2009). In open-source development, a developer's work is coordinated through a code repository or other similar artifacts, such as collaboration on revision of a source code or collaborative bug fixing. However, because of the complexity of software development, implicit collaboration also raises some critical issues in awareness. Fussell and colleagues (1998) identified four types of awareness issues for distributed software development: (1) lacking awareness of peer activities, (2) lacking awareness of each other's availability, (3) lacking awareness of the process, and (4) lacking awareness of what peers are thinking and why.

Many research efforts and technologies have been applied to address the awareness issue in distributed software development. Gutwin et al. (1996) argue that awareness contributes to effective group collaboration in a physical workspace. In distributed software development situations, software developers need to keep awareness of both the entire time and the people they need to work with (Gutwin et al. 2004), and awareness is usually realized through text-based communication, such as mailing lists and chat systems. However, Bolici et al. (2009) find that these types of communication are lacking in online software development teams: Developers are less likely to communicate with peers other than to look for traces of what others were doing and collaborate with them implicitly. Many of the awareness issues are likely to be alleviated by increasing transparency during collaboration. Dabbish et al. (2012) and Marlow et al. (2013) found that in online work production systems, developers resort to different social cues from others' profiles and activity traces to maintain awareness and coordinate work. However, the issues of awareness in the entire ecosystem are likely to be more complex, where platforms are separated from one another and there is too much information to look for.

## 10.5 THE APPLICATION OF BIG DATA TO SUPPORT THE SOFTWARE DEVELOPERS' COMMUNITY

Not only are the data produced by software developers useful for software development itself, but also, they can be applied to understand software developers' purposes, interactions, and collaborations; address the information overload and awareness issue as illustrated in Section 10.4; and satisfy software stakeholders' need for information (Hassan 2006). This chapter concludes by reviewing the literature on the data needs and applications for software developers and other related stakeholders and discusses the potential impact that Big Data brings to large communities in need of transparency.

There are two major concerns that increased transparency will address. The first is that the software developers' ecosystem is a complex entity, which contains too much information for individuals to process (Gutwin et al. 2004). The second concern is that software development is largely fragmented (Ko et al. 2007). Developers spend half of their time simply communicating with peers (Perry et al. 1994). The consequence of frequent communication is the fragmentation of time, which reduces the efficiency of the development process. A software team's frequent interruptions create a sense of a "time famine"—having too much to do and not enough time (Perlow 1999). This is where Big Data analytic techniques have the most to offer: Summarized information from a comprehensive data set can facilitate awareness in ways that are unavailable in distributed situations (Seaman and Basili 1998). Seaman and Basili (1998) term this type of facilitation *affordances of collocation*. Section 10.5 describes applying Big Data to support of the aforementioned "core" information needs of developers and analytics that can potentially be used to address these needs.

### 10.5.1 Data Generated from Core Practices

Due to the transparency of social media sites for online software practices, a vast amount of data is available through an API provided by the service providers. For example, GitHub makes activity data, user data, and repository data publicly available through both timelines and API. Timeline data contain detailed log information about which user at what time did what type of activity on which software repository (or followed whom). The types of activities can be roughly categorized into four categories (Table 10.1). Data are stored in JavaScript Object Notation (JSON) format (Crockford 2006). The log data as described here contain valuable information concerning software developers and software repositories, which is useful in tracing the full activity history of either a user or a software project.

The timeline alone creates Big Data, which contains the activities of millions of developers on millions of software repositories each day. Collecting, aggregating, and analyzing GitHub timeline data are likely to provide insights into the history, trends, and tendencies in the software industry.

Besides a timeline, GitHub also provides APIs for various inquiries, such as asking for detailed information of a specific user, like profile information, recent activities, and so forth, or a repository, such as retrieving a list of contributors and the programming language used for a specific software project. Besides GitHub, similar APIs about activity logs and user information can also be found in Stack Overflow, Bitbucket, and Twitter, which creates an opportunity to aggregate really Big Data of the software developers' ecosystem from a variety of platforms and services.

According to our discussion in Section 10.2, software developers usually perform three types of activities: (1) "core" activities that are directly related to software development; (2) activities to build or maintain social relationships; and (3) activities for professional development, such as learning. The Big Data collected from the ecosystem has opportunities to facilitate all of them. For example, the experience factory proposed by Basili (1993), which supports software development in collecting experiences from developers' projects, packaging those experiences (from examples in models), and validating and reusing experiences in future projects, can be constructed through integrating data from different sources of the ecosystem, and it can, in turn, benefit individual developers.

Software repositories use VCS to keep track of code changes and a separate bug-tracking system for raising and fixing issues and bugs in the maintenance process. A VCS is in need as soon as a project exceeds the number of one active developer. A log from VCS includes the following: (1) author, (2) date and time, (3) changed files, and (4) optional log message. On GitHub, all log data of VCS and the bug-tracking system are stored and publicly

TABLE 10.1 Types of Activities Recorded by GitHub

| Type of Activity | Description |
|---|---|
| Code | Activities that are directly related to source code, such as code commit, code review, pull request, and so forth |
| Issue | Activities that are related to bug fix, feature request, comment, and so forth |
| Wiki | Activities that are related to editing wiki pages |
| Awareness | Activities that result in receiving updates from others, such as following other users |

available through the timeline and API, which provide critical sources of information for software analytics, such as understanding the evolution of software projects and archiving the development history of the projects.

### 10.5.2 Software Analytics

Software projects are considered to be difficult for stakeholders to predict risks about (Buse and Zimmermann 2010). And in the era of Big Data and the software developers' ecosystem, more issues are raised for different stakeholders. Open-source software developers need to know on which platform to find the information they need, where and with whom to make connections, which project to participate in and to what degree, and so forth. Software users need to know whether the project is under active development and how likely the owners are to address users' feedback. Different stakeholders need a variety of information about software projects in order to make decisions. And software analytics is one possible solution for addressing this concern.

Analytics is the integration of application of analysis, data, and systematic reasoning to make decisions (Buse and Zimmermann 2012). With the help of Big Data and software analytics, stakeholders in software development can acquire information about both the big picture of the industry and the details of a specific software project. Big Data and software analytics have the potential to alleviate the information overload and awareness issue in Section 10.4 and to facilitate the decision-making process.

Analytics has revolutionized the decision-making process in many fields (Davenport et al. 2010). The goal of analytics is to assist decision makers in extracting important information and insights that would otherwise be hidden. In order to overcome the complexity of software, recently, researchers applied analytics in software development in order to better manage the process (Buse and Zimmermann 2010). They argue that software managers have high-level concerns for software projects, such as the direction of the project, the allocation of resources, the set of features, and the user experience, and analytics can help answer important questions managers ask about their projects (Buse and Zimmermann 2012). Software analytics provides the following advantages for software managers: (1) monitoring a project, (2) knowing what is working, (3) improving efficiency, (4) managing risk, (5) anticipating changes, and (6) evaluating past decisions. In the open-source community, every developer is both the manager of his/her own projects and a developer for other projects he/she participates in. Analytics can help open-source developers make various important decisions such as which software to use, which project to join, whom to collaborate with, and so forth.

Although rapid development of technologies creates potentials for software analytics, where multiple data sources in software development can be recorded and stored, the integration of these different sources and creation of meaningful analytics are still under investigation. Bachman and Bernstein (2009) reported that modern software project management systems (Jazz, Telelogic Synergy) provide full functionality needed to develop and maintain a software project. They claimed that integration of process data has multiple advantages (Bachman and Bernstein 2009): (1) predicting the location and number of future or hidden bugs (Crowston 1997) and (2) project managers being able to use such predictions to identify the critical parts of the system, limiting the gravity of their impact and allowing a better-organized plan. However, the integration of data from these standalone systems cannot be performed automatically but has to be maintained manually by developers.

With the emergence of Big Data and the ecosystem, software analytics is likely to play an even greater role. The automation integration process can be resolved because the modern platform integrates several subsystems together. For instance, GitHub integrates VCS, an issue tracker, and documentation functions for each repository. Data generated in one software project on GitHub can be identified and collected just by its unique identifier. Also, the large data set generated from the ecosystem has the potential to address the analysis types proposed by Buse and Zimmermann (2012), which include trends, alerts, forecasting, summarization, overlays, goals, modeling, benchmarking, and simulation. One application of Big Data in software analytics is integrating the context with code artifacts. For example, a popular software repository hosted on GitHub is likely to be discussed a lot on Twitter; users of the software might raise many questions on Stack Overflow; and on Hacker News, experts would address the pros and cons of the software, comparing it with other equivalents and listing application scenarios. By extracting and synthesizing information from the ecosystem, Big Data can reveal the trending of the software industry, summarizing comprehensive information about a single software repository with user feedback and questions, which provides alerts to the project stakeholders and forecasts the features and error associated with that repository. Also, as Big Data contains detailed information on the ecosystem, multiple overlays of the data set can better help developers to understand the history of a code artifact, which can be used to build models of the history, or the trending of recent programming languages. And by exploring the Big Data from the ecosystem, one can better understand users' exact needs and concerns, which facilitates the setting of goals to address users' issues and concerns. Moreover, benchmarking is important in comparing different practices, like similar code artifacts. Big Data provides a larger context for benchmarking to take place, in which more practices and more factors can be taken into consideration.

## 10.6 CONCLUSION

In this chapter, we discussed the information needs of software developers to conduct various software practices, the emergence of the software developers' ecosystem in which developers coordinate work and make social connections across social media sites, and how the interaction of these two activities raises the information overload and awareness issue. We also argued that the Big Data generated from the ecosystem has the potential to address the information overload and awareness issue by providing accurate and comprehensive information with software analytics techniques.

Software development largely becomes implicit collaboration in online space (Bolici et al. 2009), where explicit communication is lacking and mostly indirect communication happens through various cues, such as the profile information and activity traces on social media that serve as references for software developers to form impressions of others to inform collaboration (Marlow et al. 2013). While information within one platform is relatively searchable, similar and tangential information has been dispersed among several online resources, which creates the information overload and awareness issue. The Big Data techniques described in Section 10.5 have the most promise to allow for transparency to increase the efficiency and effectiveness of collaboration and innovative software development across the ecosystem.

## REFERENCES

Arora, H. (2012). Why big companies are embracing open source? IBM. Accessed on March 20, 2014. Retrieved from https://www.ibm.com/developerworks/community/blogs/6e6f6d1b-95c3-46df-8a26-b7efd8ee4b57/entry/why_big_companies_are_embracing_open_source119?lang=en.

Bachmann, A., and Bernstein, A. (2009). *Data Retrieval, Processing and Linking for Software Process Data Analysis.* Zürich, Switzerland: University of Zürich, Technical Report.

Basili, V. R. (1993). The experience factory and its relationship to other improvement paradigms. In *Software Engineering—ESEC '93*, I. Sommerville and M. Paul (Eds.) (pp. 68–83). Springer Berlin, Heidelberg.

Begel, A., DeLine, R., and Zimmermann, T. (2010, November). Social media for software engineering. In *Proceedings of the FSE/SDP Workshop on Future of Software Engineering Research* (pp. 33–38). ACM.

Biehl, J. T., Czerwinski, M., Smith, G., and Robertson, G. G. (2007, April). FASTDash: A visual dashboard for fostering awareness in software teams. In *Proceedings of the SIGCHI Conference on Human Factors in Computing Systems* (pp. 1313–1322). ACM.

Black, S., Harrison, R., and Baldwin, M. (2010). A survey of social media use in software systems development. In *Proc. of the 1st Workshop on Web 2.0 for Software Engineering* (pp. 1–5). ACM.

Black, S., and Jacobs, J. (2010). Using Web 2.0 to improve software quality. In *Proc. of the 1st Workshop on Web 2.0 for Software Engineering* (pp. 6–11). ACM.

Bolici, F., Howison, J., and Crowston, K. (2009, May). Coordination without discussion? Socio-technical congruence and stigmergy in free and open source software projects. In *Socio-Technical Congruence Workshop in Conj. Intl Conf. on Software Engineering.* Vancouver, Canada.

Bosch, J. (2009, August). From software product lines to software ecosystems. In *Proceedings of the 13th International Software Product Line Conference* (pp. 111–119). Carnegie Mellon University.

Bougie, G., Starke, J., Storey, M. A., and German, D. M. (2011, May). Towards understanding Twitter use in software engineering: Preliminary findings, ongoing challenges and future questions. In *Proceedings of the 2nd International Workshop on Web 2.0 for Software Engineering* (pp. 31–36). ACM.

Breu, S., Premraj, R., Sillito, J., and Zimmermann, T. (2010, February). Information needs in bug reports: Improving cooperation between developers and users. In *Proceedings of the 2010 ACM Conference on Computer Supported Cooperative Work* (pp. 301–310). ACM.

Buse, R. P., and Zimmermann, T. (2010, November). Analytics for software development. In *Proceedings of the FSE/SDP Workshop on Future of Software Engineering Research* (pp. 77–80). ACM.

Buse, R. P., and Zimmermann, T. (2012, June). Information needs for software development analytics. In *Proceedings of the 2012 International Conference on Software Engineering* (pp. 987–996). IEEE Press.

Campbell, P. R., and Ahmed, F. (2010, August). A three-dimensional view of software ecosystems. In *Proceedings of the Fourth European Conference on Software Architecture: Companion Volume* (pp. 81–84). ACM.

Crockford, D. (2006). The application/JSON media type for javascript object notation (JSON). RFC 4627, July 2006.

Crowston, K. (1997). A coordination theory approach to organizational process design. *Organization Science*, *8*(2), 157–175.

Dabbish, L., Stuart, C., Tsay, J., and Herbsleb, J. (2012, February). Social coding in GitHub: Transparency and collaboration in an open software repository. In *Proceedings of the ACM 2012 Conference on Computer Supported Cooperative Work* (pp. 1277–1286). ACM.

Davenport, T. H., Harris, J. G., and Morison, R. (2010). *Analytics at Work: Smarter Decisions, Better Results.* Harvard Business Press, Cambridge, MA.

Edmunds, A., and Morris, A. (2000). The problem of information overload in business organisations: A review of the literature. *International Journal of Information Management, 20*(1), 17–28.

Feeds (2014). Accessed on March 11, 2014. Retrieved from http://developer.github.com/v3/activity/feeds/.

Fussell, S. R., Kraut, R. E., Lerch, F. J., Scherlis, W. L., McNally, M. M., and Cadiz, J. J. (1998, November). Coordination, overload and team performance: Effects of team communication strategies. In *Proceedings of the 1998 ACM Conference on Computer Supported Cooperative Work* (pp. 275–284). ACM.

Gutwin, C., Greenberg, S., and Roseman, M. (1996). Workspace awareness in real-time distributed groupware: Framework, widgets, and evaluation. In *People and Computers XI*, R. J. Sasse, A. Cunningham, and R. Winder (Eds.) (pp. 281–298). Springer, London.

Gutwin, C., Penner, R., and Schneider, K. (2004, November). Group awareness in distributed software development. In *Proceedings of the 2004 ACM Conference on Computer Supported Cooperative Work* (pp. 72–81). ACM.

Hanna, R., Rohm, A., and Crittenden, V. L. (2011). We're all connected: The power of the social media ecosystem. *Business Horizons* 54(3), 265–273.

Hassan, A. E. (2006, September). Mining software repositories to assist developers and support managers. In *ICSM '06 22nd IEEE International Conference on Software Maintenance, 2006* (pp. 339–342). IEEE.

Jansen, S., Finkelstein, A., and Brinkkemper, S. (2009, May). A sense of community: A research agenda for software ecosystems. In *Software Engineering-Companion Volume, 2009. ICSE-Companion 2009* (pp. 187–190). IEEE.

Java, A., Song, X., Finin, T., and Tseng, B. (2007, August). Why we twitter: Understanding microblogging usage and communities. In *Proceedings of the 9th WebKDD and 1st SNA-KDD 2007 Workshop on Web Mining and Social Network Analysis* (pp. 56–65). ACM.

Kietzmann, J. H., Hermkens, K., McCarthy, I. P., and Silvestre, B. S. (2011). Social media? Get serious! Understanding the functional building blocks of social media. *Business Horizons 54*(3), 241–251.

Klapp, O. E. (1986). *Overload and Boredom: Essays on the Quality of Life in the Information Society* (pp. 98–99). Greenwood Press, Connecticut.

Ko, A. J., DeLine, R., and Venolia, G. (2007, May). Information needs in collocated software development teams. In *Proceedings of the 29th International Conference on Software Engineering* (pp. 344–353). IEEE Computer Society.

Kwak, H., Lee, C., Park, H., and Moon, S. (2010, April). What is Twitter, a social network or a news media? In *Proceedings of the 19th International Conference on World Wide Web* (pp. 591–600). ACM.

Lietsala, K., and Sirkkunen, E. (2008). *Social Media—Introduction to the Tools and Processes of Participatory Economy.* University of Tampere Press, Tampere, Finland.

Marlow, J., Dabbish, L., and Herbsleb, J. (2013, February). Impression formation in online peer production: Activity traces and personal profiles in GitHub. In *Proceedings of the 2013 Conference on Computer Supported Cooperative Work* (pp. 117–128). ACM.

Matthews, T., Whittaker, S., Badenes, H., and Smith, B. A. (2014). Beyond end user content to collaborative knowledge mapping: Interrelations among community social tools. In *Proceedings of the 17th ACM Conference on Computer Supported Cooperative Work and Social Computing* (pp. 900–910). ACM.

Messerschmitt, D. G., and Szyperski, C. (2005). *Software Ecosystem: Understanding an Indispensable Technology and Industry*. MIT Press Books, Cambridge, MA.

Perlow, L. A. (1999). The time famine: Toward a sociology of work time. *Administrative Science Quarterly*, *44*(1), 57–81.

Perry, D. E., Staudenmayer, N. A., and Votta, L. G. (1994). People, organizations, and process improvement. *Software, IEEE*, *11*(4), 36–45.

Reinhardt, W. (2009). Communication is the key-support durable knowledge sharing in software engineering by microblogging. In *Software Engineering (Workshops)* (pp. 329–340).

Seaman, C. B., and Basili, V. R. (1998). Communication and organization: An empirical study of discussion in inspection meetings. *IEEE Transactions on Software Engineering*, *24*(7), 559–572.

Singer, L., Figueira Filho, F., Cleary, B., Treude, C., Storey, M. A., and Schneider, K. (2013, February). Mutual assessment in the social programmer ecosystem: An empirical investigation of developer profile aggregators. In *Proceedings of the 2013 Conference on Computer Supported Cooperative Work* (pp. 103–116). ACM.

Singer, L., Figueira Filho, F., and Storey, M. A. (2014). Software engineering at the speed of light: How developers stay current using Twitter. In *Proc. ICSE 2014*.

Thung, F., Bissyande, T. F., Lo, D., and Jiang, L. (2013, March). Network structure of social coding in GitHub. In *17th European Conference on Software Maintenance and Reengineering (CSMR)*, (pp. 323–326). IEEE.8.

Tian, Y., Achananuparp, P., Lubis, I. N., Lo, D., and Lim, E. P. (2012, June). What does software engineering community microblog about? In *Mining Software Repositories (MSR), 2012 9th IEEE Working Conference on* (pp. 247–250). IEEE.

Vasilescu, B., Filkov, V., and Serebrenik, A. (2013, September). StackOverflow and GitHub: Associations between software development and crowdsourced knowledge. In *SocialCom 2013* (pp. 188–195). IEEE.

Vasilescu, B., Serebrenik, A., Devanbu, P., and Filkov, V. (2014). How social Q&A sites are changing knowledge sharing in open source software communities. In *CSCW*. ACM.

Wu, Y., Kropczynski, J., Shih, P., and Carroll, J. M. (2014). Exploring the ecosystem of software developers on GitHub and other platforms. In *CSCW '14 Interactive Posters*.

Zhao, D., and Rosson, M. B. (2009, May). How and why people Twitter: The role that micro-blogging plays in informal communication at work. In *Proceedings of the ACM 2009 International Conference on Supporting Group Work* (pp. 243–252). ACM.

# III

## Big Data Stream Techniques and Algorithms

CHAPTER 11

# Key Technologies for Big Data Stream Computing

Dawei Sun, Guangyan Zhang, Weimin Zheng, and Keqin Li

CONTENTS

## ABSTRACT

As a new trend for data-intensive computing, real-time stream computing is gaining significant attention in the Big Data era. In theory, stream computing is an effective way to support Big Data by providing extremely low-latency processing tools and massively parallel processing architectures in real-time data analysis. However, in most existing stream computing environments, how to efficiently deal with Big Data stream computing and how to build efficient Big Data stream computing systems are posing great challenges to Big Data computing research. First, the data stream graphs and the system architecture for Big Data stream computing, and some related key technologies, such as system structure, data transmission, application interfaces, and high availability, are systemically researched. Then, we give a classification of the latest research and depict the development status of some popular Big Data stream computing systems, including Twitter Storm, Yahoo! S4, Microsoft TimeStream, and Microsoft Naiad. Finally, the potential challenges and future directions of Big Data stream computing are discussed.

## 11.1 INTRODUCTION

Big Data computing is a new trend for future computing, with the quantity of data growing and the speed of data increasing. In general, there are two main mechanisms for Big Data computing, that is, Big Data stream computing (BDSC) and Big Data batch computing. BDSC is a model of straight-through computing, such as Storm [1] and S4 [2], which does for stream computing what Hadoop does for batch computing, while Big Data batch computing is a model of storing then computing, such as the MapReduce framework [3] open-sourced by the Hadoop implementation [4].

Essentially, Big Data batch computing is not sufficient for many real-time application scenarios, where a data stream changes frequently over time and the latest data are the most important and most valuable. For example, when analyzing data from real-time transactions (e.g., financial trades, e-mail messages, user search requests, sensor data tracking), a data stream grows monotonically over time as more transactions take place. Ideally, a real-time application environment can be supported by BDSC. Generally, Big Data streaming computing has the following defining characteristics [5,6]. (1) The input data stream is a real-time data stream and needs real-time computing, and the results must be updated every time the data changes. (2) Incoming data arrive continuously at volumes that far exceed the capabilities of individual machines. (3) Input streams incur multistaged computing at low latency to produce output streams, where any incoming data entry is ideally reflected in the newly generated results in output streams within seconds.

### 11.1.1 Stream Computing

Stream computing, the long-held dream of "high real-time computing" and "high-throughput computing," with programs that compute continuous data streams, has opened up the new era of future computing due to Big Data, which is a data set that is large, fast, dispersed, unstructured, and beyond the ability of available hardware and software facilities to undertake its acquisition, access, analytics, and application in a reasonable amount of time and space [7,8]. Stream computing is a computing paradigm that reads data from collections of software or hardware sensors in stream form and computes continuous data streams, where feedback results should be in a real-time data stream as well. A data stream is a sequence of data sets, a continuous stream is an infinite sequence of data sets, and parallel streams have more than one stream to be processed at the same time.

Stream computing is one effective way to support Big Data by providing extremely low-latency velocities with massively parallel processing architectures and is becoming the fastest and most efficient way to obtain useful knowledge from Big Data, allowing organizations to react quickly when problems appear or to predict new trends in the near future [9,10].

A Big Data input stream has the characteristics of high speed, real time, and large volume for applications such as sensor networks, network monitoring, microblogging, web exploring, social networking, and so on. These data sources often take the form of continuous data streams, and timely analysis of such a data stream is very important as the life cycle of most data is very short [8,11,12]. Furthermore, the volume of data is so high that there is not enough space for storage, and not all data need to be stored. Thus, the storing-then-computing batch computing model does not fit at all. Nearly all data in Big Data environments have the feature of streams, and stream computing has appeared to solve the dilemma of Big Data computing by computing data online within real-time constraints [13]. Consequently, the stream computing model will be a new trend for high-throughput computing in the Big Data era.

### 11.1.2 Application Background

BDSC is able to analyze and process data in real time to gain immediate insight, and it is typically applied to the analysis of a vast amount of data in real time and to processing them at a high speed. Many application scenarios require BDSC. For example, in financial industries, Big Data stream computing technologies can be used in risk management, marketing management, business intelligence, and so on. In the Internet, BDSC technologies can be used in search engines, social networking, and so on. In the Internet of things, BDSC technologies can be used in intelligent transportation, environmental monitoring, and so on.

Usually, a BDSC environment is deployed in a highly distributed clustered environment, as the amount of data is infinite, the rate of the data stream is high, and the results should be real-time feedback.

### 11.1.3 Chapter Organization

The remainder of this chapter is organized as follows. In Section 11.2, we introduce data stream graphs and the system architecture for BDSC and key technologies for BDSC systems. In Section 11.3, we present the system architecture and key technologies of four popular example BDSC systems, which are Twitter Storm, Yahoo! S4, Microsoft TimeStream, and Microsoft Naiad. Finally, we discuss grand challenges and future directions in Section 11.4.

## 11.2 OVERVIEW OF A BDSC SYSTEM

In this section, we first present some related concepts and definitions of directed acyclic graphs and stream computing. Then, we introduce the system architecture for stream computing and the key technologies for BDSC systems in BDSC environments.

### 11.2.1 Directed Acyclic Graph and Stream Computing

In stream computing, the multiple continuous parallel data streams can be represented by a task topology, also named a data stream graph, which is usually described by a directed acyclic graph (DAG) [5,14–16]. A measurable data stream graph view can be defined by Definition 1.

**Definition 1**

A *data stream graph* $G$ is a directed acyclic graph, which is composed of set of a vertices and a set of directed edges; has a logical structure and a special function; and is denoted as $G = (V(G), E(G))$, where $V(G) = \{v_1, v_2,\ldots,v_n\}$ is a finite set of $n$ vertices, which represent tasks, and $E(G) = \{e_{1,2}, e_{1,3},\ldots,e_{n-1,n}\}$ is a finite set of directed edges, which represent a data stream between vertices. If $\exists e_{i,j} \in E(G)$, then $v_i, v_j \in V(G)$, $v_i \neq v_j$, and $\langle v_i, v_j\rangle$ is an ordered pair, where a data stream comes from $v_i$ and goes to $v_j$.

The in-degree of vertex $v_i$ is the number of incoming edges, and the out-degree of vertex $v_i$ is the number of outgoing edges. A source vertex is a vertex whose in-degree is 0, and an end vertex is a vertex whose out-degree is 0. A data stream graph $G$ has at least one source vertex and one end vertex.

For the example data stream graph with 11 vertices shown in Figure 11.1, the set of vertices is $V = \{v_a, v_b,\ldots,v_k\}$, the set of directed edges is $E = \{e_{a,c}, e_{b,c},\ldots,e_{j,k}\}$, the source vertices

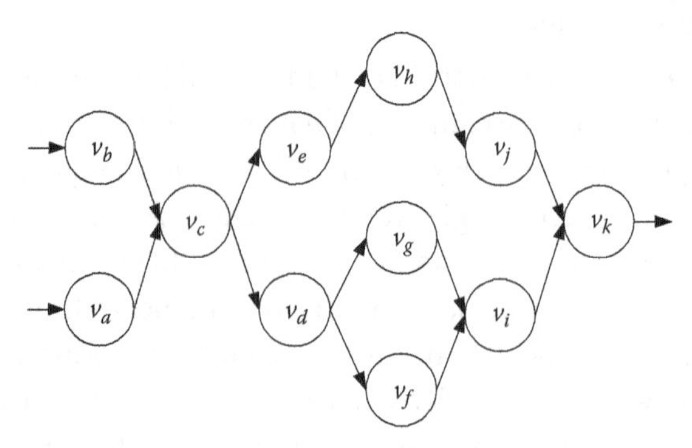


FIGURE 11.1 A data stream graph.

are $v_a$ and $v_b$, and the end vertex is $v_k$. The in-degree of vertex $v_d$ is 1, and the out-degree of vertex $v_d$ is 2.

**Definition 2**

A *subgraph sub-G* of the data stream graph G is a subgraph consisting of a subset of the vertices with the edges in between. For vertices $v_i$ and $v_j$ in the subgraph *sub-G* and any vertex $v$ in the data stream graph G, $v$ must also be in the *sub-G* if $v$ is on a directed path from $v_i$ to $v_j$, that is, $\forall v_i, v_j \in V(sub\text{-}G), \forall v \in V(G)$, and if $v \in V(p(v_i, v_j))$, then $v \in V(p(sub\text{-}G))$.

A subgraph *sub-G* is logically equivalent and can be substituted by a vertex. But reducing that subgraph to a single logical vertex would create a graph with cycle, not a DAG.

**Definition 3**

A *path* $p(v_u, v_v)$ from vertex $v_u$ to vertex $v_v$ is a subset of $E(p(v_u, v_v))$, which should meet the conditions $\exists e_{i,k} \in p(v_u, v_v)$ and $e_{k,j} \in p(v_u, v_v)$ for any directed edge $e_{k,l}$ in path $p(v_u, v_v)$ that displays the following properties: If $k \neq i$, then $\exists m$, and $e_{m,k} \in p(v_u, v_v)$; if $i \neq j$, then $\exists m$, and $e_{l,m} \in p(v_u, v_v)$.

The latency $l_p(v_u, v_v)$ of a path from vertex $v_u$ to vertex $v_v$ is the sum of latencies of both vertices and edges on the path, as given by Equation 11.1:

$$l_p(v_u, v_v) = \sum_{v_i \in V(p(v_u, v_v))} c_{v_i} + \sum_{e_{i,j} \in E(p(v_u, v_v))} c_{e_{i,j}}, c_{v_i}, c_{e_{i,j}} \geq 0. \tag{11.1}$$

A critical path, also called the longest path, is a path with the longest latency from a source vertex $v_s$ to an end vertex $v_e$ in a data stream graph G, which is also the latency of data stream graph G.

If there are $m$ paths from source vertex $v_s$ to end vertex $v_e$ in data stream graph G, then the latency $l(G)$ of data stream graph G is given by Equation 11.2:

$$l(G) = \max\left\{l_{p_i}(v_s, v_e), l_{p_2}(v_s, v_e), \ldots, l_{p_m}(v_s, v_e)\right\}, \tag{11.2}$$

where $l_{p_i}(v_s, v_e)$ is the latency of the $i$th path from vertex $v_s$ to vertex $v_e$.

**Definition 4**

In data stream graph G, if $\exists e_{i,j}$ from vertex $v_i$ to vertex $v_j$, then vertex $v_i$ is a *parent* of vertex $v_j$, and vertex $v_j$ is a *child* of vertex $v_i$.

**Definition 5**

The *throughput* $t(v_i)$ of vertex $v_i$ is the average rate of successful data stream computing in a Big Data environment and is usually measured in bits per second (bps).

We identify the source vertex $v_s$ as in the first level, the children of source vertex $v_s$ as in the second level, and so on, and the end vertex $v_e$ as in the last level.

The throughput $t(level_i)$ of the *i*th level can be calculated by Equation 11.3:

$$t(level_i) = \sum_{k=1}^{n_i} t(v_k), \tag{11.3}$$

where $n_i$ is the number of vertices in the *i*th level.

If data stream graph *G* has *m* levels, then the throughput $t(G)$ of the data stream graph *G* is the minimum throughput of all the throughput in the *m* levels, as described by Equation 11.4:

$$t(G) = \min\{t(level_1), t(level_2), \ldots, t(level_m)\}, \tag{11.4}$$

where $t(level_i)$ is the throughput of the *i*th level in data stream *G*.

**Definition 6**

A *topological sort* $TS(G) = \left(v_{x_1}, v_{x_2}, \ldots, v_{x_n}\right)$ of the vertices $V(G)$ in data stream graph *G* is a linear ordering of its vertices, such that for every directed edge $e_{x_i,x_j}\left(e_{x_i,x_j} \in E(G)\right)$ from vertex $v_{x_i}$ to vertex $v_{x_j}$, $v_{x_i}$ comes before $v_{x_j}$ in the topological ordering.

A topological sort is possible if and only if the graph has no directed cycle, that is, it needs to be a directed acyclic graph. Any directed acyclic graph has at least one topological sort.

**Definition 7**

A *graph partitioning* $GP(G) = \{GP_1, GP_2, \ldots, GP_m\}$ of the data stream graph *G* is a topological sort–based split of the vertex set $V(G)$ and the corresponding directed edges. A graph partitioning should meet the nonoverlapping and covering properties, that is, if $\forall i \neq j$, $i, j \in [1, m]$, then $GP_i \cap GP_j = \varnothing$, and $\bigcup_{i=1}^{m} GP_i = V(G)$.

### 11.2.2 System Architecture for Stream Computing

In BDSC environments, stream computing is the model of straight-through computing. As shown in Figure 11.2, the input data stream is in a real-time data stream form, all

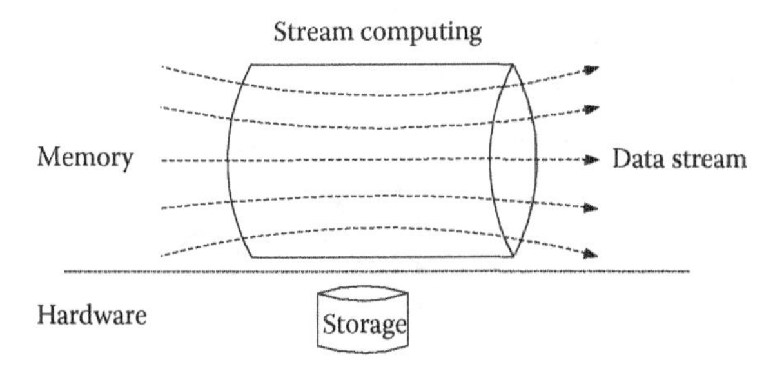


FIGURE 11.2 A Big Data stream computing environment.

continuous data streams are computed in real time, and the results must be updated also in real time. The volume of data is so high that there is not enough space for storage, and not all data need to be stored. Most data will be discarded, and only a small portion of the data will be permanently stored in hard disks.

### 11.2.3 Key Technologies for BDSC Systems

Due to data streams' distinct features of real time, volatility, burstiness, irregularity, and infinity in a Big Data environment, a well-designed BDSC system always optimizes in system structure, data transmission, application interfaces, high availability, and so on [17–19].

#### *11.2.3.1 System Structure*

Symmetric structure and master–slave structure are two main system structures for BDSC systems, as shown in Figures 11.3 and 11.4, respectively.

In the symmetric structure system, as shown in Figure 11.3, the functions of all nodes are the same. So it is easy to add a new node or to remove an unused node, and to improve the scalability of a system. However, some global functions such as resource allocation, fault tolerance, and load balancing are hard to achieve without a global node. In the S4 system, the global functions are achieved by borrowing a distributed protocol zookeeper.

In the master–slave structure system, as shown in Figure 11.4, one node is the master node, and other nodes are slave nodes. The master node is responsible for global control of the system, such as resource allocation, fault tolerance, and load balancing. Each slave node

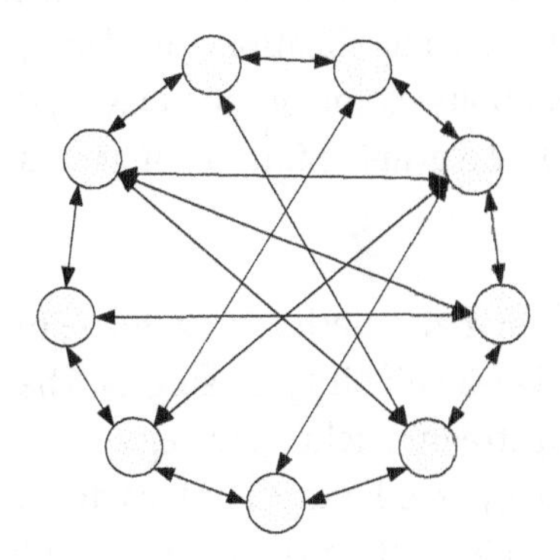

FIGURE 11.3 Symmetric structure.

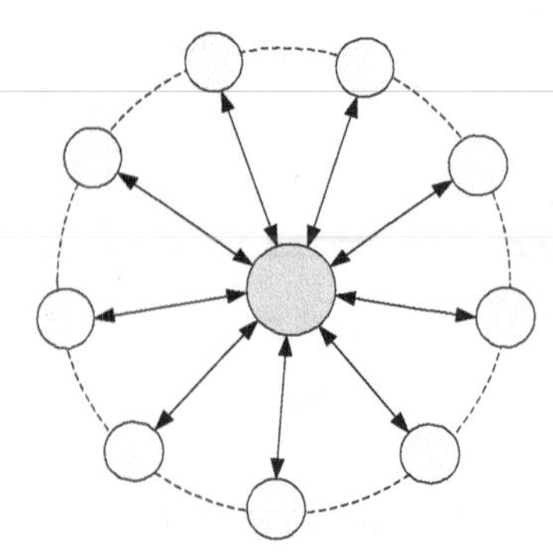

FIGURE 11.4 Master–slave structure.

has a special function, and it always receives a data stream from the master node, processes the data stream, and sends the results to the master node. Usually, the master node is the bottleneck in the master–slave structure system. If it fails, the whole system will not work.

#### *11.2.3.2 Data Stream Transmission*

Push and pull are two main data stream transmissions in a BDSC system.

In a push system, once an upstream node gets a result, it will immediately push the result data to downstream nodes. When this is done, the upstream data will be immediately sent to downstream nodes. However, if some downstream nodes are busy or fail, some data will be discarded.

In a pull system, a downstream node requests data from an upstream node. If some data need to be further processed, the upstream node will send the data to the requesting downstream node. When this is done, the upstream data will be stored in upstream nodes until corresponding downstream nodes make a request. Some data will wait a long time for further processing and may lose their timeliness.

#### *11.2.3.3 Application Interfaces*

An application interface is used to design a data stream graph, a bridge between a user and a BDSC system. Usually, a good application interface is flexible and efficient for users. Currently, most BDSC systems provide MapReduce-like interfaces; for example, the Storm system provides spouts and bolts as application interfaces, and a user can design a data stream graph using spouts and bolts. Some other BDSC systems provide structured query language (SQL)-like interfaces and graphical user interfaces.

#### *11.2.3.4 High Availability*

State backup and recovery is the main method to achieve high availability in a BDSC system. There are three main high-availability strategies, that is, passive standby strategy, active standby strategy, and upstream backup strategy.

In the passive standby strategy (see Figure 11.5), each primary node periodically sends checkpoint data to a backup node. If the primary node fails, the backup node takes over from the last checkpoint. Usually, this strategy will achieve precise recovery.

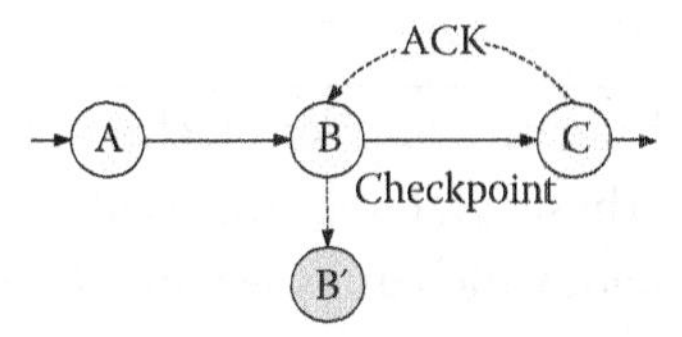


FIGURE 11.5 Passive standby. ACK, Acknowledgment.

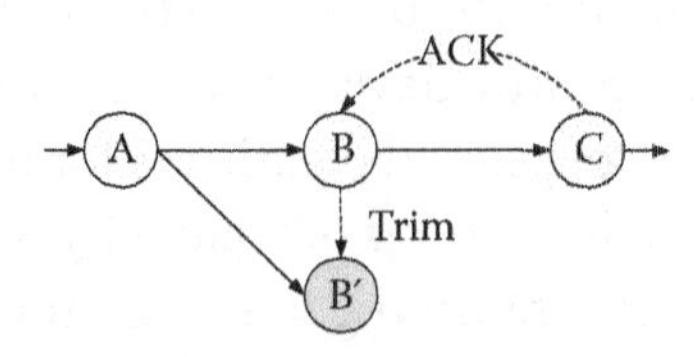


FIGURE 11.6 Active standby. ACK, Acknowledgment.

In the active standby strategy (see Figure 11.6), the secondary nodes compute all data streams in parallel with their primaries. Usually, the recovery time of this strategy is the shortest.

In the upstream backup strategy (see Figure 11.7), upstream nodes act as backups for their downstream neighbors by preserving data streams in their output queues while their downstream neighbors compute them. If a node fails, its upstream nodes replay the logged data stream on a recovery node. Usually, the runtime overhead of this strategy is the lowest.

A comparison of the three main high-availability strategies, that is, passive standby strategy, active standby strategy, and upstream backup strategy, in runtime overhead and recovery time is shown in Figure 11.8. The recovery time of the upstream backup strategy is the longest, while the runtime overhead of the passive standby strategy is the greatest.

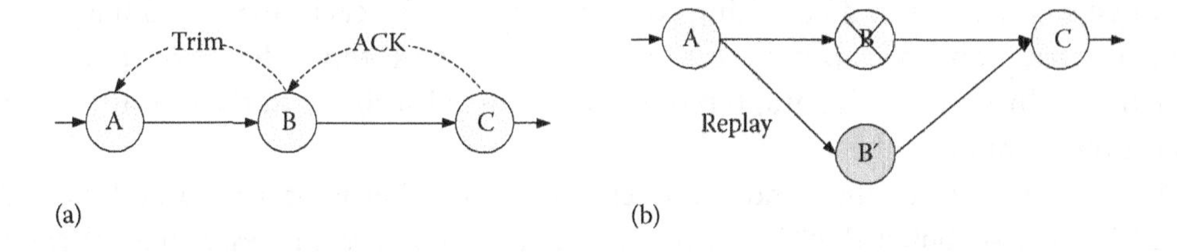


FIGURE 11.7 Upstream backup. ACK, Acknowledgment.

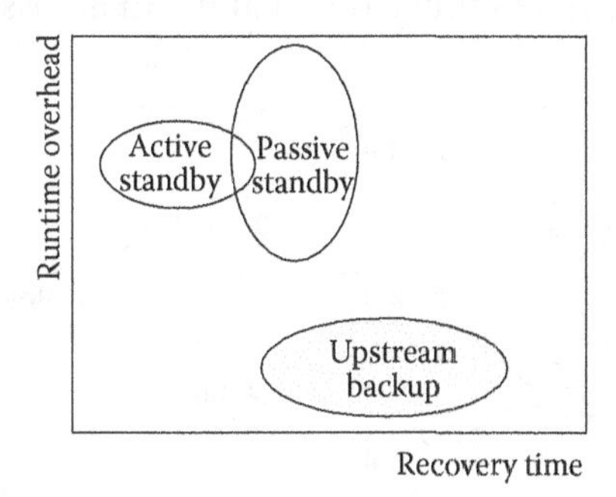


FIGURE 11.8 Comparison of high-availability strategies in runtime overhead and recovery time.

## 11.3 EXAMPLE BDSC SYSTEMS

In this section, the system architecture and key technologies of four popular BDSC system instances are presented. These systems are Twitter Storm, Yahoo! S4, Microsoft TimeStream, and Microsoft Naiad, which are specially designed for BDSC.

### 11.3.1 Twitter Storm

Storm is an open-source and distributed BDSC system licensed under the Eclipse Public License. Similar to how Hadoop provides a set of general primitives for doing batch processing, Storm provides a set of general primitives for doing real-time Big Data computing. The Storm platform has the features of simplicity, scalability, fault tolerance, and so on. It can be used with any programming language and is easy to set up and operate [1,20,21].

#### *11.3.1.1 Task Topology*

In BDSC environments, the logic for an application is packaged in the form of a task topology. Once a task topology is designed and submitted to a system, it will run forever until the user kills it.

A task topology can be described as a directed acyclic graph and comprises spouts and bolts, as shown in Figure 11.9. A spout is a source of streams in a task topology and will read data streams (in tuples) from an external source and emit them into bolts. Spouts can emit more than one data stream. The processing of a data stream in a task topology is done in bolts. Anything can be done by bolts, such as filtering, aggregations, joins, and so on. Some simple functions can be achieved by a bolt, while complex functions will be achieved by many bolts. The logic should be designed by a user. For example, transforming a stream of tweets into a stream of trending images requires at least two steps: a bolt to do a rolling count of retweets for each image and one or more bolts to stream out the top *n* images. Bolts can also emit more than one stream. Each edge in the directed acyclic graph represents a bolt subscribing to the output stream of some other spout or bolt.

A data stream is an unbounded sequence of tuples that is processed and created in parallel in a distributed BDSC environment. A task topology processes data streams in many complex ways. Repartitioning the streams between each stage of the computation is needed. Task topologies are inherently parallel and run across a cluster of machines. Any

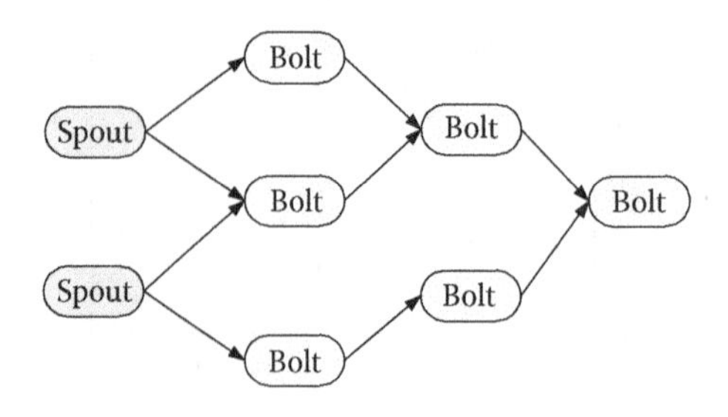


FIGURE 11.9 Task topology of Storm.

vertex in a task topology can be created in many instances. All those vertices will simultaneously process a data stream, and different parts of the topology can be allocated in different machines. A good allocating strategy will greatly improve system performance.

A data stream grouping defines how that stream should be partitioned among the bolt's tasks; spouts and bolts execute in parallel as many tasks across the cluster. There are seven built-in stream groupings in Storm, such as shuffle grouping, fields grouping, all grouping, global grouping, none grouping, direct grouping, and local or shuffle grouping; a custom stream grouping to meet special needs can also be implemented by the CustomStreamGrouping interface.

#### *11.3.1.2 Fault Tolerance*

Fault tolerance is an important feature of Storm. If a worker dies, Storm will automatically restart it. If a node dies, the worker will be restarted on another node. In Storm, Nimbus and the Supervisors are designed to be stateless and fail-fast whenever any unexpected situation is encountered, and all state information is stored in a Zookeeper server. If Nimbus or the Supervisors die, they will restart like nothing happened. This means you can kill the Nimbus and the Supervisors without affecting the health of the cluster or task topologies.

When a worker dies, the Supervisor will restart it. If it continuously fails on startup and is unable to heartbeat to Nimbus, Nimbus will reassign the worker to another machine.

When a machine dies, the tasks assigned to that machine will time out, and Nimbus will reassign those tasks to other machines.

When Nimbus or Supervisors die, they will restart like nothing happened. No worker processes are affected by the death of Nimbus or the Supervisors.

#### *11.3.1.3 Reliability*

In Storm, the reliability mechanisms guarantee that every spout tuple will be fully processed by corresponding topology. They do this by tracking the tree of tuples triggered by every spout tuple and determining when that tree of tuples has been successfully completed. Every topology has a "message timeout" associated with it. If Storm fails to detect that a spout tuple has been completed within that timeout, then it fails the tuple and replays it later.

The reliability mechanisms of Storm are completely distributed, scalable, and fault tolerant. Storm uses mod hashing to map a spout tuple ID to an acker task. Since every tuple carries with it the spout tuple IDs of all the trees they exist within, they know which acker tasks to communicate with. When a spout task emits a new tuple, it simply sends a message to the appropriate acker telling it that its task ID is responsible for that spout tuple. Then, when an acker sees that a tree has been completed, it knows to which task ID to send the completion message.

An acker task stores a map from a spout tuple ID to a pair of values. The first value is the task ID that created the spout tuple that is used later on to send completion messages. The second value is a 64-bit number called the "ack val." The ack val is a representation of the state of the entire tuple tree, no matter how big or how small. It is simply the exclusive

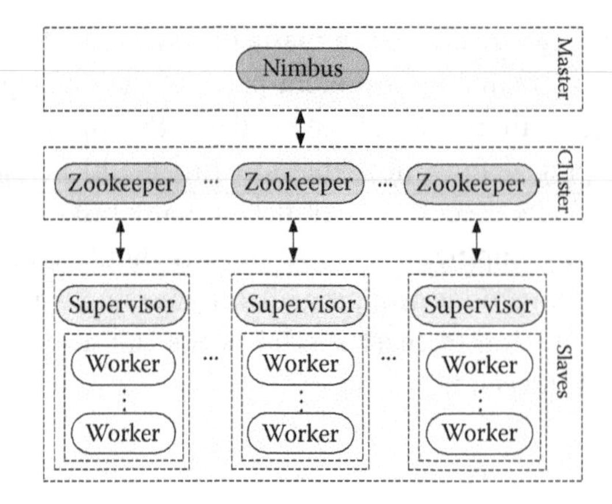


FIGURE 11.10 Storm cluster.

OR (XOR) of all tuple IDs that have been created and/or acked in the tree. When an acker task sees that an "ack val" has become 0, then it knows that the tuple tree is completed.

#### *11.3.1.4 Storm Cluster*

A Storm cluster is superficially similar to a Hadoop cluster. Whereas on Hadoop, you run "MapReduce jobs," on Storm, you run "topologies." As shown in Figure 11.10, there are two kinds of nodes on a Storm cluster, that is, the master node and the worker nodes.

The master node runs Nimbus node, which is similar to Hadoop's "JobTracker." In Storm, Nimbus node is responsible for distributing code around the cluster, assigning tasks to machines, monitoring for failures, and so on.

Each worker node runs a Supervisor node. The Supervisor listens for work assigned to its machine and starts and stops worker processes as necessary based on what Nimbus has assigned to it. Each worker process executes a subset of a topology. Usually, a running topology consists of many worker processes spread across many machines.

The coordination between Nimbus and the Supervisors is done through a Zookeeper cluster. Additionally, the Nimbus daemon and Supervisor daemons are fail-fast and stateless; all states are kept in a Zookeeper server. This means that if you kill the Nimbus or the Supervisors, they will start back up like nothing has happened.

### 11.3.2 Yahoo! S4

S4 is a general-purpose, distributed, scalable, fault-tolerant, pluggable platform that allows programmers to easily develop applications for computing continuous unbounded streams of Big Data. The core part of S4 is written in Java. The implementation is modular and pluggable, and S4 applications can be easily and dynamically combined for creating more sophisticated stream processing systems. S4 was initially released by Yahoo! Inc. in October 2010 and has been an Apache Incubator project since September 2011. It is licensed under the Apache 2.0 license [2,22–25].

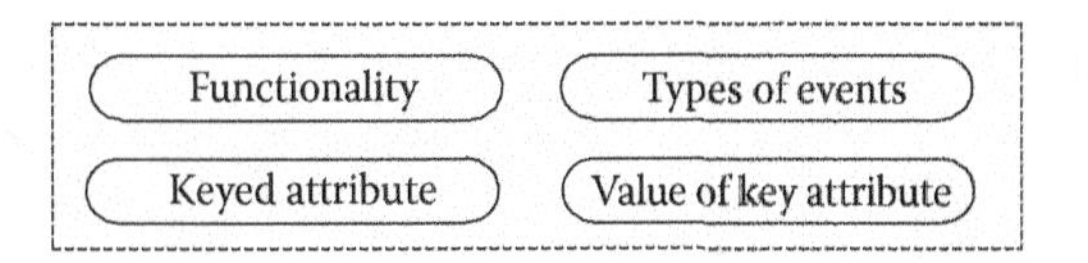


FIGURE 11.11 Processing element.

#### 11.3.2.1 Processing Element

The computing units of S4 are the *processing elements* (PEs). As shown in Figure 11.11, each instance of a PE can be identified by four components, that is, functionality, types of events, keyed attribute, and value of the keyed attribute. Each PE processes exactly those events that correspond to the value on which it is keyed.

A special class of PEs is the set of keyless PEs, with no keyed attribute or value. This type of PE will process all events of the type with which they are associated. Usually, the keyless PEs are typically used at the input layer of an S4 cluster, where events are assigned a key.

#### 11.3.2.2 Processing Nodes

*Processing nodes* (PNs) are the logical hosts to PEs. Many PEs work in a PE container, as shown in Figure 11.12. A PN is responsible for event listeners, dispatcher events, and emitter output events. In addition, the routing model, load balancing model, fail-over management model, transport protocols, and zookeeper are deployed in a communication layer.

All events will be routed to PNs by S4 according to a hash function. Every keyed PE can be mapped to exactly one PN based on the value of the hash function applied to the value of the keyed attribute of that PE. However, keyless PEs may be instantiated on every PN. The event listener model of a PN will always listen to an event from S4. If an event is allocated to a PN, it will be routed to an appropriate PE within that PN.

#### 11.3.2.3 Fail-Over, Checkpointing, and Recovery Mechanism

In S4, a fail-over mechanism will provide a high-availability environment for S4. When a node is dead, a corresponding standby node will be used. In order to minimize state loss when a node is dead, a checkpointing and recovery mechanism is employed by S4.

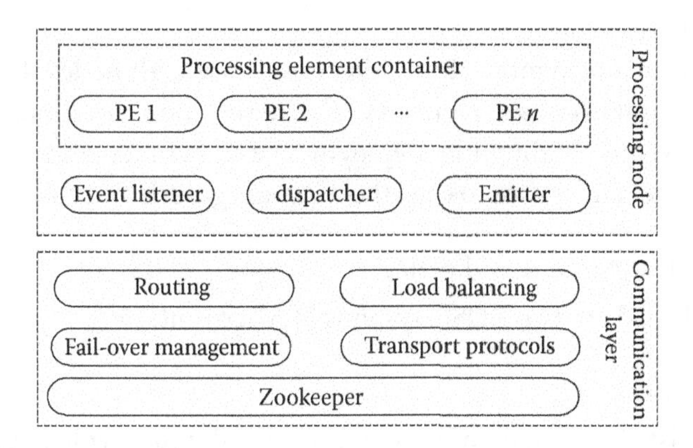


FIGURE 11.12 Processing node.

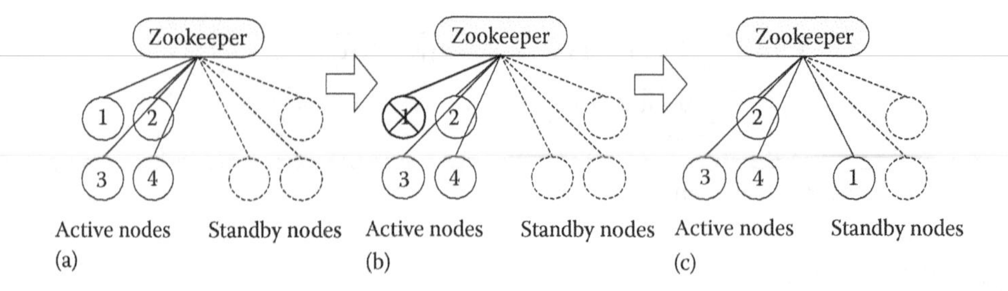


FIGURE 11.13 Fail-over mechanism. (a) In working state, (b) under failed state, and (c) after recovery state.

In order to improve the availability of the S4 system, S4 system should provide a fail-over mechanism to automatically detect failed nodes and redirect the data stream to a standby node. If you have $n$ partitions and start $m$ nodes, with $m > n$, you get $m - n$ standby nodes. For instance, if there are seven live nodes and four partitions available, four of the nodes pick the available partitions in Zookeeper. The remaining three nodes will be available standby nodes. Each active node consistently receives messages for the partition that it picked, as shown in Figure 11.13a. When Zookeeper detects that one of active nodes fails, it will notify a standby node to replace the failed node. As shown in Figure 11.13b, the node assigned with partition 1 fails. Unassigned nodes compete for a partition assignment, and only one of them picks it. Other nodes are notified of the new assignment and can reroute the data stream for partition 1, as shown in Figure 11.13c.

If a node is unreachable after a session timeout, Zookeeper will identify this node as dead. The session timeout is specified by the client upon connection and is, at minimum, twice the heartbeat specified in the Zookeeper ensemble configuration.

In order to minimize state loss when a node is dead, a checkpointing and recovery mechanism is employed by S4. The states of PEs are periodically checkpointed and stored. Whenever a node fails, the checkpoint information will be used by the recovery mechanism to recover the state of the failed node to the corresponding standby node. Most of the previous state of a failed node can be seen in the corresponding standby node.

##### *11.3.2.4 System Architecture*

In S4, a decentralized and symmetric architecture is used; all nodes share the same functionality and responsibilities (see Figure 11.14). There is no central node with specialized responsibilities. This greatly simplifies deployment and maintenance.

A pluggable architecture is used to keep the design as generic and customizable as possible.

### 11.3.3 Microsoft TimeStream and Naiad

TimeStream and Naiad are two BDSC systems of Microsoft.

##### *11.3.3.1 TimeStream*

TimeStream is a distributed system designed specifically for low-latency continuous processing of big streaming data on a large cluster of commodity machines and is based on

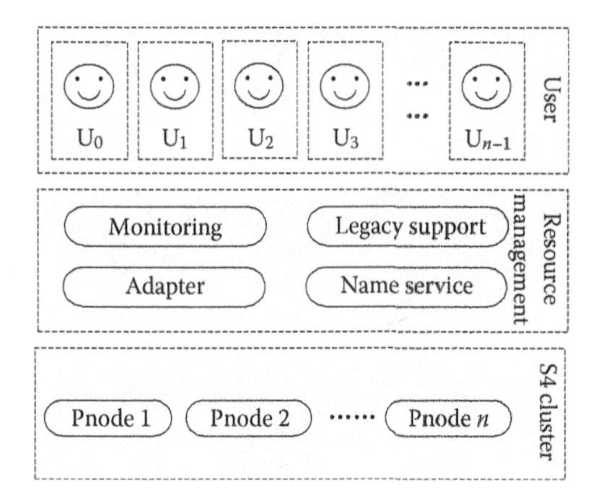


FIGURE 11.14 System architecture.

StreamInsight. TimeStream handles an online advertising aggregation pipeline at a rate of 700,000 URLs per second with a 2 s delay [5,26–29].

1. Streaming DAG

   Streaming DAG is a type of task topology, which can be dynamically reconfigured according to the loading of a data stream. All data streams in the TimeStream system will be processed in streaming DAG. Each vertex in streaming DAG will be allocated to physical machines for execution. As shown in Figure 11.15, streaming function $f_v$ of vertex $v$ is designed by the user. When input data stream $i$ is coming, streaming function $f_v$ will process data stream $i$, update $v$'s state from $\tau$ to $\tau'$, and produce a sequence $o$ of output entries as part of the output streams for downstream vertices.

   A sub-DAG is logically equivalent and can be reduced to one vertex or another sub-DAG. As shown in Figure 11.16, the sub-DAG comprised of vertices $v_2$, $v_3$, $v_4$, and $v_5$ (as well as all their edges) is a valid sub-DAG and can be reduced to a "vertex" with $i$ as its input stream and $o$ as its output stream.

2. Resilient Substitution

   Resilient substitution is an important feature of TimeStream. It is used to dynamically adjust and reconfigure streaming DAG according to the loading change of a

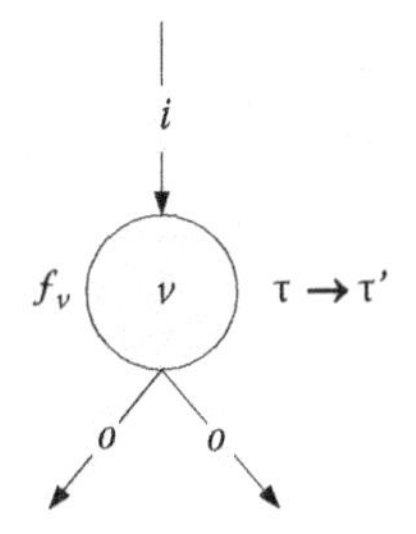


FIGURE 11.15 Streaming DAG.

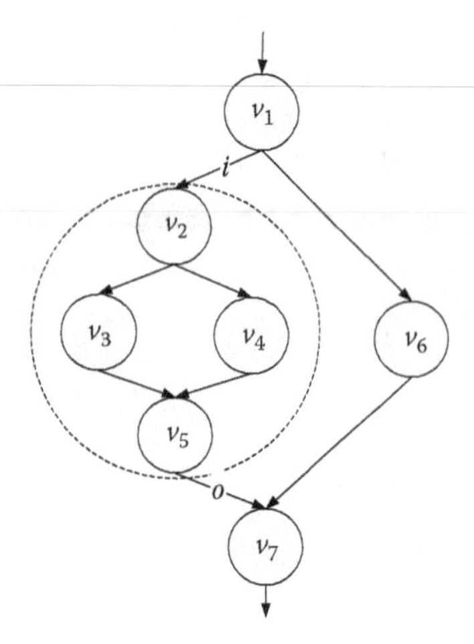


FIGURE 11.16 Streaming DAG and sub-DAG.

data stream. There are three types of resilient substitution in TimeStream. (a) A vertex is substituted by another vertex. When a vertex fails, a new corresponding standby vertex is initiated to replace the failed one and continues execution, possibly on a different machine. (b) A sub-DAG is substituted by another sub-DAG. When the number of instances of a vertex in a sub-DAG needs to be adjusted, a new sub-DAG will replace the old one. For example, as shown in Figure 11.17, a sub-DAG comprised of vertices $v_2$, $v_3$, $v_4$, and $v_5$ implements three stages: hash partitioning, computation, and union. When the load increases, TimeStream can create a new sub-DAG (shown on the left), which uses four partitions instead of two, to replace the original sub-DAG. (c) A sub-DAG is substituted by a vertex. When the load decreases, there is no need for so many steps to finish a special function, and the corresponding sub-DAG can be substituted by a vertex, as shown in Figure 11.16.

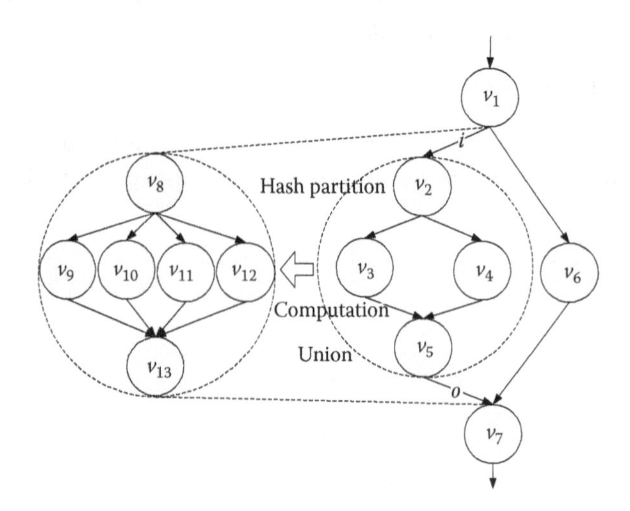


FIGURE 11.17 Resilient substitution.

#### *11.3.3.2 Naiad*

Naiad is a distributed system for executing data-parallel, cyclic dataflow programs. The core part is written in C#. It offers high throughput of batch processors and low latency of stream processors and is able to perform iterative and incremental computations. Naiad is a prototype implementation of a new computational model, timely dataflow [30].

1. Timely Dataflow

   Timely dataflow is a computational model based on directed graphs. The dataflow graph can be a directed acyclic graph, like in other BDSC environments. It can also be a directed cyclic graph; the situation of cycles in a dataflow graph is under consideration. In timely dataflow, the time stamps reflect cycle structure in order to distinguish data that arise in different input epochs and loop iterations. The external producer labels each message with an integer epoch and notifies the input vertex when it will not receive any more messages with a given epoch label.

   Timely dataflow graphs are directed graphs with the constraint that the vertices are organized into possibly nested loop contexts, with three associated system-provided vertices. Edges entering a loop context must pass through an ingress vertex, and edges leaving a loop context must pass through an egress vertex. Additionally, every cycle in the graph must be contained entirely within some loop context and include at least one feedback vertex that is not nested within any inner loop contexts. Figure 11.18 shows a single-loop context with ingress (I), egress (E), and feedback (F) vertices labeled.

2. System Architecture

   The system architecture of a Naiad cluster is shown in Figure 11.19, with a group of processes hosting workers that manage a partition of the timely dataflow vertices. Workers exchange messages locally using shared memory and remotely using TCP connections between each pair of processes.

   A program specifies its timely dataflow graph as a logical graph of stages linked by typed connectors. Each connector optionally has a partitioning function to control the exchange of data between stages. At execution time, Naiad expands the logical graph into a physical graph where each stage is replaced by a set of vertices and each connector by a set of edges. Figure 11.19 shows a logical graph and a corresponding

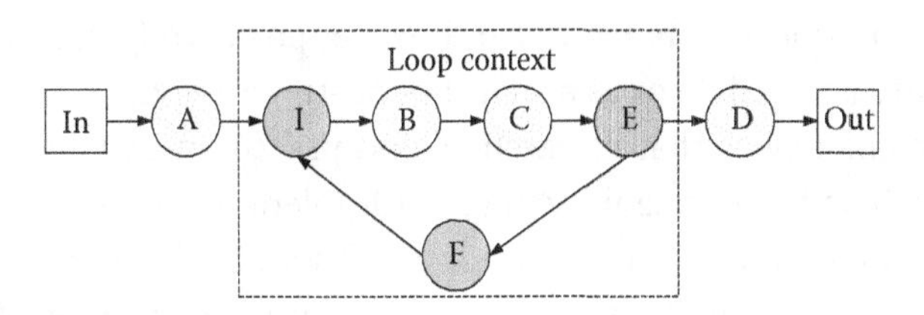


FIGURE 11.18 Timely dataflow graph.

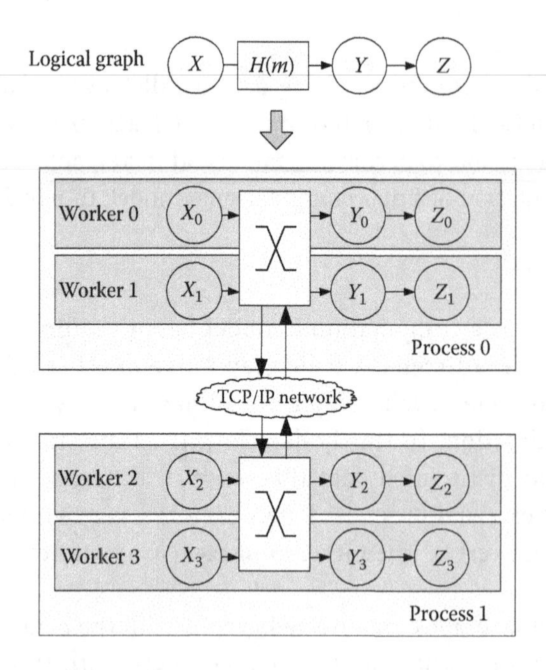


FIGURE 11.19 System architecture of a Naiad cluster.

physical graph, where the connector from $X$ to $Y$ has partitioning function $H(m)$ on typed messages $m$.

Each Naiad worker is responsible for delivering messages and notifications to vertices in its partition of the timely dataflow graph. When faced with multiple runnable actions, workers break ties by delivering messages before notifications, in order to reduce the amount of queued data.

## 11.4 FUTURE PERSPECTIVE

In this section, we focus our attention on grand challenges of BDSC and the main work we will perform in the near future.

### 11.4.1 Grand Challenges

BDSC is becoming the fastest and most efficient way to obtain useful knowledge from what is happening now, allowing organizations to react quickly when problems appear or to detect new trends helping to improve their performance. BDSC is needed to manage the data currently generated at an ever-increasing rate from such applications as log records or click-streams in web exploring, blogging, and twitter posts. In fact, all data generated can be considered as streaming data or as a snapshot of streaming data.

There are some challenges that researchers and practitioners have to deal with in the next few years, such as high scalability, high fault tolerance, high consistency, high load balancing, high throughput, and so on [6,9,31,32]. Those challenges arise from the nature of stream data, that is, data arrive at high speed and must be processed under very strict constraints of space and time.

#### *11.4.1.1 High Scalability*

High scalability of stream computing can expand to support increasing data streams and meet the quality of service (QoS) of users, or it can shrink to support decreasing data streams and improve resource utilization. In BDSC environments, it is difficult to achieve high scalability, as the change of data stream is unexpected. The key is that the software changes along with the data stream change, grows along with increased usage, or shrinks along with decreased usage. This means that scalable programs take up limited space and resources for smaller data needs but can grow efficiently as more demands are placed on the data stream.

To achieve high scalability in BDSC environments, a good scalable system architecture, a good effective resource allocation strategy, and a good data stream computing mode are required.

#### *11.4.1.2 High Fault Tolerance*

Highly fault-tolerant stream computing can enable a system to continue operating properly in the event of the failure of (or one or more faults within) some of its components. Fault tolerance is particularly sought after in high-availability or life-critical systems. In BDSC environments, it is difficult to achieve high fault tolerance, as the data stream is infinite and real time, and more importantly, most of the data are useless.

To achieve high fault tolerance in BDSC environments, a good scalable high-fault-tolerance strategy is needed, as fault tolerance provides additional resources that allow an application to continue working after a component failure without interruption.

#### *11.4.1.3 High Consistency*

Highly consistent stream computing can improve system stability and enhance system efficiency. In BDSC environments, it is difficult to achieve high consistency, as it is hard to decide which nodes should be consistent and which data are needed.

To achieve high consistency in BDSC environments, a good system structure is required. Usually, the master–slave structure is a good choice, as all data are in the master node, and it is easy to achieve highly consistent states.

#### *11.4.1.4 High Load Balancing*

Highly load-balanced stream computing can make a stream computing system self-adaptive to the changes of data streams and avoid load shedding. In BDSC environments, it is difficult to achieve high load balancing, as it is impossible to dedicate resources that cover peak loads 24 h a day, 7 days a week. Traditionally, stream computing systems use load shedding when the workload exceeds their processing. This employs a trade-off between delivering a low-latency response and ensuring that all incoming data streams are processed. However, load shedding is not feasible when the variance between peak and average workload is high, and the response should always be kept in real time for users.

To achieve high load balancing in BDSC environments, a good distributed computing environment is needed. It should provide scalable stream computing that automatically streams a partial data stream to a global computing center when local resources become insufficient.

#### *11.4.1.5 High Throughput*

High-throughput stream computing will improve data stream computing ability by running multiple independent instances of a task topology graph on multiple processors at the same time. In BDSC environments, it is difficult to achieve high throughput, as it is hard to decide how to identify the need for a replication subgraph in a task topology graph, to decide the number of replicas, and to decide the fraction of the data stream to assign to each replica.

To achieve high throughput in BDSC environments, a good multiple-instance replication strategy is needed. Usually, the data stream loading of all instances of all nodes in a task topology graph being equal is a good choice, as the computing ability of all computing nodes are efficient, and it is easy to achieve high-throughput states.

### 11.4.2 On-the-Fly Work

Future investigation will focus on the following aspects:

1. Research on new strategies to optimize a task topology graph, such as subgraph partitioning strategy, subgraph replication strategy, and subgraph allocating strategy, and to provide a high-throughput BDSC environment
2. Research on dynamic extensible data stream strategies, such that a data stream can be adjusted according to available resources and the QoS of users, and provide a highly load-balancing BDSC environment
3. Research on the impact of a task topology graph with a cycle, and a corresponding task topology graph optimize strategy and resource allocating strategy, and provide a highly adaptive BDSC environment
4. Research on the architectures for large-scale real-time stream computing environments, such as symmetric architecture and master–slave architecture, and provide a highly consistent BDSC environment
5. Develop a BDSC system with the features of high throughput, high fault tolerance, high consistency, and high scalability, and deploy such a system in a real BDSC environment

## ACKNOWLEDGMENTS

This work was supported in part by the National Natural Science Foundation of China under Grant No. 61170008 and Grant No. 61272055, in part by the National Grand Fundamental Research 973 Program of China under Grant No. 2014CB340402 in part by the National High Technology Research and Development Program of China under Grant No. 2013AA01A210, and in part by the China Postdoctoral Science Foundation under Grant No. 2014M560976.

## REFERENCES


1. Storm. Available at http://storm-project.net/ (accessed July 16, 2013).
2. Neumeyer L, Robbins B, Nair A et al. S4: Distributed stream computing platform, *Proc. 10th IEEE International Conference on Data Mining Workshops, ICDMW 2010*, Sydney, NSW, Australia, IEEE Press, December 2010, pp. 170–177.
3. Zhao Y and Wu J. Dache: A data aware caching for big-data applications using the MapReduce framework, *Proc. 32nd IEEE Conference on Computer Communications, INFOCOM 2013*, IEEE Press, April 2013, pp. 35–39.
4. Shang W, Jiang Z M, Hemmati H et al. Assisting developers of Big Data analytics applications when deploying on Hadoop clouds, *Proc. 35th International Conference on Software Engineering, ICSE 2013*, IEEE Press, May 2013, pp. 402–411.
5. Qian Z P, He Y, Su C Z et al. TimeStream: Reliable stream computation in the cloud, *Proc. 8th ACM European Conference on Computer Systems, EuroSys 2013*, Prague, Czech Republic, ACM Press, April 2013, pp. 1–14.
6. Umut A A and Yan C. Streaming Big Data with self-adjusting computation, *Proc. 2013 ACM SIGPLAN Workshop on Data Driven Functional Programming, Co-located with POPL 2013, DDFP 2013*, Rome, Italy, ACM Press, January 2013, pp. 15–18.
7. Demirkan H and Delen D. Leveraging the capabilities of service-oriented decision support systems: Putting analytics and Big Data in cloud, *Decision Support Systems*, vol. 55(1), 2013, pp. 412–421.
8. Albert B. Mining Big Data in real time, *Informatica (Slovenia)*, vol. 37(1), 2013, pp. 15–20.
9. Lu J and Li D. Bias correction in a small sample from Big Data, *IEEE Transactions on Knowledge and Data Engineering*, vol. 25(11), 2013, pp. 2658–2663.
10. Tien J M. Big Data: Unleashing information, *Journal of Systems Science and Systems Engineering*, vol. 22(2), 2013, pp. 127–151.
11. Zhang R, Koudas N, Ooi B C et al. Streaming multiple aggregations using phantoms, *VLDB Journal*, vol. 19(4), 2010, pp. 557–583.
12. Hirzel M, Andrade H, Gedik B et al. IBM streams processing language: Analyzing Big Data in motion, *IBM Journal of Research and Development*, vol. 57(3/4), 2013, pp. 7:1–7:11.
13. Dayarathna M and Toyotaro S. Automatic optimization of stream programs via source program operator graph transformations, *Distributed and Parallel Databases*, vol. 31(4), 2013, pp. 543–599.
14. Farhad S M, Ko Y, Burgstaller B et al. Orchestration by approximation mapping stream programs onto multicore architectures, *Proc. 16th International Conference on Architectural Support for Programming Languages and Operating Systems, ASPLOS 2011*, ACM Press, March 2011, pp. 357–367.
15. Schneider S, Hirzel M and Gedik B. Tutorial: Stream processing optimizations, *Proc. 7th ACM International Conference on Distributed Event-Based Systems, DEBS 2013*, ACM Press, June 2013, pp. 249–258.
16. Khandekar R, Hildrum K, Parekh S et al. COLA: Optimizing stream processing applications via graph partitioning, *Proc. 10th ACM/IFIP/USENIX International Conference on Middleware, Middleware 2009*, ACM Press, November 2009, pp. 1–20.
17. Scalosub G, Marbach P and Liebeherr J. Buffer management for aggregated streaming data with packet dependencies, *IEEE Transactions on Parallel and Distributed Systems*, vol. 24(3), 2013, pp. 439–449.
18. Malensek M, Pallickara S L and Pallickara S. Exploiting geospatial and chronological characteristics in data streams to enable efficient storage and retrievals, *Future Generation Computer Systems*, vol. 29(4), 2013, pp. 1049–1061.
19. Cugola G and Margara A. Processing flows of information: From data stream to complex event processing, *ACM Computing Surveys*, vol. 44(3), 2012, pp. 15:1–15:62.

20. Storm wiki. Available at http://en.wikipedia.org/wiki/Storm (accessed July 16, 2013).
21. Storm Tutorial. Available at https://github.com/nathanmarz/storm/wiki (accessed July 16, 2013).
22. Chauhan J, Chowdhury S A and Makaroff D. Performance evaluation of Yahoo! S4: A first look, *Proc. 7th International Conference on P2P, Parallel, Grid, Cloud and Internet Computing, 3PGCIC 2012*, Victoria, BC, Canada, IEEE Press, November 2012, pp. 58–65.
23. Simoncelli D, Dusi M, Gringoli F et al. Scaling out the performance of service monitoring applications with BlockMon, *Proc. 14th International Conference on Passive and Active Measurement, PAM 2013*, Hong Kong, China, IEEE Press, March 2013, pp. 253–255.
24. S4, distributed stream computing platform. Available at http://incubator.apache.org/s4/ (accessed July 16, 2013).
25. Stream computing StreamBase Yahoo S4 borealis comparis. Available at http://oracle-abc.wikidot.com/zh:stream-computing-streambase-yahoo-s4-borealis-comparison (accessed July 16, 2013).
26. Guo Z Y, Sean M D, Yang M et al. Failure recovery: When the cure is worse than the disease, *Proc. 14th USENIX conference on Hot Topics in Operating Systems, USENIX 2013*, Santa Ana Pueblo, NM, ACM Press, May 2013, pp. 1–6.
27. Ali M, Chandramouli B, Goldstein J et al. The extensibility framework in Microsoft StreamInsight, *Proc. IEEE 27th International Conference on Data Engineering, ICDE 2011*, Hannover, Germany, IEEE Press, April 2011, pp. 1242–1253.
28. Chandramouli B, Goldstein J, Barga R et al. Accurate latency estimation in a distributed event processing system, *Proc. IEEE 27th International Conference on Data Engineering, ICDE 2011*, Hannover, Germany, IEEE Press, April 2011, pp. 255–266.
29. Ali M, Chandramouli B, Fay J et al. Online visualization of geospatial stream data using the WorldWide telescope, *VLDB Endowment*, vol. 4(12), 2011, pp. 1379–1382.
30. Derek G M, Frank M S, Rebecca I et al. Naiad: A timely dataflow system, *Proc. the 24th ACM Symposium on Operating Systems Principles, SOSP 2013*, Pennsylvania, ACM Press, November 2013, pp. 439–455.
31. Garzo A, Benczur A A, Sidlo C I et al. Real-time streaming mobility analytics, *Proc. 2013 IEEE International Conference on Big Data, Big Data 2013*, Santa Clara, CA, IEEE Press, October 2013, pp. 697–702.
32. Zaharia M, Das T, Li H et al. R Discretized streams: Fault-tolerant streaming computation at scale, *Proc. the 24th ACM Symposium on Operating Systems Principles, SOSP 2013*, Farmington, PA, ACM Press, November 2013, pp. 423–438.

CHAPTER 12

# Streaming Algorithms for Big Data Processing on Multicore Architecture

Marat Zhanikeev

CONTENTS

## ABSTRACT

This chapter brings together three topics: hash functions, Bloom filters, and the recently emerged streaming algorithms. Hashing is the oldest of the three and is backed by much literature. Bloom filters are based on hash functions and benefit from hashing efficiency directly. Streaming algorithms use both Bloom filters and hashing in various ways but impose strict requirements on performance. This chapter views the three topics from the viewpoint of efficiency and speed. The two main performance metrics are per-unit processing time and the size of the memory footprint. All algorithms are presented as C/C++ pseudocode. Specific attention is paid to the feasibility of hardware implementation.

## 12.1 INTRODUCTION

HaDoop file system (HDFS) [1] and MapReduce are de facto standards in Big Data processing today. Although they are two separate technologies, they form a single package as far as Big Data *processing*—not just storage—is concerned. This chapter will treat them as one package. Today, Hadoop and/or MapReduce lack popular alternatives [2]. HDFS solves the practical problem of not being able to store Big Data on a single machine by distributing the storage over multiple nodes [3]. MapReduce is a framework on which one can run jobs that process the contents of the storage—also in a distributed manner—and generate statistical summaries. This chapter will show that performance improvements mostly target MapReduce [4].

There are several fundamental problems with MapReduce. First, the *map* and *reduce* operators are restricted to key–value hashes (data type, not hash function), which places a cap on usability. For example, while *data streaming* is a good alternative for Big Data processing, MapReduce fails to accommodate the necessary data types [5]. Secondly, MapReduce jobs create heterogeneous environments where jobs compete for the same resource with no guarantee of fairness [4]. Finally, MapReduce jobs, or HDFS for that matter, lack *time awareness*, while some algorithms might need to process data in their time sequence or using a time window.

The core premise of this chapter is to replace HDFS/MapReduce with a time-aware storage and processing logic. Big Data is replayed along the timeline, and all the jobs get the time-ordered sequence of data items. The major difference here is that the new method collects all the jobs in one place—the node that replays data—while HDFS/MapReduce sends jobs to remote nodes so that data can be processed locally. This architecture is chosen for the sole purpose of accommodating a wide range of data streaming algorithms and the data types they create [5].

The new method presented in this chapter can be viewed as a generic technology that exists independently of HDFS and MapReduce. This technology is one of many possible practical applications of the new method.

The generic essence of the new method is as follows. With extremely high-volume replay—like that of Big Data—one needs to partition input and process it via multiple substreams in parallel. There are two challenges here, both of which are resolved by the new method.

Firstly, it is necessary to develop a brand-new parallelization paradigm that would exploit the capacity of a multicore architecture to its fullest. In other words, parallel processing on multicore architecture should minimize overhead from parallelization and synchronization across concurrent jobs. The new method achieves this goal via a *lock-free* design, which involves a special shared memory design but also an algorithm that minimizes cross-job (interprocess) communication overhead. The need to exchange messages is also removed completely, replaced by quasi-messages in the form of asynchronous updates via shared memory that trigger changes of state. The design also involves a data construct in which data units are naturally listed in decreasing order of freshness (age), which allows for export and removal of old data units without cross-job (interprocess) synchronization.

Secondly, the process-while-replaying situation requires a new design that allows concurrent jobs not only to get access to the same stream of data but also to process the stream efficiently. Efficiency here is a complex metric consisting of *space efficiency, processing speed* expressed as per-unit overhead, and most importantly, minimization of processing rates across concurrent jobs. The last metric is extremely important in practice because the entire system is forced to advance its timeline at the rate of its *slowest* job regardless of how fast other jobs are. This aspect is referred to as *heterogeneous jobs*. This chapter introduces and implements an optimization problem that incorporates these issues. Statistical processing in each core is done using the *data streaming* concept [6].

The specific new technologies presented in this chapter are as follows. Time Aware Big Data (TABID) is the name of the method and its software implementation. This abbreviation throughout this chapter will refer to the new method and its design and software implementation. While Hadoop works with files on a file system, this chapter introduces a *timeline data store*, which is convenient when replaying data. The new design for the data store is not key to the new method, but it serves as proof that the new method is valid and can be easily implemented in software. Since many jobs run concurrently at the replay node, this chapter introduces a simple job packing heuristic that takes heterogeneity into consideration. With the packing heuristic, the new method is a *natively multicore* technology.

Analysis of the new method shows that replay-based architecture allows for more efficient use of resources, while MapReduce jobs have to read all files on all storage nodes. It is also shown that optimal packing can help maximize efficiency even for schedules with very many jobs running on commodity hardware with eight cores.

This chapter connects several seemingly unrelated topics. For example, Big Data and methods for its storage and processing currently exists as its own subject, separately from such topics as parallel processing, multicore parallelization, and others. Namely, the term *scale out*, often applied to the essence of Big Data processing, is unrelated to these other topics.

This chapter brings together the following list of other topics as well. *Big Data replay* is its own problem, which has to be solved when processing Big Data on a single machine powered by multicore technology. The multicore architecture itself poses additional new problems due to differences between traditional parallelization and that running on multicore technology. Finally, the processing method itself in its traditional form is relatively primitive, while this chapter will extend it into a more generic form known as *data streaming*. The new process not only is more generic but also can support a more statistically rigorous processing logic, while the traditional architecture based on MapReduce only accepts key–value pairs while imposing a fairly strict restraint on how the while-in-processing data can be used by user code.

## 12.2 AN UNCONVENTIONAL BIG DATA PROCESSOR

The best way to present this chapter is to announce that it is about an *unconventional Big Data processor*. This section will explain in detail what this means. The main distinction between the conventional and unconventional processes lies in the core element of the design—*distribution*—or, more precisely, in what is distributed and how. The traditional process is to distribute both data and the jobs that process the data. The unconventional processor presented in this chapter distributes only the storage part, while jobs run on a single machine.

As this point, it should be clear that the comparison between the two processes leads to a kind of trade-off. The trade-off is obvious for following reason. Distribution is always performed *over the network* where multiple machines have to *talk* to each other while processing the data. The trade-off here is in the area of throughput. Distribution is always limited by the maximum throughput that a distributed network can support. This chapter will discuss throughput limitations of Hadoop and MapReduce and will argue that the new design is strictly always better simply because large chunks of distributed storage are *streamed* via a single machine where uninterrupted (steady) streaming of large chunks of data is strictly better than multiple nodes exchanging and aggregating relatively small pieces of data.

There is a great deal of terminology in all the related areas. Section 12.2.1 is dedicated to a concise introduction of all the necessary terminology.

### 12.2.1 Terminology

Hadoop and HDFS (Hadoop file system) are used interchangeably. In fact, for simplicity, Hadoop in this section will mean *Hadoop* and *MapReduce*. A unit of action in Hadoop is *a job*.

*Streaming algorithm* and *data streaming* also denote the same technology. Here, a unit of data is *a sketch*—a statistical summary of a bulk of data. *Sketches* will function as jobs in the new method.

*Data store* is the same as *data storage*. Distributed parts of stores are called *shards* in Hadoop or *substores* in this section. Content itself is split into *records*, but in this section, we refer to them as *items*. Items in the new/alternative design are assigned in time, which is why the store is referred to as *timeline store*.

The new method runs on *multicore* technology and needs *shared memory* for the manager to communicate with *sketches* running on cores. This section employs a *ring buffer* to maintain a finite time window of items in shared memory continuously.

When estimating performance of the new method, *heterogeneity* is the main environmental parameter, while *sketchbytes* (volume of data processed by all *sketches*) and *overhead* (per data batch per core) are the two quality metrics.

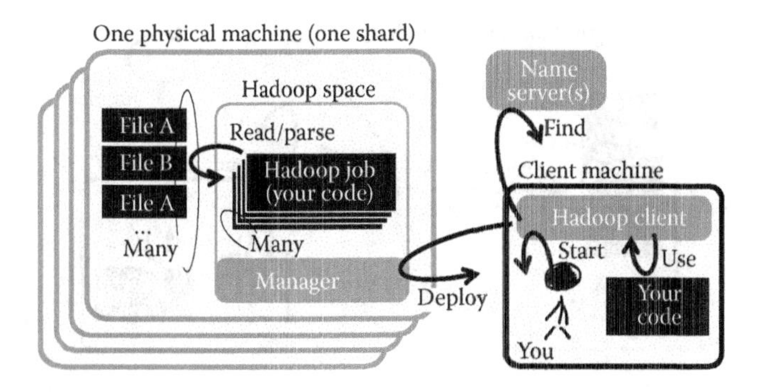


FIGURE 12.1 Architecture of a standard Hadoop/MapReduce system.

### 12.2.2 Overview of Hadoop

Figure 12.1 shows a simplified version of Hadoop architecture, which shows both the design and key processes. Gray parts denote the system, while sharp-black ones are for user files and code. The rest of this section will walk step by step through a standard MapReduce job.

*The code* for the job has to be prepared in advance. Once ready, you can *start the job* by passing it to the Hadoop (MapReduce, actually) client. The client will then *find* the shards scattered across the Hadoop cluster using a name server and *deploy* your job to each.

The manager at each shard will run your code locally. The code can then *read* and *parse files* it finds locally but, more importantly, perform the two key operations of MapReduce—*map* and *reduce*—where both operators are normally implemented by your code. The results are then sent back to the client machine, which, having collected data from all shards, will return the final result to the user.

Note that this is a simplistic view of the processes involved but contains enough information to understand the key differences in the new design presented further in this section.

Let us consider one example that is difficult to run on Hadoop—the *counting frequent items* target popular in data streaming [7], among many others [5]. To execute it on Hadoop, we need to collect all items first using the *map* operator. We can then count all items using the *reduce* operator. However, at this point, we still need to keep all items in memory, and in fact, the data have to be transmitted over the network. Memory overflow is one potential problem in this scenario. But, in a larger picture, such a solution violates the *space efficiency* objective commonly encountered in streaming algorithms [6].

Another major problem with Hadoop is that it is incompatible with data types other than *key-value* hashes, where data streaming requires much greater flexibility. For example, Hadoop cannot work with *trees* and *graph* data types, many-to-many [8] and one-to-many [9] patterns, and others.

### 12.2.3 Hadoop Alternative: Big Data Replay

Figure 12.2 presents the architecture of the new TABID method. For simplicity, the same basic layout as in Section 12.2.2 is used. Also, mirroring the narrative in Section 12.2.2, this one will also walk through the steps of a Big Data processing session. Application programming interface (APIs) pertaining to parts of the design are discussed further in Section 12.7.4.

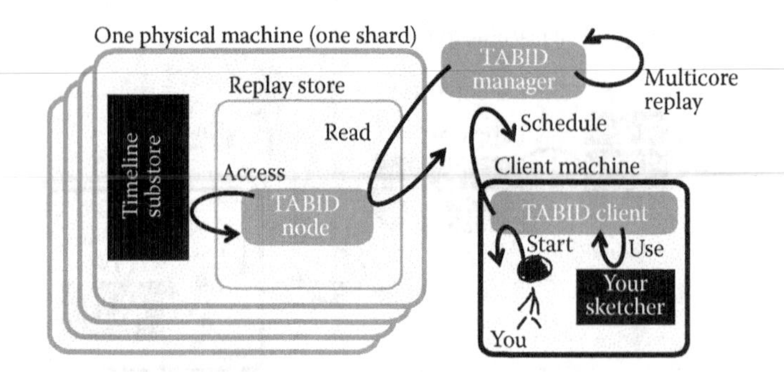


FIGURE 12.2 Architecture of the Hadoop alternative—the TABID system.

Before the process, let us consider differences in *sharding*. Sharding is used in TABID as well, but the store is ordered along the timeline. Note that such a design may not need a name server, because shards can be designed as *chains* with one shard pointing to the next. This discussion is left to further study and will be presented in future publications. Also note that as the API will show further on, data items are not key–value pairs but are arbitrary strings, thus accommodating any data type like encoded (Base64, for example) JSON [10].

The process starts with defining your *sketcher*. The format is JSON [10], and a fairly small size is expected. Unique features of sketchers are discussed at the end of this section.

When you *start* your sketch, the TABID client will *schedule* it with the TABID manager. The schedule involves optimization where sketches are packed in cores in real time, as will be shown further in this section. When a *replay* session starts, your sketch will run on one of the cores and will have access to the timeline of items for processing. When replay is over, results are *retuned* back to the client in a JSON data type with arbitrary structure.

The following are the unique features/differences of TABID. First, there is *no code*. Instead, the *sketcher* specifies which standard streaming algorithm is used and provides values for its configuration parameters. In case of a new streaming algorithm, it should be added to the library first and then referred to via the sketcher. Substore nodes in TABID are dumb storage devices and do not run the code, although it should not be difficult to create a version of TABID that would migrate to shards to replay data locally. This version with its performance evaluation will be studied in future publications.

## 12.3 PUTTING THE PIECES TOGETHER

Earlier sections presented a number of small parts; this section puts the pieces together but also maps them to a set of scientific problems that can be used to achieve the ultimate practical goals presented in this chapter. The scientific problems are *hashing*, *Bloom filters*, *data streaming* and related statistics [11,12], and finally, parallelization on *multicore* architecture.

Also, this section helps to solidify the scientific scope covered by this chapter. This section also gives an overview of the literature covering the scope.

### 12.3.1 More on the Scope of the Problem

This chapter exists in the space created by three separate (although somewhat related) topics—*hashing, Bloom filters*, and *data streaming*. The last term is not fully established in the literature, having been created relatively recently—within the last decade or so—which is why it appears under various code names in literature, some of which are *streaming algorithms*, *data streaming*, and *data streams*. The title of this chapter clearly shows that this author prefers the term *data streaming*. The first two topics are well known and have been around in both practice and theory for many years.

The two major objectives posed in this chapter are as follows.

- *Objective 1*: Fast hashing—how to calculate hash functions of arbitrary length data using as few CPU cycles as possible
- *Objective 2*: Efficient lookup—how to find items in structures of arbitrary size and complexity with the highest achievable efficiency [13]

These objectives do not necessarily complement each other. In fact, they can be conflicting under certain circumstances. For example, faster hash functions may be inferior and cause more key collisions on average [11,12]. Such collisions have to be resolved by the lookup algorithm, which should be designed to allow multiple records under the same hash key [14]. The alternative of not using collision resolution is a bad design because it directly results in loss of valuable information.

*Data streaming* as a topic has appeared in the research community relatively recently. The main underlying reason is a fundamental change in how large volumes of data had to be handled. The traditional way to handle large data (*Big Data* may be a better term), which is still used in many places today, is to store the data in a database and analyze it later—where the *later* is normally referred to as *offline* [15]. As the Big Data problem—the problem of having to deal with an extremely large volume of data—starts to appear in many areas, storing data in any kind of database has become difficult and, in some cases, impossible. Specifically, it is pointless to store Big Data if its arrival rate exceeds processing capacity, by the way of logic.

Hence the *data streaming* problem, which is defined as a process that extracts all the necessary information from the input raw data stream without having to store it. The first obvious logical outcome from this statement is that such processing has to happen in *real time*. In view of this major design feature, the need for both *fast hashing* and *efficient lookup* should be obvious.

There is a long list of practical targets for data streaming. Some common targets are

- Calculating a median of all the values in the arrival stream
- Counting all the distinct items
- Detecting a longest increasing or decreasing sequence of values

It should be obvious that the first two targets in this list would be trivial to achieve without any special algorithm had they come without the conditionals. For example, it is easy to calculate *the average*—simply sum up all the values and divide them by the total number of values. The same goes for the counting of distinct items. This seemingly small detail makes for the majority of the complexity in data streaming.

In addition to the above list of practical targets, the following describes *the catch* of data streaming:

- There is limited space for storing the current state—otherwise, we would revert back to the traditional database-oriented design.
- Data have to be accessed in their natural arrival sequence, which is the obvious side effect of a real-time process—again a major change from the database-backed processes, which can access any record in the database.
- There is an upper limit on per-unit processing cost, which, if violated, would break the continuity of a data streaming algorithm (arguably, buffering can help smoothen out temporary spikes in arrival rate).

### 12.3.2 Overview of Literature

Although Hadoop is unchallenged in practice today [2], the technology is known to be inefficient in several respects. Maximum achievable throughput for the HDFS system itself is found to be around 50 Mbps [3]. HDFS is also found to be inefficient when content is split into a large number of small files [16], where the default block size used by HDFS is 64 MB and above.

Recently, much attention in literature has been paid to the performance of Hadoop and its improvement. Statistics from a real commercially operated cluster are presented in Reference 17, while workload modeling and synthesis are proposed in Reference 18. Performance improvement can be split into the following two groups. Parallel processing on multicore technology is proposed in Reference 4 and requires a major rewrite of the architecture. Research in Reference 2 proposes using more of local processing in RAM rather than distribution processing over the network.

There is some literature on improvement of the HDFS technology without the MapReduce component. Research in Reference 19 proposes creating a searchable version of HBase—a simple, nonsearchable spreadsheet data type on top of HDFS. Another method in Reference 20 creates an entirely new form of data storage on top of HDFS, arguably an alternative to HBase. Given that resources on substores are shared by all jobs running in parallel, optimizing heterogeneous access to HDFS is also a subject of interest [21].

Data streaming is a relatively new method but solves the same problem as MapReduce—high-volume real-time Big Data processing. Simply put, the main premise of data streaming is the ability to compress large volumes of data into small statistical summaries. Reference 5 is an excellent 100+ page introduction to the topic. Practical algorithms in literature are already applied to traffic analysis [6]. There is research on space efficiency designs and optimizations [22]. Temporal features of data streaming and sliding windows are considered in Reference 23.

Well-known practical data streaming targets are *frequent item discovery* [7], aggregation of one-to-many records such as found in quickly spreading viruses [9], and generic aggregation of many-to-many records [8]. However, the tool kit of data streaming is extremely flexible and can theoretically accommodate any practical target.

Data streaming is the main reason for the new design in this section. Time awareness and processing during replay are the features of the new design that are implemented with data streaming in mind. Coincidentally, the resulting architecture can benefit from running on multicore hardware.

*Multicore* technology is a special case of parallel processing [24]. It has already been applied to MapReduce in Reference 4, which improves its performance but does not solve the other problems discussed in this section. Most existing multicore methods implement traditional scheduling-based parallelization [4,24,25], which requires intensive message exchange across processes. This chapter uses a special case of parallelization, which features minimum overhead due to a lock-free design. More details are presented further in this section.

## 12.4 THE DATA STREAMING PROBLEM

This section is specific to data streaming. It may sound like data streaming is only about methods that are used to *stream* data, but in fact, the topic is extremely rigorous scientifically. The rigor is the information theory that underlies the topic. But many formulations in data streaming are demanded in new areas of data processing and therefore require somewhat new-looking formulations. Although data streaming is commonly written with a great deal of mathematical notation, this section and this chapter as a whole will stick to the bare minimum of math, strictly with the purpose of providing basic understanding of the numeric essence of formulations.

A second round of terminology is presented in Section 12.4.1.

### 12.4.1 Data Streaming Terminology

As was mentioned before, the terms *data streams*, *data streaming*, and *streaming algorithms* all refer to the same class of methods.

*Bloom filter* and Bloom structure both refer to a space in memory that stores the current state of the filter, which normally takes the form of a bit string. However, the Bloom filter itself is not only the data it contains but also the methods used to create and maintain the state.

A *Double-Linked List (DLL)* is also a kind of structure. However, DLLs are arguably exclusively used in C/C++ [14,26]. This is not 100% true—in reality, even this author has been able to use DLLs in other programming languages like PHP or JavaScript—but C/C++ programs can benefit the most from DLLs because of the nature of pointers in C/C++ versus those in any other programming language. DLLs refer to a memory structure (*struct* in C/C++) alternatively to traditional lists—vectors, stacks, and so forth. In DLLs, each item/element is linked to its neighbors via raw C/C++ pointers, thus forming a chain that can be traversed in either direction. DLLs are revisited later in this chapter and are part of the practical application discussed at the end of the chapter.

The terms *word*, *byte*, and *digest* are specific to hashing. The word is normally a 32-bit (4-B) integer on 32-bit architectures. A hashing method *digests* an arbitrary length input at the grain of a *byte* or *word* and outputs a hash key. Hashing often involves *bitwise* operations where individual bits of a word are manipulated.

This chapter promised a minimum of mathematics. However, some basic notation is necessary. Sets of variables $a$, $b$, and $c$ are written as $\{a, b, c\}$ or $\{a\}_n$ if the set contains $n$ values of a parameter $a$. In Information Theory, the term *universe* can be expressed as a set.

Sequences of $m$ values of variable $b$ are denoted as $\langle b \rangle_m$. Sequences are important for data streaming where sequential arrival of input is one of the environmental conditions. In case of sequences, $m$ can also be interpreted as *window size*, given that arrival is normally continuous.

Operators are denoted as functions, that is, the *minimum of a set* is denoted as $\min\{a, b\}$.

### 12.4.2 Related Information Theory and Formulations

We start with a universe of size $n$. In data streaming, we do not have access to the entire universe; instead, we are limited to the current window of size $m$. The ultimate real-time streaming is when input is read and processed one item at a time, that is, $m = 1$.

Using the complexity notation, the upper bound for the space (memory, etc.) that is required to maintain the state is

$$S = o(\min\{m,n\}).$$

If we want to build a robust and sufficiently generic method, it would pay to design it in such a way that it would require roughly the same space for a wide range of $n$ and $m$, that is,

$$S = O(\log(\min\{m,n\})).$$

When talking about *space efficiency*, the closest concept in traditional Information Theory is *channel capacity* (see the work of Shannon [27] for the original definition). Let us put function $f(\{a\}_n)$ as the cost (time, CPU cycles, etc.) of operation for each item in the input stream. The cost can be aggregated into $F(\{a\}_n)$ to denote the entire output. It is possible to judge the quality of a given data streaming method by analyzing the latter metric. The analysis can extend into other efficiency metrics like memory size and so forth, simply by changing the definition of a per-unit processing cost.

A simple example is in order. Let us discuss the unit cost, defined as

$$f(\{a\}_n) = f\{i{:}a_i = C\}, \quad i \in 1, \ldots, n.$$

The unit cost in this case is the cost of defining—for each item in the arrival stream—if it is equal to a given constant $C$. Although it sounds primitive, the same exact formulation can be used for much more complicated unit functions.

Here is one example of a slightly higher complexity. This time, let us phrase the unit cost as the following condition. Upon receiving item $a_i$, update a given record $f_j \leftarrow f_j + C$. This time, prior to updating a record, we need to find the record in the current state. Since it is common that $i$ and $j$ have no easily calculable relation between them, finding the $j$ efficiently can be a challenge.

Note that these formulations may make it look like *data streaming is similar to traditional hashing*, where the latter also needs to update its state on every item in the input. This is a gross misrepresentation of what data streaming is all about. Yes, it is true that some portion of the state is potentially updated on each item in the arrival stream. However, in hashing, the method always knows which part of the state is to be updated, given that the state itself is often just a single 32-bit word. In data streaming, the state is normally much larger, which means that it takes at least a calculation or an algorithm to find a spot in the state that is to be updated.

The best way to describe the relation between data streaming and hashing is to state that data streaming uses hashing as one of its primitive operations. Another primitive operation is blooming.

The term *sketch* is often used in relation to data streaming to describe the entire state, that is, the $\{f\}_m$ set, at a given point of time. Note that $f$ here denotes the value obtained from the unit function $f()$, using the same name for convenience.

### 12.4.3 Practical Applications and Designs

Early proposals related to data streaming were abstract methodologies without any specific application. For example, Reference 5 contains several practical examples referred to as *puzzles* without any overlying theme. Regardless, all the examples in early proposals were based on realistic situations. In fact, all data streaming targets known today were established in the very early works. For example, counting frequent items in streams [7] and algorithms working with and optimizing the size of sliding windows [23] are both topics introduced in early proposals.

Data streaming was also fast to catch up with the older area of packet traffic processing [28,29]. Early years have seen proposals on data streaming methods in Internet traffic and content analysis [6] as well as detection of complex communication patterns in traffic [8,9]. Reference 8 specifically is an earlier work by this author and is also this author's particular interest as far as application of data streaming to packet traffic is concerned. The particular problem of detecting complex communication patterns is revisited several times in this chapter.

Data streaming has been applied to other areas besides traffic. For example, Reference 22 applies the discipline to detection of triangles in large graphs with obvious practical applications in social networks among many other areas where graphs can be used to describe underlying topology.

## 12.5 PRACTICAL HASHING AND BLOOM FILTERS

Having understood the data streaming technology, we now can move on to a lower layer of underlying methods, such as hashing, blooming (using Bloom filters), and others [30]. Some small volume of math is used in this section as well.

### 12.5.1 Bloom Filters: Store, Lookup, and Efficiency

Reference 31 provides a good theoretical background on the notion of Bloom filters. This section is a brief overview of commonly available information before moving on to the more advanced features actively discussed in research today.

Some of this section will discuss the traditional design presented by this figure but then replace it with a more modern design. From earlier in this chapter, we know that blooming is performed by calculating one or more hash keys and updating the value of the filter by OR-ing each hash key with its current state. This is referred to as the *insert* operation. The *lookup* operation is done by taking the bitwise operation AND between a given hash key and current state of the filter. The decision making from this point on can go in one of the following two directions:

- The result of AND is the same as the value of the hash key—this is either true positive (TP) or false positive (FP), with no way to tell between the two.
- The result of AND is not the same as the value of the hash key—this is a 100% reliable true negative.

One common way to describe this lookup behavior of Bloom filters is to describe the filter as a person with memory who can only answer the question "Have you seen this item before?" reliably [11]. This is not to underestimate the utility of the filter, as the answer to this exact question is precisely what is needed in many practical situations.

Let us look at the Bloom filter design from the viewpoint of hashing, especially given that the state of the filter is gradually built by adding more hash keys onto its state.

Let us put $n$ as the number of items and $m$ as the bit length of hash keys and, therefore, the filter. We know from before that each bit in the hash key can be set to 1 with 50% probability. Therefore, omitting details, the optimal number of hash function can be calculated as

$$k = \ln 2\left(\frac{m}{n}\right) \approx 0.6\frac{m}{n}.$$

If each hash function is perfectly independent of all others, then the probability of a bit remaining 0 after $n$ elements is

$$p = \left(1 - \frac{1}{m}\right)^{kn} \approx e^{\frac{-kn}{m}}.$$

The FP—an important performance metric of a Bloom filter—is then

$$p_{FP} = (1-p)^k \approx \left(1 - e^{\frac{-kn}{m}}\right)^k \approx \frac{1}{2^k},$$

for the optimal $k$. Note with that increasing $k$, the probability of an FP actually is supposed to decrease, which is an unintuitive outcome because one would expect the filter to get filled up with keys earlier.

Let us analyze $k$. For the majority of cases, $m \ll n$, which means that the optimal number of hash functions is 1. Two functions are feasible only with $m > 2.5n$. In most realistic cases, this is almost never so, because $n$ is normally huge, while $m$ is something practical like 24 or 32 (bits).

### 12.5.2 Unconventional Bloom Filter Designs for Data Streams

Based on Section 12.5.1, the obvious problem in Bloom filters is how to improve their flexibility. As a side note, such Bloom filters are normally referred to as *dynamic*.

Figure 12.3 shows the generic model, which applies to most of the proposals of dynamic Bloom filters. The simple idea is to replace a simple bit string with a richer data structure (the change in the Bloom filter in the figure). Each bit in the filter now simply is a pointer to a structure that supports dynamic operations.

The other change that ensues is that the OR operation is no longer applicable. Instead, a nontrivial manipulation has to be performed on each bit of the value that was supposed to be OR-ed in the traditional design. Naturally, this incurs a considerable overhead on performance.

The following classes of dynamic Bloom filters are found in the literature.

- *Stop additions filter.* This filter will stop accepting new keys beyond a given point. Obviously, this is done in order to keep the FP beyond a given target value.
- *Deletion filter.* This filter is tricky to build, but if accomplished, it can revert to a given previous state by forgetting the change introduced by a given key.
- *Counting filter.* This filter can count both individual bits of potential occurrences and entire values—combinations of bits. This particular class of filters obviously can find practical applications in data streaming. In fact, the existing example of the d-left hashing method in Reference 32 uses a kind of counting Bloom filter [33]. Another example can be found in Reference 34, where it is used roughly for the same purpose.

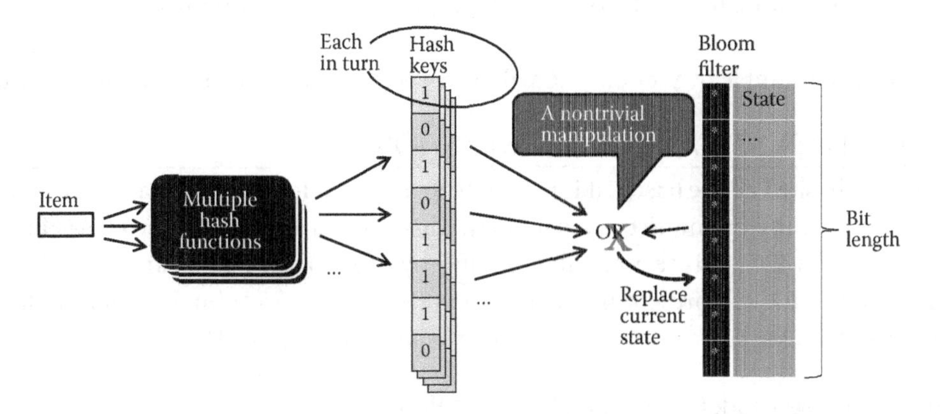


FIGURE 12.3 A generic model representing Bloom filters with dynamic functionality. The two main changes are (1) extended design of the Bloom filter structure itself, which is not a bit string anymore, and (2) nontrivial manipulation logic dictated by the first change—simply put, one cannot simply use logical ORs between hashes and Bloom filter states.

There are other kinds of unconventional designs. Reference 35 declares that it can do with fewer hash functions while providing the same blooming performance. Bloom filters specific to perfect hashing are proposed in Reference 36.

### 12.5.3 Practical Data Streaming Targets

Besides the relatively simple (you can call them traditional) data streaming targets, there are several interesting practical targets that need a higher level of algorithmic complexity. This section lists only the problems and leaves the search for solutions to the reader. In fact, some solutions are the subject of active discussion in the research community today. Pointers to such research are provided.

- *Example 1: Finding heavy hitters (beyond the min-count sketch).* Find $k$ most frequently accessed items in a list. One algorithm is proposed in Reference 7. Generally, more sound algorithms for sliding windows can be found in Reference 23.
- *Example 2: Triangle detection.* Detect triangles defined as A talks to B, B talks to C, and C talks to A (other variants are possible as well) in the input stream. An algorithm is proposed in Reference 22.
- *Example 3: Superspreaders.* Detect items that access or are accessed by exceedingly many other items. Related research can be found in Reference 9.
- *Example 4: Many-to-many patterns.* This is a more generic case of heavy hitters and superspreaders, but in this definition, the patterns are not known in advance. Earlier work by this author in Reference 8 is one method. However, the subject is popular with several methods such as many-to-many broadcasting [37], and various many-to-many memory structures [38,39] and data representations (like the graph in Reference 40) are proposed—all outside of the concept of data streaming. The topic is of high intrinsic value because it has direct relevance to group communications where one-to-many and many-to-many are the two popular types of group communications [41–44].

A practical example in Section 12.6 will be based on a many-to-many pattern capture.

## 12.6 BIG DATA STREAMING OPTIMIZATION

Having established all the basics, it is now time to use them for the core practical use case—Big Data streaming on multicore technology. This content is split into two parts between this and Section 12.7. This section talks specifically about optimization and related theory, while Section 12.7 will present implementation details. Software implementation for fast hashing and blooming parts of the technology is made public [45,46].

### 12.6.1 A Simple Model of a Data Streaming Process

Figure 12.4 shows the generic model of a data streaming situation. The parameters are *arrival rate*, *record size*, *record count*, and the *index*, where the last term is a replacement term for data streaming. Note that only arrival rate is important because departure rate, by

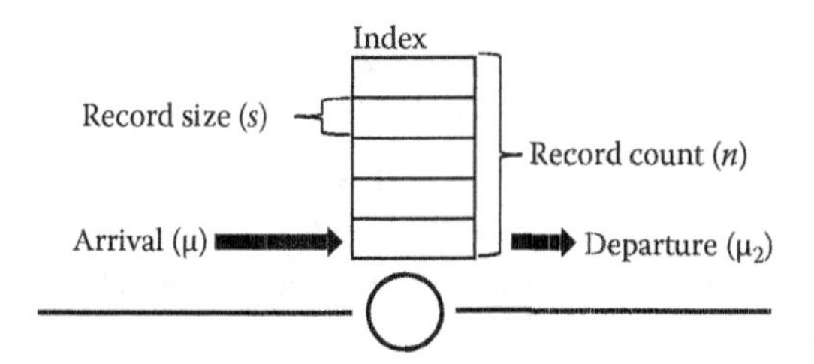


FIGURE 12.4 Common components shared by all data streaming applications.

definition, cannot be higher than the arrival rate. On the other hand, arrival rate is important because a data streaming method has to be able to support a given rate of arrival in order to be feasible in practice.

Arrival rate is also the least popular topic in related research. In fact, per-unit processing time is not discussed much in literature, which instead focuses on efficient hashing or blooming methods [30,32,47,48]. Earlier work by this author in Reference 49 shows that per-unit processing cost is important—the study specifically shows that too much processing can have a major impact on throughput.

The topic of arrival rate is especially important in packet traffic [28,29,50]. With constantly increasing transmission rates as well as traffic volume, a higher level of efficiency is demanded of switching equipment. A Click router is one such technology [51], which is in the active development phase with recent achievements of billion packets per second (pps) processing rates [52]. The same objectives are pursued by Open vSwitch—a technology in network virtualization. It is interesting that research in this area uses roughly the same terminology as is found in data streaming. For example, References 52 and 53 talks about space efficiency of data structures used to support per-packet decision making. Such research often uses Bloom filters to improve search and lookup efficiency [13,32]. In general, this author predicts with high probability that high-rate packet processing research in the near future will discover the topic of data streaming and will greatly benefit from the discovery [50].

### 12.6.2 Streaming on Multicore

A method that packs $n$ sketches into $m$ cores is a well-known *bin packing* problem. The method presented in this section performs one (blind) initial packing but also repacks every time there is a change in state. Given that each sketch has *start* and *end times* that can differ from those of the replayed Big Data, changes in state can be frequent. With convergence speed in mind, this section and Section 12.7 discuss a simple packing heuristic. Given that all sketches on all cores share one time-ordered stream of items, the main objective of the heuristic is to minimize variance in processing *cursors* (current position in stream) across sketches. Intuitively, if all sketches with heavy-duty processing are packed into one core and all light-duty ones are packed into another, the replay will move at a greatly diminished speed. The new heuristic aspires to avoid such situations, while the design itself facilitates parallel processing without cores passing messages or competing for memory locks.

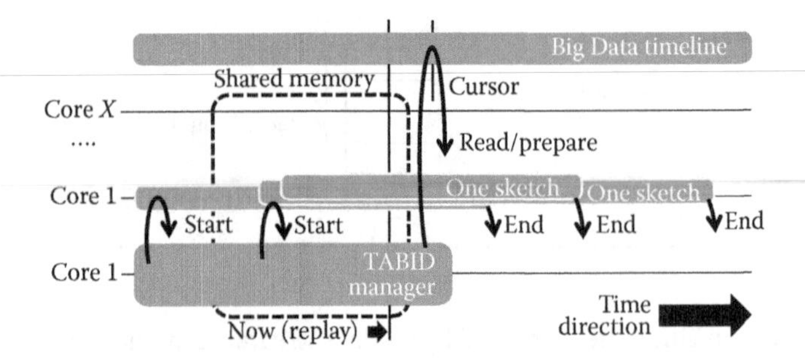


FIGURE 12.5 Big Data timeline and sketches running in groups on multiple cores.

Figure 12.5 shows the design of the TABID manager node. The manager is running on one core and is in charge of starting and ending sketches as per its current schedule. Note that this point alone represents a much higher flexibility compared to MapReduce, which has no scheduling component. Read/write access to the time-ordered stream, although remaining asynchronous, is collision-free by ensuring that writing and reading cursors never point to the same position in the stream. Sketches read the data at the *now* cursor while the manager is writing to a position further along the stream, thus creating a collision-safe buffer.

### 12.6.3 Performance Metrics

This section discusses parameters and metrics used for analysis. First, let us define the concept of a *configuration tuple,*

$$\langle v_{\min}, v_{\max}, a \rangle,$$

which specifies a distribution of values between $v_{\min}$ and $v_{\max}$ configured by $a$. The tuple can be used to define *heterogeneity*:

$$y = \{(\exp^{-ax})^{-1}, \quad \forall x \in \{1, 2, \ldots, 100\}\}, a > 0 v_{\max}, \quad a = 0, \tag{12.1}$$

where $a$ is the exponent. Note that the case of $a = 0$ is a special *homogenous* case. In analysis, values for $a$ are randomly chosen from the list 0, 0.7, 0.3, 0.1, 0.05, 0.01, where $a = 0$ is a homogenous distribution (horizontal line), $a$ between 0.7 and 0.3 creates distributions with a majority of large values, $a = 0.1$ is almost a linear trend, and $a$ between 0.05 and 0.01 outputs distributions where most values are small. The last two cases are commonly found in natural systems and can therefore be referred to as *realistic*. Note that this selection only appears arbitrary; in reality, it covers the entire range of all possible distributions.

These distributions apply to *sketch life span* and *per-unit overhead*—the two practical metrics that directly affect performance of a TABID system.

The packing heuristic is defined as follows. $C$ denotes a set of item counts for all sketches in all cores, one value per sketch. $M$ is a set of per-core item counts, one value per core. The optimization objective is then (var and max are operators)

$$\text{minimize } \operatorname{var}(C) + \max(M).$$

For simplicity, values for $C$ and $M$ are normalized within a time window to avoid dealing with different units of measure in the two terms. In analysis, the problem is solved using a genetic algorithm (GA), for which this objective serves as a *fitness function*. A detailed description of the GA is omitted due to limited space.

### 12.6.4 Example Analysis

This section presents results in two settings: performance in terms of *sketchbytes* versus *overhead*. The former is evaluated by the ratio of *sketchbytes* between traditional Hadoop jobs and TABID sketches. The latter is a *worst-case* analysis only for TABID, which shows what item throughput can be expected given heterogeneity and number of sketches.

The specific tuples used for overhead are ⟨100, 10000, 0⟩; ⟨100, 10000, 0.7⟩; ⟨100, 10000, 0.3⟩; ⟨100, 10000, 0.1⟩; ⟨100, 10000, 0.05⟩; and ⟨100, 10000, 0.01⟩—where the unit of measure is *microseconds*. The specific life span tuples are ⟨100, 2500, 0⟩; ⟨100, 2500, 0.7⟩; ⟨100, 2500, 0.3⟩; ⟨100, 2500, 0.1⟩; ⟨100, 2500, 0.05⟩; and ⟨100, 2500, 0.01⟩—where the unit of measure is *minutes*. It is therefore assumed that the example Big Data used for analysis is *2500 m* long (used for replay). With the commonly found practical throughput of about 1 GB per minute, the Big Data in question is 2500 GB in size [3]. There can be between 10 and 1000 parallel sketches and eight cores (assuming commodity eight-core hardware).

Figure 12.6 shows *sketchbytes* performance expressed as Hadoop/TABID ratio. In cases when heterogeneity $a \geq 0.1$, the ratio is around 1000. For smaller (more realistic) $a$, the ratio does not saturate but grows continuously up to 5–6 orders of magnitude (log scale). Note that this gap in *sketchbytes* is solely due to heterogeneity of sketch life spans. If to account for the fact that in TABID, the stream is read/replayed only once, the gap widens further.

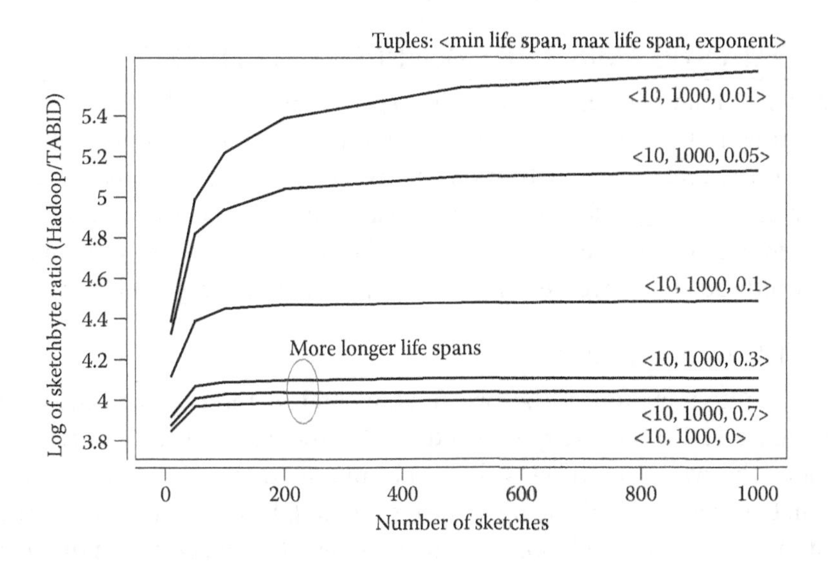


FIGURE 12.6 Performance of *sketchbytes* metric over a wide range of the number of sketches and several heterogeneity setups.

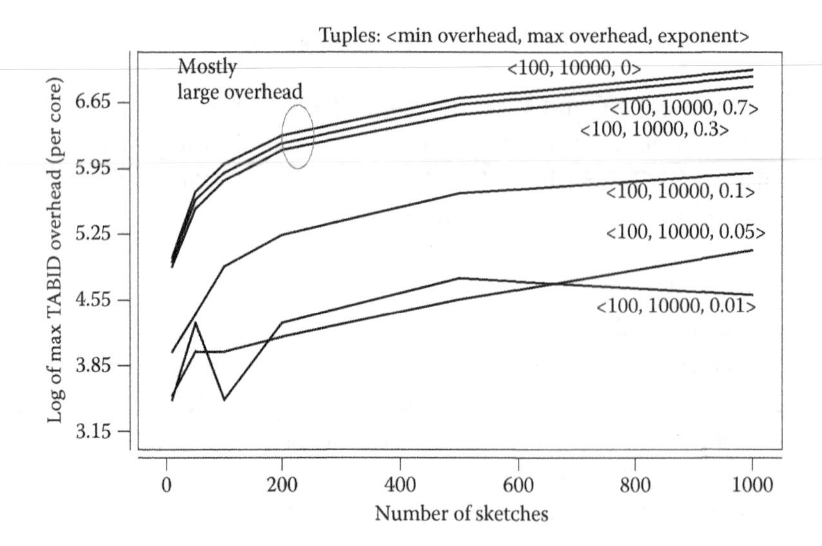


FIGURE 12.7 Performance of *max per-core overhead* metric over a wide range of the number of sketches and several heterogeneity setups.

Figure 12.7 shows the performance of this packing heuristic. For $a \geq 0.1$, it is difficult to constrain a continuous increase in maximum overhead (per core, not per sketch). However, for the realistic cases of $a = 0.05$ or, even better, for $a = 0.01$, there is a near-saturation trend, which means that performance is about the same for few sketches as for very many sketches. The worst case is about 5.4 *ms* (see parameters at the head of this Section 12.6.4) per item (10 ms is the configured max), which means that the slowest core can still process roughly 200 items per second per sketch. Note that the total throughput of the system is much greater because of multiple sketches and multiple cores.

## 12.7 BIG DATA STREAMING ON MULTICORE TECHNOLOGY

Parallelization in multicore environments is a special kind of parallel processing. This section will start with the very basic theory and practice of parallel processing but will go on with a brand-new kind of parallelization referred to as a *lock-free process. Lock-free* is a self-explanatory term and refers to situations when multiple processes running in parallel do not have to use locks in memory they share. Clearly, a lock-free process incurs zero (or very little) overhead, while overhead in traditional locking cannot be neglected.

### 12.7.1 Parallel Processing Basics

A very simple interpretation of the two fundamental methods in parallel processing is offered in Figure 12.8. There are *synced* and *syncless* methods, where *sync* is short for synchronization. The *synced* approach will be split into *locking* and *message passing* later in this section, but there can only be one *syncless* method, by definition. This is the same as to say that there can be multiple solutions to a problem but the best (and unique) possible solution is to remove the problem. The *syncless* approach does just that and describes a

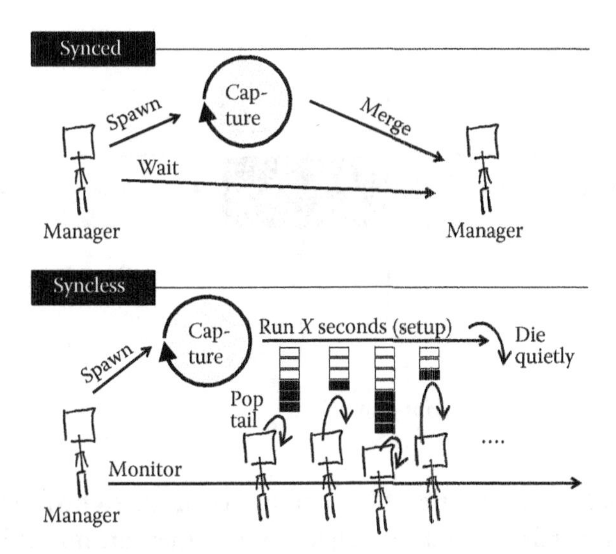


FIGURE 12.8 Two distinct parallel processes, one (above) that requires synchronization and the other (below) that can work without it.

method or a design that avoids having locks of having to pass messages while processing the same data stream in parallel.

A design shown later in this section will avoid *locks*, which is why it will be referred to as *lock-free*. However, *syncless* remains the umbrella term since both locking and message passing are means of synchronization.

It may be useful to provide more background on the main problem from having locks in shared memory or having to pass messages between processes. The problem is that both these methods are applied in an *asynchronous* environment, where multiple processes run continuously in parallel but have to exchange data across each other occasionally. This, in fact, is the core of the problem. Because the exchange is asynchronous, by definition, it means that one process never knows when the other process is going for contact. Naturally, having a synchronous process would entirely solve the problem but would render the parallelization itself useless in practice.

The key point of a lock-free process is to reduce the time communicating processes having to wait on each other. Existing research shows that multicore technology requires such a novel approach simply because having locks greatly reduces the overall throughput of the system. With four- and eight-core architectures becoming common today, such a hard limitation on throughput is undesirable.

### 12.7.2 DLL

Figure 12.9 shows a four-way DLL. Strictly speaking, there is only big DLL and many small sideways DLLs for each element in the big one. But it is convenient to call it *four-way DLL*.

Four-way DLL is an excellent solution to collisions of keys in hash functions. Since collision literally means that several elements contend for the same spot in an index table (not DLL

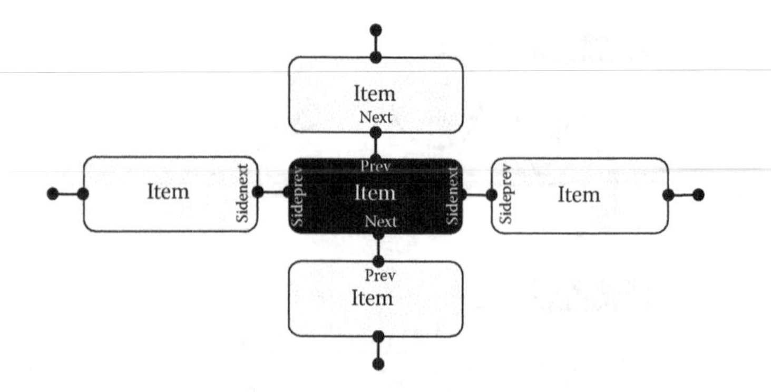


FIGURE 12.9 Design of a standard four-way DLL.

itself!), sideways DLLs help to connect the multiple elements so that they can be searched. Note that collisions are relatively rare, which means that there are normally few of those.

### 12.7.3 Lock-Free Parallelization

Figure 12.10 shows one simple method that makes it possible to avoid locks in shared memory. For clarity, let us concentrate on one manager–core pair of processes. Note that, as was stated earlier, if a lock-free process works for one pair, then it should work for any number of such pairs. Also note that the manager is the same process in all pairs, thus making for a one-to-many interprocess communication.

The manager (Figure 12.11, right side) logic is as follows. Since it is helpful to avoid per-item signaling (via shared memory), data are read in batches within the physical limits of

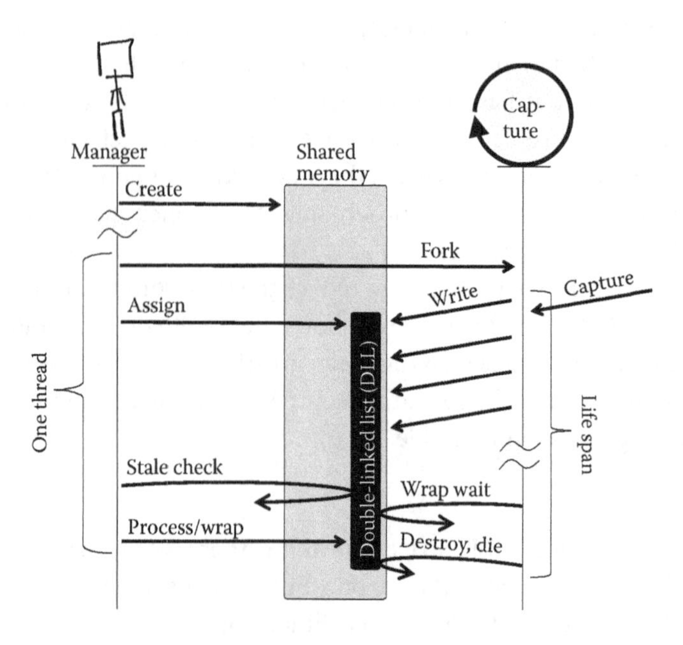


FIGURE 12.10 One method of avoiding locks in shared memory.

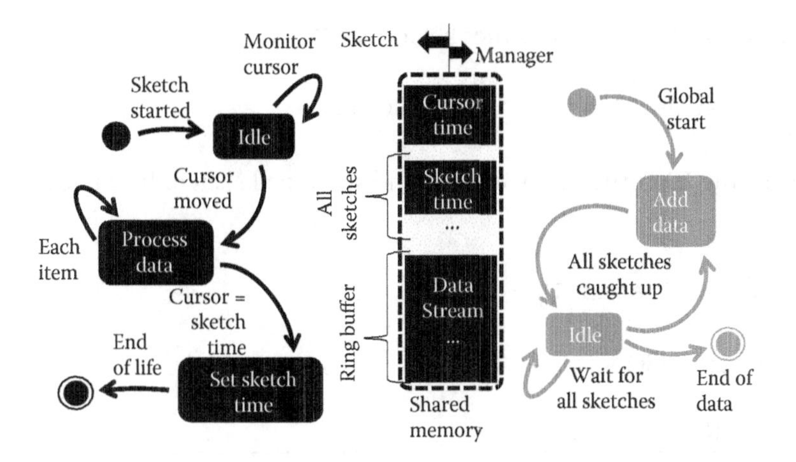


FIGURE 12.11 Shared memory design and logic followed by each sketch (left) and TABID manager (right).

the *ring buffer*, which in turn depends on the size of the shared memory. When a new batch has been read into the ring buffer, the global *now* cursor is updated, which allows for all sketches to start processing new items. The manager also monitors all *sketch times*—cursors for individual sketches—and can change packing configuration based on collected statistics.

The sketch (Figure 12.11, left side) follows the following logic. It waits for the *now* cursor to advance beyond its own cursor, which is stored independently for each sketch in the shared memory. When it detects change, the sketch processes all newly arrived items. Once its cursor reaches the global *now*, the sketch returns to the idle state and starts polling for new change at a given time interval.

Note that the ring buffer is accessed via a C/C++ library call [54] rather than directly by each party. This is a useful convention because it removes the need for each accessing party to maintain its current position in the ring buffer. The buffer only appears to be continuous, while in reality, it has finite size and has to wrap back to the head when its tail is reached.

### 12.7.4 Software APIs

Although far from an exhaustive list, the APIs in this section can fully describe TABID functionality. All APIs are executed over hypertext transfer protocol (HTTP), even if they occur within the same physical machine.

To append a data item to the timeline store, one can send

```
POST tabidStorePut(timestamp, 'string'),
```

where *string* is completely freetype. It is common practice to send Base64-encoded JSON in this manner.

To schedule a sketch, one sends the sketcher JSON as

```
POST tabidSketchAdd(startTime, JSON),
```

which returns the *ID* of the newly created and scheduled sketch. Its current status can be verified using

```
GET tabidSketchStatus(ID),
```

which returns status JSON.

## 12.8 SUMMARY

This chapter presented a brand-new paradigm for Big Data processing. The new paradigm abandons the "let us distribute everything" approach in search of a more flexible technology. Flexibility is in several parts of the new paradigm—it is in how storage is distributed, how Big Data is processed by user jobs, and finally, how user jobs are distributed across cores in a multicore architecture.

Setting all that aside, it can be said that the new paradigm is an interesting formulation simply because it brings together several currently active areas of research. These are specifically *efficient hashing* technology, parallel processing *on multicore* technology, and, finally, the scientifically rigorous data streaming approach to creating statistical sketches from raw data. This collection of topics creates an interesting platform for dynamic algorithms and optimizations aimed at improving the performance of Big Data processing as a whole.

The *unconventional Big Data processor* was introduced early in this chapter in order to avoid having to provide a unified background for each separate part of the technology. This early introduction also presented several major parts that are required for the technology to work in the first place. A separate section was then dedicated to putting all these parts together into a complete practical/technical solution.

Having established the background, the chapter continued with specifics. A separate section was dedicated to issues related to hashing, using Bloom filters, and the related efficiency. Another section fully focused on optimization related to assigning jobs in multicore environments. Finally, a section was dedicated to the issue of data streaming on top of a multicore architecture, yet on top of a real-time Big Data replay process.

It should be noted that real-time replay of Big Data in itself is a brand-new way to view Big Data. While the traditional approach distributes both the storage and processing of data, this chapter discussed a more orderly, and therefore more controlled, process, where Big Data is replayed on a single multicore machine and tasks are packed into separate cores in order to process a replayed stream of data in real time. This chapter hinted that such a design allows for much higher flexibility given that the design is based on a natural trade-off. Although this chapter does not provide examples of various settings that would showcase the dynamic nature of the design, it is clear from the contents which parameters can be used to change the balance of the technology.

The issue of scientific rigor when referring to the data streaming method itself was covered in good detail, where several examples of practical statistic targets were shown. It was also shown that such a level of statistical rigor cannot be supported by traditional MapReduce technology, which only supports key–value pairs and also imposes strict rules on how they can be used—the map and reduce operators themselves. When the data streaming paradigm is used, the jobs are relieved of any such restraint and can use any data structure necessary for processing. This single feature in this chapter is a huge factor in Big Data, where authenticity of conclusions drawn from Big Data has to be defended.

Unfortunately, given the large set of topics in this chapter, several topics could not be expanded with sufficient detail. However, enough references are provided for recent research on each of the specific scientific problems for the reader to be able to follow up.

## REFERENCES

1. Apache Hadoop. Available at: http://hadoop.apache.org/.
2. A. Rowstron, S. Narayanan, A. Donnelly, G. O'Shea, A. Douglas, "Nobody Ever Got Fired for Using Hadoop on a Cluster," *1st International Workshop on Hot Topics in Cloud Data Processing*, April 2012.
3. K. Shvachko, "HDFS Scalability: The Limits to Growth," *The Magazine of USENIX*, vol. 35, no. 2, pp. 6–16, 2012.
4. R. Chen, H. Chen, B. Zang, "Tiled-MapReduce: Optimizing Resource Usages of Data-Parallel Applications on Multicore with Tiling," *19th International Conference on Parallel Architectures and Compilation Techniques (PACT)*, pp. 523–534, 2010.
5. S. Muthukrishnan, "Data Streams: Algorithms and Applications," *Foundations and Trends in Theoretical Computer Science*, vol. 1, no. 2, pp. 117–236, 2005.
6. M. Sung, A. Kumar, L. Li, J. Wang, J. Xu, "Scalable and Efficient Data Streaming Algorithms for Detecting Common Content in Internet Traffic," *ICDE Workshop*, 2006.
7. M. Charikar, K. Chen, M. Farach-Colton, "Finding Frequent Items in Data Streams," *29th International Colloquium on Automata, Languages, and Programming*, 2002.
8. M. Zhanikeev, "A Holistic Community-Based Architecture for Measuring End-to-End QoS at Data Centres," *Inderscience International Journal of Computational Science and Engineering (IJCSE)*, (in print) 2013.
9. S. Venkataraman, D. Song, P. Gibbons, A. Blum, "New Streaming Algorithms for Fast Detection of Superspreaders," *Distributed System Security Symposium (NDSS)*, 2005.
10. JSON Format. Available at: http://www.json.org.
11. D. MacKey, *Information Theory, Inference, and Learning Algorithms*. Cambridge University Press, West Nyack, NY, 2003.
12. A. Konheim, *Hashing in Computer Science: Fifty Years of Slicing and Dicing*. Wiley, Hoboken, NJ, 2010.
13. G. Antichi, A. Pietro, D. Ficara, S. Giordano, G. Procissi, F. Vitucci, "A Heuristic and Hybrid Hash-Based Approach to Fast Lookup," *International Conference on High Performance Switching and Routing (HPSR)*, pp. 1–6, June 2009.
14. M. Zadnik, T. Pecenka, J. Korenek, "NetFlow Probe Intended for High-Speed Networks," *International Conference on Field Programmable Logic and Applications*, pp. 695–698, 2005.
15. R. Kimball, M. Ross, W. Thornthwaite, J. Mundy, B. Becker, *The Data Warehouse Lifecycle Toolkit*. John Wiley and Sons, Indianapolis, IN, 2008.
16. Small File Problem in Hadoop (blog). Available at: http://amilaparanawithana.blogspot.jp/2012/06/small-file-problem-in-hadoop.html.

17. Z. Ren, X. Xu, J. Wan, W. Shi, M. Zhou, "Workload Characterization on a Production Hadoop Cluster: A Case Study on Taobao," *IEEE International Symposium on Workload Characterization*, pp. 3–13, 2012.
18. Y. Chen, A. Ganapathi, R. Griffith, R. Katz, "The Case for Evaluating MapReduce Performance Using Workload Suites," *19th International Symposium on Modeling, Analysis and Simulation of Computer and Telecommunication Systems (MASCOTS)*, pp. 390–399, July 2011.
19. X. Gao, V. Nachankar, J. Qiu, "Experimenting with Lucene Index on HBase in an HPC Environment," *1st Annual Workshop on High Performance Computing Meets Databases (HPCDB)*, pp. 25–28, 2012.
20. S. Das, Y. Sismanis, K. Beyer, R. Gemulla, P. Haas, J. McPherson, "Ricardo: Integrating R and Hadoop," *SIGMOD*, pp. 987–999, June 2010.
21. A. Rasooli, D. Down, "COSHH: A Classification and Optimization Based Scheduler for Heterogeneous Hadoop Systems," Technical Report of McMaster University, Canada, 2013.
22. Z. Bar-Yossef, R. Kumar, D. Sivakumar, "Reductions in Streaming Algorithms, with an Application to Counting Triangles in Graphs," *13th ACM-SIAM Symposium on Discrete Algorithms (SODA)*, January 2002.
23. M. Datar, A. Gionis, P. Indyk, R. Motwani, "Maintaining Stream Statistics over Sliding Windows," *SIAM Journal on Computing*, vol. 31, no. 6, pp. 1794–1813, 2002.
24. M. Aldinucci, M. Torquati, M. Meneghin, "FastFlow: Efficient Parallel Streaming Applications on Multi-Core," Technical Report no. TR-09-12, Universita di Pisa, Italy, September 2009.
25. R. Brightwell, "Workshop on Managed Many-Core Systems," *1st Workshop on Managed Many-Core Systems*, 2008.
26. K. Michael, *The Linux Programming Interface*. No Starch Press, San Francisco, 2010.
27. C. Shannon, "A Mathematical Theory of Communication," *The Bell System Technical Journal*, vol. 27, pp. 379–423, 1948.
28. M. Zhanikeev, Y. Tanaka, "Popularity-Based Modeling of Flash Events in Synthetic Packet Traces," *IEICE Technical Report on Communication Quality*, vol. 112, no. 288, pp. 1–6, 2012.
29. M. Zhanikeev, Y. Tanaka, "A Graphical Method for Detection of Flash Crowds in Traffic," *Springer Telecommunication Systems Journal*, vol. 57, no. 1, pp. 91–105, 2014.
30. S. Heinz, J. Zobel, H. Williams, "Burst Tries: A Fast, Efficient Data Structure for String Keys," *ACM Transactions on Information Systems (TOIS)*, vol. 20, no. 2, pp. 192–223, 2002.
31. F. Putze, P. Sanders, J. Singler, "Cache-, Hash- and Space-Efficient Bloom Filters," *Journal of Experimental Algorithmics (JEA)*, vol. 14, no. 4, 2009.
32. F. Bonomi, M. Mitzenmacher, R. Panigrahy, S. Singh, G. Varghese, "Bloom Filters via d-Left Hashing and Dynamic Bit Reassignment," *44th Allerton Conference on Communication, Control, and Computing*, 2006.
33. F. Bonomi, M. Mitzenmacher, R. Panigrahi, S. Singh, G. Vargrese, "An Improved Construction for Counting Bloom Filters," *14th Conference on Annual European Symposium (ESA)*, vol. 14, pp. 684–695, 2006.
34. H. Song, S. Dharmapurikar, J. Turner, J. Lockwood, "Fast Hash Table Lookup Using Extended Bloom Filter: An Aid to Network Processing," *SIGCOMM*, 2005.
35. A. Kirsch, M. Mitzenmacher, "Less Hashing, Same Performance: Building a Better Bloom Filter," *Wiley Inderscience Journal on Random Structures and Algorithms*, vol. 33, no. 2, pp. 187–218, 2007.
36. G. Antichi, D. Ficara, S. Giordano, G. Procissi, F. Vitucci, "Blooming Trees for Minimal Perfect Hashing," *IEEE Global Telecommunications Conference (GLOBECOM)*, pp. 1–5, December 2008.
37. C. Bhavanasi, S. Iyer, "M2MC: Middleware for Many to Many Communication over Broadcast Networks," *1st International Conference on Communication Systems Software and Middleware*, pp. 323–332, 2006.

38. D. Digby, "A Search Memory for Many-to-Many Comparisons," *IEEE Transactions on Computers*, vol. C22, no. 8, pp. 768–772, 1973.
39. M. Hattori, M. Hagiwara, "Knowledge Processing System Using Multidirectional Associative Memory," *IEEE International Conference on Neural Networks*, vol. 3, pp. 1304–1309, 1995.
40. Y. Keselman, A. Shokoufandeh, M. Demirci, S. Dickinson, "Many-to-Many Graph Matching via Metric Embedding," *IEEE Conference on Computer Vision and Pattern Recognition (CVPR)*, pp. 850–857, 2003.
41. D. Lorenz, A. Orda, D. Raz, "Optimal Partition of QoS Requirements for Many-to-Many Connections," *International Conference on Computers and Communications (ICC)*, pp. 1670–1680, 2003.
42. V. Dvorak, J. Jaros, M. Ohlidal, "Optimum Topology-Aware Scheduling of Many-to-Many Collective Communications," *6th International Conference on Networking (ICN)*, pp. 61–66, 2007.
43. A. Silberstein, J. Yang, "Many-to-Many Aggregation for Sensor Networks," *23rd International Conference on Data Engineering*, pp. 986–995, 2007.
44. M. Saleh, A. Kamal, "Approximation Algorithms for Many-to-Many Traffic Grooming in WDM Mesh Networks," *INFOCOM*, pp. 579–587, 2010.
45. Source Code for this Chapter. Available at: https://github.com/maratishe/fasthash4datastreams.
46. Stringex Project Repository. Available at: https://github.com/maratishe/stringex.
47. M. Ramakrishna, J. Zobel, "Performance in Practice of String Hashing Functions," *5th International Conference on Database Systems for Advanced Applications*, April 1997.
48. D. Lemire, O. Kaser, "Strongly Universal String Hashing is Fast," Cornell University Technical, Ithaca, NY, Report arXiv:1202.4961, 2013.
49. M. Zhanikeev, "Experiments with Practical On-Demand Multi-Core Packet Capture," *15th Asia-Pacific Network Operations and Management Symposium (APNOMS)*, September 2013.
50. T. Benson, A. Akella, D. Maltz, "Network Traffic Characteristics of Data Centers in the Wild," *Internet Measurement Conference (IMC)*, pp. 202–208, November 2010.
51. E. Kohler, R. Morris, B. Chen, J. Jannotti, M. Kaashoek, "The Click Modular Router," *ACM Transactions on Computer Systems (TOCS)*, vol. 18, no. 3, pp. 263–297, 2000.
52. M. Zec, L. Rizzo, M. Mikuc, "DXR: Towards a Billion Routing Lookups per Second in Software," *ACM SIGCOMM Computer Communication Review*, vol. 42, no. 5, pp. 30–36, 2012.
53. J. Zhang, X. Niu, J. Wu, "A Space-Efficient Fair Packet Sampling Algorithm," *Asia-Pacific Network Operation and Management Symposium (APNOMS)*, Springer LNCS vol. 5297, pp. 246–255, September 2008.
54. MCoreMemory Project Page. Available at: https://github.com/maratishe/mcorememory.

CHAPTER 13

# Organic Streams

## *A Unified Framework for Personal Big Data Integration and Organization Towards Social Sharing and Individualized Sustainable Use*

Xiaokang Zhou and Qun Jin

CONTENTS



ABSTRACT


This chapter describes a unified framework for dynamically integrating and meaningfully organizing personal and social Big Data. With the rapid development of emerging computing paradigms, we have been continuously experiencing a change in work, life, playing, and learning in the highly developed information society, which is a kind of seamless integration of the real physical world and cyber digital space. More and more people have been accustomed to sharing their personal contents across the social networks due to the high accessibility of social media along with the increasingly widespread adoption of wireless mobile computing devices. User-generated information has spread more widely and quickly and provided people with opportunities to obtain more knowledge and information than ever before, which leads to an explosive increase of data scale, containing big potential value for individual, business, domestic, and national economy development. Thus, it has become an increasingly important issue to sustainably manage and utilize personal Big Data, in order to mine useful insight and real value to better support information seeking and knowledge discovery. To deal with this situation in the Big Data era, a unified approach to aggregation and integration of personal Big Data from life logs in accordance with individual needs is considered essential and effective, which can benefit the sustainable information sharing and utilization process in the social networking environment. In this chapter, a new concept of *organic stream*, which is designed as a flexibly extensible data carrier, is introduced and defined to provide a simple but efficient means to formulate, organize, and represent personal Big Data. As an abstract data type, organic streams can be regarded as a logic metaphor, which aims to meaningfully process the raw stream data into an associatively and methodically organized form, but no concrete implementation for physical data structure and storage is defined. Under the conceptual model of organic streams, a heuristic method is proposed and applied to extract diversified individual needs from the tremendous amount of social stream data through social media. And an integrated mechanism is developed to aggregate and integrate the relevant data together based on individual needs in a meaningful way, in which personal data can be physically stored and distributed in private personal clouds and logically represented and processed by a set of newly introduced metaphors named *heuristic stone*, *associative drop*, and *associative ripple*. The architecture of the system with the foundational modules is described, and the prototype implementation with the experiment's result is presented to demonstrate the usability and effectiveness of the framework and system.

## 13.1 INTRODUCTION

With the continuous development of social media (such as Twitter and Facebook), more and more populations have been involved in this social networking revolution, which leads to a tremendous increase of data scale, ranging from daily text data to multimedia data that describe different aspects of people's lives. Websites no longer change in weeks or days but in hours, minutes, or even seconds, as people are willing to share their personal contents with each other at all times and places. For instance, every month, Facebook deals with 570 billion page views, stores 3 billion new photos, and manages 25 billion pieces of content [1]. Big Data, which "includes data sets with sizes beyond the ability of current technology, method and theory to capture, manage, and process the data within a tolerable elapsed time" [2], is an example of "high-volume, high-velocity, and/or high-variety information assets that require new forms of processing to enable enhanced decision making, insight discovery and process optimization" [3]. It has become a big challenge to process such massive amounts of complex Big Data, which has attracted a lot of attention from academia, industry, and government as well.

Life logs, a kind of personal Big Data, have also attracted increasing attention in recent years. Life logs include a variety of data, such as text, location, sound, images, and videos, which are dynamically produced from multiple sources and with different data structures or even no data structures. Consequently, it is difficult for an individual (end user) to utilize the useful information hidden in these records, such as a user's personal experience, or the social knowledge shared in a community, by simply processing the raw data from life logs. Therefore, it is essential to effectively integrate and mine this kind of personal Big Data in both cyberspace and the real world, in order to provide people with valuable individualized services.

Following this discussion, we refer to either stream data in the cyber world or life log data from the physical world, which represent different aspects of people's information behaviors and social activities, *social streams*. On one hand, the amount of stream data that people can obtain is growing at a stupendous rate. On the other hand, it is difficult for one person to remember such a tremendous amount of information and knowledge that are mixed together with fewer relations simultaneously, especially when he/she faces the situation where newer information emerges continuously. Thus, we try to find a new way by which the massive amount of raw stream data can be meaningfully and methodically organized into a pre-prepared form and the related information and knowledge hidden in the stream data can be associatively provided to users when they match users' personal intentions, in order to benefit the individualized information-seeking and retrieval process.

In a previous study, a graph model with a set of metaphors was introduced and presented, in order to represent a variety of social stream data with a hierarchical structure on an abstract level [4]. A mechanism was developed to assist users' information seeking that could best fit users' current needs and interests with two improved algorithms [5]. Based on these, the organized stream data have further been utilized to facilitate the enrichment of the user search experience [6]. Moreover, the so-called organic stream with formal descriptions was introduced and defined, which could discover and represent the potential relations among data [7]. And a mechanism was proposed to organize the raw stream data into the methodically pre-prepared form, in order to assist the data aggregation and integration process [8].

In this chapter, an effective approach to aggregating and integrating personal Big Data in accordance with individual needs is introduced, which can further benefit the information fusion and knowledge discovery process and facilitate the sustainable data utilization process as well. We describe the extended and refined definition of organic stream in order to make it an extensible data carrier to formulize and organize personal Big Data from life logs in a meaningful and flexible way, which can assist the transformation from data to information, from information to knowledge, and finally, from knowledge to asset iteratively.

## 13.2 OVERVIEW OF RELATED WORK

Big Data has drawn more and more researchers in recent years [9–14]. Alsubaiee et al. [9] built a parallel database system called ASTERIX, which tried to combine time-tested principles from parallel database systems with web-scale computing community, in order to deal with the "Big Data" management challenges. Berkovich and Liao [10] developed a tool to cluster diverse information items in a data stream mode, which can enhance the available information-processing resources for on-the-fly clusterization from diverse sources. Hoi et al. [11] proposed a feature selection algorithm in order to solve the online feature selection (OFS) problem. Zhang et al. [12] presented a parallel rough set–based approach using MapReduce for knowledge acquisition on the basis of the characteristics of data, in order to mine knowledge from Big Data. Menon [13] focused on the data warehousing and analytics platform of Facebook to provide support for batch-oriented analytics applications. Han et al. [14] introduced a Big Data model that used social network data for information recommendation related to a variety of social behaviors.

There are many analyses, as well as applications, that focus on life logs [15–20]. Yamagiwa et al. [15] proposed a system to achieve an ecological lifestyle at home, in which sensors closely measure a social life log by the temperature, humidity, and intensity of illumination related to human action and living environment. Hori and Aizawa [16] developed a context-based video retrieval system that utilized various sensor data to provide video browsing and retrieval functions for life log applications. Hwang and Cho [17] developed a machine learning method for life log management, in which a probabilistic network model was employed to summarize and manage human experiences by analyzing various kinds of log data. Kang et al. [18] defined metadata to save and search life log media, in order to deal with problems such as high-capacity memory space and long search time cost. Shimojo et al. [19] reengineered the life log common data model (LLCDM) and life log mash-up API (LLAPI) with the relational database MySQL and web services, which can help access the standardized data, in order to support the integration of heterogeneous life log services. Nakamura and Nishio [20] proposed a method to infer users' temporal preference according to their interests by analyzing web browsing logs.

As for data aggregation [21–26], Ozdemir [21] presented a reliable data aggregation and transmission protocol based on the concept of functional reputation, which can improve the reliability of data aggregation and transmission. Jung et al. [22] proposed and developed two data aggregation mechanisms based on hybrid clustering, a combination of the clustering-based data aggregation mechanism and adaptive clustering-based data aggregation, in order to improve both data aggregation and energy efficiency. Iftikhar and Pedersen [23] developed a rule-based tool to maintain data at different levels of granularity, which features automatic gradual data aggregation. Rahman et al. [24] proposed a privacy-preserving data aggregation scheme named REBIVE (reliable private data aggregation scheme), which considers both data accuracy maintenance and privacy protection, in order to provide a privacy preservation technique to maintain data accuracy for realistic environments. Wei et al. [25] proposed three prediction-based data aggregation approaches, Grey Model–based Data Aggregation (GMDA), Kalman Filter–based Data Aggregation (KFDA), and Combined Grey model and Kalman Filter Data Aggregation (CoGKDA), which can help reduce redundant data communications. Ren et al. [26] proposed an attribute-aware data aggregation (ADA) scheme, which consists of a packet-driven timing algorithm and a special dynamic routing protocol, to improve data aggregation efficiency and further benefit data redundancy elimination.

Research has also been done on data integration. Getta [27] exploited a system for online data integration based on the binary operations of relational algebra. Atkinson et al. [28] developed a framework of data mining, access, and integration, to deal with the scale-up issue of data integration and data mining across heterogeneous and distributed data resources and data mining services. Sarma et al. [29] described approximation algorithms to solve the cost-minimization and maximum-coverage problems for data integration over a large number of dependent sources. Tran et al. [30] developed a platform for distributed data integration and mining, in which data are processed as streams by processing elements connected together into workflows. Gong et al. [31] introduced the architecture and prototype implementation to provide a dynamic solution for integration of heterogeneous data sources.

Research has been done on processing stream data [32,33]. The Chronicle data model [32] is the first model proposed to capture and maintain stream data. Babcock et al. [33] examined models and issues in stream query language and query processing. A host of research has been done to make use of stream data and create Social Semantic Microblogs or use Semantic Webs to link and reuse stream data across Web 2.0 platforms [34–39]. Ebner [34] showed how a microblog can be used during a presentation to improve the situation through instant discussions by the individuals in a classroom. In addition to traditional conference tools, Reinhardt et al. [35] utilized the microblog to enhance knowledge among a group by connecting a diverse online audience. Ebner et al. [36] indicated that a microblog should be seen as a completely new form of communication that can support informal learning beyond classrooms. Studies have also been done to create a prototype for distributed semantic microblogging [37]. Passant et al. [38] developed a platform for open, semantic, and distributed microblogs by combining Social Web principles and state-of-the-art Semantic Web and Linked Data technologies. Bojars et al. [39] used the Semantic Web to link and reuse distributed stream data across Web 2.0 platforms.

## 13.3 ORGANIC STREAM: DEFINITIONS AND ORGANIZATIONS

### 13.3.1 Metaphors and Graph Model

Due to the diverse nature of the data carried within social networking sites, these contents posted in Social Networking Service (SNS) can only be viewed as raw data streams. In order to obtain meaningful information from these raw data streams, the data from these social networking sites must be organized systematically. Therefore, we introduce the metaphors for data streams as follows [4]:

- *Drop*: A drop is a minimum unit of data streams, such as a message posted to a microblog (e.g., Twitter) by a user or a status change in SNS (e.g., Facebook).
- *Stream*: A stream is a collection of drops in a timeline, which contains the messages, activities, and actions of a user.
- *River*: A river is a confluence of streams from different users that is formed by following or subscribing to their followers/friends. It could be extended to followers' followers.
- *Ocean*: An ocean is a combination of all the streams.

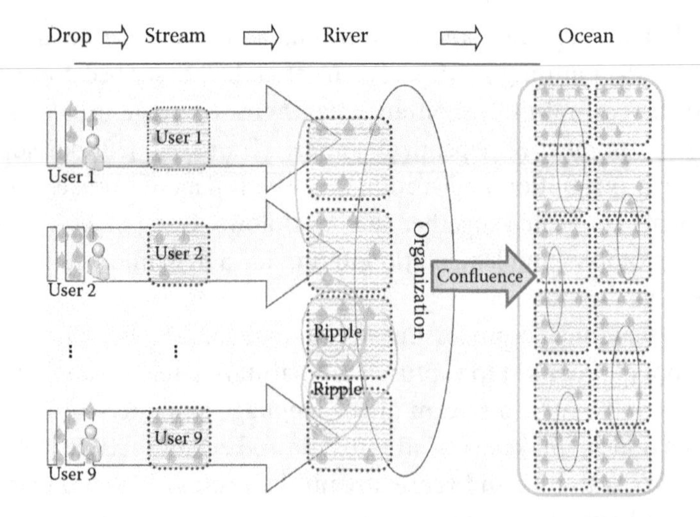


FIGURE 13.1 Graph model for social streams. (With kind permission from Springer Science+ Business Media: *Multimedia Tools and Applications*, "Enriching User Search Experience by Mining Social Streams with Heuristic Stones and Associative Ripples," 63, 2013, 129–144, Zhou, X.K. et al.)

Thus, the personal data posted by each user can be seen as a drop, and the drops coming from one user can converge together to form a stream. Then the streams of the user and his/her friends can form a river. Finally, all the streams can come together to form an ocean. All these processes are shown in Figure 13.1.

The following definitions are used for seeking related information that satisfies users' current needs and interests [5].

- *Heuristic stone*: It represents one of a specific user's current interests, which may be changed dynamically.
- *Associative ripple*: It is a meaningfully associated collection of the drops related to some topics of a specific user's interests, which are formed by the heuristic stone in the river.
- *Associative drop*: It is the drop distributed in an associative ripple, which is related to one specific heuristic stone and can be further collected into the organic stream.

### 13.3.2 Definition of Organic Stream

In a previous study [4], an organic stream is considered as a collection of drops and streams, in which they refer or link to each other by perceiving the inherent logicality among them, so that they may contain the dynamic and potential knowledge and can further benefit the information-seeking process. In this chapter, the refined the organic stream and the descriptions of the relations among them are introduced and defined as follows.

*Organic stream*: An organic stream is a dynamically changed carrier of organized social streams that may contain hidden knowledge and potential information. An organic stream can be expressed as in Equation 13.1,

$$OS = \Phi\ (Hs, Ad, R) \tag{13.1}$$

where

$Hs = \{Hs[u_1, t_1], Hs[u_2, t_2], \ldots, Hs[u_m, t_n]\}$: a nonempty set of heuristic stones in accordance with different users' intentions, in which each $Hs[u_i, t_j]$ indicates a specific heuristic stone of a specific user, $u_i$ indicates the owner of this heuristic stone, and $t_j$ indicates the time slice this heuristic stone belongs to;

$Ad = \{Ad_1, Ad_2, \ldots, Ad_n\}$: a collection of associative drops, which can refer or link to each other based on inherent or potential logicality; and

$R$: the relations among heuristic stones and associative drops in the organic stream.

Following these definitions, the relation $R$ in the organic stream can be categorized into three major types: relation between each heuristic stone, relation between each associative drop, and relation between heuristic stones and associative drops. The details are addressed as follows.

*Relation between heuristic stone and associative drop*: This type of relation identifies the relationships between one heuristic stone and a series of associative drops, which can be represented as *heuristic stone* × *associative drop*. It is the basic relation in the organic stream. In detail, due to different heuristic stones, those related drops can be connected together in the organic stream.

*Relation between heuristic stone and heuristic stone*: This type of relation identifies the relationships among the heuristic stones in the organic stream, which can be represented as *heuristic stone* × *heuristic stone*. Note that the expression of heuristic stones has two parameters, $u_i$ and $t_j$; thus, this kind of relation can further be categorized into two subtypes, which can be addressed as follows.

$Hs[u_i, t_x] \leftrightarrow Hs[u_i, t_y]$: This relation identifies the relationships of the heuristic stones within a specific user $u_i$. That is, this relation is used to describe those internal relationships or changes for a specific user's intentions. In detail, given a series of heuristic stones belonging to a specific user $u_i$, represented as $\{Hs[u_i, t_1], Hs[u_2, t_2], \ldots, Hs[u_i, t_n]\}$, the differences from $Hs[u_i, t_1]$ to $Hs[u_i, t_n]$ changed in a sequence can demonstrate the transitions of this user's interests or needs in a specific period, which can be employed to infer his/her further intention.

$Hs[u_i, t_x] \leftrightarrow Hs[u_j, t_y]$: This relation identifies the relationships of the heuristic stones between two different users, $u_i$ and $u_j$. That is, this relation is used to describe those external relationships among different users' intentions. In detail, given two heuristic stones, represented as $Hs[u_i, t_a]$ and $Hs[u_j, t_b]$ for two different users, the relationship can demonstrate the potential connection between these two users in accordance with their dynamic interests or needs.

*Relation between associative drop and associative drop*: This type of relation identifies the relationships among those drops that are clustered to the associative ripples and further selected into the organic stream, which can be represented as *associative drop* × *associative drop*. In detail, the drops connected together based on this relation can represent the whole trend as well as its changes following the timeline.

According to its definition, the organic stream is defined as a carrier of hidden knowledge and potential information, in which three major relations are constructed to describe the inherent relationship and logicality, in order to organize the raw stream data into meaningful and methodic content. In other words, the organic stream is a pre-prepared collection that contains diversified and dynamically changed information, which shall also be extended in time, so that it can associatively provide users with various services in accordance with their different requirements.

### 13.3.3 Organization of Social Streams

As defined in Section 13.3.1, the heuristic stone is defined to represent a specific user's current interest or need, which can be discovered and extracted from his/her own streams. The associative ripple can then be generated in accordance with the heuristic stones, which will further contribute to the generation of the organic stream. The conceptual generation process is shown in Figure 13.2. That is, in this study, each piece of personal data posted by users could be viewed as the drop. The key words in these drops will be extracted to generate the heuristic stone. The Term Frequency-Inverse Document Frequency (TF-IDF)–based method is further employed to calculate the weight of each key word, which can be expressed in Equation 13.2,

$$w_i = F(i,t)^* \frac{T}{\sum_{j=1}^{n} In(i,t_j)} + \frac{\sum_{j=1}^{n} F(i,t_j)}{M} \tag{13.2}$$

where

$$In(i,t) = \begin{cases} 1 & \text{if the keyword } i \text{ exists in this period } t \\ 0 & \text{else} \end{cases}$$

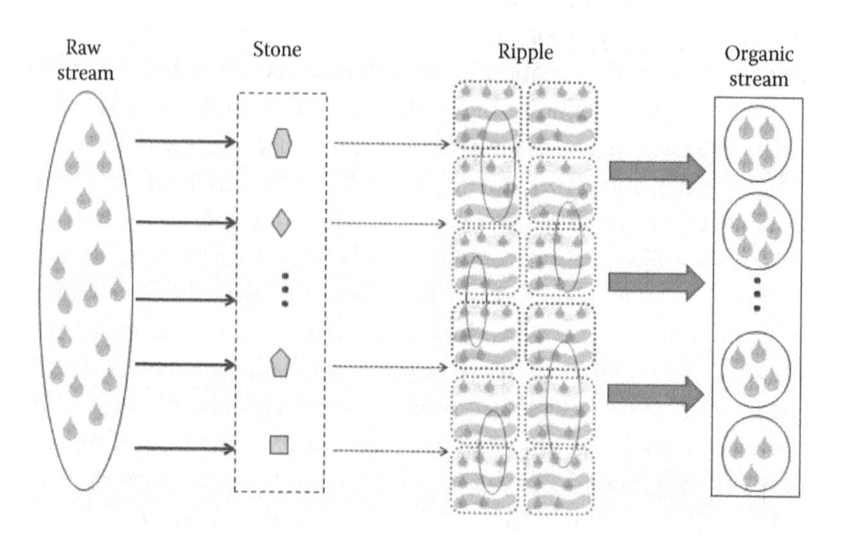


FIGURE 13.2 Image of the generation process of organic streams. (From Zhou, X.K. et al., "Organic Stream: Meaningfully Organized Social Stream for Individualized Information Seeking and Knowledge Mining," Proc. the 5th IET International Conference on Ubi-Media Computing [U-Media2012], Xining, China, Aug. 16–18, 2012.)

In Equation 13.2, $F(i, t)$ indicates the frequency of key word $i$ in a specific period $t$ in which the heuristic stone will be extracted, while $T$ indicates the whole interval in which the organic stream will be generated. For example, if $T$ indicates 1 month, then $t$ can be set as 1 day. $M$ indicates the total amount of the key words in period $T$. That is, the former part $F(i,t)^* \frac{T}{\sum_{j=1}^{n} In(i,t_j)}$ is employed to calculate the transilient interest, while the latter part $\frac{\sum_{j=1}^{n} F(i,t_j)}{M}$ is employed to calculate the durative interest. Finally, the key word with the weight that is higher than a given threshold δ will be extracted as the heuristic stone.

Based on these, those related drops can be selected into the associative ripples with the corresponding heuristic stone and go further to generate the organic stream according to the relation $R$. The organization algorithm is shown in Figure 13.3.

As a summary, in order to capture an individual's time-changing needs, the whole time period should be divided into several time slices (e.g., 1 day, 1 week), in which the developed TF-IDF method can be employed to extract data to represent individual needs, concerns, and interests within this specific time slice. After that, the data related to the extracted individual need can be selected and aggregated according to the *heuristic stone* × *associative drop* relation in a heuristic way. Based on these, the associative ripples can be generated, in which the selected data shall be integrated and organized following the *associative drop* × *associative drop* relation. Finally, a series of associative ripples will compose the organic stream according to the *heuristic stone* × *heuristic stone* relation in an extensible way.

**Input:** A specific user's post set $Z_s$ and all users' post $G_s$
**Output:** The heuristic stone set *Hs* and the associative drops set *Ad*

**Step 1:** Input $Z_1$, $G_s$, initialize the duration $T$ and $t$
**Step 2:** For each key word $k_i$ in each $Z_i$

Calculate $w_i = F(i,t)^* \frac{T}{\sum_{j=1}^{n} \text{In}(i,t_j)} + \frac{\sum_{j=1}^{n} F(i,t_j)}{M}$

if $w_i >$ threshold δ
add $k_i$ into *Hs*

**Step 3:** For each $Hs[u_i, t_j]$ and $G_i$
if $Q(G_i, Hs[u_i, t_j]) \rightarrow G_i \in r_x$
add $G_i$ into the associative ripple $r_x$ generated in each $t$
**Step 4:** For each $r_x$, select high-ranked $G_i$ into the associative drop set *Ad*, record the *Heuristic Stone* × *Associative Drop* relation
**Step 5:** Calculate the other two kinds of relations among elements in set *Hs* and *Ad*
**Step 6:** Return the set *Hs* and *Ad* with the relation *R*

FIGURE 13.3 Algorithm for generating organic stream. (From Zhou, X.K. et al., "Organic Stream: Meaningfully Organized Social Stream for Individualized Information Seeking and Knowledge Mining," Proc. the 5th IET International Conference on Ubi-Media Computing [U-Media2012], Xining, China, Aug. 16–18, 2012.)

## 13.4 EXPERIMENTAL RESULT AND ANALYSIS

### 13.4.1 Functional Modules

The architecture to realize the individual need-based data integration and organization in organic streams is shown in Figure 13.4, which consists of five major components: data collector, user need extractor, data aggregator, data integrator, and data relation analyzer.

As shown in Figure 13.4, the data collector, which is the fundamental function module in this system, is used to collect the raw data from individuals and connect with the major database, which saves the life logs of all users. The user need extractor is employed to extract individual needs, concerns, and interests from a user's life logs to generate the heuristic stones. In addition, the data aggregator works to aggregate the data related to the extracted individual needs, while the data integrator is responsible for integrating and organizing the heuristic stones with the associative drops. Finally, the data relation analyzer analyzes the major relations among these data and further organizes them to generate the associative ripples.

### 13.4.2 Experiment Analysis

Twitter data have been utilized to conduct our experiment in order to demonstrate the feasibility of the proposed framework and method.

We obtained the Twitter data set from a list of Twitter users selected from a famous Twitter list named "twitter," as well as some of their followers. The majority of the tweets that were collected for our experiment were published in April 2013. Finally, a total of 320,000 tweets were collected. In order to illuminate our method, in this case, during the time period from April 4 to 16, the key word "Boston" was extracted as the heuristic stone using the TF-IDF method, which could indicate the interest or need within this group of users. And then those data that were related to "Boston" were aggregated to form the associative drops. An example of experiment analysis results is shown in Figure 13.5.

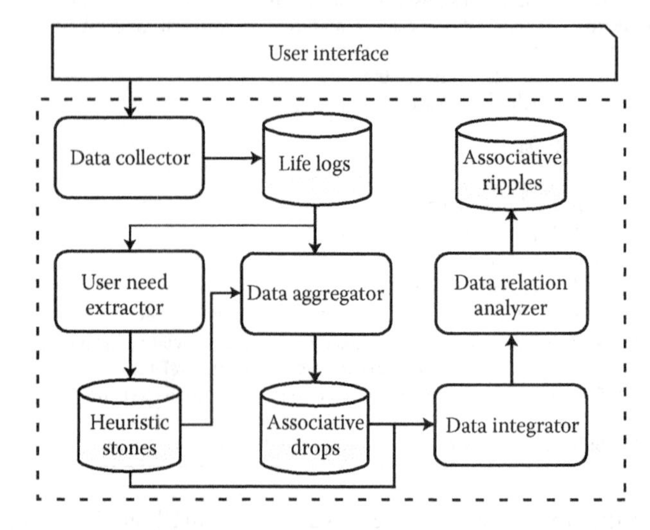


FIGURE 13.4 Algorithm for generating organic stream. (From Zhou, X.K. et al., "Organic Streams: Data Aggregation and Integration Based on Individual Needs," Proc. the 6th IEEE International Conference on Ubi-Media Computing [U-Media2013], Aizu-Wakamatsu, Japan, Nov. 2–4, 2013.)

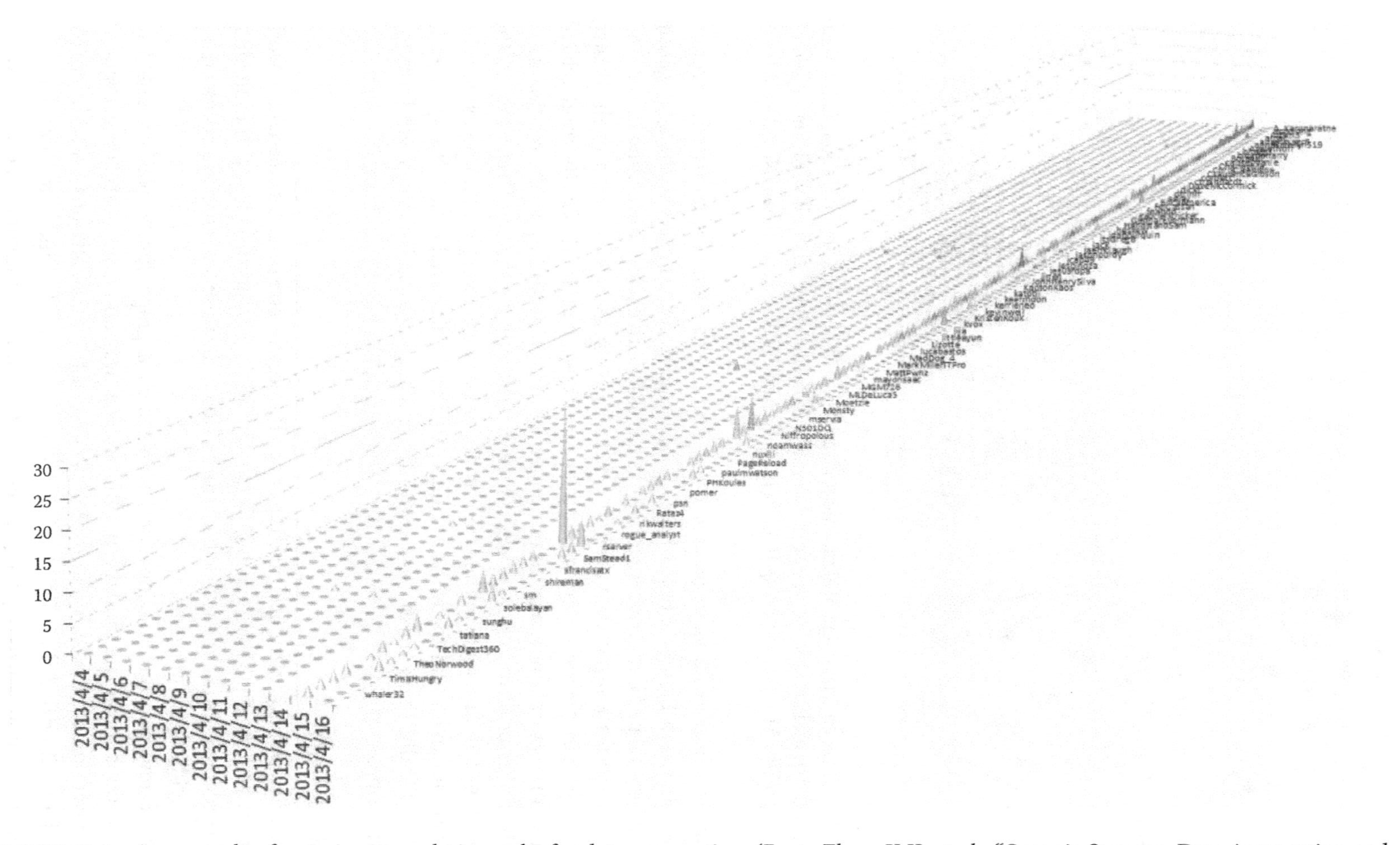


FIGURE 13.5 An example of experiment analysis results for data aggregation. (From Zhou, X.K. et al., "Organic Streams: Data Aggregation and Integration Based on Individual Needs," Proc. the 6th IEEE International Conference on Ubi-Media Computing [U-Media2013], Aizu-Wakamatsu, Japan, Nov. 2–4, 2013.)

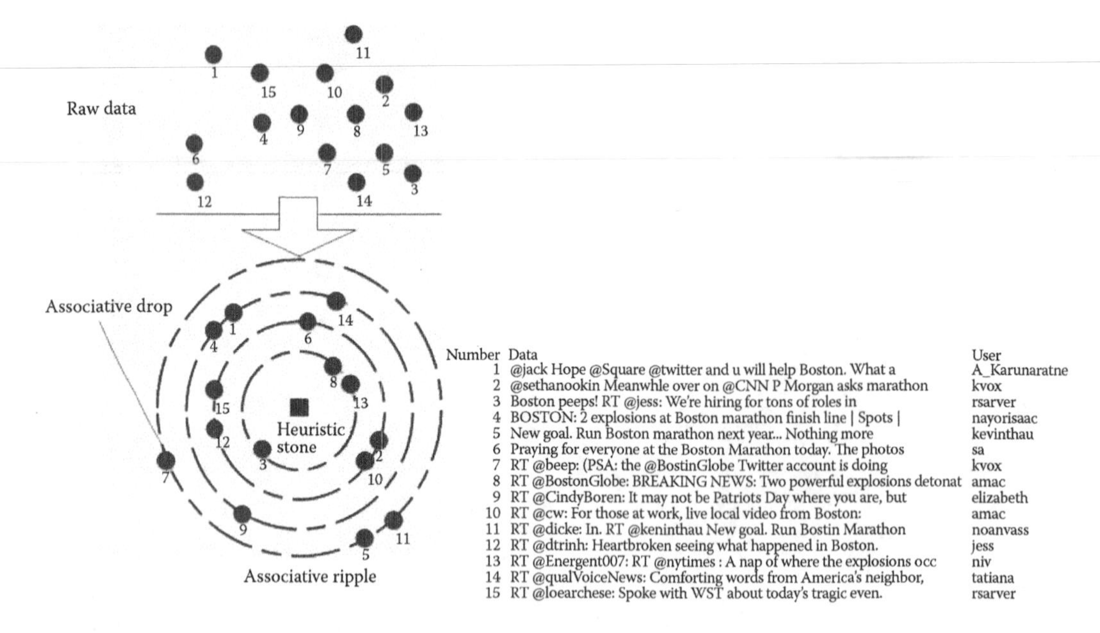


| Number | Data | User |
|---|---|---|
| 1 | @jack Hope @Square @twitter and u will help Boston. What a | A_Karunaratne |
| 2 | @sethanookin Meanwhle over on @CNN P Morgan asks marathon | kvox |
| 3 | Boston peeps! RT @jess: We're hiring for tons of roles in | rsarver |
| 4 | BOSTON: 2 explosions at Boston marathon finish line \| Spots \| | nayorisaac |
| 5 | New goal. Run Boston marathon next year... Nothing more | kevinthau |
| 6 | Praying for everyone at the Boston Marathon today. The photos | sa |
| 7 | RT @beep: (PSA: the @BostinGlobe Twitter account is doing | kvox |
| 8 | RT @BostonGlobe: BREAKING NEWS: Two powerful explosions detonat | amac |
| 9 | RT @CindyBoren: It may not be Patriots Day where you are, but | elizabeth |
| 10 | RT @cw: For those at work, live local video from Boston: | amac |
| 11 | RT @dicke: In. RT @keninthau New goal. Run Bostin Marathon | noanvass |
| 12 | RT @dtrinh: Heartbroken seeing what happened in Boston. | jess |
| 13 | RT @Energent007: RT @nytimes : A nap of where the explosions occ | niv |
| 14 | RT @qualVoiceNews: Comforting words from America's neighbor, | tatiana |
| 15 | RT @loearchese: Spoke with WST about today's tragic even. | rsarver |

FIGURE 13.6 Conceptual image of data integration. (From Zhou, X.K. et al., "Organic Streams: Data Aggregation and Integration Based on Individual Needs," Proc. the 6th IEEE International Conference on Ubi-Media Computing [U-Media2013], Aizu-Wakamatsu, Japan, Nov. 2–4, 2013.)

We further integrate the aggregated data according to the relation $R$ discussed in Section 13.3.2. As shown in Figure 13.6, the selected heuristic stone has become the aggregating center, and the related data converged to it as the associative drop. According to the relation *heuristic stone* × *associative drop*, the distance from associative drop to the center, namely, the heuristic stone, describes the relevance between them, that is, the more related associative drops would be closer to the center. Moreover, according to the relation *associative drop* × *associative drop*, the associative drops that have the same relevance to the heuristic stone will distribute in the same layer. For instance, in Figure 13.6, all associative drops distribute in four layers, while in each layer, the more related data will stay closer to each other. Finally, the heuristic stone with these related associative drops will form the associative ripple.

## 13.5 SUMMARY

In this chapter, we have introduced and described the organic stream as a unified framework to integrate and organize personal Big Data from life logs in accordance with individual needs, in order to benefit the sustainable information utilization and sharing process in the social networking environment.

We have reviewed our previous works, in which a graph model was presented to describe various social stream data in a hierarchical way and a set of stream metaphors was employed to describe the organization process of the social streams. In addition, we described and discussed the refined organic stream as a flexible and extensible data carrier, to aggregate and integrate personal Big Data according to individual needs, in which the raw stream data can be associatively and methodically organized into a meaningful form. We introduced in detail the heuristic stone, associative drop, and associative ripple to represent individual need, related data, and an organized data set, respectively. Three major relations, *heuristic stone* × *associative drop*, *heuristic stone* × *heuristic stone*, and *associative drop* × *associative drop*, were defined and proposed to discover and describe the intrinsic and potential relationships among the aggregated data. Based on these, an integrated mechanism was developed to extract individuals' time-varying needs and further aggregate and integrate personal Big Data into an associatively organized form, which can further support the information fusion and knowledge discovery process. Finally, we presented the system architecture with the major functional modules, and the experimental analysis results illuminated the feasibility and effectiveness of the proposed framework and approach.

## REFERENCES


1. C.Q. Ji, Y. Li, W.M. Qiu, U. Awada, and K.Q. Li, "Big Data processing in cloud computing environments," in *Proc. 12th International Symposium on Pervasive Systems, Algorithms and Networks (ISPAN)*, December 13–15, 2012, pp. 17–23.
2. "Big Data: Science in the petabyte era," *Nature* vol. 455, no. 7209, p. 1, 2008.
3. L. Douglas, *The Importance of "Big Data": A Definition*, Gartner, Stamford, CT, 2008.
4. H. Chen, X.K. Zhou, H.F. Man, Y. Wu, A.U. Ahmed, and Q. Jin, "A framework of organic streams: Integrating dynamically diversified contents into ubiquitous personal study," in *Proc. of 2nd International Symposium on Multidisciplinary Emerging Networks and Systems*, Xi'an, China, pp. 386–391, 2010.
5. X.K. Zhou, H. Chen, Q. Jin, and J.M. Yong, "Generating associative ripples of relevant information from a variety of data streams by throwing a heuristic stone," in *Proc. of 5th International Conference on Ubiquitous Information Management and Communication (ACM ICUIMC '11)*, Seoul, Korea, 2011.
6. X.K. Zhou, N.Y. Yen, Q. Jin, and T.K. Shih, "Enriching user search experience by mining social streams with heuristic stones and associative ripples," *Multimed. Tools Appl. (Springer)* vol. 63, no. 1, pp. 129–144, 2013.
7. X.K. Zhou, J. Chen, Q. Jin, and T.K. Shih, "Organic stream: Meaningfully organized social stream for individualized information seeking and knowledge mining," in *Proc. the 5th IET International Conference on Ubi-Media Computing (U-Media2012)*, Xining, China, August 16–18, 2012.
8. X.K. Zhou, Q. Jin, B. Wu, W. Wang, J. Pan, and W. Zheng, "Organic streams: Data aggregation and integration based on individual needs," in *Proc. the 6th IEEE International Conference on Ubi-Media Computing (U-Media2013)*, Aizu-Wakamatsu, Japan, November 2–4, 2013.
9. S. Alsubaiee, Y. Altowim, H. Altwaijry, A. Behm, V. Borkar, Y. Bu, M. Carey, R. Grover, Z. Heilbron, Y. Kim, C. Li, N. Onose, P. Pirzadeh, R. Vernica, and J. Wen, "ASTERIX: An open source system for 'Big Data' management and analysis (demo)," *Proc. VLDB Endow.* vol. 5, no. 12, pp. 1898–1901, 2012.
10. S. Berkovich, and D. Liao, "On clusterization of 'Big Data' streams," in *Proc. the 3rd International Conference on Computing for Geospatial Research and Applications (COM.Geo '12)*, ACM, New York, Article 26, 6 pp, 2012.
11. S.C.H. Hoi, J. Wang, P. Zhao, and R. Jin, "Online feature selection for mining Big Data," in *Proc. the 1st International Workshop on Big Data, Streams and Heterogeneous Source Mining: Algorithms, Systems, Programming Models and Applications (BigMine '12)*, ACM, New York, pp. 93–100, 2012.

12. J. Zhang, T. Li, and Y. Pan, "Parallel rough set based knowledge acquisition using MapReduce from Big Data," in *Proc. the 1st International Workshop on Big Data, Streams and Heterogeneous Source Mining: Algorithms, Systems, Programming Models and Applications (BigMine '12)*, ACM, New York, pp. 20–27, 2012.
13. A. Menon, "Big Data @ Facebook," in *Proc. the 2012 Workshop on Management of Big Data Systems (MBDS '12)*, ACM, New York, pp. 31–32, 2012.
14. X. Han, L. Tian, M. Yoon, and M. Lee, "A Big Data model supporting information recommendation in social networks," in *Proc. the Second International Conference on Cloud and Green Computing (CGC)*, November 1–3, 2012, pp. 810–813.
15. M. Yamagiwa, M. Uehara, and M. Murakami, "Applied system of the social life log for ecological lifestyle in the home," in *Proc. International Conference on Network-Based Information Systems (NBIS '09)*, August 19–21, 2009, pp. 457–462.
16. T. Hori, and K. Aizawa, "Capturing life-log and retrieval based on contexts," in *Proc. IEEE International Conference on Multimedia and Expo (ICME '04)*, June 27–30, 2004, pp. 301–304.
17. K.S. Hwang, and S.B. Cho, "Life log management based on machine learning technique," in *Proc. IEEE International Conference on Multisensor Fusion and Integration for Intelligent Systems (MFI)*, August 20–22, 2008, pp. 691–696.
18. H.H. Kang, C.H. Song, Y.C. Kim, S.J. Yoo, D. Han, and H.G. Kim, "Metadata for efficient storage and retrieval of life log media," in *Proc. IEEE International Conference on Multisensor Fusion and Integration for Intelligent Systems (MFI)*, August 20–22, 2008, pp. 687–690.
19. A. Shimojo, S. Matsumoto, and M. Nakamura, "Implementing and evaluating life-log mashup platform using RDB and web services," in *Proc. the 13th International Conference on Information Integration and Web-based Applications and Services (iiWAS '11)*, ACM, New York, pp. 503–506, 2011.
20. A. Nakamura, and N. Nishio, "User profile generation reflecting user's temporal preference through web life-log," in *Proc. the 2012 ACM Conference on Ubiquitous Computing (UbiComp '12)*, ACM, New York, pp. 615–616, 2012.
21. S. Ozdemir, "Functional reputation based reliable data aggregation and transmission for wireless sensor networks," *Comput. Commun.* vol. 31, no. 17, pp. 3941–3953, 2008.
22. W.S. Jung, K.W. Lim, Y.B. Ko, and S.J. Park, "Efficient clustering-based data aggregation techniques for wireless sensor networks," *Wirel. Netw.* vol. 17, no. 5, pp. 1387–1400, 2011.
23. N. Iftikhar, and T.B. Pedersen, "A rule-based tool for gradual granular data aggregation," in *Proc. the ACM 14th International Workshop on Data Warehousing and OLAP (DOLAP '11)*, ACM, New York, pp. 1–8, 2011.
24. F. Rahman, E. Hoque, and S.I. Ahamed, "Preserving privacy in wireless sensor networks using reliable data aggregation," *ACM SIGAPP Appl. Comput. Rev.* vol. 11, no. 3, pp. 52–62, 2011.
25. G. Wei, Y. Ling, B. Guo, B. Xiao, and A.V. Vasilakos, "Prediction-based data aggregation in wireless sensor networks: Combining grey model and Kalman filter," *Comput. Commun.* vol. 34, no. 6, pp. 793–802, 2011.
26. F. Ren, J. Zhang, Y. Wu, T. He, C. Chen, and C. Lin, "Attribute-aware data aggregation using potential-based dynamic routing in wireless sensor networks," *IEEE Trans. Parallel Distrib. Syst.* vol. 24, no. 5, pp. 881–892, 2013.
27. J.R. Getta, "Optimization of online data integration," in *Proc. 7th International Baltic Conference on Databases and Information Systems*, pp. 91–97, 2006.
28. M.P. Atkinson, J.I. Hemert, L. Han, A. Hume, and C.S. Liew, "A distributed architecture for data mining and integration," in *Proc. the Second International Workshop on Data-Aware Distributed Computing (DADC '09)*, ACM, New York, pp. 11–20, 2009.
29. A.D. Sarma, X.L. Dong, and A. Halevy, "Data integration with dependent sources," in *Proc. the 14th International Conference on Extending Database Technology (EDBT/ICDT '11)*, A. Ailamaki, S. Amer-Yahia, J. Pate, T. Risch, P. Senellart, and J. Stoyanovich (Eds.), ACM, New York, pp. 401–412, 2011.

30. V. Tran, O. Habala, B. Simo, and L. Hluchy, "Distributed data integration and mining," in *Proc. the 13th International Conference on Information Integration and Web-Based Applications and Services (iiWAS '11)*, ACM, New York, pp. 435–438, 2011.
31. P. Gong, I. Gorton, and D.D. Feng, "Dynamic adapter generation for data integration middleware," in *Proc. the 5th International Workshop on Software Engineering and Middleware (SEM '05)*, ACM, New York, pp. 9–16, 2005.
32. H.V. Jagadish, I.S. Mumick, and A. Silberschatz, "View maintenance issues for the chronicle data model," in *Proc. ACM/PODS 1995*, pp. 113–124, 1995.
33. B. Babcock, S. Babu, M. Datar, R. Motwani, and J. Widom, "Models and issues in data stream systems," in *Proc. of 21st ACM SIGMOD-SIGACT-SIGART Symposium on Principles of Database Systems*, Madison, WI, 2002.
34. M. Ebner, "Introducing live microblogging: How single presentations can be enhanced by the mass," *J. Res. Innov. Teach.* vol. 2, no. 1, pp. 108–119, 2009.
35. W. Reinhardt, M. Ebner, G. Beham, and C. Costa, "How people are using Twitter during conferences," in *Proc. of 5th EduMedia Conference*, Salzburg, Austria, pp. 145–156, 2009.
36. M. Ebner, C. Lienhardt, M. Rohs, and I. Meyer, "Microblogs in higher education—A chance to facilitate informal and process-oriented learning?," *Comput. Educ.* vol. 55, no. 1, pp. 92–100, 2010.
37. A. Passant, T. Hastrup, U. Bojars, and J. Breslin, "Micro-blogging: A semantic web and distributed approach," in *Proc. of ESWC/SFSW 2008*, Tenerife, Spain, 2008.
38. A. Passant, U. Bojars, J.G. Breslin, T. Hastrup, M. Stankovic, and P. Laublet, "An overview of SMOB 2: Open, semantic and distributed micro-blogging," in *Proc. of AAAI/ICWSM 2010*, 2010.
39. U. Bojars, J. Breslin, A. Finn, and S. Decker, "Using the Semantic Web for linking and reusing data across Web 2.0 communities," *J. Web Semant.* vol. 6, no. 1, pp. 21–28, 2008.

CHAPTER 14

# Managing Big Trajectory Data

## *Online Processing of Positional Streams*

Kostas Patroumpas and Timos Sellis

CONTENTS



ABSTRACT

As smartphones and GPS-enabled devices proliferate, *location-based services* become all the more important in social networking, mobile applications, advertising, traffic monitoring, and many other domains. Managing the locations and trajectories of numerous people, vehicles, vessels, commodities, and so forth must be efficient and robust, since this information must be processed online and should provide answers to users' requests in real time. In this *geostreaming* context, such long-running continuous queries must be repeatedly evaluated against the most recent positions relayed by moving objects, for instance, reporting which people are now moving in a specific area or finding friends closest to the current location of a mobile user. In essence, modern processing engines must cope with huge amounts of streaming, transient, uncertain, and heterogeneous spatiotemporal data, which can be characterized as *big trajectory data*. In this chapter, we examine Big Data processing techniques over frequently updated locations and trajectories of moving objects. Rapidly evolving trajectory data pose several research challenges with regard to their acquisition, storage, indexing, analysis, discovery, and interpretation in order to be really useful for intelligent, cost-effective decision making. Indeed, the Big Data issues regarding volume, velocity, variety, and veracity also arise in this case. Thus, we foster a close synergy between the established stream processing paradigm and spatiotemporal properties inherent in motion features. Taking advantage of the spatial locality and temporal timeliness that characterize each trajectory, we present methods and heuristics from our recent research results that address such problems. We highlight certain aspects of big trajectory data management through several case studies. Regarding *volume*, we suggest single-pass algorithms that can summarize each object's course into succinct, reliable representations. To cope with *velocity*, an amnesic trajectory approximation structure may offer fast, multiresolution synopses by dropping details from obsolete segments. Detection of objects that travel together can lead to trajectory multiplexing, hence reducing the *variety* inherent in raw positional data. As for *veracity*, we discuss a probabilistic method for continuous range monitoring against user locations with varying degrees of uncertainty, due to privacy concerns in geosocial networking. Last, but not least, as we are heading toward a next-generation framework in trajectory data management, we point out interesting open issues that may provide rich opportunities for innovative research and applications.

## 14.1 INTRODUCTION

Nowadays, hundreds of millions of GPS-enabled devices are in use globally. The amount of information exchanged on a daily basis is in the order of terabytes for several social networks (like Facebook or Twitter) or financial platforms (e.g., New York Stock Exchange). Billions of radio-frequency identification (RFID) tags are generated per day, and millions of sensors collect measurements in diverse application domains, ranging from meteorology and biodiversity to energy consumption, battlefield monitoring, and many more. These staggering figures are expected to double every 2 years over the next decade, as the number of Internet users, smartphone holders, online customers, and networked sensors is growing at a very fast rate. Quite often, this information also includes a *spatial aspect*, either explicitly in the form of coordinates (from GPS, RFID, or global system for mobile communications [GSM]) or implicitly (via geo-tagged photos or addresses). As smartphones and GPS-enabled devices proliferate and related platforms penetrate into the market, managing the bulk of rapidly accumulating traces of objects' movement becomes all the more crucial in modern, high-availability applications. Such *location-based services* (LBSs) are capable of identifying the geographical location of a mobile device and then providing services based on that location. Platforms for traffic surveillance and fleet management, mobile applications in tourism or advertising, notification systems for natural resources and hazards, tracing services for stolen objects or elder people, and so forth must cope with huge amounts of such flowing, uncertain, and heterogeneous spatiotemporal data.

These data sets can certainly be characterized as *Big Data* [1], not only because of their increasingly large volumes but also due to their volatility and complexity, which make their processing and analysis quite difficult, if not impossible, through traditional data management tools and applications. Obviously, it is too hard for a centralized server to sustain massive amounts of positional updates from a multitude of people, vehicles, vessels, containers, and so forth, which may also affect network traffic and load balancing.

The main challenge is how such fluctuating, transient, and possibly unbounded *positional data streams* can be processed in online fashion. In such a geostreaming context, it is important to provide incremental answers to *continuous queries* (CQs), that is, to various and numerous user requests that remain active for long and require their results to be refreshed upon changes in the incoming data [2]. There are several types of *location-aware* CQs [3,4] that examine spatial relationships against the *current locations* of numerous moving objects. For instance, a *continuous range search* must report which objects are now moving in a specific area of interest (e.g., vehicles in the city center). A mobile user who wishes to find *k* of his/her friends that are closest to his/her current location is an example of a *continuous k-nearest neighbor* (*k*-NN) search. As the monitored objects are moving and their relative positions are changing, new results must be emitted, thus cancelling or modifying previous ones (e.g., 5 min after the previous answer to this *k*-NN CQ, the situation has changed, and another friend is now closer to him/her).

In the course of time, many locations per object are being accumulated and can actually be used to keep track of its trace. Of course, it is not realistic to maintain a continuous trace of each object, since practically, point samples can only be collected from the respective data source at distinct time instants (e.g., every few seconds, a car sends its location measured by a GPS device). Still, these traces can be a treasure trove for innovative data analysis and intelligent decision making. In fact, data exploration and trend discovery against such collections of *evolving trajectories* are also crucial, beyond their effective storage or the necessity for timely response to user requests. From detection of flocks [5] or convoys [6] in fleet management to clustering [7] for wildlife protection, similarity joins [8] in vehicle traces for carpooling services, or even identification of frequently followed routes [9,10] for effective traffic control, the prospects are enormous.

The core challenges regarding Big Data [11] are usually described through the well-known three *V*s, namely, *volume*, *velocity*, and *variety*, although in Reference 12, it was proposed to include a fourth one for *veracity*. In the particular case of *big trajectory data* collected from moving objects, these challenges may be further specified as follows:

- *Volume*. If numerous vehicles, parcels, smartphones, animals, and so forth are monitored and their locations get relayed very frequently, then large amounts of positional data are being captured. Hence, the processing mechanism must be scalable, as the sheer volume of positional data may be overwhelming and could exceed several terabytes per day. Further, processing every single position incurs some overhead but does not necessarily convey significant movement changes (e.g., if a vessel moves along a straight line for some time), so opportunities exist for data compression because of similar redundancies in the input.

- *Velocity*. As data are generated continuously at various rates and are inherently streaming in a time-varying fashion, they must be handled and processed in real time. However, typical spatiotemporal queries like range, *k*-NN, and so forth are costly, and their results should not be computed from scratch upon admission of fresh input. Ideally, evaluation of the current batch of positional data must be completed before the next one arrives, in order to provide timely results and also keep in pace with the incoming streaming locations.

- *Variety.* Data can actually come from multiple, possibly heterogeneous sources and may be represented in various types or formats, either structured (e.g., as relational tuples, sensor readings) or unstructured (logs, text messages, etc.). Location could have differing semantics, for example, positions may come from both GPS devices and GSM cells; hence, accuracy may vary widely, up to hundreds of meters. Besides, these dynamic positional data might also require interaction with static data sets, for example, for matching vehicle locations against the underlying road network. Handling the intricacies of such data, eliminating noise and errors (e.g., positional outliers), and interpreting latent motion patterns are nontrivial tasks and may be subject to assumptions at multiple stages of the analysis.
- *Veracity.* Due to privacy concerns, hiding a user location and not just his/her identity is important. Hence, positional data may be purposely noisy, obfuscated, or erroneous in order to avoid malicious or inappropriate use (e.g., prevent locating of people or linking them to certain groups in terms of social, cultural, or market habits). But handling such uncertain, incomplete information adds too much complexity to processing; hence, results cannot be accurate and are usually associated with a probabilistic confidence margin.

To address such challenging research issues related to trajectory data collection, maintenance, and analytics at a very large scale, there have been several recent initiatives, each proposing methods that take advantage of the unique characteristics of these data. In this chapter, we focus on results from our own work regarding real-time processing techniques over frequently updated locations and trajectories of moving objects. We foster a close synergy between the stream processing paradigm and spatiotemporal properties inherent in motion features, in line with modern trends regarding mobility of data, fast data access, and summarization. Thus, we highlight certain case studies on big trajectory data, especially regarding online data reduction of streaming trajectories and approximate query answering against scalable data volumes. In particular,

- In Section 14.2, we present fundamental notions about moving objects and trajectory representation. We also stress certain characteristics that arise in modeling and querying in a geostreaming context, such as the necessity of timestamps in the incoming locations, the use of sliding windows, and the most common types of online analytics.
- In Section 14.3, we discuss single-pass approximation techniques based on sampling, which take advantage of the spatial locality and temporal timeliness inherent in trajectory streams. The objective is to tackle *volume* and maintain a concise yet quite reliable summary of each object's movement, avoiding any superfluous details and reducing processing complexity and communication cost.

- In Section 14.4, we present a hierarchical tree structure that reserves more precision for the most recent object positions, while tolerating increasing error for gradually aging stream portions. Intended to cope with rapidly updated locations (i.e., *velocity*), this time-decaying approach can effectively provide online trajectory approximations at multiple resolutions and may also offer affordable estimates when counting distinct moving objects.
- In Section 14.5, we describe a methodology for probabilistic range monitoring over privacy-preserving user locations of limited *veracity*. Assuming a continuous uncertainty model for streaming positional updates, novel pruning heuristics based on spatial and probabilistic properties of the data are employed so as to offer reliable response with quality guarantees.
- In Section 14.6, we outline an online framework for detecting groups of moving objects with approximately similar routes over the recent past. Thanks to a flexible encoding scheme, this technique can cope with the *variety* in motion features, by synthesizing an indicative trajectory that collectively represents motion patterns pertaining to objects in the same group.
- Finally, in Section 14.7, we point out several interesting open issues that may provide rich opportunities for innovative research and applications involving big trajectory and spatiotemporal data.

## 14.2 TRAJECTORY REPRESENTATION AND MANAGEMENT

In the sequel, we assume a three-dimensional model for tracing entities (e.g., people, vehicles, commodities, etc.) of known identities that are moving in two spatial and one temporal dimensions. In general, moving objects have spatial extent (i.e., shape), which might be varying over time. Our principal concern is to capture movement characteristics of objects, rather than other changes (appearance, disappearance, etc.) occurring during their lifetime. Without loss of generality, we restrict our interest to *point* entities (not regions or lines) moving on the Euclidean plane across time, as illustrated in Figure 14.1. As a convention, for a given object *id*, its successive positional samples $p$ are pairs of geographic coordinates $(x, y)$, measured at discrete, totally ordered timestamps $\tau$ from a given time domain. This *time domain* may be considered as an ordered, infinite set of discrete, primitive time instants (e.g., seconds). Successive positions of an individual object are assigned ever-increasing timestamp values that denote the time at which each recording took place. This time-stamping actually establishes a unique ordering reference among all positional elements. Of course, not all objects may be relaying a new position concurrently. Indeed, sampling rates may not be identical (e.g., a vehicle could report every few seconds, while a pedestrian, once every minute) and may be also varying even for a single object (e.g., less point samples when moving at steady speed along a straight course).

When a large number of objects are being monitored, their relayed time-stamped locations effectively constitute a *positional stream* of tuples $< id, p, \tau >$ that keep arriving into

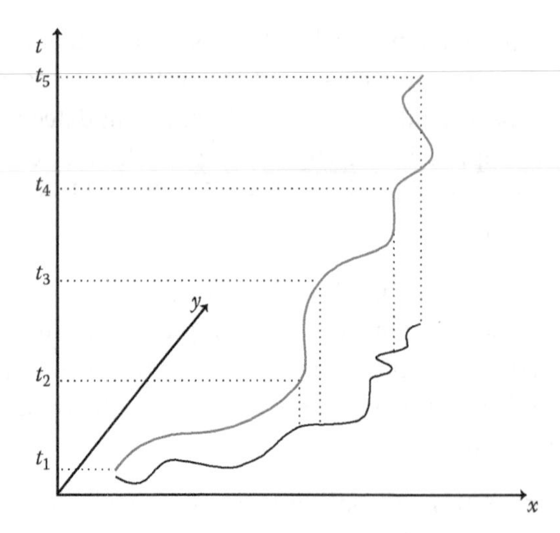


FIGURE 14.1 A three-dimensional trajectory and its two-dimensional trace on the Euclidean plane.

the processing engine from each moving source. However, no deletions or updates are allowed to already-registered locations, so that coherence is preserved among *append-only* positional items. On the basis of such current locations, the processor may evaluate location-aware queries, for example, detect if a vehicle has just entered into a designated region or if the closest friend has changed. The sequential nature in each object's path is also very significant in order to inspect movement patterns over a period of time, as well as interactions between objects (e.g., objects traveling together) or with stationary entities (e.g., crossing region boundaries). In essence, this sequence of point locations traces the movement of this object across time. More specifically, *trajectory T* of a point object identified as *id* and moving over the Euclidean plane is a possibly unbounded sequence of items consisting of its position $p$ recorded at timestamp $\tau$. Thus, each trajectory is approximated as an evolving sequence of point samples collected from the respective data source at distinct time instants (e.g., a GPS reading taken every few seconds).

Considering that numerous objects may be monitored, this information creates a *trajectory stream* of locations concurrently evolving in space and time. It is worth mentioning two characteristics inherent in such streaming trajectories. First, *time monotonicity* dictates a strict ordering of spatial positions taken by a moving object all along its trajectory. Thus, items of the trajectory stream are in increasing temporal order as time advances. Hence, positional tuples referring to the same trajectory can be ordered by simply comparing their time indications. Second, *locality* in an object's movement should be expected, assuming that its next location will be recorded somewhere close to the current one. Therefore, it is plausible to anticipate consistent object paths in space, with the possible exception of discontinuities in movement due to communication failures or noise.

Several models and algorithms have been proposed for managing continuously moving objects in spatiotemporal databases. In Reference 13, an abstract data model and a query language were developed toward a foundation for implementing a spatiotemporal Data Base

Management System (DBMS) extension where trajectories are considered as moving points. Based on that infrastructure, the SECONDO prototype [14] offered several built-in and extensible operations. Due to obvious limitations in storage and communication bandwidth, it is not computationally efficient to maintain a continuous trace of an object's movement. As a trade-off between performance and positional accuracy, a discrete model is usually assumed for movement, by simply collecting point samples along its course. The discrete model proposed in Reference 15 decomposes temporal development into fragments and then uses a simple function to represent movement along every such "time slice." For trajectories, each time slice is a line segment connecting two consecutively recorded locations of a given object, as a trade-off between performance and positional accuracy. In moving-object databases [16], *interpolation* techniques can be used to estimate intermediate positions between recorded samples and thus approximately reconstruct an acceptable trace of movement. In addition, spline methods have also been suggested as a means to provide synchronized, curve-based trajectory representations by resampling original positional measurements at a fixed rate.

In practice, a trajectory is often reconstructed as a *"polyline" vector*, that is, a sequence of line segments for each object that connect each observed location to the one immediately preceding it. Therefore, from the original stream of successive positions taken by the moving points, a stream of append-only line segments may be derived online, maintaining progressively the trace of each object's movement up to its most recent position. Again, no updates are allowed to segments already registered, in order to avoid any modification to historical information and thus preserve coherence among streaming trajectory segments. We adhere to such representation of trajectories as point sequences across time, implying that locations are linearly connected. Also, note that trajectories are viewed from a historical perspective, not dealing at all with predictions about future object positions as in Reference 17. Therefore, robust indexing structures are needed in order to support complex evaluations in both space and time [18].

In LBS applications, several types of *location-aware CQs* can be expressed, so as to examine spatial interactions among moving objects or with specific static geometries (including spatial containment, range or distance search). According to the classification in Reference 18, two types of spatiotemporal queries can be distinguished:

1. *Coordinate-based* queries. In this case, it is the *current position* of objects that matters most in typical range [19] and nearest-neighbor search [3].

2. *Trajectory-based* queries, involving the topology of trajectories (*topological* queries) and derived information, such as speed or heading of objects (*navigational* queries). Such queries may be used to examine any interactions among *trajectories*, like joins [8] to identify those that exhibit some kind of similarity (e.g., vehicles following similar routes after 8 a.m.) or nearest-neighbor queries [20]. Clustering operations can also be employed to detect convoys [6], flocks [5], or swarms [21].

In addition, this model conveniently enables calculation of derived attributes, such as speed or acceleration. This information may be produced from location updates after some processing or by simply manipulating time-stamped locations per object. For example, as soon as a positional update is received, the average or instantaneous speed of the respective moving object can be calculated at once. Results may be propagated to a secondary data stream that maintains the speed for each moving object. Note that this kind of *derived stream* for speed, acceleration, traveled distance, and so forth can be suitably time-stamped with respect to the time indications of incoming tuples, but they do not necessarily retain any spatial features from the original positional stream.

Given that a possibly large but always finite number of location updates arrive for processing at each timestamp $\tau$, the evaluation mechanism must maintain an ordered sequence of all data elements accumulated thus far, practically representing the historical movement of objects. Due to the sheer volume of positional updates and the necessity to answer CQs in real time, it is hardly feasible to deal with lengthy, ever-growing trajectories that represent every detail of the entire history of movement. Instead, it becomes imperative to examine lightweight yet connected motion paths for a limited time period close to the present. Stream processing is mainly achieved by employing CQs, whose results are always refreshed with the most recently arrived positional updates. Such requests usually refer to the most recent portion of objects' paths, rather than remote ones, since users are primarily concerned with the current or most recent situation. So, they specify evolving periods of interest for their computations (e.g., examine locations received during the last hour), instead of dealing with potentially unbounded object traces piling up during long periods.

This inherent and valuable characteristic, along with time monotonicity, has a subtle effect: After a while, older positions may be considered as obsolete, so they may be either discarded or archived. The semantics of *sliding windows* [22] against such positional streams are an ideal choice, as trajectories always evolve steadily along the temporal dimension. Such windows actually restrict the amount of inspected data into temporary yet finite chunks. They are declared in user requests against the stream through properties inherent in the data, mostly thanks to timestamps in the incoming items. Typically, users express interest in a recent time range $\omega$ (e.g., positions relayed during past 10 min), which gets frequently refreshed every $\beta$ units (e.g., each minute), so that the window slides forward to keep in pace with newly arriving items. In effect, such windows can naturally abstract the recent portion of trajectories and thus provide synchronized and compact subsequences without temporal gaps.

## 14.3 ONLINE TRAJECTORY COMPRESSION WITH SPATIOTEMPORAL CRITERIA

One way to deal with the overwhelming *volume* of streaming positions is to maintain a concise yet quite reliable *summary* of each object's movement, avoiding any superfluous details and reducing processing complexity. A suitable algorithm could be based on heuristics, taking advantage of the spatial locality and temporal timeliness that characterize each trajectory, and thus distinguish locations that should be preserved in the compressed paths. As a rule in data stream processing, *single-pass algorithms* are the most adequate means to effectively summarize massive data items into concise *synopses*, by examining each incoming point only once. Essentially, there is always a trade-off between approximation quality and both time and space complexity. In the special case of trajectory streams, an additional requirement is posed: Not only exploit the timely spatiotemporal information but also take into account and preserve the sequential nature of the data. Therefore, in order to efficiently maintain trajectories online, there is no other way but to apply *compression* techniques, thus not only reducing drastically the overall data size but also speeding up query answering, for example, identifying pairs of trajectories that have recently been moving close together. By intelligently dropping some points with negligible influence on the general movement of an object, a simplified yet quite reliable trajectory representation may be obtained. Such a procedure may be used as a filter over the incoming spatiotemporal updates, essentially controlling the stream arrival rate by discarding items before passing them to further processing. Besides, since each object is considered in isolation, such item-at-a-time filtering could be applied directly at the sources, with substantial savings both in communication bandwidth and in computation overhead at the processing engine.

Clearly, there is a need for techniques that can successfully maintain an online summary consisting of the most significant trajectory locations, with minimal process cost per point. Such algorithms should take advantage of the spatiotemporal features that characterize movement, successfully detecting changes in speed and orientation in order to produce a representative synopsis as close as possible to the original route. Our key intuition in the techniques proposed in Reference 23 is that a point should be taken into the sample as long as it reveals a relative change in the course of a trajectory. If the location of an incoming point can be safely predicted (e.g., via interpolation) from the current movement pattern, then this point contributes little information and hence can be discarded without significant loss in accuracy. As a result, retained points may be used to approximately reconstruct the trajectory; discarded locations could be derived via linear interpolation with small error. It is important to note that locations are examined not on the basis of spatial positions only but, rather, on *velocity* considerations (e.g., speed changes, turns, etc.), such that the sample may catch any significant alterations at the known pattern of movement.

The first algorithm is called *threshold-guided sampling* because a new point is appended to the retained trajectory sample once a specified threshold is exceeded for an incoming location. A decision to accept or reject a point is taken according to user-defined rules specifying the allowed tolerance ("threshold") for changes in speed and orientation. In order to decide whether the current location can be safely predicted from the recent past and is thus superfluous, this approach takes under consideration both the mean and instantaneous velocity of the object. As illustrated in Figure 14.2, the mean velocity $V_T$ comes from the last two locations ($A$, $B$) stored in the current trajectory sample, while the instantaneous velocity $V_C$ is derived from the last two observed locations ($B$, $C$). Also note that small variations in the orientation of the predicted course are tolerable; hence, deviation by a few degrees is allowed, as long as it does not change the overall direction of movement by more than $d\phi$ (threshold parameter in degrees). Actually, the corresponding two loci derived by these criteria are the ring sections $SA_T$ and $SA_C$, called "safe areas," where the object is normally expected to be found next, according to the mean and current instantaneous velocities, respectively. As justified in Reference 23, it is not sufficient to use only one of these loci as a criterion to drop locations, because critical points may be missed

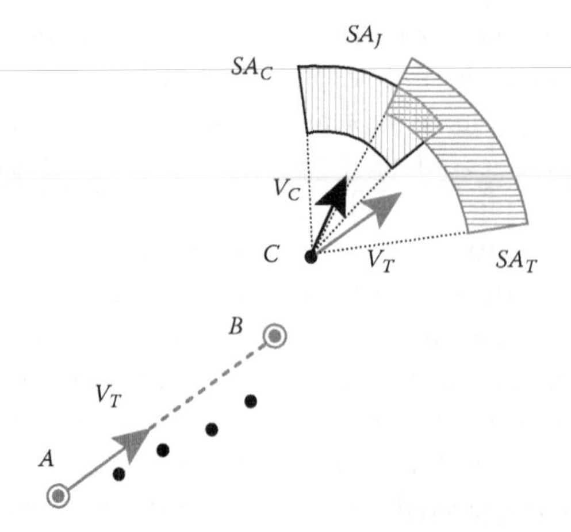


FIGURE 14.2 Finding the joint safe area ($SA_J$) in threshold-guided sampling.

and errors may be propagated and hence lead to distortions in the compressed trajectories. Instead, taking the intersection $SA_J$ of the two loci shown in Figure 14.2 as a *joint safe area* is a more secure policy. As time goes by, this area is more likely to shrink as the number of discarded points increases after the last insertion into the sample, so the probability of missing any critical points is diminishing. As soon as no intersection of these loci is found, an insertion will be prompted into the compressed trajectory, regardless of the current object location. The only problem with this scheme is that the total amount of items in the trajectory sample keeps increasing without eliminating any point already stored, so it is not possible to accommodate under a fixed memory space allocated to each trajectory.

To remedy this downside, the second algorithm introduced in Reference 23 and called *STTrace* is more tailored for streaming environments. The intuition behind STTrace is to use an insertion scheme based on the recent movement features but at the same time allowing deletions from the sample to make room for the newly inserted points without exceeding allocated memory per trajectory. However, a point candidate for deletion is chosen not randomly over the current sample contents but according to its significance in trajectory preservation. In order to quantify the importance of each point in the trajectory, we used a metric based on the notion of *synchronous Euclidean distance* (*SED*) [24]. As shown in Figure 14.3, for any point *B* in the retained sample (i.e., the compressed trajectory), SED is the distance between its actual location and its time-aligned position *B'* estimated via

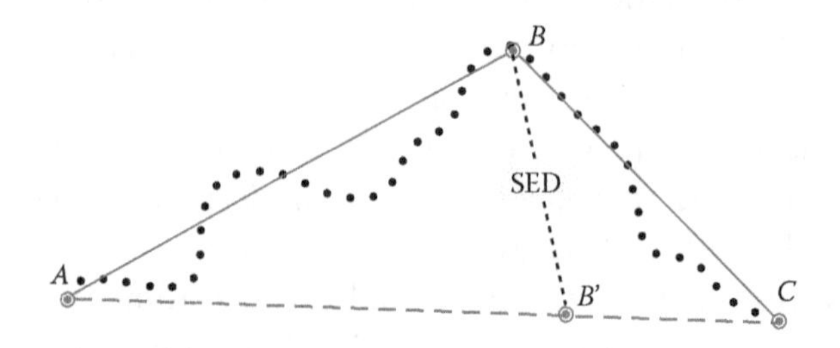


FIGURE 14.3 The notion of synchronous Euclidean distance as used in the STTrace algorithm.

interpolation between its predecessor $A$ and successor point $C$ in the sample. Since this is essentially the distance between the actual point and its *spatiotemporal trace* along the line segment that connects its immediate neighbors in the sequence, this sampling algorithm was named STTrace. Admittedly, it is better to discard a point that will produce the least distortion to the current trajectory synopsis. As soon as the allocated memory gets exhausted and a new point is examined for possible insertion, the compressed representation is searched for the item with the lowest SED. That point represents the least possible loss of information in case it gets discarded. But this comes at a cost: For every new insertion, the most appropriate candidate point must be searched for deletion over the entire sample with $O(N)$ worst-case cost, where $N$ is the actual size of the compressed trajectory. Nevertheless, as $N$ is expectedly very small and the sampled points may be maintained in an appropriate data structure (e.g., a binary balanced tree) with logarithmic cost for operations (search, insert, delete), normally, this is an affordable trade-off.

There have been several other attempts to compress moving-object traces, either through load shedding policies on incoming locations like in References 25 through 27 or by taking also into account trajectory information as in References 28 and 29. Still, based on a series of experiments reported in Reference 23, threshold-guided sampling emerges as a robust mechanism for semantic load shedding *for trajectories*, filtering out negligible locations with minor computational overhead. The actual size of the sample it provides is a rough indication of the complexity of each trajectory, and the parameters can be fine-tuned according to trajectory characteristics and memory availability. Besides, it can be applied close to the data sources instead of a central processor, sparing both transmission cost and processing power. Regarding efficiency, STTrace always manages to maintain a small representative sample for a trajectory of unknown size. It outperforms threshold-guided sampling for small compression rates since it is not easy to define suitable threshold values in this case. Empirical results show that STTrace incurs some overhead in maintaining the minimum synchronous distance and in-memory adjustment of the sampled points. However, this cost can be reduced if STTrace is applied in a batch mode, that is, executed at consecutive time intervals.

## 14.4 AMNESIC MULTIRESOLUTION TRAJECTORY SYNOPSES

It is apparent that the significance of each isolated location is inherently *time decaying*, since any recorded position of an object will be soon outdated by forthcoming ones. This motivates a policy for *dropping detail with age*: The older a location gets, the coarser its representation could become in a progressive fashion, implying that greater precision should be reserved for the most recent positions, as exemplified in Figure 14.4. Effectively, this *amnesic* treatment of trajectories can cope with the frequent positional updates gathered per object, known as the *velocity* challenge in big trajectory data.

With respect to approximation of one-dimensional time series, a wide range of amnesic functions was identified in Reference 30, useful in controlling the amount of error tolerated for every single point in the time series. In our work in Reference 31, we presented a *multiple-granularity* framework based on an *amnesic tree* structure (*AmTree*), which accepts streaming items and maintains summaries over hierarchically organized levels

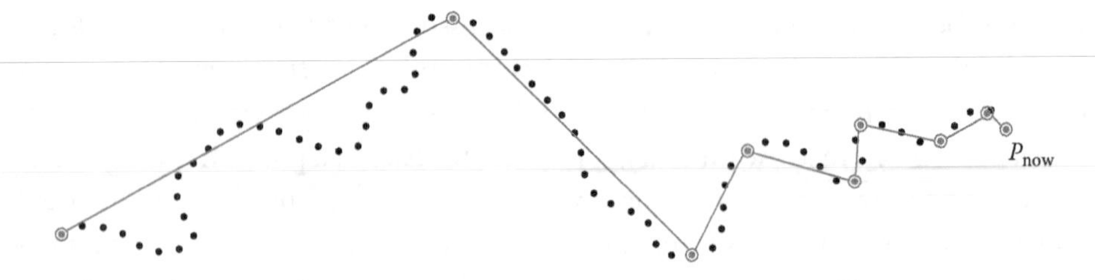


FIGURE 14.4 Amnesic approximation of a trajectory, as projected on the Euclidean plane.

of precision, essentially realizing an amnesic behavior over stream portions. In addition, different levels of abstraction are inherent in semantics related to multiscale representation of spatiotemporal features, allowing progressive refinements of their evolution. This time-aware scheme distantly resembles SWAT [32], but this latter is intrinsically bound to wavelet transform of scalar stream values and cannot handle multidimensional points with user-specified approximation functions.

Actually, AmTree can be easily calibrated to work with a varying number of *time granules* (i.e., nodes) per level. In Figure 14.5, we present several consecutive snapshots of this amnesic processing scheme that manipulates pairs of items at every level. For simplicity, we utilize one-dimensional numeric values and not two-dimensional positional ones, and we assume that a time granule at each level spans two granules half its temporal span at the level beneath. Except for the root, each level $i$ of the tree consists of a right ($R_i$) and a left node ($L_i$). At the lowest level, node $R_0$ accepts data with reference to the finest time granularity (e.g., seconds), which characterizes every timestamp attached to incoming tuples. Each node at level $i$ contains information across twice as many timestamps as a node at level $i - 1$. Hence, a node at level $i$ contains information characterizing $2^i$ timestamps, and it is being updated accordingly.

Once this structure starts consuming streaming items, node updates are performed in a bottom-up fashion. A user-defined mapping $f$ is applied over the batch of fresh positions with current timestamp value $\tau$ and transforms them into a single tuple that can become the content of a tree node. As illustrated in Figure 14.5, where $f$ is a simple summation of numeric items, the resulting content is assigned to node $R_0$, while the previous content of

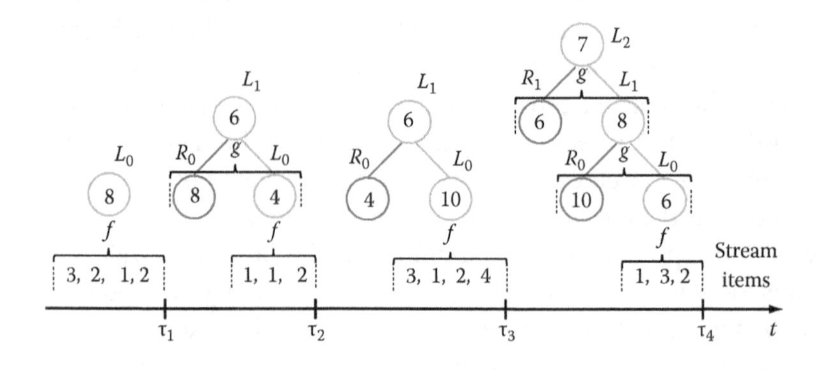


FIGURE 14.5 Four snapshots of an example operation of AmTree against numeric streaming items.

$R_0$ is shifted to node $L_0$. As time goes by and new data come in, the contents of each level are merged using a function $g$ (average in the example of Figure 14.5) and propagated higher up in the tree, thus retaining less detail. Note that node $R_0$ is the only entry point to the synopsis maintained by the AmTree. As justified in Reference 31, AmTree updates can be carried out online in $O(1)$ amortized time per location with only logarithmic requirements in memory storage.

This framework is best suited for summarizing streams of sequential features, and it has been applied to create amnesic synopses concerning singleton trajectories. As an alternative representation to a time series of points, the trajectory of a moving object can be represented with a polyline composed of consecutive displacements. Every such line segment connects a pair of successive point locations recorded for this object, eventually providing a continuous, though approximate, trace of its movement. With respect to compressing a single trajectory, an AmTree instantiation manipulates all successive displacement tuples relayed by this object. In direct correspondence to the generic AmTree functionality, mapping $f$ converts every current object position into a displacement tuple with respect to its previous location. This displacement is then inserted into the $R_0$ node, possibly triggering further updates higher up in the AmTree. When the contents of level $i$ must be merged to produce a coarser representation, a simple concatenation function $g$ is used to combine the successive displacements stored in nodes $L_i$ and $R_i$. After eliminating the common articulation point of the two original segments, a single line segment is produced and then stored in node $R_{i+1}$. As a result, end points of all displacements stored in AmTree nodes correspond to original positional updates, while consecutive displacements remain connected to each other at every level. Evidently, an amnesic behavior is achieved for trajectory segments through levels of gradually less detail in this bottom-up tree maintenance. As long as successive displacements are preserved, the movement of a particular object can be properly reconstructed by choosing points in descending temporal order, starting from its most recent position and going steadily backward in time. Any trajectory reconstruction process can be gradually refined by combining information from multiple levels and nodes of the tree, leading to a *multiresolution approximation* for a given trajectory, as depicted in Figure 14.4.

This structure can also provide unbiased estimates for the number of objects that are moving in an area of interest during a specified time interval. When each object must be counted only once, the problem is known as *distinct counting* [33]. We consider a regular decomposition of the two-dimensional Euclidean plane into equal-area grid cells, which serve as a simplified spatial reference for the moving objects instead of their exact coordinates. Thus, each cell corresponds to a separate AmTree, which maintains gradually aging information concerning the number of moving objects inside that cell. Query-oriented compression is achieved using Flajolet-Martin (FM) sketches [34], which are based on Probabilistic Counting with Stochastic Averaging (FM_PCSA). Each node of an AmTree retains $m$ bitmap vectors utilized by the FM_PCSA sketch, as illustrated in Figure 14.6. Hence, we avoid enumeration of objects, as we are satisfied with an acceptable estimate of their distinct count (*DC*) given by the sketching algorithm. In order to estimate the number of distinct objects moving within a given area $a$ during a time interval $\Delta\tau$, we first identify

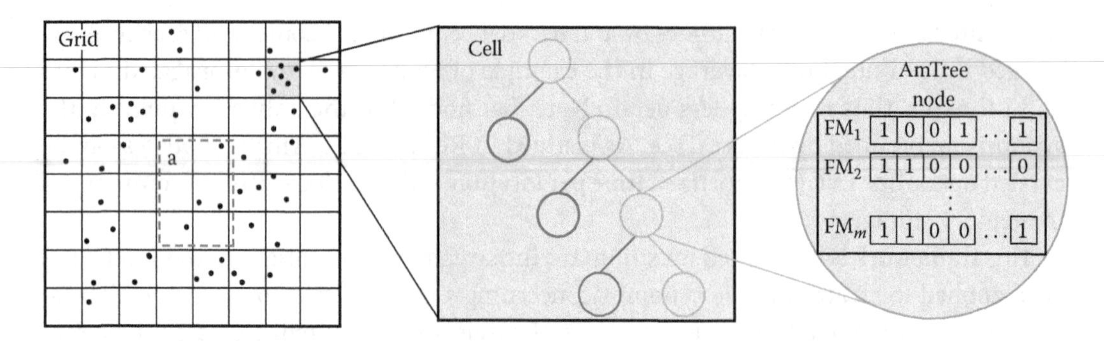


FIGURE 14.6 The three-tier FM-AmTree structure used for approximate distinct counting of moving objects.

the grid cells that completely cover region $a$. Those cells determine the group of qualifying AmTree structures that maintain the aggregates. For each such tree, we need to locate the set of nodes that overlap time period $\Delta\tau$ specified by the query; these nodes are identical for each qualifying tree. By taking the union of the sketches attached to these nodes (i.e., an OR operation over the respective bitmaps), we can finally provide an approximate answer to the DC query.

The experimental study conducted in Reference 31 confirms that recent trajectory segments always remain more accurate, while overall error largely depends on the temporal extent of the query in the past. Even for heavily compressed trajectories, accuracy is proven quite satisfactory for answering spatiotemporal range queries. With respect to DC queries, it was observed that a finer grid partitioning incurs more processing time at the expense of increased accuracy.

Overall, the suggested AmTree framework is a modular amnesic structure with exponential decay characteristics. It is especially tailored to cope with streaming positional updates, and it can retain a compressed outline of entire trajectories, always preserving *contiguity* among successive segments for each individual object. Effectively, such a policy constructs a multiresolution path, offering finer motion paths for the recent past and keeping gradually less and less details for aging parts of the trajectory.

In conjunction with FM sketches, AmTree can further be used in spatiotemporal aggregation for providing good-quality estimates to DC queries over locations of moving objects.

## 14.5 CONTINUOUS RANGE SEARCH OVER UNCERTAIN LOCATIONS

Recently, the increasingly popular social networking applications have been enhanced with location-aware features toward *geosocial networking services* [35], thus allowing mobile users to interact relative to their current positions. Platforms like Facebook Places [36], Google Latitude [37], and Foursquare [38] enable users to pinpoint friends on a map and share their whereabouts and preferences with the followers they choose. Despite their attraction, such features may put people's privacy at risk, revealing sensitive information about everyday habits, political affiliations, cultural interests, and so forth. Mobile users are aware of their own exact location, but they may not wish to disclose it to third parties; instead, they consent to relay just a *cloaked* indication of their whereabouts [39]; hence, the *veracity* of this information is purposely limited.

In Reference 40, we suggested a framework for real-time processing of *continuous range queries* against such imprecise locations, so that a user may receive instant notifications when a friend appears *with sufficient probability* within his/her area of interest (e.g., "it is more than 75% probable that one of my friends is in my neighborhood"). Each position is obfuscated as an *uncertainty region* $r_o$ with *bivariate Gaussian* characteristics, enclosing (but apparently not centered at) the user's current location. Depending on the variance, the density of a Gaussian random variable is rapidly diminishing with increasing distances from the mean. Thanks to its inherent simplicity, the uncertainty region can be truncated in a natural way on the server side, so the user itself does not need to specify a bounded area explicitly. It suffices that an object sends the origin $(\mu_x, \mu_y)$ of its own probability density function (*pdf*) and the standard deviation $\sigma$ (common along both spatial dimensions). As illustrated in Figure 14.7a, there is 99.73% probability that the actual location is found somewhere within a radius $3\sigma$ from the origin. Uncertainty parameters are expressed in distance units (e.g., meters) and can be changed dynamically as the object is moving or adjusting its degree of privacy. Larger $\sigma$ values indicate that an object's location can be hidden in a greater area around its indicated mean $(\mu_x, \mu_y)$. Based on such massive, transient, imprecise data, the server attempts to give a response to multiple search requests. Each CQ $q$ may dynamically modify its spatial range (a moving *rectangular extent* $r_q$) and its *cutoff probability threshold* $\theta$. Evaluation takes place periodically at execution cycles, upon reception of a batch of updates from objects and changes in queries. Query $q$ identifies any object $o$ registered as a member in its *contact list* $L_q$ (e.g., friends or followers of $q$) and is currently within range $r_q$ with an appearance probability of at least $\theta$.

The problem is that Gaussian distributions cannot be integrated analytically, whereas numerical methods like Monte Carlo are prohibitively costly for getting a fair estimation for appearance probabilities in online fashion. Given the mobility and mutability of objects and queries alike, in Reference 40, we approximated each uncertainty region with its rectilinear circumscribed square of side $6\sigma$ around the origin of the pdf. As illustrated in Figure 14.7a, the cumulative probability of this minimum bounding box (MBB) is greater than 99.73% and tends asymptotically to 1, although its area is larger than the circle of

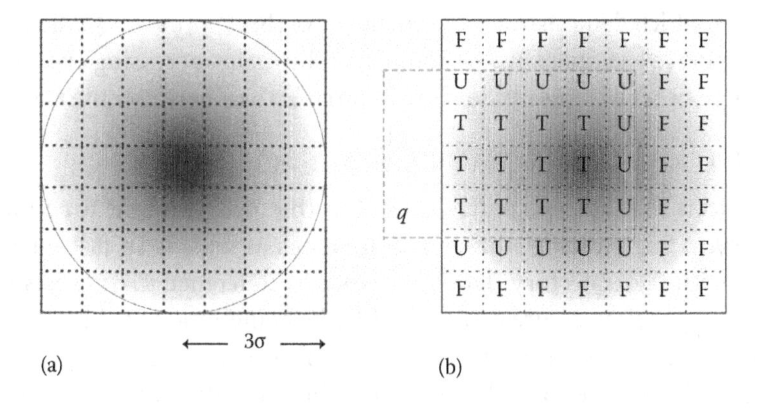


FIGURE 14.7 (a) Standard bivariate Gaussian distribution $N(\mathbf{0}, \mathbf{1})$ as a model for an uncertain location. (b) Verifier with $7 \times 7$ elementary boxes for checking an object against range query $q$.

radius $3\sigma$. If this MBB were uniformly subdivided into $\lambda \times \lambda$ *elementary boxes*, then each box would represent diverse cumulative probabilities, indicated by the differing shades in Figure 14.7a. However, for a known $\sigma$, the cumulative probability in each elementary box is independent of the parameters of the applied bivariate Gaussian distribution. Once precomputed (e.g., by Monte Carlo), these probabilities can be retained in a lookup table.

The rationale behind this subdivision is that it may be used as a *discretized verifier* $V$ when probing uncertain Gaussians. Consider the case of query $q$ against an object, shown as a shaded rectangle in Figure 14.7b. Depending on its topological relation with the given query, each elementary box of $V$ can be easily characterized by one of three possible states: (1) $T$ is assigned to elementary boxes totally within query range; (2) $F$ signifies disjoint boxes, that is, those entirely outside the range; and (3) $U$ marks boxes partially overlapping with the specified query range. Then, summing up the respective cumulative probabilities for each subset of boxes returns three *indicators* $p_T$, $p_F$, and $p_U$ suitable for object validation against the cutoff threshold $\theta$. The confidence margin in the results equals the overall cumulative probability of the $U$-boxes, which depends entirely on granularity $\lambda$. Indeed, a small $\lambda$ can provide answers quickly, which is critical when coping with numerous objects. In contrast, the finer the subdivision into elementary boxes (i.e., a larger $\lambda$), the less the uncertainty in the emitted results.

As a trade-off between timeliness and answer quality in a geostreaming context, in Reference 40, we turned this range search into an ($\varepsilon$, $\delta$) *approximation problem*, where $\varepsilon$ quantifies *the error margin* in the allowed overestimation when reporting a qualifying object and $\delta$ specifies the tolerance that an invalid answer may be given (i.e., a false positive). We introduced several optimizations based on inherent probabilistic and spatial properties of the uncertain streaming data; details can be found in Reference 40.

In a nutshell, this technique can quickly determine whether an item possibly qualifies for the query or safely prune examination of nonqualifying cases altogether. Inevitably, such a probabilistic treatment returns approximate answers, along with confidence margins as a measure of their quality. Simulations over large synthetic data sets indicated that, compared with an exhaustive Monte Carlo evaluation, about 15% of candidates were eagerly rejected, while another 25% were pruned. Most importantly, false negatives were less than 0.1% in all cases, which demonstrates the efficiency of this approach. Qualitative results are similar for varying $\lambda$, but finer subdivisions naturally incur increasing execution costs, yet always at least an order of magnitude more affordable than naïve Monte Carlo evaluation.

## 14.6 MULTIPLEXING OF EVOLVING TRAJECTORIES

Identifying objects that approximately travel together over a recent time interval would also be important for reducing storage requirements as well as in fast query answering when dealing with big trajectory data. Our work in Reference 41 suggests a framework that intends to cope with the great *variety* of paths being monitored in applications that handle massive motion data, as in traffic monitoring (e.g., carpooling services); fleet management (logistics); airspace control using radars; biodiversity protection (e.g., for tracking wildlife); and so forth. The key idea is that a symbolic encoding for sequences of trajectory segments may offer a rough yet succinct abstraction of their concurrent evolution.

Intuitively, a processing scheme could take advantage of inherent spatiotemporal properties, such as heading, speed, and current position, in order to continuously report groups of objects with similar motion traces. Then, an indicative *"delegate" path* could be regularly constructed per detected group, since it could actually epitomize spatiotemporal features shared by its participating objects.

The main objective of this scheme is to detect groups of objects according to the following constraints:

- *Similarity*: pairwise similar trajectory segments must be within distance $\varepsilon$.
- *Simultaneity*: positions are checked over a recent *sliding window* [22] with temporal range $\omega$.
- *Timeliness*: groups are adjusted periodically every $\beta$ time units (i.e., once the window slides), in order to reflect any changes in objects' movement.

Hence, pairs of concurrently recorded locations from each object should not deviate more than $\varepsilon$ during interval $\omega$. This notion of similarity is confined within the recent past (window $\omega$) and does not extend over the entire history of movement. Moreover, it can be easily generalized for multiple objects with pairwise similar trajectory segments (Figure 14.8). In addition, a *threshold* is applied when incrementally creating "delegate" traces from such trajectory groupings; a synthetic trace $T$ is returned only if the respective group currently includes more than $n$ objects. Note that no original location in such a group can deviate more than the given tolerance $\varepsilon$ from its delegate path $T$.

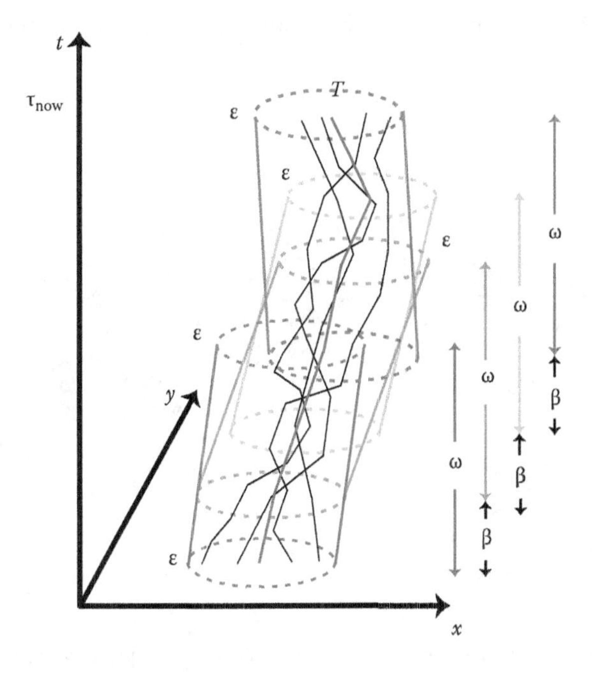


FIGURE 14.8 Similarly evolving trajectories and their delegate path $T$.

Checking the similarity of trajectory segments according to their time-stamped positions soon becomes a bottleneck for escalating numbers of moving objects or wider window ranges. To avoid this, in Reference 41, we opt for an approximate representation of traces based on consecutive velocity vectors that end up at the current location of each respective object (Figure 14.9). Every vector is characterized by a symbol that signifies the orientation of movement using the familiar notion of a *compass*, which roughly exemplifies an object's course between successive position messages. Effectively, *compass resolution* $\alpha$ determines the degree of motion smoothing; when $\alpha = 4$, orientation symbols $\{N, S, E, W\}$ offer just a coarse indication, but finer representations are possible with $\alpha = 16$ symbols (Figure 14.9) or more. Typically, once the window slides at the next execution cycle, an additional symbol (approximating motion during the latest $\beta$ timestamps) will be appended at the tail of this first-in, first-out (FIFO) sequence, while the oldest one (i.e., at the head) gets discarded. Thus, instead of the bulk of original positions, only a series of few symbols and speed measures need be maintained per trajectory, thus offering substantial memory savings.

Symbolic sequences are more amenable to similarity checks since they act as *motion signatures*. In Reference 41, we identified objects with common signatures through a hash table. Objects with identical symbolic sequences might have almost "parallel" courses but can actually be very distant from each other. The crux of our approach is that objects with identical signatures and currently within distance $\varepsilon$ from each other most probably have followed similar paths recently. Therefore, identifying groups of at least $n$ objects with a common signature can be performed against their current locations through a point clustering technique. We provisionally make use of the Density-Based Spatial Clustering of Applications with Noise (DBSCAN) algorithm [42] to detect groups of similar trajectories with proximal current positions. Afterward, a delegate path $T$ per discovered cluster with sufficient membership ($>n$ objects) can be easily created by simply starting from the most

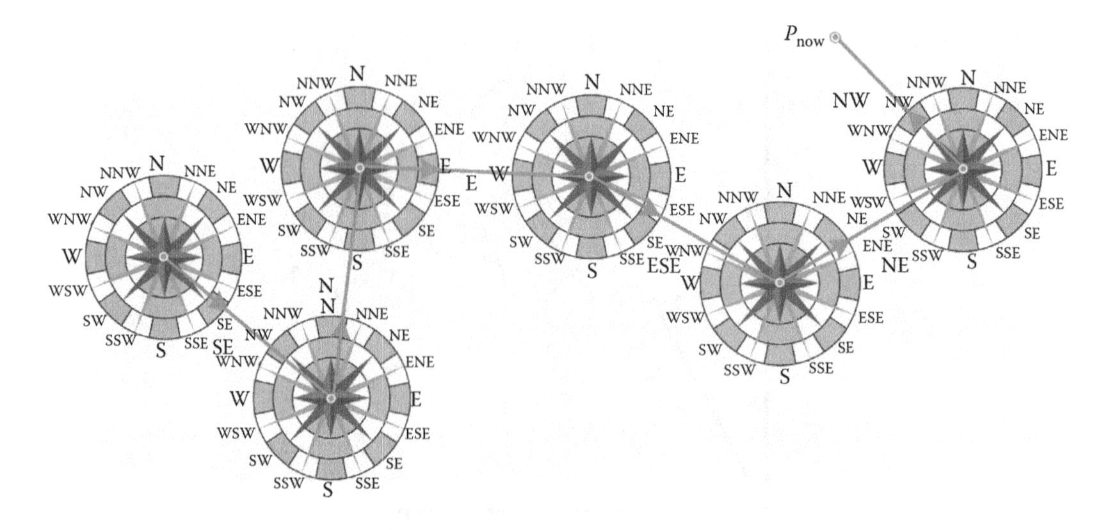


FIGURE 14.9 Orientation-based encoding of an evolving trajectory using a compass with 16 symbols.

recent location and rewinding the symbolic sequence backward, that is, reversely visiting all samples retained within the sliding window frame.

Simulations against synthetic trajectories have shown that this framework has great potential for data reduction and timely detection of motion trends, without hurting performance or approximation quality. Some preliminary results regarding traffic trends on the road network of Athens are available in Reference 41.

Overall, we believe that this ongoing work fuses ideas from trajectory clustering [7] and path simplification [23] but proceeds further beyond. Operating in a geostreaming context, not only can it identify important motion patterns in online fashion, but it may also provide concise summaries without resorting to sophisticated spatiotemporal indexing. Symbolic representation of routes was first proposed in Reference 8 for filtering against trajectory databases. Yet, our encoding differs substantially, as it attempts to capture evolving spatiotemporal vectors using a versatile alphabet of tunable object headings instead of simply compiling time-stamped positions in a discretized space. In practice, processed features may serve various needs, such as the following: (i) *data compression*, by collectively representing traces of multiple objects with a single "delegate" that suitably approximates their common recent movement; (ii) *data discovery*, by detecting trends or motion patterns from real-time location feeds; (iii) *data visualization*, by estimating the significance of each multiplexed group of trajectories and illustrating its mutability across time; and (iv) *query processing*, if multiplexed traces are utilized at the filtering stage when evaluating diverse queries (range, *k*-NN, aggregates, etc.) instead of the detailed, bulky trajectories.

## 14.7 TOWARD NEXT-GENERATION MANAGEMENT OF BIG TRAJECTORY DATA

Processing trajectory data sets with a focus on their inherent spatiotemporal features has been widely studied in the last decade. However, big spatiotemporal data brings in new research challenges and the need for effective and efficient ways to process such information and extract useful knowledge.

Undoubtedly, the processing infrastructure should be able to tackle scalability and load balancing. Stream processing in the cloud may offer flexible, highly distributed resource allocation, as data emanates from multimodal devices and flows through heterogeneous networks. For instance, Hadoop-GIS [43] is a scalable and high-performance data warehousing system specifically for spatial data and queries. Running on Hadoop, it supports several types of queries (including spatial join, containment, aggregation, etc.) through partitioning and multilevel spatial indexing.

Novel *index* structures are also needed to support a variety of queries regarding spatial proximity, navigation, recommendation, and so forth, as well as interaction with trajectories. TrajStore [44] is a recently proposed scheme for indexing, clustering, and storing trajectory data, which is optimized for retrieving data about many trajectories passing through a particular location. The suggested index dynamically identifies both spatially and temporally adjacent segments in order to compress them on disk.

Exploring *dynamic motion patterns* across time, like flocks [5], convoys [6], traveling companions [45], or swarms [21], is a very attractive topic and may have many applications, for example, in fleet management, carpooling services, biodiversity protection, and so forth. These concepts aim to identify groups of moving objects traveling together for a certain time period. The recent notion of gathering [46] is able to model a variety of additional group events or incidents. Examples include celebrations, parades, large-scale business promotions, protests, traffic jams, and other public congregations.

Regarding more specific *analytics*, the methodology in Reference 47 identifies time period–based most-frequent paths over large trajectory data sets. The proposed two-step strategy first constructs a graph with the frequencies of the candidate paths and then executes a graph search to find the results. The SimpleFleet platform [48] aggregates online data from vehicle trajectories and provides live traffic analytics for large metropolitan areas, including services for route searching; statistics; as well as map visualization of paths, traffic conditions, and isochrones. This may be useful in car navigation, in order to recommend routes according to actual traffic patterns. vTrack [49] is a prototype system that uses mobile phones in order to accurately estimate road travel times from a sequence of inaccurate position samples. For example, if GPS readings are unavailable or erroneous (e.g., in "urban canyons"), it is possible to use alternative, but noisier, sensors (like Wi-Fi) so as to estimate both a user's trajectory and travel times. The techniques proposed in Reference 50 are based on a multicost, time-dependent, uncertain graph model of a road network according to GPS positions from vehicles. Thus, optimal routes may consider multiple costs and time-dependent uncertainty for a given origin–destination pair and a start time.

Rapid proliferation of popular, location-based social networking (e.g., Foursquare, Facebook) has also given rise to large amounts of trajectories associated with activity information. Such *activity trajectories* [51] essentially record not only the spatiotemporal aspects of movement but textual descriptions of users' activities as well. This gives rise to interesting problems such as similarity retrieval: Given a sequence of query locations, each associated with a set of desired activities, an activity trajectory similarity query returns $k$ trajectories that cover the query activities and yield the shortest minimum match distance [51]. It is expected that more interesting problems around *spatiotextual* descriptions of activities in time will emerge, including personalized aspects of similarity retrieval [52].

Similarly, interpreting *human lifestyles* [53] not only is essential to many scientific disciplines but also has a profound business impact on geomarketing applications. Devising algorithms that integrate data from multiple social network accounts of millions of individuals along with their publicly available heterogeneous behavioral data as stored in location-based social networks could lead to interesting classification schemes. The results can be precious for personalized recommendation and targeted advertising.

Another increasingly common trend is the advent of geographic information created by voluntary communities, known as *volunteered geographic information* (VGI) or crowd-sourced geospatial data [54]. The OpenStreetMap project [55] is one of the most successful VGI efforts, which aims to develop a collaborative, freely available, digital map of the planet with contributions from volunteers. Another form of crowdsourcing is location sharing in social media, such as annotated Twitter posts [56] or geo-tagged photos in Flickr [57]. Beyond issues in privacy [58], extra processing is required on these social media data to retrieve any spatial information [59]. Due to the nonauthoritative nature, VGI efforts have often been questioned regarding the quality and reliability of the information collected. Preprocessing and noise removal of VGI seems a promising research direction toward addressing uncertainty in these spatiotemporal data.

Last, but not least, next-generation platforms should offer advanced functionality to users in order to better deal with the peculiarities in trajectory data and better interpret hidden properties. Interactive mapping tools should sustain the bulk of data, by leveraging representation detail according to the actual map scale, the amount of distinct objects, the complexity of their traces, and so forth. In application development, APIs for fine-grain control over complex events that could correlate dynamic positional data against stationary data sets (e.g., transportation networks or administrative boundaries) or other trajectories would be valuable as well. Finally, dealing with inherent stream imperfections like disorder or noise would offer the ability for dynamic revision of query results, thus increasing the quality of answers.

## REFERENCES


1. Agrawal, D., P. Bernstein, E. Bertino, S. Davidson, U. Dayal, M. Franklin, J. Gehrke, L. Haas, A. Halevy, J. Han, H.V. Jagadish, A. Labrinidis, S. Madden, Y. Papakonstantinou, J.M. Patel, R. Ramakrishnan, K. Ross, C. Shahabi, D. Suciu, S. Vaithyanathan, and J. Widom. 2012. Challenges and Opportunities with Big Data—A Community White Paper Developed by Leading Researchers across the United States. Available at http://www.cra.org/ccc/files/docs/init/bigdatawhitepaper.pdf (accessed April 30, 2014).
2. Stonebraker, M., U. Cetintemel, and S. Zdonik. 2005. The 8 Requirements of Real-Time Stream Processing. *ACM SIGMOD Record*, 34(4):42–47.
3. Mouratidis, K., M. Hadjieleftheriou, and D. Papadias. 2005. Conceptual Partitioning: An Efficient Method for Continuous Nearest Neighbor Monitoring. In *Proceedings of the 24th ACM SIGMOD International Conference on Management of Data*, pp. 634–645.
4. Hu, H., J. Xu, and D.L. Lee. 2005. A Generic Framework for Monitoring Continuous Spatial Queries over Moving Objects. In *Proceedings of the 24th ACM SIGMOD International Conference on Management of Data*, pp. 479–490.
5. Vieira, M., P. Bakalov, and V.J. Tsotras. 2009. On-line Discovery of Flock Patterns in Spatio-Temporal Data. In *Proceedings of the 17th ACM SIGSPATIAL International Conference on Advances in Geographic Information Systems (ACM GIS)*, pp. 286–295.
6. Jeungy, H., M.L. Yiu, X. Zhou, C.S. Jensen, and H.T. Shen. 2008. Discovery of Convoys in Trajectory Databases. *Proceedings of Very Large Data Bases (PVLDB)*, 1(1):1068–1080.
7. Lee, J., J. Han, and K. Whang. 2007. Trajectory Clustering: A Partition-and-Group Framework. In *Proceedings of the 26th ACM SIGMOD International Conference on Management of Data*, pp. 593–604.
8. Bakalov, P., M. Hadjieleftheriou, E. Keogh, and V.J. Tsotras. 2005. Efficient Trajectory Joins Using Symbolic Representations. In *MDM*, pp. 86–93.
9. Chen, Z., H.T. Shen, and X. Zhou. 2011. Discovering Popular Routes from Trajectories. In *Proceedings of the 27th International Conference on Data Engineering (ICDE)*, pp. 900–911.
10. Sacharidis, D., K. Patroumpas, M. Terrovitis, V. Kantere, M. Potamias, K. Mouratidis, and T. Sellis. 2008. Online Discovery of Hot Motion Paths. In *Proceedings of the 11th International Conference on Extending Database Technology (EDBT)*, pp. 392–403.
11. Manyika, J., M. Chui, B. Brown, J. Bughin, R. Dobbs, C. Roxburgh, and A.H. Byers. 2011. *Big Data: The Next Frontier for Innovation, Competition, and Productivity*. McKinsey Global Institute, New York. Available at http://www.mckinsey.com/insights/business_technology/big_data_the_next_frontier_for_innovation (accessed April 30, 2014).

12. Ferguson, M. 2012. Architecting a Big Data Platform for Analytics. Whitepaper Prepared for IBM.
13. Güting, R.H., M.H. Böhlen, M. Erwig, C.S. Jensen, N.A. Lorentzos, M. Schneider, and M. Vazirgiannis. 2000. A Foundation for Representing and Querying Moving Objects. *ACM Transactions on Database Systems*, 25(1):1–42.
14. Güting, R.H., T. Behr, and C. Duentgen. 2010. SECONDO: A Platform for Moving Objects Database Research and for Publishing and Integrating Research Implementations. *IEEE Data Engineering Bulletin*, 33(2):56–63.
15. Forlizzi, L., R.H. Güting, E. Nardelli, and M. Schneider. 2000. A Data Model and Data Structures for Moving Objects Databases. In *Proceedings of the 19th ACM SIGMOD International Conference on Management of Data*, pp. 319–330.
16. Pfoser, D., and C.S. Jensen. 1999. Capturing the Uncertainty of Moving-Object Representations. In *Proceedings of the 6th International Symposium on Advances in Spatial Databases (SSD)*, pp. 111–132.
17. Sistla, A.P., O. Wolfson, S. Chamberlain, and S. Dao. 1997. Modeling and Querying Moving Objects. In *Proceedings of the 13th International Conference on Data Engineering (ICDE)*, pp. 422–432.
18. Pfoser, D., C.S. Jensen, and Y. Theodoridis. 2000. Novel Approaches to the Indexing of Moving Object Trajectories. In *Proceedings of 26th International Conference on Very Large Data Bases (VLDB)*, pp. 395–406.
19. Mokbel, M., X. Xiong, and W. Aref. 2004. SINA: Scalable Incremental Processing of Continuous Queries in Spatiotemporal Databases. In *Proceedings of the 23rd ACM SIGMOD International Conference on Management of Data*, pp. 623–634.
20. Frentzos, E., K. Gratsias, N. Pelekis, and Y. Theodoridis. 2007. Algorithms for Nearest Neighbor Search on Moving Object Trajectories. *GeoInformatica*, 11(2):159–193.
21. Zhenhui, L., D. Bolin, H. Jiawei, and R. Kays. 2010. Swarm: Mining Relaxed Temporal Moving Object Clusters. *Proceedings on Very Large Data Bases (PVLDB)*, 3(1–2):723–734.
22. Patroumpas, K., and T. Sellis. 2011. Maintaining Consistent Results of Continuous Queries under Diverse Window Specifications. *Information Systems*, 36(1):42–61.
23. Potamias, M., K. Patroumpas, and T. Sellis. 2006. Sampling Trajectory Streams with Spatiotemporal Criteria. In *Proceedings of the 18th International Conference on Scientific and Statistical Data (SSDBM)*, pp. 275–284.
24. Meratnia, N., and R. de By. 2004. Spatiotemporal Compression Techniques for Moving Point Objects. In *Proceedings of the 9th International Conference on Extending Database Technology (EDBT)*, pp. 765–782.
25. Nehme, R., and E. Rundensteiner. 2007. ClusterSheddy: Load Shedding Using Moving Clusters over Spatio-Temporal Data Streams. In *Proceedings of the 12th International Conference on Database Systems for Advanced Applications (DASFAA)*, pp. 637–651.
26. Gedik, B., L. Liu, K. Wu, and P.S. Yu. 2007. Lira: Lightweight, Region-Aware Load Shedding in Mobile CQ Systems. In *Proceedings of the 23rd International Conference on Data Engineering (ICDE)*, pp. 286–295.
27. Mokbel, M.F., and W.G. Aref. 2008. SOLE: Scalable On-line Execution of Continuous Queries on Spatio-Temporal Data Streams. *VLDB Journal*, 17(5):971–995.
28. Cao, H., O. Wolfson, and G. Trajcevski. 2006. Spatio-Temporal Data Reduction with Deterministic Error Bounds. *VLDB Journal*, 15(3):211–228.
29. Lange, R., F. Dürr, and K. Rothermel. 2011. Efficient Real-Time Trajectory Tracking. *VLDB Journal*, 25:671–694.
30. Palpanas, T., M. Vlachos, E. Keogh, D. Gunopulos, and W. Truppel. 2004. Online Amnesic Approximation of Streaming Time Series. In *Proceedings of the 20th International Conference on Data Engineering (ICDE)*, pp. 338–349.

31. Potamias, M., K. Patroumpas, and T. Sellis. 2007. Online Amnesic Summarization of Streaming Locations. In *Proceedings of the 10th International Symposium on Spatial and Temporal Databases (SSTD)*, pp. 148–165.
32. Bulut, A., and A.K. Singh. 2003. SWAT: Hierarchical Stream Summarization in Large Networks. In *Proceedings of the 19th International Conference on Data Engineering (ICDE)*, pp. 303–314.
33. Tao, Y., G. Kollios, J. Considine, F. Li, and D. Papadias. 2004. Spatio-Temporal Aggregation Using Sketches. In *Proceedings of the 20th International Conference on Data Engineering (ICDE)*, pp. 214–226.
34. Flajolet, P., and G.N. Martin. 1985. Probabilistic Counting Algorithms for Database Applications. *Journal of Computer and Systems Sciences*, 31(2):182–209.
35. Mascetti, S., D. Freni, C. Bettini, X.S. Wang, and S. Jajodia. 2011. Privacy in Geo-Social Networks: Proximity Notification with Untrusted Service Providers and Curious Buddies. *VLDB Journal*, 20(4):541–566.
36. Facebook. Available at http://facebook.com/about/location (accessed April 30, 2014).
37. Google Latitude. Available at http://google.com/latitude (accessed August 1, 2013).
38. Foursquare. Available at https://foursquare.com/ (accessed April 30, 2014).
39. Chow, C.-Y., M.F. Mokbel, and W.G. Aref. 2009. Casper*: Query Processing for Location Services without Compromising Privacy. *ACM Transactions on Database Systems*, 34(4):24.
40. Patroumpas, K., M. Papamichalis, and T. Sellis. 2012. Probabilistic Range Monitoring of Streaming Uncertain Positions in GeoSocial Networks. In *Proceedings of the 24th International Conference on Scientific and Statistical Data (SSDBM)*, pp. 20–37.
41. Patroumpas, K., K. Toumbas, and T. Sellis. 2012. Multiplexing Trajectories of Moving Objects. In *Proceedings of the 24th International Conference on Scientific and Statistical Data (SSDBM)*, pp. 595–600.
42. Ester, M., H.-P. Kriegel, J. Sander, and X. Xu. 1996. A Density-Based Algorithm for Discovering Clusters in Large Spatial Databases with Noise. In *Proceedings of the 2nd International Conference on Knowledge Discovery and Data Mining (KDD)*, pp. 226–231.
43. Aji, A., F. Wang, H. Vo, R. Lee, Q. Liu, X. Zhang, and J. Saltz. 2013. Hadoop-GIS: A High Performance Spatial Data Warehousing System over MapReduce. *Proceedings of Very Large Data Bases (PVLDB)*, 6(11):1009–1020.
44. Cudré-Mauroux, P., E. Wu, and S. Madden. 2010. TrajStore: An Adaptive Storage System for Very Large Trajectory Data Sets. In *Proceedings of the 26th International Conference on Data Engineering (ICDE)*, pp. 109–120.
45. Tang, L., Y. Zheng, J. Yuan, J. Han, A. Leung, C. Hung, and W. Peng. 2012. On Discovery of Traveling Companions from Streaming Trajectories. In *Proceedings of the 29th International Conference on Data Engineering (ICDE)*, pp. 186–197.
46. Zheng, K., Y. Zheng, N. Jing Yuan, and S. Shang. 2013. On Discovery of Gathering Patterns from Trajectories. In *Proceedings of the 29th IEEE International Conference on Data Engineering (ICDE 2013)*, pp. 242–253.
47. Luo, W., H. Tan, L. Chen, and L.M. Ni. 2013. Finding Time Period-Based Most Frequent Path in Big Trajectory Data. In *Proceedings of the 33rd ACM SIGMOD International Conference on Management of Data*, pp. 713–724.
48. Efentakis, A., S. Brakatsoulas, N. Grivas, G. Lamprianidis, K. Patroumpas, and D. Pfoser. 2013. Towards a Flexible and Scalable Fleet Management Service. In *Proceedings of the 6th ACM SIGSPATIAL International Workshop on Computational Transportation Science (IWCTS)*, p. 79.
49. Thiagarajan, A., L. Ravindranath, K. LaCurts, S. Madden, H. Balakrishnan, S. Toledo, and J. Eriksson. 2009. vTrack: Accurate, Energy-Aware Road Traffic Delay Estimation Using Mobile Phones. *In Proceedings of the 7th ACM Conference on Embedded Networked Sensor Systems (SenSys)*, pp. 85–98.

50. Yang, B., C. Guo, C.S. Jensen, M. Kaul, and S. Shang. 2014. Multi-Cost Optimal Route Planning under Time-Varying Uncertainty. In *Proceedings of the 30th IEEE International Conference on Data Engineering (ICDE)*.
51. Zheng, K., S. Shang, N. Jing Yuan, and Y. Yang. 2013. Towards Efficient Search for Activity Trajectories. In *Proceedings of the 29th International Conference on Data Engineering (ICDE)*, pp. 230–241.
52. Shang, S., R. Ding, K. Zheng, C.S. Jensen, P. Kalnis, and X. Zhou. 2014. Personalized Trajectory Matching in Spatial Networks. *VLDB Journal*, 23(3): 449–468.
53. Yuan, N.J., F. Zhang, D. Lian, K. Zheng, S. Yu, and X. Xie. 2013. We Know How You Live: Exploring the Spectrum of Urban Lifestyles. In *Proceedings of the 1st ACM Conference on Online Social Networks (COSN)*, pp. 3–14.
54. Goodchild, M.F. 2007. Citizens as Voluntary Sensors: Spatial Data Infrastructure in the World of Web 2.0. *International Journal of Spatial Data Infrastructures Research (IJSDIR)*, 2(1):24–32.
55. OpenStreetMap Project. Available at http://www.openstreetmap.org (accessed April 30, 2014).
56. Twitter. Available at https://twitter.com (accessed April 30, 2014).
57. Flickr. Available at https://www.flickr.com (accessed April 30, 2014).
58. Damiani, M.L., C. Silvestri, and E. Bertino. 2011. Fine-Grained Cloaking of Sensitive Positions in Location-Sharing Applications. *Pervasive Computing*, 10(4):64–72.
59. Zheng, Y.-T., Z.-J. Zha, and T.-S. Chua. 2012. Mining Travel Patterns from Geotagged Photos. *ACM Transactions on Intelligent Systems and Technology (TIST)*, 3(3):56.

# IV

## Big Data Privacy

CHAPTER 15

# Personal Data Protection Aspects of Big Data

Paolo Balboni

CONTENTS



ABSTRACT

New technologies have significantly changed both the way that we live and the ways in which we do business. The advent of the Internet and the vast amounts of data disseminated across global networks and databases has brought with it both significant advantages in terms of scientific understanding and business opportunities, and challenges in terms of competition and privacy concerns. More data are currently available than ever before. Neelie Kroes, vice president of the European Commission responsible for the Digital Agenda, defined our era as "the data gold rush." What is Big Data? In its most general sense, the term *Big Data* refers to large digital data sets often held by corporations, governments, and other organizations that are subsequently analyzed by way of computer algorithms. In a more practical sense, "Big Data means big money!" The most significant challenge in harnessing the value of Big Data is to rightly balance business aspects with social and ethical implications, especially when consumers, or data subjects, possess a limited understanding of how their personal data are being collected and what becomes of the same. This chapter provides an analysis of applicable data protection provisions and their impact on both businesses and consumers/data subjects. More precisely, it will be determined whether and to what extent (1) data protection law applies and (2) personal Big Data can be (further) processed (e.g., by way of analytic software programs). It is vital to have a strategic and accurate approach to data protection compliance in order to collect personal data in a way that enables further lawful processing activities. The difference for a company between dying buried under personal data and harnessing their value is directly related to privacy compliance management.

## 15.1 INTRODUCTION

New technologies have significantly changed both the way that we live and the ways in which we do business. The advent of the Internet and the vast amounts of data disseminated across global networks and databases has brought with it both significant advantages in terms of scientific understanding and business opportunities, and challenges in terms of competition and privacy concerns. More data are currently available than ever before. As Neelie Kroes, vice president of the European Commission responsible for the Digital Agenda, pointed out in her March 19, 2014, speech entitled "The Data Gold Rush," not only do we have more data at our disposal than ever before, but also, we have access to a multitude of ways to manipulate, collect, manage, and make use of it [1]. The key, however, as Kroes wisely pointed out, is to find the value amid the multiplying mass of data: "The right infrastructure, the right networks, the right computing capacity and, last but not least, the right analysis methods and algorithms help us break through the mountains of rock to find the gold within" [1].

What is Big Data? In its most general sense, the term *Big Data* refers to large digital data sets often held by corporations, governments, and other organizations that are subsequently analyzed by way of computer algorithms [2, p. 35]. While the way that Big Data is defined depends largely on your point of reference, Big Data is comprised of aggregated and anonymous data, the personal data that are generated by the 369 million European Union citizens utilizing online platforms and services including but not limited to apps, games, social media, e-commerce, and search engines [3, p. 9]. The personal data that are collected, for example, in order to subscribe to an online service, usually include name, sex, e-mail address, location, IP address, Internet search history, and personal preferences, which are used to help companies provide more personalized and enhanced services and even to reach out to potential customers [3, p. 9].

The worldwide digital economy is "marked by strong, dynamic growth, a high turnover of new services" [3, p. 9], and is characterized by the presence of few dominant industry members. The services that these few large players offer often appear to be "free" but, in all reality, are paid for by the granting of personal data, estimated to be worth 300 billion euro [3, p. 8], a value that is only expected to grow in the coming years. Big Data means big money. Estimates suggest that 4 zettabytes of data were generated across the globe in 2013 [4]. That is a staggering amount of information, which continues to grow by the minute, presenting possible innovations and new ways to generate value from the same. The power of Big Data and the influence that it has on the global marketplace is therefore immense as it becomes increasingly intertwined with our daily lives as we make use of such new technological devices and services.

The ways in which such data are used by both business and government are in a process of constant evolution. In the past, data were collected in order to provide a specific service. In the age of Big Data, however, the genuine value of such data can be understood in terms of their potential (re)uses, and in fact, the estimated digital value that EU customers place on their data is forecast to rise to 1 trillion euro by 2020 [3, p. 9]. Big Data offers endless potential in terms of bettering our lives, from allowing for an increased provision and efficiency of services, to monitoring climate change, health trends, and disease epidemics, to allowing for increased monitoring of government fraud, abuse, and waste [5, p. 37]. In this way, we can understand Big Data not only as a technological tool but also as a public good, an expression of individual identity, or even property, a new form of value [5, pp. 37–49]. For the purposes of this chapter, however, we will focus on the value generated by personal data and the vital balance of adequate privacy and security safeguards with business opportunity.

In the European Data Protection Supervisor's (EDPS's) preliminary opinion of March 2014, the authority stressed the challenges that Big Data represents in the balancing act between competition, data protection, and consumer protection [3, p. 2]. Indeed, the most significant challenge in harnessing the value of Big Data must also take social and ethical aspects into consideration, especially when consumers, or data subjects, possess a limited understanding of how their personal data are being collected and what becomes of the same. Privacy and security measures, however, need not focus on limitations and protections alone, in the words of Neelie Kroes:

> Data generates value, and unlocks the door to new opportunities: you don't need to 'protect' people from their own assets. What you need is to empower people, give them control, give them a fair share of that value. Give them rights over their data—and responsibilities too, and the digital tools to exercise them. And ensure that the networks and systems they use are affordable, flexible, resilient, trustworthy, secure. [1]*

Under President Obama, the United States government has committed itself to promoting the free flow of data and the growth of the digital economy, calling on stakeholders of the private and public sectors in order to "harness the power of data in ways that boost productivity, improve lives, and serve communities" [5, p. 9]. The analysis of transactional and operational data has allowed manufacturers better management of warranties and equipment, logistics optimization, behavioral advertising, customized pricing, and even better fraud detection for banks [5, pp. 37–49]. In this sense, Big Data provides the possibility for improvement, minimization of waste, new customers, and therefore, higher monetary returns.

---

* See also The World Economic Forum (2014) *Rethinking Personal Data: Trust and Context in User-Centred Data Ecosystems*, p. 3. Available at: http://www3.weforum.org/docs/WEF_RethinkingPersonalData_TrustandContext_Report_2014.pdf. "To thrive, the growing number of economies that depend on the potential of "Big Data" must earn the trust of individuals, and be centred on empowering those individuals by respecting their needs."

The European Union has also affirmed the increase in productivity, innovation, and services as a result of cloud computing and Big Data [1]. Further economic growth as a result of new technologies, however, must be fostered by embracing new technologies, bettering infrastructure, and providing adequate safeguards for citizens. In this respect, the World Economic Forum rightly stressed the fundamental task of regulation, which "must keep pace with new technology and protect consumers without stifling innovation or deterring uptake" [6, p. 3].

### 15.1.1 Topic and Aim

The first six months of 2014 were characterized by the publication of US and EU (preliminary) positions on Big Data. This demonstrates that Big Data lies at the top their agendas from both the economic and regulatory perspectives. It is possible to preliminarily state that the United States has taken a clearer position on the topic than the European Union, which is still considering a number of aspects. Both powers have identified the numerous economic benefits of Big Data and an equal number of possible threats to consumers' privacy and data protection rights. The goal is the same for the United States and the European Union, as they are both determined to solve the identified issues. However, it seems that the United States is considering a review of the traditional principles of privacy and data protection, these being notice and consent,* whereas the European Union is not questioning traditional principles but is looking at how to adapt them to Big Data.

In March 2014, the EDPS published a preliminary opinion:

> *Preliminary Opinion of the European Data Protection: Supervisor Privacy and Competitiveness in the Age of Big Data: The Interplay between Data Protection, Competition Law and Consumer Protection in the Digital Economy* (henceforth *EDPS Preliminary Opinion on Big Data*). [3]

Moreover, the Article 29 Data Protection Working Party issued a relevant opinion on April 10, 2014, on anonymization techniques [7], which, together with Opinion 3/2013 on purpose limitation [2] and Opinion 4/2007 on the concept of personal data [8], represents the fundamental tools to interpret the ways that the current legal privacy framework applies to Big Data.

In May 2014, the White House published two very detailed reports:

*Big Data: Seizing Opportunities, Preserving Values* [5]

*Report to the President: Big Data and Privacy: A Technological Perspective* [5, p. ix]

---

* "The notice and consent is defeated by exactly the positive benefits that Big Data enables: new, non-obvious, unexpectedly powerful uses of data." The White House (2014) *Report to the President Big Data and Privacy: A Technological Perspective*, p. 38. Available at: http://www.whitehouse.gov/sites/default/files/microsites/ostp/PCAST/pcast_big_data_and_privacy_-_may_2014.pdf. See also "What really matters about Big Data is what it does. Aside from how we define Big Data as a technological phenomenon, the wide variety of potential uses for Big Data analytics raises crucial questions about whether our legal, ethical, and social norms are sufficient to protect privacy and other values in a Big Data world." The White House (2014) *Big Data: Seizing Opportunities, Preserving Values*, p. 3. Available at: http://www.whitehouse.gov/sites/default/files/docs/big_data_privacy_report_may_1_2014.pdf.

In this chapter, I will take into consideration both the US and the EU perspectives on Big Data; however, being a European lawyer, I will focus my analysis on applicable EU data protection provisions and their impact on both businesses and consumers/data subjects. More precisely, I will utilize a methodology to determine whether (1) data protection law applies and (2) personal Big Data can be (further) processed (e.g., by way of analytic software programs).

### 15.1.2 Note to the Reader, Structure, and Arguments

This chapter is intended for readers who are familiar with the basic principles of European privacy and data protection law. Consequently, I will not explain the general principles but will instead focus on specific aspects directly relevant to the current analysis. For those unfamiliar with the basics, I suggest that you first read Section 3.1 of the *EDPS Preliminary Opinion on Big Data* [3, p. 11ss].

Section 15.2 will deal with diverse aspects of data protection, providing an understanding of Big Data from the perspective of personal data protection using the Organization for Economic Co-operation and Development's (OECD's) four-step life cycle of personal data along the value chain, paying special attention to the concept of compatible use. Specifically, Section 15.2.2 will shed light on the development of the concept of personal data and its relevance in terms of data processing. Further focus will be placed on aspects such as pseudonymization, anonymous data, and reidentification. Section 15.2.3 instead will specifically deal with purpose limitation. Finally, my conclusions and recommendations focusing on the privacy and data implications of Big Data processing and the importance of data protection compliance management will be illustrated in Section 15.3.

## 15.2 DATA PROTECTION ASPECTS

In a paper discussing Big Data, the World Economic Forum pointed out that "[a]s ecosystem players look to use (mobile-generated) data, they face concerns about violating user trust, rights of expression, and confidentiality. Privacy and security concerns must be addressed before firms, governments, and individuals can be convinced to share data more openly" [6, p. 5]. Let us therefore try to understand what exactly Big Data means from the personal data protection point of view.

### 15.2.1 Big Data and Analytics in Four Steps

In order to explain what Big Data means from a strictly personal data protection perspective, I will borrow the four-step life cycle of personal data along the value chain described by the OECD as follows:

- Step 1: Collection/access
- Step 2: Storage and aggregation
- Step 3: Analysis and distribution
- Step 4: Usage [9]

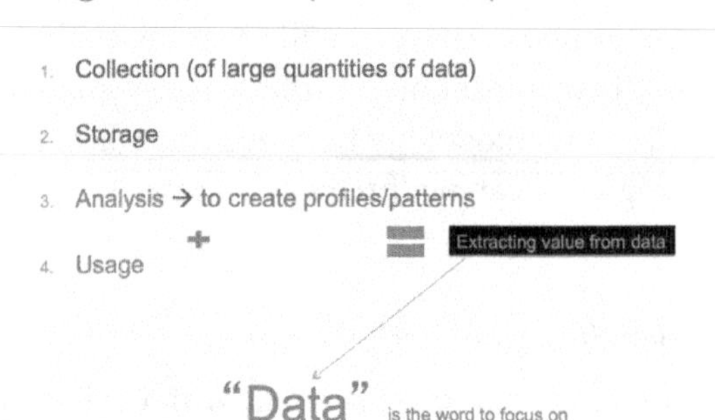


FIGURE 15.1 Big Data and analytics in four steps.

As the Article 29 Data Protection Working Party rightly stressed, the value of Big Data rests in the increasing ability of technology to meaningfully and thoroughly analyze them* in order to make better and more informed decisions:

> 'Big Data' refers to the exponential growth in availability and automated use of information: it refers to gigantic digital datasets held by corporations, governments and other large organisations, which are then extensively analysed using computer algorithms. Big Data relies on the increasing ability of technology to support the collection and storage of large amounts of data, but also to analyse, understand and take advantage of the full **value** of data (in particular using analytics applications). The expectation from Big Data is that it may ultimately lead to better and more informed decisions. [2, p. 45]

Therefore, going back to the OECD scheme, the crucial steps for Big Data are Step 3, analysis, and Step 4, usage, as shown in Figure 15.1.

If "[c]omputational capabilities now make 'finding a needle in a haystack' not only possible, but practical" [5, p. 6], for businesses, two fundamental privacy-related questions that data protection officers of companies should ask themselves are (1) whether the "haystack of data" has been lawfully built and (2) whether analytic software can be lawfully run on that big set of data to find the needle.

---

* This has been stressed also in the recent US reports on Big Data: "Big Data is big in two different senses. It is big in the quantity and variety of data that are available to be processed. And, it is big in the scale of analysis (termed "analytics") that can be applied to those data, ultimately to make inferences and draw conclusions. By data mining and other kinds of analytics, non-obvious and sometimes private information can be derived from data that, at the time of their collection, seemed to raise no, or only manageable, privacy issues." The White House (2014) *Report to the President Big Data and Privacy: A Technological Perspective*, p. ix. Available at: http://www.whitehouse.gov/sites/default/files/microsites/ostp/PCAST/pcast_big_data_and_privacy_-_may_2014.pdf; and "Computational capabilities now make "finding a needle in a haystack" not only possible, but practical. In the past, searching large datasets required both rationally organized data and a specific research question, relying on choosing the right query to return the correct result. Big Data analytics enable data scientists to amass lots of data, including unstructured data, and find anomalies or patterns." The White House (2014) *Big Data: Seizing Opportunities, Preserving Values*, p. 6. Available at: http://www.whitehouse.gov/sites/default/files/docs/big_data_privacy_report_may_1_2014.pdf.

The first question can be rephrased as follows: Have the personal data been lawfully collected?

The second question can be rephrased as follows: Is (further) processing of personal data by way of analytics compatible with the purposes for which the data were collected (so-called compatible use)?

Given the limited space in this chapter, I will focus only on the second question—which, in my opinion, represents the core question that the entire Big Data phenomenon poses—assuming that the personal data have been lawfully collected.

However, before answering the "compatible use" question, it is first necessary to understand if the processing operations concern personal data at all. We therefore need to focus on the concept of personal data.

### 15.2.2 Personal Data

It is of fundamental importance to understand whether personal data are processed. This is because if they are not processed, the principles of data protection do not apply. It is important, however, to stress that "[c]ompanies may consider most of their data to be non-personal datasets, but in reality it is now rare for data generated by user activity to be completely and irreversibly anonymised" [3, p. 9].* A description of what is considered personal data and anonymous data is therefore paramount. The distinction between pseudonymous and anonymous data is also briefly discussed in Sections 15.2.2.2 and 15.2.2.3 for the purpose of consistency and complete understanding.

The concept of personal data is established in Directive 95/46/EC.† The definition according to the same directive provides guidelines regarding personal data in addition to briefly outlining anonymous data. Article 2(a) of Directive 95/46/EC states that "'personal data' shall mean any information relating to an identified or identifiable natural person ('data subject'); an identifiable person is one who can be identified, directly or indirectly, in particular by reference to an identification number or to one or more factors specific to his physical, physiological, mental, economic, cultural or social identity."

* The same position was taken by The White House (2014) *Big Data: Seizing Opportunities, Preserving Values*, p. 6 and p. xi. Available at: http://www.whitehouse.gov/sites/default/files/docs/big_data_privacy_report_may_1_2014.pdf. "As techniques like data fusion make Big Data analytics more powerful, the challenges to current expectations of privacy grow more serious. When data is initially linked to an individual or device, some privacy-protective technology seeks to remove this linkage, or 'de-identify' personally identifiable information—but equally effective techniques exist to pull the pieces back together through 're-identification.' Moreover, it is stated in the study that "[a]nonymization is increasingly easily defeated by the very techniques that are being developed for many legitimate applications of Big Data. In general, as the size and diversity of available data grows, the likelihood of being able to re-identify individuals (that is, re-associate their records with their names) grows substantially. While anonymization may remain somewhat useful as an added safeguard in some situations, approaches that deem it, by itself, a sufficient safeguard need updating."

† Directive 95/46/EC of the European Parliament and of the Council of 24 October 1995 on the protection of individuals with regard to the processing of personal data and on the free movement of such data. *Official Journal L* 281, 23/11/1995, pp. 0031–0050.

In 2007, the Article 29 Working Party issued an opinion on the concept of personal data [8], elaborating four specific and fundamental elements of the definition:

1. Any information
2. Relating to
3. Identified or identifiable
4. Natural person

    1. Any information

    The Article 29 Working Party noted that this element underscores the rather broad approach illustrated in Directive 95/46/EC. All information relevant to a person is included, regardless of the "position or capacity of those persons (as consumer, patient, employee, customer, etc.)" [8, p. 7]. In this case, the information can be objective or subjective and does not necessarily have to be true or proven.

    The words "any information" also imply information of any form, audio, text, video, images, and so forth. Importantly, the manner in which the information is stored is irrelevant. The Article 29 Working Party expressly mentions biometric data as a special case [8, p. 8], as such data can be considered as information content as well as a link between the individual and the information. Because biometric data are unique to an individual, they can also be used as identifiers.

    2. Relating to

    Information related to an individual is information about that individual. The relationship between data and an individual is often self-evident, an example of which is when the data are stored in an individual employee file or in a medical record. This is, however, not always the case, especially when the information relates to objects. Such objects belong to individuals, but additional meanings or information is required to create the link to the individual [8, p. 9].

    At least one of the following three elements should be present in order to consider information to be related to an individual: content, purpose, or result. An element of *content* is present when the information is in reference to an individual, regardless of the (intended) use of the information. The *purpose* element instead refers to whether the information is used or is likely to be used "with the purpose to evaluate, treat in a certain way or influence the status or behaviour of an individual" [8, p. 10]. A *result* element is present when the use of the data is likely to have an impact on a certain person's rights and interests [8, p. 11]. These elements are alternatives and are not cumulative, implying that one piece of data can relate to different individuals based on diverse elements.

3. Identified or identifiable

   "A natural person can be 'identified' when, within a group of persons, he or she is 'distinguished' from all other members of the group" [8, p. 12]. When identification has not occurred but is possible, the individual is considered to be "identifiable."

   In order to determine whether those with access to the data are able to identify the individual, all reasonable means likely to be used either by the controller or by any other person should be taken into consideration. The cost of identification, the intended purpose, the way the processing is structured, the advantage expected by the data controller, the interest at stake for the data subjects, and the risk of organizational dysfunctions and technical failures should be taken into account in the evaluation [8, p. 15].

4. Natural person

   Directive 95/46/EC is applicable to the personal data of natural persons, a broad concept that calls for protection wholly independent from the residence or nationality of the data subject.

   The concept of personality is understood as "the capacity to be the subject of legal relations, starting with the birth of the individual and ending with his death" [8, p. 22]. Personal data thus relate to identified or identifiable living individuals. Data concerning deceased persons or unborn children may, however, indirectly be subject to protection in particular cases. When the data relate to other living persons, or when a data controller makes no differentiation in his/her documentation between living and deceased persons, it may not be possible to ascertain whether the person the data relate to is living or deceased; additionally, some national laws consider deceased or unborn persons to be protected under the scope of Directive 95/46/EC [8, pp. 22–23].

   Legal persons are excluded from the protection provided under Directive 95/46/EC. In some cases, however, data concerning a legal person may relate to an individual, such as when a business holds the name of a natural person. Some provisions of Directive 2002/58/EC* (amended by Directive 2009/136/EC)† extend the scope of Directive 95/46/EC to legal persons.‡

* Directive 2002/58/EC of the European Parliament and of the Council of 12 July 2002 concerning the processing of personal data and the protection of privacy in the electronic communications sector (Directive on privacy and electronic communications) [2002]. *Official Journal L* 31/07/2002, pp. 0037–0047.

† Directive 2009/136/EC of the European Parliament and of the Council of 25 November 2009 amending Directive 2002/22/EC on universal service and users' rights relating to electronic communications networks and services, Directive 2002/58/EC concerning the processing of personal data and the protection of privacy in the electronic communications sector and Regulation (EC) No 2006/2004 on cooperation between national authorities responsible for the enforcement of consumer protection laws Text with EEA relevance [2006]. *Official Journal L* 337, 18/12/2009, pp. 0011–0036.

‡ In the EDPS Preliminary Opinion on Big Data, it is also expected that: "[c]ertain national jurisdictions (Austria, Denmark, Italy and Luxembourg) extend some protection to legal persons." European Data Protection Supervisor (2014) *Preliminary Opinion of the European Data Protection Supervisor Privacy and Competitiveness in the Age of Big Data: The Interplay between Data Protection, Competition Law and Consumer Protection in the Digital Economy*, p. 13, footnote 31. Available at: https://secure.edps.europa.eu/EDPSWEB/webdav/site/mySite/shared/Documents/Consultation/Opinions/2014/14-03-26_competitition_law_big_data_EN.pdf.

*15.2.2.1 Profiling Activities on Personal Data*

Article 15 of Directive 95/46/EC establishes that "Member States shall grant the right to every person not to be subject to a decision which produces legal effects concerning him or significantly affects him and which is based solely on automated processing of data intended to evaluate certain personal aspects relating to him, such as his performance at work, creditworthiness, reliability, conduct, etc."

The Council of Europe's "Recommendation on Profiling" (henceforth "Recommendation")* considers that "profiling an individual may result in unjustifiably depriving her or him from accessing certain goods or services and thereby violate the principle of non-discrimination." Inclusion or exclusion alone can therefore be considered a legal effect. The "Recommendation" indicates that different contexts should be maintained separately when data are collected and analyzed for profiling purposes because the protection of personal data and the fundamental right to privacy "entails the existence of different and independent spheres of life where each individual can control the use she or he makes of her or his identity."*

Examining the current European data protection reform, the "Proposal for a General Data Protection Regulation" (henceforth "Proposal GDPR"),† Article 20, builds on Article 15 of Directive 95/46/EC with several modifications and additional safeguards. In this way, every natural person has the right "to not be subject"/"to object" (depending on whether we consider the European Commission Proposal or the compromise version approved by the European Parliament on March 12, 2014) to profiling, which leads to measures "producing legal effects concerning the data subject or does similarly significantly affect the interests, rights or freedoms of the concerned data subject." In this case, profiling "means any form of automated processing of personal data intended to evaluate certain personal aspects relating to a natural person or to analyse or predict in particular that natural person's performance at work, economic situation, location, health, personal preferences, reliability or behaviour" (Article 4 of the compromise version approved by the European Parliament on March 12, 2014).

*15.2.2.2 Pseudonymization*

Personal data are data that directly or indirectly identify an individual. Identification should be understood broadly; reference by way of a unique number is such an example. It is not necessary to know the name of a person; recognition is sufficient. In other words, it is possible to single out an individual in a group based on the data. The recent Opinion 05/2014 on anonymization techniques of the Article 29 Data Protection Working Party specifies that "pseudonymisation is not a method of anonymisation. It merely reduces the linkability of a dataset with the original identity of a data subject,

* Recommendation CM/Rec (2010) 13 of the Committee of Ministers to member states on the protection of individuals with regard to automatic processing of personal data in the context of profiling (Adopted by the Committee of Ministers on 23 November 2010 at the 1099th meeting of the Ministers' Deputies).

† Proposal for a Regulation of the European Parliament and of the Council on the protection of individuals with regard to the processing of personal data and on the free movement of such data (General Data Protection Regulation) [2012] OJ COM (2012) 11 final. Available at: http://ec.europa.eu/justice/data-protection/document/review2012/com_2012_11_en.pdf.

and is accordingly a useful security measure" [7, p. 20]. Pseudonyms indirectly identify the individual and are therefore considered personal data.* The compromise version of the GDPR approved by the EU Parliament on March 12, 2014, Recital 58(a), states that "[p]rofiling based solely on the processing of pseudonymous data should be presumed not to significantly affect the interests, rights or freedoms of the data subject. Where profiling, whether based on a single source of pseudonymous data or on the aggregation of pseudonymous data from different sources, permits the controller to attribute pseudonymous data to a specific data subject, the processed data should no longer be considered to be pseudonymous."

#### *15.2.2.3 Anonymous Data*

Anonymous data are data that cannot be directly or indirectly connected to an identified or identifiable individual originally or after processing.† The distinction between personal data and anonymous data depends on a possible connection, that is, on the possibility to reassociate the data to a specific data subject. Aggregated data concerning a large group of people are anonymous "only if the data controller would aggregate the data to a level where the individual events are no longer identifiable" [7].

Anonymizing data is sufficient in the case that the data cannot be reversed to the original identifying data. Data are considered anonymous when an unreasonable effort (amount of time and manpower) is required to (re)turn the data into personally identifiable data. The likelihood of making a connection between data and a data subject is thus measured in relation to the time, cost, and technical means necessary to do so.‡ The test of identifiability must also consider state-of-the-art technology at the time of processing. The technical threshold at which sensitive information is considered identifiable is lower, as it warrants a higher level of protection [10, p. 525].

---

* In this respect, it is interesting to notice that in the 17.12.12 Draft report on the proposal for a regulation of the European Parliament and of the Council on the protection of individual with regard to the processing of personal data and on the free movement of such data (General Data Protection Regulation) (COM(2012)0011–C7-0025/2012–2012/0011(COD)) Committee on Civil Liberties, Justice and Home Affairs, Rapporteur: Jan Philipp Albrecht ('Albrecht's 17.12.12 Draft report on GDPR'), "the rapporteur encourages the pseudonymous and anonymous use of services. For the use of pseudonymous data, there could be alleviations with regard to obligations for the data controller (Articles 4(2)(a), 10), Recital 23)," p. 211. Available at: http://www.europarl.europa.eu/RegData/commissions/libe/projet_rapport/2012/501927/LIBE_PR(2012)501927_EN.doc. Moreover, on pseudonymization, see the recent European Privacy Association paper by Rosario Imperiali: Pseudonymity and legitimate interest: two solutions to reduce data protection impact? Available at: http://www.academia.edu/2308794/Part_1_Pseudonymity_and_legitimate_interest_two_solutions_to_reduce_data_protection_impact.

† This definition of anonymous data is used also in Section 4.1.n of the Italian Personal Data Protection Code Legislative Decree No. 196 of June 30, 2003: "'anonymous data' shall mean any data that either in origin or on account of its having been processed cannot be associated with any identified or identifiable data subject." See also Albrecht's 17.12.12 Draft report on GDPR, "This Regulation should not apply to anonymous data, meaning any data that can not be related, directly or indirectly, alone or in combination with associated data, to a natural person or where establishing such a relation would require a disproportionate amount of time, expense, and effort, taking into account the state of the art in technology at the time of the processing and the possibilities for development during the period for which the data will be processed," p. 15.

‡ Council Recommendation (EC) R97/5 of February 13, 1997, on the protection of medical data, Article 29 Working Party, Opinion 04/2007 on the concept of personal data, 01248/07/EN, WP 136. See also Albrecht's 17.12.12 Draft report on GDPR, p. 15.

Data are considered unintelligible when personal data are securely encrypted with a standardized secure encryption algorithm, or when hashed with a standardized cryptographic keyed hash function and the key used to encrypt or hash the data has not been compromised in any security breach or was generated in a way that it cannot be guessed by exhaustive key searches using technological means [11].

Data protection experts and authorities agree that when a data controller anonymizes personal data, even though it holds the "key" that would allow for reidentification, the publication of such data does not amount to a disclosure of personal data [12, p. 13]. In this case, then, personal data protection laws do not apply to the relative disclosed information.

Anonymity of personal data provides important benefits for the protection of personal data and can be seen as a way to advance one's right to privacy [10, p. 527]. Anonymized data transfers ensure a high level of privacy and are particularly beneficial in the processing of sensitive personal data [13]. Additionally, anonymization partially enforces the principle of data minimization [12, p. 12]. The anonymization techniques eliminate personal data or identifying data from the processing if the purpose sought in the individual case can be achieved by using anonymous data.

However, as the Article 29 Data Protection Working Party rightly stated in Opinion 05/2014 on anonymization techniques, "data controllers should consider that an anonymised dataset can still present residual risks to data subjects. Indeed, on the one hand, anonymisation and reidentification are active fields of research and new discoveries are regularly published, and on the other hand even anonymised data, like statistics, may be used to enrich existing profiles of individuals, thus creating new data protection issues. Thus, anonymisation should not be regarded as a one-off exercise and the attending risks should be reassessed regularly by data controllers" [7, p. 9].*

The European data protection legislation does not expressly provide a definition of "anonymous data." The European data protection reform, the "Proposal GDPR" compromise version approved by the European Parliament on March 12, 2014, does not provide a definition of anonymous data either, but reference to anonymous data is made in Recital 23: "The principles of data protection should apply to any information concerning an identified or identifiable natural person. To determine whether a person is identifiable, account should be taken of all the means reasonably likely to be used either by the controller or by any other person to identify or single out the individual directly or indirectly. To ascertain whether means are reasonably likely to be used to identify the individual, account should be taken of all objective factors, such as the costs of and the amount of time required for identification, taking into consideration both available technology at the time of the processing and technological development. The principles of data protection should therefore

* This view have also been confirmed by The White House (2014) *Report to the President Big Data and Privacy: A Technological Perspective*, p. xi. Available at: http://www.whitehouse.gov/sites/default/files/microsites/ostp/PCAST/pcast_big_data_and_privacy_-_may_2014.pdf: "Anonymization is increasingly easily defeated by the very techniques that are being developed for many legitimate applications of Big Data. In general, as the size and diversity of available data grows, the likelihood of being able to re-identify individuals (that is, re-associate their records with their names) grows substantially. While anonymization may remain somewhat useful as an added safeguard in some situations, approaches that deem it, by itself, a sufficient safeguard need updating."

not apply to anonymous data, which is information that does not relate to an identified or identifiable natural person. This Regulation does therefore not concern the processing of such anonymous data, including for statistical and research purposes."

Considering the legislative framework in force, the notion of anonymous data can be derived from some recitals of Directive 95/46/EC and Directive 2002/58/EC (amended by Directive 2009/136/EC). Recital 26 of the Preamble of Directive 95/46/EC refers to the concept of anonymous data: "[t]he principles of protection shall not apply to data rendered anonymous in such a way that the data subject is no longer identifiable." Codes of conduct within the meaning of Article 27 of Directive 95/46/EC may be useful for understanding the ways in which data can be rendered anonymous and retained in a form in which identification of the data subject is not possible.

Further reference to anonymous data is included in Recitals 9, 26, 28, and 33 of Directive 2002/58/EC. Specifically, Recital 9 states that "[t]he Member States (...) should cooperate in introducing and developing the relevant technologies where this is necessary to apply the guarantees provided for by this Directive and taking particular account of the objectives of minimizing the processing of personal data and of using anonymous or pseudonymous data where possible."

#### *15.2.2.4 Reidentification*

In order to determine that personal data are appropriately anonymized, it is important to assess whether reidentification of data subjects is possible and whether anyone can perform such an operation.* While data protection legislation does not currently provide guidance in these terms, data protection authorities apply several tests in practice to determine whether the data are appropriately anonymized. For this purpose, the UK Information Commissioner (ICO) has developed a "motivated intruder" test as part of a risk assessment process [12, pp. 22–24]. This test, according to the ICO, should be applied primarily in the context of anonymized information disclosure to the public or a third party and is considered a significant official benchmark.

The test itself is based on consideration of whether a person, without any prior knowledge, but with a desire to achieve reidentification of the individuals the information relates to (a motivated intruder), would be successful. In the test, the motivated intruder represents the general public and is thus a reasonably competent individual who has access to publicly available sources, including the Internet or libraries, but has no special knowledge, such as computer hacking skills or equipment [12, p. 23].

Section 15.2.2 of this chapter examined the concept of personal data and its relevance in terms of data processing. I have highlighted the importance of fully understanding whether or not the principles of data protection are applicable, an aspect that is paramount in the discussion and examination of personal data protection in the field of Big Data. A deep understanding of anonymization, pseudonimization, and reidentification has been

* See generally on reidentification Article 29 Data Protection Working Party Opinion 05/2014 on anonymisation techniques. Adopted on April 10, 2014. Available at: http://ec.europa.eu/justice/data-protection/article-29/documentation/opinion-recommendation/files/2014/wp216_en.pdf.

provided insofar as they are strictly related to the discussion. It is of utmost importance that the implications of different types of data are adequately understood as they have highly relevant consequences on the legality of plausible processing activities. Section 15.2.3 will explore purpose limitation, compatibility assessments, and the concept of "functional separation," which must, however, consider that the risk of reidentification is increasingly present due to the development of new technologies.

### 15.2.3 Purpose Limitation

On April 2, 2013, the Article 29 Data Protection Working Party published an opinion on the principle of purpose limitation [2], as outlined by Directive 95/46/EC. I will briefly provide the fundamental elements of this opinion.

According to Directive 95/46/EC, the Article 29 Data Protection Working Party outlines the fundamental elements that characterize the purpose limitation principle. The concept of purpose limitation has two primary building blocks:

- Personal data must be collected for specified, explicit, and legitimate purposes (so-called purpose specification) [2, p. 11].
- Personal data must not be further processed in a way incompatible with those purposes (so-called compatible use) [2, p. 12].

Compatible or incompatible use needs are to be assessed—compatibility assessment—on a case-by-case basis, according to the following key factors:

- The relationship between the purposes for which the personal data have been collected and the purposes of further processing [2, p. 23]
- The context in which the personal data have been collected and the reasonable expectations of the data subjects as to their further use [2, p. 24]
- The nature of the personal data and the impact of the processing on the data subjects [2, p. 25]
- The safeguards adopted by the controller to ensure fair processing and to prevent any undue impact on the data subjects [2, p. 26]

The purpose limitation principle can only be restricted subject to the conditions set forth in Article 13 of Directive 95/46/EC (i.e., national security, defense, public security, or protection of the data subject or of the rights and freedoms of others).

In this opinion, the Article 29 Data Protection Working Party deals with Big Data [2, p. 45ss]. More precisely, the Article 29 Data Protection Working Party specifies that, in order to lawfully process Big Data, in addition to the four key factors of the compatibility assessment to be fulfilled, additional safeguards must be assessed to ensure fair processing and to prevent any undue impact. The Article 29 Data Protection Working Party considers two scenarios to identify such additional safeguards:

1. In the first one, the organizations processing the data want to detect trends and correlations in the information.
2. In the second one, the organizations are interested in individuals (...) [as they specifically want] to analyse or predict personal preferences, behaviour and attitudes of individual customers, which will subsequently inform 'measures or decisions' that are taken with regard to those customers [2, p. 46].

In the first scenario, so-called functional separation plays a major role in deciding whether further use of data may be considered compatible. Examples of functional separation are "full or partial anonymisation, pseudonymisation, or aggregation of the data, privacy enhancing technologies, as well as other measures to ensure that the data cannot be used to take decisions or other actions with respect to individuals" [2, p. 27; see also Section 2.3].

In the second scenario, prior consent of customers/data subjects (i.e., free, specific, informed, and unambiguous opt-in) would be required for further use to be considered compatible. In this respect, the Article 29 Data Protection Working Party specifies that "such consent should be required, for example, for tracking and profiling for purposes of direct marketing, behavioural advertisement, data-brokering, location-based advertising or tracking-based digital market research" [2, p. 46]. Furthermore, as a prerequisite for consent to be informed and for ensuring transparency, data subjects must have access (1) to their profiles, (2) to the algorithm that develops the profiles, and (3) to the source of data that led to the creation of the profiles [2, p. 47]. Furthermore, data subjects should be effectively granted the right to correct or update their profiles. Last but not least, the Article 29 Data Protection Working Party recommends allowing "data portability": "safeguards such as allowing data subjects/customers to have access to their data in a portable, user-friendly and machine readable format [as a way] to enable businesses and data-subjects/consumers to maximise the benefit of Big Data in a more balanced and transparent way" [2].*

This section dealt with the implications of the Article 29 Data Protection Working Party opinion regarding the principle of purpose limitation as outlined in Directive 95/46/EC and explored the elements that characterize the principle of limitation, including purpose specification and compatibility use. Specifically, the compatibility assessment, to be carried out on a case-by-case basis, and the necessary additional safeguards for fair processing were highlighted. The concept of functional separation was outlined in terms of its importance in the lawfulness of further processing activities.

## 15.3 CONCLUSIONS AND RECOMMENDATIONS

This chapter has outlined the relevance of the data revolution and the importance it will continue to have in the future, especially in terms of personal data protection compliance. The importance of Big Data is growing at an accelerated pace, a notion that is exemplified

* "For example, access to information about energy consumption in a user-friendly format could make it easier for households to switch tariffs and get the best rates on gas and electricity, as well as enabling them to monitor their energy consumption and modify their lifestyles to reduce their bills as well as their environmental impact."

by the increasing attention that the United States and the European Union have placed on both the regulation and further exploration of the phenomenon.

Section 15.2.1 looked at the OECD's four-step life cycle of personal data along the value chain, giving special importance to Steps 3 and 4, analysis and usage, respectively. The section also dealt with the question of compatible use or whether further personal data processing by way of analytics is compatible with the purposes for which the data were collected. The importance of identifying whether or not a company is processing personal Big Data is highlighted in Section 15.2.2 and is paramount to this discussion. The same section therefore focused on the definition of personal data and the importance of understanding what type of data is being processed, a determining factor in the applicability of data protection law. Section 15.2.2 further explored the definitions of pseudonymization, anonymization, and reidentification and the relative consequences of further activities. Finally, Section 15.2.3 provided insights with respect to the fundamental elements of the principle of purpose limitation and the vitality of the compatibility assessment, bringing other notions such as functional separation into the argument. Indeed, anonymization techniques and the aggregation of data are to be considered as key elements in determining compatibility. The inclusion of the Article 29 Data Protection Working Party Opinion 3/2013 therefore underlined the criticality of purpose limitation for companies to understand the conditions for further use for processing to be lawful.

Proliferation of data and increasing computational resources represent a great business opportunity for companies. This chapter has provided a strategic look at the privacy and data protection implications of Big Data processing. The importance of adequate and rigorous data protection compliance management throughout the entire data life cycle is key and was discussed at length. In fact, companies are usually advised on how to store, protect, and analyze a large amount of data and turn them into valuable information to improve their businesses. However, data—unless anonymized*—can only be processed for the purposes they were collected for, or those that are compatible with them.† It is therefore vital to have a strategic and accurate approach to data protection compliance in order to collect personal data in a way that enables further lawful processing activities. The

* As Article 29 Data Protection Working Party rightly stated in Opinion 05/2014 on anonymisation techniques: "data controllers should consider that an anonymised dataset can still present residual risks to data subjects. Indeed, on the one hand, anonymisation and re-identification are active fields of research and new discoveries are regularly published, and on the other hand even anonymised data, like statistics, may be used to enrich existing profiles of individuals, thus creating new data protection issues. Thus, anonymisation should not be regarded as a one-off exercise and the attending risks should be reassessed regularly by data controllers." Article 29 Data Protection Working Party Opinion 05/2014 on anonymisation techniques. Adopted on April 10, 2014, p. 9. Available at: http://ec.europa.eu/justice/data-protection/article-29/documentation/opinion-recommendation/files/2014/wp216_en.pdf. This view has also been confirmed by The White House (2014) *Report to the President Big Data and Privacy: A Technological Perspective*, p. xi. Available at: http://www.whitehouse.gov/sites/default/files/microsites/ostp/PCAST/pcast_big_data_and_privacy_-_may_2014.pdf: "Anonymization is increasingly easily defeated by the very techniques that are being developed for many legitimate applications of Big Data. In general, as the size and diversity of available data grows, the likelihood of being able to re-identify individuals (that is, re-associate their records with their names) grows substantially. While anonymization may remain somewhat useful as an added safeguard in some situations, approaches that deem it, by itself, a sufficient safeguard need updating."

† See extensively Section 2.3.

difference for a company between dying buried under personal data and harnessing their value is directly related to privacy compliance management.

## REFERENCES


1. Kroes, N. (2014) "The data gold rush." European Commission. Athens, March 19. Speech. Available at: http://europa.eu/rapid/press-release_SPEECH-14-229_en.htm.
2. Article 29 Data Protection Working Party (2013) Opinion 03/2013 on purpose limitation. Adopted on April 2, 2013. Available at: http://ec.europa.eu/justice/data-protection/article-29/documentation/opinion-recommendation/files/2013/wp203_en.pdf.
3. European Data Protection Supervisor (2014) *Preliminary Opinion of the European Data Protection Supervisor Privacy and Competitiveness in the Age of Big Data: The Interplay between Data Protection, Competition Law and Consumer Protection in the Digital Economy*. Available at: https://secure.edps.europa.eu/EDPSWEB/webdav/site/mySite/shared/Documents/Consultation/Opinions/2014/14-03-26_competitition_law_big_data_EN.pdf.
4. Meeker, M. and Yu, L. (2013) *Internet Trends*. Kleiner Perkins Caulfield Byers. Available at: http://www.slideshare.net/kleinerperkins/kpcb-internet-trends-2013.
5. The White House (2014) *Big Data: Seizing Opportunities, Preserving Values*. Available at: http://www.whitehouse.gov/sites/default/files/docs/big_data_privacy_report_may_1_2014.pdf.
6. The World Economic Forum (2012) *Big Data, Big Impact: New Possibilities for International Development*. Available at: http://www3.weforum.org/docs/WEF_TC_MFS_BigDataBigImpact_Briefing_2012.pdf.
7. Article 29 Data Protection Working Party (2014) Opinion 05/2014 on anonymisation techniques. Adopted on April 10, 2014. Available at: http://ec.europa.eu/justice/data-protection/article-29/documentation/opinion-recommendation/files/2014/wp216_en.pdf.
8. Article 29 Working Party (2007) Opinion 4/2007 on the concept of personal data. Adopted on June 20. Available at: http://ec.europa.eu/justice/policies/privacy/docs/wpdocs/2007/wp136_en.pdf.
9. Organisation for Economic Co-operation and Development (2013) *OECD Digital Economy Papers No. 220 Exploring the Economics of Personal Data a Survey of Methodologies for Measuring Monetary Value*, pp. 4–39.
10. Kerr, I., Lucock, C. and Steeves, V. (Eds.) (2009) *Lessons from the Identity Trail: Anonymity, Privacy, and Identity in a Networked Society*. Oxford: Oxford University Press.
11. ENISA (2012) *Recommendations on Technical Implementation Guidelines of Article 4*, April, p. 17. Available at: http://www.enisa.europa.eu.
12. ICO (2012) *Anonymisation: Managing Data Protection Risk Code of Practice*, November 20. Available at: http://www.ico.gov.uk/news/latest_news/2012/~/media/documents/library/Data_Protection/Practical_application/anonymisation_code.ashx.
13. Nicoll, C., Prins, J. E. J. and van Dellen, M. J. M. (Eds.) (2003) *Digital Anonymity and the Law: Tensions and Dimensions, Information Technology and Law Series*. The Hague: T.M.C. Asser Press, p. 149.

CHAPTER 16

# Privacy-Preserving Big Data Management

## *The Case of OLAP*

Alfredo Cuzzocrea

CONTENTS

## ABSTRACT

This chapter explores the emerging context of privacy-preserving OLAP over Big Data, a novel topic that is playing a critical role in actual Big Data research, and proposes an innovative framework for supporting *intelligent techniques for computing privacy-preserving OLAP aggregations on data cubes.* The proposed framework originates from the evidence stating that state-of-the-art *privacy-preserving OLAP* approaches lack *strong theoretical bases* that provide solid foundations to them. In other words, there is not a theory underlying such approaches, but rather an *algorithmic vision* of the problem. A class of methods that clearly confirm to us the trend above is represented by the so-called *perturbation-based techniques*, which propose to alter the target data cube cell-by-cell to gain privacy-preserving query processing. This approach exposes us to clear limits, whose lack of *extendibility* and *scalability* is only the tip of an enormous iceberg. With the aim of fulfilling this critical drawback, this chapter describes and experimentally assesses a *theoretically-sound accuracy/privacy-constrained framework for computing privacy-preserving data cubes in OLAP environments.* The benefits derived from our proposed framework are twofold. First, we provide and meaningfully exploit *solid theoretical foundations* to the privacy-preserving OLAP problem that pursue the idea of obtaining privacy-preserving data cubes via *balancing accuracy and privacy of cubes* by means of *flexible sampling methods.* Second, we ensure the *efficiency* and the *scalability* of the proposed approach, as confirmed to us by our experimental results, thanks to the idea of leaving the algorithmic vision of the privacy-preserving OLAP problem.

## 16.1 INTRODUCTION

One among the most challenging topics in *Big Data research* is represented, without doubt, by the issue of ensuring the *security and privacy of Big Data repositories* (e.g., References 1 and 2). To become convinced about this, consider the case of *cloud systems* [3,4], which are very popular now. Here, cloud nodes are likely to exchange data very often. Therefore, the *privacy breach risk* arises, as distributed data repositories can be accessed from a node to another one, and hence, *sensitive information* can be inferred.

Another relevant data management context for Big Data research is represented by the issue of effectively and efficiently supporting *data warehousing and online analytical processing (OLAP) over Big Data* [5–7], as multidimensional data analysis paradigms are likely to become an "enabling technology" for *analytics over Big Data* [8,9], a collection of models, algorithms, and techniques oriented to extract useful knowledge from cloud-based Big Data repositories for decision-making and analysis purposes.

At the convergence of the three axioms introduced here (i.e., security and privacy of Big Data, data warehousing and OLAP over Big Data, analytics over Big Data), a critical research challenge is represented by the issue of *effectively and efficiently computing privacy-preserving OLAP data cubes over Big Data* [10,11]. It is easy to foresee that this problem will become more and more important in future years, as it not only involves relevant theoretical and methodological aspects, not all explored by actual literature, but also regards significant modern scientific applications, such as *biomedical tools over Big Data* [12,13], *e-science and e-life Big Data applications* [14,15], *intelligent tools for exploring Big Data repositories* [16,17], and so forth.

Inspired by these clear and evident trends, in this chapter, we focus the attention on privacy-preserving OLAP data cubes over Big Data, and we provide two kinds of contributions:

- We provide a complete survey of privacy-preserving OLAP approaches available in literature, with respect to both *centralized* and *distributed environments.*
- We provide an innovative framework that relies on flexible sampling-based data cube compression techniques for computing privacy-preserving OLAP aggregations on data cubes.

### 16.1.1 Problem Definition

Given a multidimensional range *R* of a data cube *A*, an *aggregate pattern* over *R* is defined as an aggregate value extracted from *R* that is capable of providing a "description" of data stored in *R*. In order to capture the privacy of aggregate patterns, in this chapter, we introduce a *novel notion of privacy OLAP.* According to this novel notion, given a data cube *A*, *the privacy preservation of A is modeled in terms of the privacy preservation of aggregate patterns defined on multidimensional data stored in A*. Therefore, we say that a data cube *A* is privacy preserving iff aggregate patterns extracted from *A* are privacy preserving. Contrary to our innovative privacy OLAP notion, previous privacy-preserving OLAP proposals totally neglect this even-relevant theoretical aspect and, inspired by well-established techniques that focus on the privacy preservation of relational tuples [18,19], mostly focus on the privacy preservation of data cells (e.g., Reference 20) accordingly. Despite this, OLAP deals with aggregate data and neglects individual information. Therefore, it makes more sense to deal with the privacy preservation of aggregate patterns rather than the privacy preservation of data cube cells.

Given a multidimensional range *R* of a data cube *A*, AVG(*R*), that is, the average value of data cells in *R*, is the simplest aggregate pattern one could think about. It should be noted that this pattern could be inferred with a high degree of accuracy starting from (1) the knowledge about the summation of items in *R*, SUM(*R*) (which could be public for some reason—e.g., the total amount of salaries of a certain corporate department) and (2) the estimation of the number of items in *R*, COUNT(*R*) (which, even if not disclosed to external users, could be easily estimated—e.g., the total number of employees of the department). In turn, basic aggregate patterns like AVG can be meaningfully combined to progressively discover (1) more complex aggregate patterns (e.g., Reference 21 describes a possible methodology for deriving complex OLAP aggregates from elementary ones) to be exploited for trend analysis of sensitive aggregations and prediction purposes or, contrary to this, (2) aggregations of coarser hierarchical data cube levels until the privacy of individual data cells is breached. From the latter amenity, it follows that the results of Reference 20, which conventionally focuses on the privacy preservation of data cube cells, are covered by our innovative privacy OLAP notion.

Inspired by these considerations, in this chapter, we provide a pioneering framework that encompasses *flexible sampling-based data cube compression techniques for computing privacy-preserving OLAP aggregations on data cubes while allowing approximate answers to be efficiently evaluated over such aggregations.* This framework addresses an application scenario where, given a multidimensional data cube $A$ stored in a *producer* data warehouse server, a collection of multidimensional portions of $A$ defined by a given (range) *query workload QWL* of interest must be published online for *consumer* OLAP client applications. Moreover, once published, the collection of multidimensional portions is no longer connected to the data warehouse server, and updates are handled from scratch at each new online data delivery. The query workload *QWL* is cooperatively determined by the data warehouse server and OLAP client applications, mostly depending on OLAP analysis goals of client applications and other parameters such as business processes and requirements, frequency of accesses, and locality. OLAP client applications wish to retrieve summarized knowledge from $A$ via adopting a *complex multiresolution query model* whose components are (1) queries of *QWL* and (2) for each query $Q$ of *QWL*, *subqueries* of $Q$ (i.e., in a multiresolution fashion). To this end, for each query $Q$ of *QWL*, an *accuracy grid* $\mathcal{G}(Q)$, whose cells model subqueries of interest, is defined. While aggregations of (authorized) queries and (authorized) subqueries in *QWL* are disclosed to OLAP client applications, it must be avoided that, by meaningfully combining aggregate patterns extracted from multidimensional ranges associated with queries and subqueries in *QWL*, malicious users could infer sensitive knowledge about other multidimensional portions of $A$ that, due to privacy reasons, are hidden to unauthorized users. Furthermore, in our reference application scenario, target data cubes are also massive in size, so that data compression techniques are needed in order to efficiently evaluate queries, yet introducing *approximate answers* (e.g., References 22 and 23) having a certain *degree of approximation* that is perfectly tolerable for OLAP analysis goals [24]. In our proposal, the described application scenario with accuracy and privacy features is accomplished by means of the so-called *accuracy/privacy contract*, which determines the *accuracy/privacy constraint* under which client applications must access and process multidimensional data. In this contract, the data warehouse server and client OLAP applications play the role of mutual subscribers.

### 16.1.2 Chapter Organization

The remaining part of this chapter is organized as follows. In Section 16.2, we provide a comprehensive overview on actual proposals that focus on privacy-preserving OLAP over Big Data, both in centralized and distributed environments. In Section 16.3, we provide the fundamental definitions and formal tools of our proposed privacy-preserving OLAP framework. Section 16.4 contains results on managing overlapping query workloads defined over privacy-preserving OLAP data cubes. Section 16.5 defines metrics for modeling and measuring the accuracy of privacy-preserving OLAP data cubes, whereas Section 16.6 defines metrics for modeling and measuring the privacy of privacy-preserving OLAP data cubes as well. Section 16.7 introduces the accuracy and privacy thresholds of the proposed privacy-preserving OLAP framework. In Section 16.8, we focus our attention on so-called accuracy grids and multiresolution accuracy grids, a family of conceptual tools for handing accuracy and privacy of privacy-preserving OLAP data cubes. Section 16.9 contains the core algorithm of our proposed privacy-preserving OLAP framework. In Section 16.10, we provide a comprehensive experimental assessment and analysis of algorithms embedded in our proposed privacy-preserving OLAP framework. Finally, in Section 16.11 we derive conclusions from our research and set a basis for future efforts in the investigated area.

## 16.2 LITERATURE OVERVIEW AND SURVEY

In our proposed research, two contexts are relevant, namely, privacy-preserving OLAP in centralized environments and privacy-preserving OLAP in distributed environments. Here, we provide a survey on both areas.

### 16.2.1 Privacy-Preserving OLAP in Centralized Environments

Despite the above-discussed in Section 16.1 privacy-preserving issues in OLAP, today's OLAP server platforms lack effective countermeasures to face relevant-in-practice limitations deriving from privacy breaches. Contrary to this actual trend, privacy-preserving issues in statistical databases, which represent the theoretical foundations of privacy-preserving OLAP, have been deeply investigated during past years [25], and a relevant number of techniques developed in this context are still waiting to be studied, extended, and integrated within the core layer of OLAP server platforms. Basically, privacy-preserving techniques for statistical databases can be classified into two main classes: *restriction-based techniques* and *perturbation-based techniques*. The first ones propose restricting the number of classes of queries that can be posed to the target database (e.g., Reference 26); the second ones propose adding random noise at various levels of the target database, ranging from schemas [27] to query answers [28]. *Auditing query techniques* aims at devising intelligent methodologies for detecting *which* queries must be forbidden, in order to preserve privacy. Therefore, these approaches have been studied in the broader context of restriction-based privacy-preserving techniques. Recently, some proposals on auditing techniques in OLAP appeared. Among these proposals, noticeable ones are Reference 29, which makes use of an *information theoretic approach*, and Reference 30, which exploits *integer linear programming* (ILP) techniques. Nevertheless, due to different, specific motivations, both restriction-based and perturbation-based techniques are not effective and efficient in OLAP. Restriction-based techniques are quite ineffective in OLAP since the nature of OLAP analysis is intrinsically *interactive* and based on a *wide* set of operators and query classes. Perturbation-based techniques, which process one data cube cell at a time, are quite inefficient in OLAP since they introduce excessive computational overheads when executed on massive data cubes. It should be noted that the latter drawback derives from the lack of a proper notion of privacy OLAP, as highlighted in Section 16.1.1.

More recently, Reference 31 proposes a cardinality-based inference control scheme that aims at finding sufficient conditions for obtaining *safe data cubes*, that is, data cubes such that the number of known values is under a tight bound. In line with this research, Reference 32 proposes a privacy-preserving OLAP approach that combines access and inference control techniques [28]: (1) the first one is based on the hierarchical nature of data cubes modeled in terms of *cuboid lattices* [33] and multiresolution of data, and (2) the second one is based on *directly* applying *restriction* to coarser aggregations of data cubes and then *removing* remaining inferences that can be still derived. References 31 and 32 are not properly comparable with our work, as they basically combine a technique inspired from statistical databases with an access control scheme, which are both outside the scope of this chapter. Reference 34 extends results of Reference 32 via proposing the algorithm *FMC*, which still works on the cuboid lattice to hide sensitive data that cause inference. Finally, Reference 20 proposes a *random data distortion technique*, called *zero-sum method*, for preserving the privacy of data cells while providing accurate answers to range queries. To this end, Reference 20 *iteratively* alters the values of data cells of the target data cube in such a way as to maintain the *marginal sums* of data cells along rows and columns of the data cube equal to zero. According to motivations given in Section 16.1.1, when applied to massive data cubes, Reference 20 clearly introduces excessive overheads, which are not comparable with low computational requirements due to sampling-based techniques like ours. In addition to this, in our framework, we are interested in preserving the privacy of aggregate patterns rather than the one of data cells, which, however, can be still captured by introducing aggregate patterns at the coarser degree of aggregation of the input data cube, as stated in Section 16.1.1. In other words, Reference 20 does not introduce a proper notion of privacy OLAP but only restricts the analysis to the privacy of data cube cells. Despite this, we observe that, from the client-side perspective, Reference 20 (1) solves the same problem we investigate, that is, providing privacy-preserving (approximate) answers to OLAP queries against data cubes, and (2) contrary to References 32 and 35, adopts a "data-oriented" approach, which is similar in nature to ours. For these reasons, in our experimental analysis, we test the performance of our framework against the one of Reference 20, which, apart from being the state-of-the-art perturbation-based privacy-preserving OLAP technique, will be hereby considered as the comparison technique for testing the effectiveness of the privacy-preserving OLAP technique we propose.

### 16.2.2 Privacy-Preserving OLAP in Distributed Environments

While *privacy-preserving distributed data mining* has been widely investigated, and a plethora of proposals exists (e.g., References 36–40), the problem of effectively and efficiently supporting privacy-preserving OLAP over distributed collections of data repositories, which is relevant in practice, has been neglected so far.

Distributed privacy preservation techniques over OLAP data cubes solve the problem of making privacy-preserving distributed OLAP data cubes or, under an alternative interpretation, making a privacy-preserving OLAP data cube model over distributed data sources. Deriving problems is similar but different in nature. As regards the first problem, to the best of our knowledge, in the literature, there does not exist any proposal that deals with it, whereas concerning the second problem, Reference 35 is the state-of-the-art result existent in literature.

By looking at the active literature, while a plethora of initiatives focusing on privacy-preserving distributed data mining [41] exists, References 36–40 being some noticeable ones, to the best of our knowledge, only References 35, 42, and 43 deal with the yet-relevant problem of effectively and efficiently supporting privacy-preserving OLAP over distributed data sources, specifically falling in the second scientific context according to the provided taxonomy. Reference 35 defines a privacy-preserving OLAP model over data partitioned across multiple clients using a *randomization approach*, which is implemented by the so-called *retention replacement perturbation* algorithm, on the basis of which (1) clients perturb tuples with which they participate to the partition in order to gain *row-level privacy* and (2) a server is capable of evaluating OLAP queries against perturbed tables via *reconstructing* original distributions of attributes involved by such queries. In Reference 35, the authors demonstrate that the proposed distributed privacy-preserving OLAP model is safe against privacy breaches. Reference 42 is another distributed privacy-preserving OLAP approach that is reminiscent of ours. More specifically, Reference 42 pursues the idea of obtaining a privacy-preserving OLAP data cube model from *distributed data sources across multiple sites* via applying perturbation-based techniques on *aggregate data* that are retrieved from each singleton site as a baseline step of the main (distributed) OLAP computation task. Finally, Reference 43 focuses the attention on the significant issue of providing efficient data aggregation while preserving privacy over *wireless sensor networks*. The proposed solution is represented by two privacy-preserving data aggregation schemes that make use of innovative *additive aggregation functions*, these schemes being named *cluster-based private data aggregation* (CPDA) and *slice-mix-aggregate* (SMART). Proposed aggregation functions fully exploit topology and dynamics of the underlying wireless sensor network and bridge the gap between collaborative data collection over such networks and data privacy needs.

## 16.3 FUNDAMENTAL DEFINITIONS AND FORMAL TOOLS

A *data cube* $A$ defined over a relational data source $\mathcal{S}$ is a tuple $A = \langle D, \mathcal{F}, \mathcal{H}, \mathcal{M} \rangle$, such that (1) $D$ is the data domain of $A$ containing (OLAP) data cells, which are the basic aggregations of $A$ computed over relational tuples stored in $\mathcal{S}$; (2) $\mathcal{F}$ is the set of *dimensions* of $A$, that is, the *functional attributes* with respect to which the underlying OLAP analysis is defined (in other words, $\mathcal{F}$ is the set of attributes along which tuples in $\mathcal{S}$ are aggregated); (3) $\mathcal{H}$ is the set of *hierarchies* related to the dimensions of $A$, that is, hierarchical representations of the functional attributes shaped in the form of general trees; (4) $\mathcal{M}$ is the set of *measures* of $A$, that is, the *attributes of interest* for the underlying OLAP analysis (in other words, $\mathcal{M}$ is the set of attributes taken as argument of SQL aggregations whose results are stored in data cells of $A$). Given these definitions, (1) $|\mathcal{F}|$ denotes the number of dimensions of $A$, (2) $d \in \mathcal{F}$ a generic dimension of $A$, (3) $|d|$ the cardinality of $d$, and (4) $H(d) \in \mathcal{H}$ the hierarchy related to $d$. Finally, for the sake of simplicity, we assume that we are dealing with data cubes having a single measure (i.e., $|\mathcal{M}| = 1$). However, extending schemes, models, and algorithms proposed in this chapter to deal with data cubes having *multiple measures* (i.e., $|\mathcal{M}| > 1$) is straightforward.

Given an $|\mathcal{F}|$-dimensional data cube $A$, an *m-dimensional range query* $Q$ against $A$, with $m \leq |\mathcal{F}|$, is a tuple $Q = \langle R_{k_0}, R_{k_1}, \ldots, R_{k_{m-1}}, \mathcal{A} \rangle$, such that (1) $R_{k_i}$ denotes a *contiguous* range defined on the dimension $d_{k_i}$ of $A$, with $k_i$ belonging to the range $[0, |\mathcal{F}| - 1]$, and (2) $\mathcal{A}$ is an SQL aggregation operator. The evaluation of $Q$ over $A$ returns the $\mathcal{A}$-based aggregation computed over the set of data cells in $A$ contained within the multidimensional subdomain of $A$ bounded by the ranges $R_{k_0}, R_{k_1}, \ldots, R_{k_{m-1}}$ of $Q$. Range-SUM queries, which return the SUM of the involved data cells, are trendy examples of range queries. In our framework, we take into consideration range-SUM queries, as SUM aggregations are very popular in OLAP and efficiently support summarized knowledge extraction from massive amounts of multidimensional data as well as other SQL aggregations (e.g., COUNT, AVG, etc). Therefore, our framework can be straightforwardly extended to deal with other SQL aggregations different from SUM. However, the latter research aspect is outside the scope of this chapter and thus left as future work.

Given a query $Q$ against a data cube $A$, the *query region* of $Q$, denoted by $R(Q)$, is defined as the subdomain of $A$ bounded by the ranges $R_{k_0}, R_{k_1}, \ldots, R_{k_{m-1}}$ of $Q$.

Given an $m$-dimensional query $Q$, the accuracy grid $\mathcal{G}(Q)$ of $Q$ is a tuple $\mathcal{G}(Q) = \langle \Delta\ell_{k_0}, \Delta\ell_{k_1}, \ldots, \Delta\ell_{k_{m-1}} \rangle$, such that $\Delta\ell_{k_i}$ denotes the range partitioning $Q$ along the dimension $d_{k_i}$ of $A$, with $k_i$ belonging to $[0, |\mathcal{F}|-1]$, in a $\Delta\ell_{k_i}$-based (one-dimensional) partition. By combining the one-dimensional partitions along *all* the dimensions of $Q$, we finally obtain $\mathcal{G}(Q)$ as a *regular multidimensional partition* of $R(Q)$. From Section 16.1.1, recall that the elementary cell of the accuracy grid $\mathcal{G}(Q)$ is implicitly defined by subqueries of $Q$ belonging to the query workload $QWL$ against the target data cube. An example of an accuracy grid is depicted in Figure 16.1: Each elementary data cell corresponds to a subquery in $QWL$.

Based on the latter definitions, in our framework, we consider the broader concept of *extended range query* $Q^+$, defined as a tuple $Q^+ = \langle Q, \mathcal{G}(Q) \rangle$, such that (1) $Q$ is a "classical" range query, $Q = \langle R_{k_0}, R_{k_1}, \ldots, R_{k_{m-1}}, \mathcal{A} \rangle$, and (2) $\mathcal{G}(Q)$ is the accuracy grid associated with $Q$, $\mathcal{G}(Q) = \langle \Delta\ell_{k_0}, \Delta\ell_{k_1}, \ldots, \Delta\ell_{k_{m-1}} \rangle$, with the condition that each interval $\Delta\ell_{k_i}$ is defined on the *corresponding* range $R_{k_i}$ of the dimension $d_{k_i}$ of $Q$. For the sake of simplicity, here and in the remaining part of the chapter, we assume $Q \equiv Q^+$.

Given an $n$-dimensional data domain $D$, we introduce the *volume* of $D$, denoted by $||D||$, as follows: $||D|| = |d_0| \times |d_1| \times \ldots \times |d_{n-1}|$, such that $|d_i|$ is the cardinality of the dimension $d_i$ of $D$. This definition can also be extended to a multidimensional data cube $A$, thus introducing the volume of $A$, $||A||$, and to a multidimensional range query $Q$, thus introducing the volume of $Q$, $||Q||$.

Given a data cube $A$, a range query workload $QWL$ against $A$ is defined as a *collection* of (range) queries against $A$, as follows: $QWL = \{Q_0, Q_1, \ldots, Q_{|QWL|-1}\}$, with $R(Q_k) \subseteq R(A)\ \forall\ Q_k \in QWL$. An example query workload is depicted in Figure 16.1.

Given a query workload $QWL = \{Q_0, Q_1, \ldots, Q_{|QWL|-1}\}$, we say that $QWL$ is *nonoverlapping* if there do not exist two queries $Q_i$ and $Q_j$ belonging to $QWL$ such that $R(Q_i) \cap R(Q_j) \neq \emptyset$. Given a query workload $QWL = \{Q_0, Q_1, \ldots, Q_{|QWL|-1}\}$, we say that $QWL$ is *overlapping* if

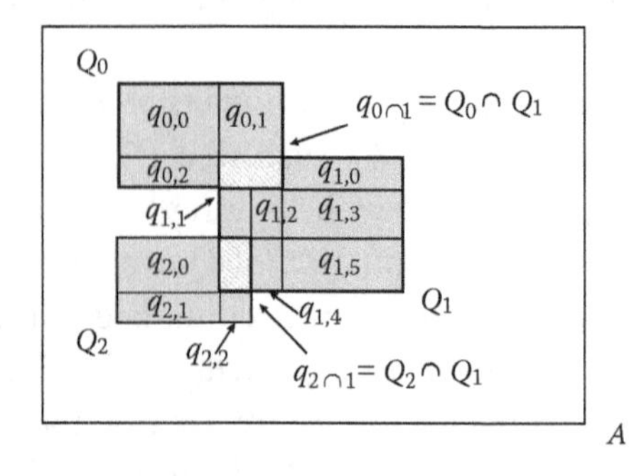


FIGURE 16.1 Building the nonoverlapping query workload (plain lines) from an overlapping query workload (bold lines).

there exist two queries $Q_i$ and $Q_j$ belonging to $QWL$ such that $R(Q_i) \cap R(Q_j) \neq \varnothing$. Given a query workload $QWL = \{Q_0, Q_1, \ldots, Q_{|QWL|-1}\}$, the *region set* of $QWL$, denoted by $R(QWL)$, is defined as the *collection* of regions of queries belonging to $QWL$, as follows: $R(QWL) = \{R(Q_0), R(Q_1), \ldots, R(Q_{|QWL|-1})\}$.

## 16.4 DEALING WITH OVERLAPPING QUERY WORKLOADS

When overlapping query workloads are considered, we adopt a *unifying strategy* that allows us to handle nonoverlapping and overlapping query workloads in the same manner. As we will better discuss in Section 16.9, this approach involves several benefits in all the phases of our privacy-preserving OLAP technique. In this respect, given an overlapping query workload $QWL$, our aim is to obtain a nonoverlapping query workload $QWL'$ from $QWL$ such that $QWL'$ is *equivalent* to $QWL$, that is, (1) $QWL'$ provides the *same information content* as $QWL$, and (2) $QWL$ can be totally reconstructed from $QWL'$. To this end, for each query $Q_i$ of $QWL$ without any intersection with other queries in $QWL$, we simply add $Q_i$ to $QWL'$ and remove $Q_i$ from $QWL$. Contrary to this, for each query $Q_j$ of $QWL$ having at least one non-null intersection with other queries in $QWL$, we (1) extract from $Q_j$ two new subsets of queries (defined next), denoted by $\tau_1(Q_j)$ and $\tau_2(Q_j)$; (2) add $\tau_1(Q_j)$ and $\tau_2(Q_j)$ to $QWL'$; and (3) remove $Q_j$ from $QWL$. Specifically, $\tau_1(Q_j)$ contains the subqueries of $Q_j$ defined by intersection regions of $Q_j$, and $\tau_2(Q_j)$ contains the subqueries of $Q_j$ defined by regions of $Q_j$ obtained via prolonging the ranges of subqueries in $\tau_1(Q_j)$ along the dimensions of $Q_j$. As an example, consider Figure 16.1, where the nonoverlapping query workload $QWL'$ extracted from an overlapping query workload $QWL = \{Q_0, Q_1, Q_2\}$ is depicted. Here, we have the following: (1) $\tau_1(Q_0) = \{q_{0\cap 1}\}$—note that $q_{0\cap 1} \equiv Q_0 \cap Q_1$; (2) $\tau_2(Q_0) = \{q_{0,0}, q_{0,1}, q_{0,2}\}$; (3) $\tau_1(Q_1) = \{q_{0\cap 1}, q_{2\cap 1}\}$—note that $q_{2\cap 1} \equiv Q_2 \cap Q_1$; (4) $\tau_2(Q_1) = \{q_{1,0}, q_{1,1}, q_{1,2}, q_{1,3}, q_{1,4}, q_{1,5}\}$; (5) $\tau_1(Q_2) = \{q_{2\cap 1}\}$; and (6) $\tau_2(Q_2) = \{q_{2,0}, q_{2,1}, q_{2,2}\}$. Therefore, $QWL' = \tau_1(Q_0) \cup \tau_2(Q_0) \cup \tau_1(Q_1) \cup \tau_2(Q_1) \cup \tau_1(Q_2) \cup \tau_2(Q_2)$.

## 16.5 METRICS FOR MODELING AND MEASURING ACCURACY

As accuracy metrics for answers to queries of the target query workload $QWL$, we make use of the *relative query error* between exact and approximate answers, which is a measure of quality for approximate query answering techniques in OLAP that is well recognized in literature (e.g., Reference 24).

Formally, given a query $Q_k$ of $QWL$, we denote as $A(Q_k)$ the exact answer to $Q_k$ (i.e., the answer to $Q_k$ evaluated over the original data cube $A$) and as $\tilde{A}(Q_k)$ the approximate answer to $Q_k$ (i.e., the answer to $Q_k$ evaluated over the synopsis data cube $A'$). Therefore, the relative query error $E_Q(Q_k)$ between $A(Q_k)$ and $\tilde{A}(Q_k)$ is defined as follows: $E_Q(Q_k) = \frac{|A(Q_k) - \tilde{A}(Q_k)|}{\max\{A(Q_k), 1\}}$.

$E_Q(Q_k)$ can be extended to the whole query workload $QWL$, thus introducing the *average relative query error* $\bar{E}_Q(QWL)$ that takes into account the contributions of relative query errors of all the queries $Q_k$ in $QWL$, each of them weighted by the volume of the query, $\|Q_k\|$, with respect to the whole volume of queries in $QWL$, that is, the *volume of* $QWL$, $\|QWL\|$. $\|QWL\|$ is defined as follows: $\|QWL\| = \sum_{k=0}^{|QWL'|-1} \|Q_k\|, Q_k \in QWL$.

Based on the previous definition of $||QWL||$, the average relative query error $\bar{E}_Q(QWL)$ for a given query workload $QWL$ can be expressed as a *weighted linear combination* of relative query errors $E_Q(Q_k)$ of all the queries $Q_k$ in $QWL$, as follows: $\bar{E}_Q(QWL) = \sum_{k=0}^{|QWL|-1} \frac{\|Q_k\|}{\|QWL\|} \cdot E_Q(Q_k)$, that is, $\bar{E}_Q(QWL) = \sum_{k=0}^{|QWL|-1} \frac{\|Q_k\|}{\sum_{j=0}^{|QWL|-1} \|Q_j\|} \cdot \frac{|A(Q_k) - \tilde{A}(Q_k)|}{\max\{A(Q_k), 1\}}$, under the constraint $\sum_{k=0}^{|QWL|-1} \frac{\|Q_k\|}{\|QWL\|} = 1$.

## 16.6 METRICS FOR MODELING AND MEASURING PRIVACY

Since we deal with the problem of ensuring the privacy preservation of OLAP aggregations, our privacy metrics take into consideration how sensitive knowledge can be discovered from aggregate data and try to limit this possibility. On a theoretical plane, this is modeled by the privacy OLAP notion introduced in Section 16.1.1.

To this end, we first study how sensitive aggregations can be discovered from the target data cube $A$. Starting from the knowledge about $A$ (e.g., range sizes, OLAP hierarchies, etc.) and the knowledge about a given query $Q_k$ belonging to the query workload $QWL$ [i.e., the volume of $Q_k$, $||Q_k||$ and the exact answer to $Q_k$, $A(Q_k)$], it is possible to infer knowledge about sensitive ranges of data contained within $R(Q_k)$. For instance, it is possible to derive the average value of the contribution throughout which each basic data cell of $A$ within $R(Q_k)$ contributes to $A(Q_k)$, which we name *singleton aggregation* $I(Q_k)$. $I(Q_k)$ is defined as follows: $I(Q_k) = \frac{A(Q_k)}{\|Q_k\|}$.

In order to provide a clearer practical example of privacy breaches deriving from the proposed singleton aggregation model in real-life OLAP application scenarios, consider a three-dimensional OLAP data cube $A$ characterized by the set of dimensions $\mathcal{F} = \{Region, Product, Time\}$ and the set of measures $\mathcal{M} = \{Sale\}$. Assume that $A$ stores data about sales performed in stores located in Italy. In the running example, for the sake of simplicity, consider again the SQL aggregation operator AVG. If malicious users know the AVG value of the range $R$ of $A$ : $R$ = ⟨*SouthItaly*, [*ElectricProducts* : *OfficeProducts*], [*2008:2009*]⟩, say AVG($R$) = 23,500K€, and the volume of $R$, say $||R||$ = 1000, they will easily infer that, during the time interval between the 2008 and 2009, each store located in South Italy has originated a volume of sales of about 23,500€ (suppose that data are uniformly distributed) for electric and office products. This sensitive knowledge, also combined with the knowledge about the hierarchies defined on dimensions of $A$, for example, *H*(*Region*): *Italy* ← {*North Italy*, *Central Italy*, *South Italy*, *Insular Italy*} ← … ← *South Italy* ← {*Abruzzo*, *Molise*, *Campania*, *Puglia*, *Basilicata*, *Calabria*} ← …, can allow malicious users to infer sensitive sale data about specific individual stores located in Calabria related to specific individual classes of products and specific individual days, just starting from aggregate values of sale data of stores located in the whole of Italy during 2008 and 2009 (!).

Coming back to the singleton aggregation model, it is easy to understand that, starting from the knowledge about $I(Q_k)$, it is possible to *progressively* discover aggregations of a larger range of data within $R(Q_k)$, rather than the one stored within the basic data cell, thus inferring sensitive knowledge that is even more useful. Also, by exploiting OLAP hierarchies and the well-known ROLL-UP operator, it is possible to discover aggregations of ranges of data at higher degrees of such hierarchies. It should be noted that the singleton aggregation model, $I(Q_k)$, indeed represents an *instance* of our privacy OLAP notion targeted to the problem of preserving the privacy of range-SUM queries (the focus of our chapter). As a consequence, $I(Q_k)$ is essentially based on the conventional SQL aggregation operator AVG. Despite this, the underlying theoretical model we propose is general enough to be straightforwardly extended to deal with more sophisticated privacy OLAP notion instances, depending on the particular class of OLAP queries considered. Without loss of generality, given a query $Q_k$ belonging to an OLAP query class $C$, in order to handle the privacy preservation of $Q_k$, we only need to define the formal expression of the related singleton aggregation $I(Q_k)$ (like the previous one for the specific case of range-SUM queries). Then, the theoretical framework we propose works in the same way.

Secondly, we study how OLAP client applications can discover sensitive aggregations from the knowledge about approximate answers and, similarly to the previous case, from the knowledge about data cube and query metadata. Starting from the knowledge about the synopsis data cube $A'$ and the knowledge about the answer to a given query $Q_k$ belonging to the query workload $QWL$, it is possible to derive an *estimation* on $I(Q_k)$, denoted by $\tilde{I}(Q_k)$, as follows: $\tilde{I}(Q_k)=\frac{\tilde{A}(Q_k)}{S(Q_k)}$, such that $S(Q_k)$ is the *number of samples* effectively extracted from $R(Q_k)$ to compute $A'$ (note that $S(Q_k) < ||Q_k||$). The relative difference between $I(Q_k)$ and $\tilde{I}(Q_k)$, named *relative inference error* and denoted by $E_I(Q_k)$, gives us metrics for the privacy of $\tilde{A}(Q_k)$, which is defined as follows: $E_I(Q_k)=\frac{|I(Q_k)-\tilde{I}(Q_k)|}{\max\{I(Q_k),1\}}$.

Indeed, while OLAP client applications are aware about the definition and metadata of both the target data cube and queries of the query workload $QWL$, the number of samples $S(Q_k)$ (for each query $Q_k$ in $QWL$) is not disclosed to them. As a consequence, in order to model this aspect of our framework, we introduce the *user-perceived singleton aggregation*, denoted by $\tilde{I}_U(Q_k)$, which is the *effective* singleton aggregation *perceived* by external applications based on the knowledge made available to them. $\tilde{I}_U(Q_k)$ is defined as follows: $\tilde{I}_U(Q_k)=\frac{\tilde{A}(Q_k)}{\|Q_k\|}$.

Based on $\tilde{I}_U(Q_k)$, we derive the definition of the *relative user-perceived inference error* $E_I^U(Q_k)$, as follows: $E_I^U(Q_k)=\frac{|I(Q_k)-\tilde{I}_U(Q_k)|}{\max\{I(Q_k),1\}}$.

Since $S(Q_k) < ||Q_k||$, it is trivial to demonstrate that $\tilde{I}_U(Q_k)$ provides a better estimation of the singleton aggregation of $Q_k$ than that provided by $\tilde{I}(Q_k)$, as $\tilde{I}_U(Q_k)$ is evaluated with respect to *all* the items contained within $R(Q_k)$ (i.e., $||Q_k||$), whereas $\tilde{I}(Q_k)$ is evaluated with respect to the effective number of samples extracted from $R(Q_k)$ [i.e., $S(Q_k)$]. In other words, $\tilde{I}_U(Q_k)$ is an *upper bound* for $\tilde{I}(Q_k)$. Therefore, in our framework, we consider $\tilde{I}(Q_k)$ to compute the synopsis data cube, whereas we consider $\tilde{I}_U(Q_k)$ to model inference issues on the OLAP client application side.

$E_I^U(Q_k)$ can be extended to the whole query workload $QWL$ by considering the *average relative inference error* $\bar{E}_I(QWL)$ that takes into account the contributions of relative inference errors $E_I(Q_k)$ of all the queries $Q_k$ in $QWL$. Similarly to what was done for the average relative query error $\bar{E}_Q(QWL)$, we model $\bar{E}_I(QWL)$ as follows: $\bar{E}_I(QWL) = \sum_{k=0}^{|QWL|-1} \frac{\|Q_k\|}{\|QWL\|} \cdot E_I(Q_k)$, that is, $\bar{E}_I(QWL) = \sum_{k=0}^{|QWL|-1} \frac{\|Q_k\|}{\sum_{j=0}^{|QWL|-1} \|Q_j\|} \cdot \frac{|I(Q_k) - \tilde{I}_U(Q_k)|}{\max\{I(Q_k),1\}}$, under the constraint $\sum_{k=0}^{|QWL|-1} \frac{\|Q_k\|}{\|QWL\|} = 1$.

Note that, $\bar{E}_I(QWL)$ is defined in dependence on $\tilde{I}_U(Q_k)$ rather than $\tilde{I}(Q_k)$. For the sake of simplicity, here and in the remaining part of the chapter, we assume $E_I(Q_k) \equiv E_I^U(Q_k)$.

The concepts and definitions allow us to introduce the *singleton aggregation privacy-preserving model* $\mathcal{X} = \langle I(\cdot), \tilde{I}(\cdot), \tilde{I}_U(\cdot) \rangle$, which is a fundamental component of the privacy-preserving OLAP framework we propose. $\mathcal{X}$ properly realizes our privacy OLAP notion.

Given a query $Q_k \in QWL$ against the target data cube $A$, in order to preserve the privacy of $Q_k$ under our privacy OLAP notion, we must *maximize the inference error* $E_I(Q_k)$ *while minimizing the query error* $E_Q(Q_k)$. While the definition of $E_Q(Q_k)$ can be reasonably considered as an *invariant* of our theoretical model, the definition of $E_I(Q_k)$ strictly depends on $\mathcal{X}$. Therefore, given a particular class of OLAP queries $C$, in order to preserve the privacy of queries of kind $C$, we only need to *appropriately* define $\mathcal{X}$. This nice amenity states that the privacy-preserving OLAP framework we propose is orthogonal to the particular class of queries considered and can be straightforwardly adapted to a large family of OLAP query classes.

## 16.7 ACCURACY AND PRIVACY THRESHOLDS

Similarly to related proposals that appeared in literature recently [20], in our framework, we introduce the accuracy threshold $\Phi_Q$ and the privacy threshold $\Phi_I$. $\Phi_Q$ and $\Phi_I$ give us an *upper bound* for the average relative query error $\bar{E}_Q(QWL)$ and a *lower bound* for the average relative inference error $\bar{E}_I(QWL)$ of a given query workload $QWL$ against the synopsis data cube $A'$. As stated in Section 16.1.1, $\Phi_Q$ and $\Phi_I$ allow us to meaningfully model and treat the accuracy/privacy constraint by means of rigorous mathematical/statistical models.

In our application scenario, $\Phi_Q$ and $\Phi_I$ are cooperatively negotiated by the data warehouse server and OLAP client applications. The issue of determining how to set these parameters is a nontrivial engagement. Intuitively enough, regarding the accuracy of answers, it is possible to (1) refer to the widely accepted *query error threshold* belonging to the interval [15, 20]% that, according to results of a plethora of research experiences in the context of approximate query answering techniques in OLAP (e.g., see Reference 44), represents the current state of the art and (2) use it as a baseline to trade off the parameter $\Phi_Q$. Regarding the privacy of answers, there are no immediate guidelines to be considered since privacy-preserving techniques for advanced data management (like OLAP) are relatively new; hence, we cannot refer to any widely accepted threshold like what happens with approximate query answering techniques. As a result, the parameter $\Phi_I$ can be set according to a *two-step approach* where *first* the accuracy constraint is accomplished in the dependence of $\Phi_Q$, and *then* $\Phi_I$ is *consequently* set by trying to maximize it (i.e., augmenting the privacy of answers) *as much as possible, thus following a best-effort approach.*

## 16.8 ACCURACY GRIDS AND MULTIRESOLUTION ACCURACY GRIDS: CONCEPTUAL TOOLS FOR HANDLING ACCURACY AND PRIVACY

From Sections 16.1.1 and 16.7, it follows that, in our framework, the synopsis data cube $A'$ stores OLAP aggregations satisfying the accuracy/privacy constraint with respect to queries of the target query workload *QWL*. Hence, computing $A'$ via sampling the input data cube $A$ is the most relevant task of the privacy-preserving OLAP framework we propose.

To this end, we adopt the *strategy of sampling query regions according to the partitioned representation defined by their accuracy grids.* This strategy is named *accuracy grid–constrained sampling.* On the basis of this strategy, *samples are extracted from cells of accuracy grids,* according to *a vision that considers the elementary cell of accuracy grids as the atomic unit of our reasoning.* This assumption is well founded under the evidence of noticing that, given a query $Q_k$ of *QWL* and the collection of its subqueries $q_{k,0}, q_{k,1}, \ldots, q_{k,m-1}$ defined by the accuracy grid $\mathcal{G}(Q_k)$ of $Q_k$, sampling the (sub)query regions $R(q_{k,0}), R(q_{k,1}), \ldots, R(q_{k,m-1})$ of $R(Q_k)$ allows us to (1) efficiently answer subqueries $q_{k,0}, q_{k,1}, \ldots, q_{k,m-1}$, as sampling is accuracy grid constrained, and at the same time, (2) efficiently answer the super-query $Q_k$, being the answer to $Q_k$ given by the summation of the answers to $q_{k,0}, q_{k,1}, \ldots, q_{k,m-1}$ (recall that we consider range-SUM queries). It is a matter of fact to note that the alternative solution of sampling the super-query $Q_k$ directly, which we name *region-constrained sampling,* would expose us to the flaw of being unable to efficiently answer the subqueries $q_{k,0}, q_{k,1}, \ldots, q_{k,m-1}$ of $Q_k$, since there could exist the risk of having (sub)regions of $Q_k$ characterized by *high density* of samples and (sub)regions of $Q_k$ characterized by *low density* of samples.

It is important to further highlight that, similarly to the privacy OLAP notion (see Section 16.7), the proposed sampling strategy depends on the particular class of OLAP queries considered, that is, range-SUM queries. If different OLAP queries must be handled, different sampling strategies must be defined accordingly.

Similarly to what was done with the accuracy of answers, we exploit amenities offered by accuracy grids to *accomplish the privacy constraint as well.* In other words, just like accuracy, privacy of answers is handled by means of the granularity of accuracy grids and still considering the elementary cell of accuracy grids as the atomic unit of our reasoning.

To become convinced of the benefits coming from our sampling strategy, consider Figure 16.2, where a data cube $A$ and a query $Q_k$ are depicted along with 63 samples (represented by blue points) extracted from $R(Q_k)$ by means of two different strategies: (1) accuracy grid–constrained sampling (Figure 16.2a) and (2) region-constrained sampling (Figure 16.2b). As shown in Figure 16.2, accuracy grid–constrained sampling allows us to avoid "favorite" regions in the synopsis data cube $A'$ being obtained, that is, regions for which the total amount of allocated space (similarly, the total number of extracted samples) is much greater than that of other regions in $A'$. It is easy to understand that the latter circumstance,

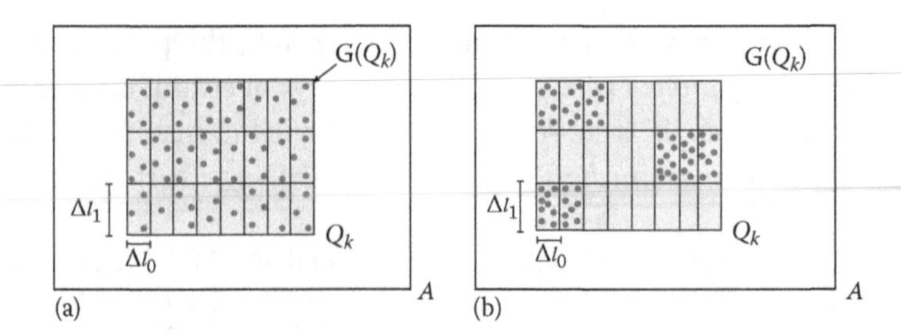


FIGURE 16.2 Two sampling strategies: (a) accuracy grid–constrained sampling and (b) region-constrained sampling.

which could be caused by the alternative strategy (i.e., region-constrained sampling), arbitrarily originates regions in $A'$ for which the accuracy error is low and the inference error is high (which is a desiderata in our framework), and regions for which the accuracy error is very high and the inference error is very low (which are both undesired effects in our framework). The final, global effect of such a scenario results in a limited capability of answering queries by satisfying the accuracy/privacy constraint. Contrary to the latter scenario, accuracy grid–constrained sampling aims at obtaining a *fair* distribution of samples across $A'$, so that a *large* number of queries against $A'$ can be accommodated by satisfying the accuracy/privacy constraint.

When overlapping query workloads are considered, intersection (query) regions pose the issue of *dealing with the overlapping of different accuracy grids*, which we name *multiresolution accuracy grids*, meaning that such grids partition the *same* intersection region of multiple queries by means of *cells at different granularities*. As much as granularities of such cells are different, we obtain "problematic" settings to be handled where subqueries have very *different volumes*, so that, due to geometrical issues, handling both accuracy and privacy of answers as well as dealing with the sampling phase become more questioning. It should be noted that, contrary to what happens with overlapping queries, accuracy grids of nonoverlapping queries originate subqueries having volumes that are equal one to another, so that we obtain facility at both modeling and reasoning tasks.

To overcome issues deriving from handling multiresolution accuracy grids, we introduce an innovative solution that consists in *decomposing nonoverlapping and overlapping query workloads in sets of appropriately selected subqueries*, thus achieving the amenity of treating both kinds of query workloads in a unified manner. The baseline operation of this process is represented by the decomposition of a query $Q_k$ of the target query workload $QWL$. Given an $I_{0,k} \times I_{1,k}$ query $Q_k$ and its accuracy grid $\mathcal{G}(Q_k) = \langle \Delta\ell_{0,k}, \Delta\ell_{1,k} \rangle$, the query decomposition process generates a set of subqueries $\zeta(Q_k)$ on the basis of the nature of $Q_k$. In more detail, if $Q_k$ is nonoverlapping, then $Q_k$ is decomposed in $\frac{I_{0,k}}{\Delta\ell_{0,k}} \cdot \frac{I_{1,k}}{\Delta\ell_{1,k}}$ subqueries given by cells in $\mathcal{G}(Q_k)$, such that each subquery has volume equal to $\Delta\ell_{0,k} \times \Delta\ell_{1,k}$. Otherwise, if $Q_k$ is overlapping, that is, there exists another query $Q_h$ in $QWL$ such that $R(Q_k) \cap R(Q_h) \neq \emptyset$, the query decomposition process works as follows: (1) The intersection region of $Q_k$ and $Q_h$, denoted by $R^I(Q_k,Q_h)$, is decomposed in the set of subqueries $\zeta(Q_k,Q_h)$ given by the overlapping of $\mathcal{G}(Q_k)$ and $\mathcal{G}(Q_h)$. (2) Let $\pi(Q_k)$ be the set of subqueries in $\zeta(Q_k)$ (computed considering $Q_k$ as nonoverlapping) that partially overlap $R^I(Q_k,Q_h)$, and $\pi(Q_h)$ be the analogous set for $Q_h$; $\pi(Q_k)$ and $\pi(Q_h)$ are decomposed in sets of subqueries obtained by considering portions of subqueries in $\pi(Q_k)$ and $\pi(Q_h)$ that are completely contained by the regions $R(Q_k) - R^I(Q_k,Q_h)$ and $R(Q_h) - R^I(Q_k,Q_h)$, respectively, thus obtaining the sets of *border queries* $\mu_1(Q_k)$ and $\mu_1(Q_h)$. (3) Let $\omega(Q_k)$ be the set of subqueries in $\zeta(Q_k)$ that are completely contained by the region $R(Q_k) - R^I(Q_k,Q_h) - \mu_1(Q_k)$, and $\omega(Q_h)$ be the analogous set for $Q_h$; $\omega(Q_k)$ and $\omega(Q_h)$ are decomposed in sets of subqueries given by $\mathcal{G}(Q_k)$ and $\mathcal{G}(Q_h)$, respectively, thus obtaining the sets $\mu_2(Q_k)$ and $\mu_2(Q_h)$. (4) Finally, the set of subqueries originated by the decomposition of the overlapping queries $Q_k$ and $Q_h$, denoted by $\zeta^I(Q_k,Q_h)$, is obtained as follows: $\zeta^I(Q_k,Q_h) = \zeta(Q_k,Q_h) \cup \mu_1(Q_k) \cup \mu_1(Q_h) \cup \mu_2(Q_k) \cup \mu_2(Q_h)$. Given a query workload *QWL*, *QWL* is decomposed by iteratively decomposing its queries $Q_0, Q_1, \ldots, Q_{|QWL|-1}$ according to the query decomposition process described for a given (singleton) query.

## 16.9 AN EFFECTIVE AND EFFICIENT ALGORITHM FOR COMPUTING SYNOPSIS DATA CUBES

From the Sections 16.3–16.8, it follows that our privacy-preserving OLAP technique, which is finally implemented by greedy algorithm `computeSynDataCube`, encompasses three main phases: (1) allocation of the input storage space *B*, (2) sampling of the input data cube *A*, and (3) refinement of the synopsis data cube *A′*. In this section, we present in detail these phases and finally conclude with algorithm `computeSynDataCube`.

### 16.9.1 Allocation Phase

Given the input data cube *A*, the target query workload *QWL*, and the storage space *B*, in order to compute the synopsis data cube *A′*, *the first issue to be considered is how to allocate B across query regions of QWL*. Given a query region $R(Q_k)$, allocating an amount of storage space to $R(Q_k)$, denoted by $B(Q_k)$, corresponds to assigning to $R(Q_k)$ a certain number of samples that can be extracted from $R(Q_k)$, denoted by $N(Q_k)$. To this end, during the allocation phase of algorithm `computeSynDataCube`, we *assign more samples to those query regions of QWL having skewed (i.e., irregular and asymmetric) data distributions (e.g., Zipf) and fewer samples to those query regions having uniform data distributions*. The idea underlying such an approach is that few samples are enough to "describe" uniform query regions, due to the fact that data distributions of such regions are "regular," whereas we need more samples to "describe" skewed query regions, due to the fact that data distributions of such regions are, contrary to the previous case, not "regular." Specifically, we face the deriving allocation problem by means of a *proportional storage space allocation scheme*, which allows us to efficiently allocate *B* across query regions of *QWL* via assigning a *fraction* of *B* to each region. This allocation scheme has been preliminarily proposed in Reference 45 for the different context of approximate query answering techniques for two-dimensional OLAP data cubes, and in this work, it is extended to deal with multidimensional data cubes and (query) regions.

First, if *QWL* is overlapping, we compute its corresponding nonoverlapping query workload *QWL′* (see Section 16.4). Hence, in both cases (i.e., *QWL* is overlapping and not), a set of regions $R(QWL) = \{R(Q_0), R(Q_1), \ldots, R(Q_{|QWL|1})\}$ is obtained. Let $R(Q_k)$ be a region belonging to $R(QWL)$; the amount of storage space allocated to $R(Q_k)$, $B(Q_k)$, is determined according to a proportional approach that considers (1) the nature of the data distribution of $R(Q_k)$ and geometrical issues of $R(Q_k)$ and (2) the latter parameters of $R(Q_k)$ in proportional comparison with the same parameters of all the regions in $R(QWL)$, as follows: $B(Q_k) = \left\lfloor \frac{\varphi(R(Q_k)) + \Psi(R(Q_k)) \cdot \xi(R(Q_k))}{\sum_{h=0}^{|QWL|-1} \varphi(R(Q_k)) + \sum_{h=0}^{|QWL|-1} \Psi(R(Q_k)) \cdot \xi(R(Q_k))} \cdot B \right\rfloor$, such that [45] (1) $\Psi(R)$ is a Boolean *characteristic function* that, given a region $R$, allows us to decide if data in $R$ are uniform or skewed; (2) $\varphi(R)$ is a factor that captures the *skewness* and the *variance* of $R$ in a combined manner; and (3) $\xi(R)$ is a factor that provides the ratio between the skewness of $R$ and its standard deviation, which, according to Reference 46, allows us to estimate the *skewness degree* of the data distribution of $R$. The previous formula can be extended to handle the overall allocation of $B$ across regions of $QWL$, thus achieving the formal definition of our proportional storage space allocation scheme, denoted by $\mathcal{W}(A, R(Q_0), R(Q_1), \ldots, R(Q_{|QWL|-1}), B)$, via the following system:

$$\begin{cases} B(Q_0) = \left\lfloor \dfrac{\varphi(R(Q_0)) + \Psi(R(Q_0)) \cdot \xi(R(Q_0))}{\sum_{k=0}^{|QWL|-1} \varphi(R(Q_k)) + \sum_{k=0}^{|QWL|-1} \Psi(R(Q_k)) \cdot \xi(R(Q_k))} \cdot B \right\rfloor \\ \ldots \\ B(Q_{|QWL|-1}) = \left\lfloor \dfrac{\varphi(R(Q_{|QWL|-1})) + \Psi(R(Q_{|QWL|-1})) \cdot \xi(R(Q_{|QWL|-1}))}{\sum_{k=0}^{|QWL|-1} \varphi(R(Q_k)) + \sum_{k=0}^{|QWL|-1} \Psi(R(Q_k)) \cdot \xi(R(Q_k))} \cdot B \right\rfloor \\ \sum_{k=0}^{|QWL|-1} B(Q_k) \leq B \end{cases} \tag{16.1}$$

In turn, for each query region $R(Q_k)$ of $R(QWL)$, we further allocate the amount of storage space $B(Q_k)$ across the subqueries of $Q_k$, $q_{k,0}, q_{k,1}, \ldots, q_{k,m-1}$, obtained by decomposing $Q_k$ according to our decomposition process (see Section 16.8), via using the *same* allocation scheme (Equation 16.1). Overall, this approach allows us to obtain a storage space allocation for each *subquery* $q_{k,i}$ of $QWL$ in terms of the maximum sample number $N(q_{k,i}) = \left\lfloor \frac{B(q_{k,i})}{32} \right\rfloor$ that can be extracted from $q_{k,i}$,* $B(q_{k,i})$ being the amount of storage space allocated to $q_{k,i}$.

* Here, we are assuming that an integer is represented in memory by using 32 bits.

It should be noted that the described approach allows us to achieve an extremely accurate level of detail in handling accuracy/privacy issues of the final synopsis data cube $A'$. To become convinced of this, recall that the granularity of OLAP client applications is *that one of queries* (see Section 16.1.1), which is *much greater* than that one of subqueries (specifically, the latter depends on the degree of accuracy grids) we use as the atomic unit of our reasoning. Thanks to this difference between granularity of input queries and accuracy grid cells, which, in our framework, is made "conveniently" high, we finally obtain a crucial *information gain* that allows us to efficiently accomplish the accuracy/privacy constraint.

### 16.9.2 Sampling Phase

Given an instance of our proportional allocation scheme (Equation 16.1), $\mathcal{W}$, during the second phase of algorithm `computeSynDataCube`, we sample the input data cube $A$ in order to obtain the synopsis data cube $A'$, in such a way as to satisfy the accuracy/privacy constraint with respect to the target query workload $QWL$. To this end, we apply a different strategy in dependence on the fact that query regions characterized by uniform or skewed distributions are handled, according to similar insights that have inspired our allocation technique (see Section 16.9.1). Specifically, for a skewed region $R(q_{k,i})$, given the maximum number of samples that can be extracted from $R(q_{k,i})$, $N(q_{k,i})$, we *sample the* $N(q_{k,i})$ *outliers of* $q_k$. It is worth noticing that, for skewed regions, the *sum of outliers represents an accurate estimation of the sum of all the data cells contained within such regions*. Also, it should be noted that this approach allows us to gain advantages with respect to approximate query answering as well as the privacy preservation of sensitive ranges of multidimensional data of skewed regions. Contrary to this, for a uniform region $R(q_{k,i})$, given the maximum number of samples that can be extracted from $R(q_{k,i})$, $N(q_{k,i})$, let (1) $\bar{C}_{R(q_{k,i})}$ be the average of values of data cells contained within $R(q_{k,i})$; (2) $\mathcal{U}(R(q_{k,i}),\bar{C}_{R(q_{k,i})})$ be the set of data cells $C$ in $R(q_{k,i})$ such that $value(C) > \bar{C}_{R(q_{k,i})}$, where $value(C)$ denotes the value of $C$; and (3) $\bar{C}^{\uparrow}_{R(q_{k,i})}$ be the average of values of data cells in $\mathcal{U}(R(q_{i,k}),\bar{C}_{R(q_{k,i})})$. We adopt the strategy of extracting $N(q_{k,i})$ samples from $R(q_{k,i})$ by selecting them as the $N(q_{k,i})$ *closer-to-*$\bar{C}^{\uparrow}_{R(q_{k,i})}$ *data cells* $C$ *in* $R(q_{k,i})$ *such that value* $(C) > \bar{C}_{R(q_{k,i})}$. Just like previous considerations given for skewed regions, it should be noted that the described sampling strategy for uniform regions allows us to meaningfully trade off the need for efficiently answering range-SUM queries against the synopsis data cube and the need for limiting the number of samples to be stored within the synopsis data cube.

In order to satisfy the accuracy/privacy constraint, the sampling phase aims at accomplishing (decomposed) accuracy and privacy constraints *separately*, based on a two-step approach. Given a query region $R(Q_k)$, we *first* sample $R(Q_k)$ in such a way as to satisfy the accuracy constraint, and *then*, we check if samples extracted from $R(Q_k)$ *also* satisfy, beyond the accuracy constraint, the privacy constraint. As mentioned in Section 16.7, this strategy follows a best-effort approach aiming at minimizing computational overheads due to computing the synopsis data cube, and it is also the conceptual basis of guidelines for setting the thresholds $\Phi_Q$ and $\Phi_I$.

Moreover, our sampling strategy aims at obtaining a *tunable* representation of the synopsis data cube $A'$, which can be *progressively refined* until the accuracy/privacy constraint is satisfied as much as possible. This means that, given the input data cube $A$, we first sample $A$ in order to obtain the *current* representation of $A'$. If such a representation satisfies the accuracy/privacy constraint, then the *final* representation of $A'$ is achieved and used at query time to answer queries instead of $A$. Otherwise, if the current representation of $A'$ does not satisfy the accuracy/privacy constraint, then we perform "corrections" on the current representation of $A'$, thus refining such representation in order to obtain a final representation that satisfies the constraint, on the basis of a best-effort approach. What we call the *refinement process* (described in Section 16.9.3) is based on a greedy approach that *"moves"* samples from regions of QWL whose queries satisfy the accuracy/privacy constraint to regions of QWL whose queries do not satisfy the constraint, yet ensuring that the former do not violate the constraint.*

Given a query $Q_k$ of the target query workload $QWL$, we say that $Q_k$ satisfies the accuracy/privacy constraint iff the following inequalities simultaneously hold: $\begin{cases} E_Q(Q_k) \leq \Phi_Q \\ E_I(Q_k) \geq \Phi_I \end{cases}$.

In turn, given a query workload $QWL$, we decide about its *satisfiability* with respect to the accuracy/privacy constraint by inspecting the satisfiability of queries that compose $QWL$. Therefore, we say that $QWL$ satisfies the accuracy/privacy constraint iff the following inequalities simultaneously hold: $\begin{cases} \overline{E}_Q(QWL) \leq \Phi_Q \\ \overline{E}_I(QWL) \geq \Phi_I \end{cases}$.

Given the target query workload $QWL$, the criterion of our greedy approach used during the refinement process is the *minimization* of the average relative query error, $\overline{E}_Q(QWL)$, and the *maximization* of the average relative inference error, $\overline{E}_I(QWL)$, within the *minimum* number of movements that allows us to accomplish both the goals simultaneously [i.e., minimizing $\overline{E}_Q(QWL)$ and maximizing $\overline{E}_I(QWL)$]. Furthermore, the refinement process is bounded by a *maximum occupancy of samples moved across queries of* $QWL$, which we name *total buffer size* and denote as $\mathcal{L}_{A',QWL}$. $\mathcal{L}_{A',QWL}$ depends on several parameters, such as the size of the buffer, the number of sample pages moved at each iteration, the overall available swap memory, and so forth.

### 16.9.3 Refinement Phase

In the refinement process, the third phase of algorithm `computeSynDataCube`, given the current representation of $A'$ that does *not* satisfy the accuracy/privacy constraint with respect to the target query workload $QWL$, we try to obtain an alternative representation of $A'$ that satisfies the constraint, according to a best-effort approach. To this end, the refinement process encompasses the following steps: (1) sort queries in $QWL$ according to their "distance" from the satisfiability condition, thus obtaining the ordered query set $QWL^{\mathcal{P}}$; (2) select from $QWL^{\mathcal{P}}$ a pair of queries $Q^{\mathcal{T}}$ and $Q^{\mathcal{F}}$ such that (*ii.j*) $Q^{\mathcal{T}}$ is the query of $QWL^{\mathcal{P}}$ having the *greater positive distance* from the satisfiability condition, that is, $Q^{\mathcal{T}}$ is the query of $QWL^{\mathcal{P}}$ that has the greater *surplus* of samples that can be moved toward queries in $QWL^{\mathcal{P}}$ that do

* In Section 16.9.3, we describe in detail the meaning of "moving" samples between query regions.

not satisfy the satisfiability condition, and (*ii.jj*) $Q^{\mathcal{F}}$ is the query of $QWL^{\mathcal{P}}$ having the *greater negative distance* from the satisfiability condition, that is, $Q^{\mathcal{F}}$ is the query of $QWL^{\mathcal{P}}$ that is in most need of new samples; (3) move enough samples from $Q^{\mathcal{T}}$ to $Q^{\mathcal{F}}$ in such a way as to satisfy the accuracy/privacy constraint on $Q^{\mathcal{F}}$ while, at the same time, ensuring that $Q^{\mathcal{T}}$ does not violate the constraint; and (4) repeat steps 1, 2, and 3 until the current representation of $A'$ satisfies, as much as possible, the accuracy/privacy constraint with respect to $QWL$, within the maximum number of iterations bounded by $\mathcal{L}_{A',QWL}$. Regarding step 3, moving $\rho$ samples from $Q^{\mathcal{T}}$ to $Q^{\mathcal{F}}$ means (1) removing $\rho$ samples from $R(Q^{\mathcal{T}})$, thus obtaining an *additional* space, named $B(\rho)$; (2) allocating $B(\rho)$ to $R(Q^{\mathcal{F}})$; and (3) resampling $R(Q^{\mathcal{F}})$ by considering the additional number of samples that have become available—in practice, this means extracting from $R(Q^{\mathcal{F}})$ further $\rho$ samples.

Let $S^*(Q_k)$ be the number of samples of a query $Q_k \in QWL$ satisfying the accuracy/privacy constraint. From the formal definitions of $E_Q(Q_k)$ (see Section 16.5), $I(Q_k)$, $\tilde{I}(Q_k)$, and $E_I(Q_k)$ (see Section 16.6), and the satisfiability condition, it could be easily demonstrated that $S^*(Q_k)$ is given by the following formula: $S^*(Q_k) = \frac{(1-\Phi_Q)}{(1-\Phi_I)} \cdot \|Q_k\|$.

Let $S_{\text{eff}}(Q^{\mathcal{F}})$ and $S_{\text{eff}}(Q^{\mathcal{T}})$ be the numbers of samples *effectively* extracted from $R(Q^{\mathcal{F}})$ and $R(Q^{\mathcal{T}})$, respectively, during the previous sampling phase. Note that $S_{\text{eff}}(Q^{\mathcal{F}}) < S^*(Q^{\mathcal{F}})$ and $S_{\text{eff}}(Q^{\mathcal{T}}) \geq S^*(Q^{\mathcal{T}})$. It is easy to prove that the number of samples to be moved from $Q^{\mathcal{T}}$ to $Q^{\mathcal{F}}$ such that $Q^{\mathcal{F}}$ satisfies the accuracy/privacy constraint and $Q^{\mathcal{T}}$ does not violate the constraint, denoted by $S_{\text{mov}}(Q^{\mathcal{T}}, Q^{\mathcal{F}})$, is finally given by the following formula: $S_{\text{mov}}(Q^{\mathcal{T}}, Q^{\mathcal{F}}) = S^*(Q^{\mathcal{F}}) - S_{\text{eff}}(Q^{\mathcal{F}})$, under the constraint $S_{\text{mov}}(Q^{\mathcal{T}}, Q^{\mathcal{F}}) < S_{\text{eff}}(Q^{\mathcal{T}}) - S^*(Q^{\mathcal{T}})$.

Without going into detail, it is possible to demonstrate that, given (1) an *arbitrary* data cube $A$, (2) an *arbitrary* query workload $QWL$, (3) an arbitrary pair of thresholds $\Phi_Q$ and $\Phi_I$, and (4) an *arbitrary* storage space $B$, it is not always possible to make $QWL$ satisfiable via the refinement process. From this evidence, our idea of using a best-effort approach makes sense perfectly.

### 16.9.4 The computeSynDataCube Algorithm

Main greedy algorithm `computeSynDataCube` is described by the pseudocode listed in Figure 16.3, wherein (1) `allocateStorageSpace` implements the proportional storage space allocation scheme (Equation 16.1), (2) `sampleDataCube` implements the sampling strategy, (3) `check` tests the satisfiability of the target query workload against the current representation of the synopsis data cube, and (4) `refineSynDataCube` is in charge of refining the current representation of the synopsis data cube.

```
computeSynDataCube(A,QWL,Φ_Q,Φ_I,B,L_A',QWL)
W ← allocateStorageSpace(A,QWL,B)
A' ← sampleDataCube(A,QWL,W)
while (!check(A',QWL,Φ_Q,Φ_I) && L_A',QWL > 0){
⟨A',currSwapMemorySize⟩ ← refineSynDataCube(A',QWL,Φ_Q,Φ_I,W,L_A',QWL)
L_A',QWL ← L_A',QWL – currSwapMemorySize
}
return A'
```

FIGURE 16.3 Algorithm `computeSynDataCube`.

## 16.10 EXPERIMENTAL ASSESSMENT AND ANALYSIS

In order to test the effectiveness of our framework throughout studying the performance of algorithm `computeSynDataCube`, we conducted an experimental evaluation where we tested how the relative query error (similarly, the accuracy of answers) and the relative inference error (similarly, the privacy of answers) due to the evaluation of populations of randomly generated queries, which model query workloads of our framework, over the synopsis data cube range with respect to the volume of queries. The latter is a relevant parameter costing computational requirements of any query processing algorithm (also referred as *selectivity*—e.g., Reference 44). According to motivations given in Section 16.2.1, we considered the zero-sum method [20] as the comparison technique.

In our experimental assessment, we engineered three classes of two-dimensional data cubes: synthetic, benchmark, and real-life data cubes. For all these data cubes, we limited the cardinalities of both dimensions to a threshold equal to 1000, which represents a reliable value modeling significant OLAP applications (e.g., Reference 44). In addition to this, data cubes of our experimental framework expose different *sparseness coefficients s*, which measures the percentage number of non-null data cells with respect to the total number of data cells of a data cube. As has been widely known since early experiences in OLAP research [47], the sparseness coefficient holds a critical impact on every data cube processing technique, thus including privacy-preserving data cube computation as well.

In particular, synthetic data cubes store two kinds of data: uniform data and skewed data, the latter being obtained by means of a Zipf distribution [48]. The benchmark data cube we considered has been built from the *TPC-H* data set [49], the real-life one from the *Forest CoverType* (FCT) data set [50]. Both data sets are well known in the data warehousing and OLAP research community. The final sparseness of the TPC-H and FCT data cube, respectively, has been easily *artificially* determined within the same OLAP data cube aggregation routine. The benefits derived from using different kinds of data cubes are manifold, among which we recall the following: (1) The algorithm can be tested against *different* data distributions, thus stressing the reliability of the collection of techniques we propose (i.e., allocation, sampling, refinement), which, as described in Section 16.9, inspect the nature of input data to compute the final synopsis data cube. (2) Parameters of data distributions characterizing the data cubes can be controlled easily, thus obtaining a reliable experimental evaluation. Selectivity of queries has been modeled in terms of a percentage value of the overall volume of synthetic data cubes, and for each experiment, we considered queries with selectivity increasing in size, in order to stress our proposed techniques under the ranging of an increasing input.

Regarding compression issues, we imposed a *compression ratio r*, which measures the percentage occupancy of the synopsis data cube $A'$, $size(A')$, with respect to the occupancy of the input data cube $A$, $size(A)$, equal to 20%, which is a widely accepted threshold for data cube compression techniques (e.g., Reference 44). To simplify, we set the accuracy and privacy thresholds in such a way as not to trigger the refinement process. This is also because Reference 20 does not support any "dynamic" computational feature (e.g., tuning of the quality of the random data distortion technique), so that it would have been particularly difficult to compare the two techniques under completely different experimental settings. On the other hand, this aspect puts in evidence the innovative characteristics of our privacy-preserving OLAP technique with respect to Reference 20, which is indeed a state-of-the-art proposal in perturbation-based privacy-preserving OLAP techniques.

Figure 16.4 shows experimental results concerning relative query errors of synopsis data cubes built from uniform, skewed, TPC-H, and FCT data, and for several values of $s$. Figure 16.5 shows instead the results concerning relative inference errors on the same data cubes. In both figures, our approach is labeled as G, whereas the approach in Reference 20, is labeled as Z. Obtained experimental results confirm the effectiveness of our algorithm, also in comparison with Reference 20, according to the following considerations. First, relative query and inference errors decrease as selectivity of queries increases, that is, the accuracy of answers increases and the privacy of answers decreases as selectivity of queries increases. This is because the more the data cells involved by a given query $Q_k$, the more the samples extracted from $R(Q_k)$ are able to "describe" the original data distribution of $R(Q_k)$ (this also depends on the proportional storage space allocation scheme [Equation 16.1]), so that accuracy increases. At the same time, more samples cause a decrease in privacy, since they provide *accurate* singleton aggregations and, as a consequence, the inference error decreases. Secondly, when $s$ increases, we observe a higher query error (i.e., accuracy of answers decreases) and a higher inference error (i.e., privacy of answers increases). In other

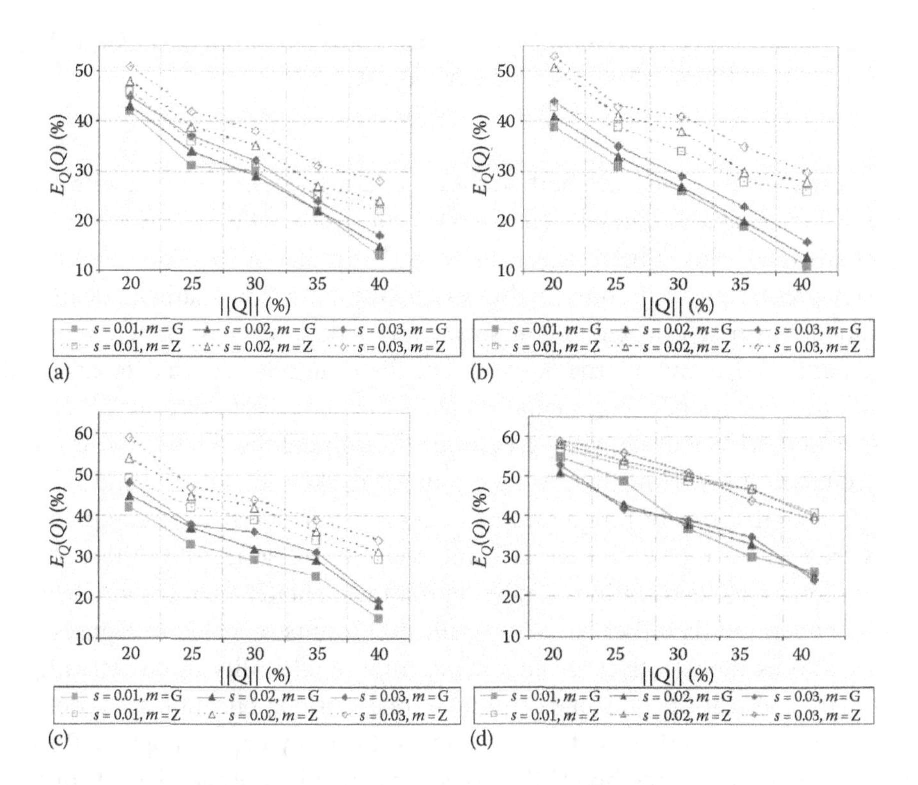


FIGURE 16.4 Relative query errors of synopsis data cubes built from (a) uniform, (b) skewed, (c) TPC-H, and (d) FCT data cubes for several values of $s$ ($r$ = 20%).

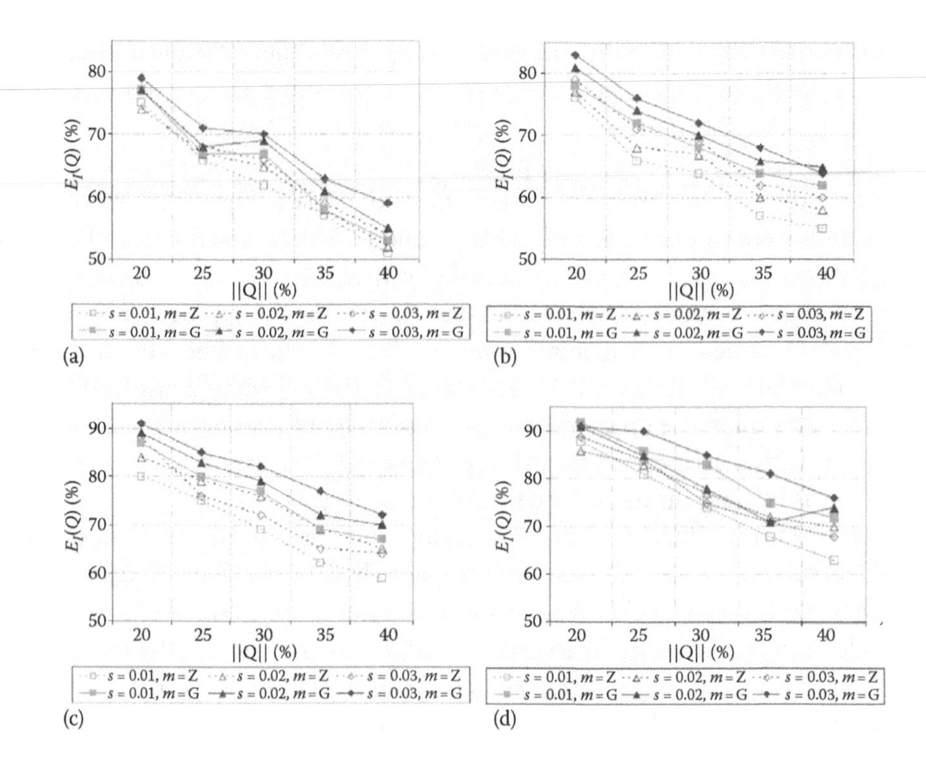


FIGURE 16.5 Relative inference errors of synopsis data cubes built from (a) uniform, (b) skewed, (c) TPC-H, and (d) FCT data cubes for several values of $s$ ($r = 20\%$).

words, data sparseness influences both accuracy and privacy of answers, with a negative effect in the first case (i.e., accuracy of answers) and a positive effect in the second case (i.e., privacy of answers). This is because, similarly to the results of Reference 20, we observe that privacy-preserving techniques, being essentially based on mathematical/statistical models and tools, *strongly* depend on the sparseness of data, since the latter, in turn, influences the *nature* and, above all, the *shape* of data distributions kept in databases and data cubes. Both these pieces of experimental evidence further corroborate our idea of trading off accuracy and privacy of OLAP aggregations to compute the final synopsis data cube. Also, by comparing experimental results on uniform, skewed, TPC-H, and FCT (input) data, we observe that our technique works better on uniform data, as expected, while it decreases the performance on benchmark and real-life data gracefully. This is due to the fact that uniform data distributions can be approximated better than skewed, benchmark, and real-life ones. On the other hand, experimental results reported in Figures 16.4 and 16.5 confirm to us the effectiveness and, above all, the reliability of our technique even on benchmark and real-life data one can find in real-world application scenarios. Finally, Figures 16.4 and 16.5 clearly state that our proposed privacy-preserving OLAP technique outperforms the zero-sum method [20]. This achievement is another relevant contribution of our research.

## 16.11 CONCLUSIONS AND FUTURE WORK

In this chapter, we have discussed principles and fundamentals of privacy-preserving OLAP over Big Data, both in centralized and distributed environments, and we have proposed a complete framework for efficiently supporting privacy-preserving OLAP aggregations on data cubes, along with its experimental assessment. We rigorously presented theoretical foundations as well as intelligent techniques for processing data cubes and queries, and algorithms for computing the final synopsis data cube whose aggregations balance, according to a best-effort approach, accuracy and privacy of retrieved answers. An experimental evaluation conducted on several classes of data cubes has clearly demonstrated the benefits deriving from the privacy-preserving OLAP technique we propose, also in comparison with a state-of-the-art proposal.

Future work is mainly oriented toward extending the actual capabilities of our framework in order to encompass the following features: (1) *intelligent update management techniques* (e.g., what happens when a query workload's characteristics change dynamically over time), perhaps inspired by well-known principles of *self-tuning databases*; (2) assuring the robustness of the framework with respect to *coalitions of attackers*; (3) integrating the proposed framework within the core layer of *next-generation data warehousing and data mining server systems*; and (4) building useful real-life case studies in emerging application scenarios (e.g., analytics over complex data [51,52]).

## REFERENCES


1. C. Wu and Y. Guo, "Enhanced User Data Privacy with Pay-by-Data Model," in: *Proceedings of BigData Conference*, 53–57, 2013.
2. M. Jensen, "Challenges of Privacy Protection in Big Data Analytics," in: *Proceedings of BigData Congress*, 235–238, 2013.
3. M. Li et al., "MyCloud: Supporting User-Configured Privacy Protection in Cloud Computing," in: *Proceedings of ACSAC*, 59–68, 2013.
4. S. Betgé-Brezetz et al., "End-to-End Privacy Policy Enforcement in Cloud Infrastructure," in: *Proceedings of CLOUDNET*, 25–32, 2013.
5. M. Weidner, J. Dees and P. Sanders, "Fast OLAP Query Execution in Main Memory on Large Data in a Cluster," in: *Proceedings of BigData Conference*, 518–524, 2013.
6. A. Cuzzocrea, R. Moussa and G. Xu, "OLAP*: Effectively and Efficiently Supporting Parallel OLAP over Big Data," in: *Proceedings of MEDI*, 38–49, 2013.
7. A. Cuzzocrea, L. Bellatreche and I.-Y. Song, "Data Warehousing and OLAP Over Big Data: Current Challenges and Future Research Directions," in: *Proceedings of DOLAP*, 67–70, 2013.
8. A. Cuzzocrea, "Analytics over Big Data: Exploring the Convergence of Data Warehousing, OLAP and Data-Intensive Cloud Infrastructures," in: *Proceedings of COMPSAC*, 481–483, 2013.
9. A. Cuzzocrea, I.-Y. Song and K.C. Davis, "Analytics over Large-Scale Multidimensional Data: The Big Data Revolution!," in: *Proceedings of DOLAP*, 101–104, 2011.
10. A. Cuzzocrea and V. Russo, "Privacy Preserving OLAP and OLAP Security," J. Wang (ed.), *Encyclopedia of Data Warehousing and Mining*, 2nd ed., IGI Global, Hershey, PA 1575–1581, 2009.
11. A. Cuzzocrea and D. Saccà, "Balancing Accuracy and Privacy of OLAP Aggregations on Data Cubes," in: *Proceedings of the 13th ACM International Workshop on Data Warehousing and OLAP*, 93–98, 2010.

12. X. Chen et al., "OWL Reasoning over Big Biomedical Data," in: *Proceedings of BigData Conference*, 29–36, 2013.
13. M. Paoletti et al., "Explorative Data Analysis Techniques and Unsupervised Clustering Methods to Support Clinical Assessment of Chronic Obstructive Pulmonary Disease (COPD) Phenotypes," *Journal of Biomedical Informatics 42(6)*, 1013–1021, 2009.
14. Y.-W. Cheah et al., "Milieu: Lightweight and Configurable Big Data Provenance for Science," in: *Proceedings of BigData Congress*, 46–53, 2013.
15. A.G. Erdman, D.F. Keefe and R. Schiestl, "Grand Challenge: Applying Regulatory Science and Big Data to Improve Medical Device Innovation," *IEEE Transactions on Biomedical Engineering 60(3)*, 700–706, 2013.
16. D. Cheng et al., "Tile Based Visual Analytics for Twitter Big Data Exploratory Analysis," in: *Proceedings of BigData Conference*, 2–4, 2013.
17. N. Ferreira et al., "Visual Exploration of Big Spatio-Temporal Urban Data: A Study of New York City Taxi Trips," *IEEE Transactions on Visualization and Computer Graphics 19(12)*, 2149–2158, 2013.
18. L. Sweeney, "k-Anonymity: A Model for Protecting Privacy," *International Journal on Uncertainty Fuzziness and Knowledge-based Systems 10(5)*, 557–570, 2002.
19. A. Machanavajjhala et al., "L-diversity: Privacy beyond k-Anonymity," *ACM Transactions on Knowledge Discovery from Data 1(1)*, art. no. 3, 2007.
20. S.Y. Sung et al., "Privacy Preservation for Data Cubes," *Knowledge and Information Systems 9(1)*, 38–61, 2006.
21. J. Han et al., "Efficient Computation of Iceberg Cubes with Complex Measures," in: *Proceedings of ACM SIGMOD*, 1–12, 2001.
22. A. Cuzzocrea and W. Wang, "Approximate Range-Sum Query Answering on Data Cubes with Probabilistic Guarantees," *Journal of Intelligent Information Systems 28(2)*, 161–197, 2007.
23. A. Cuzzocrea and P. Serafino, "LCS-Hist: Taming Massive High-Dimensional Data Cube Compression," in: *Proceedings of the 12th International Conference on Extending Database Technology*, 768–779, 2009.
24. A. Cuzzocrea, "Overcoming Limitations of Approximate Query Answering in OLAP," in: *IEEE IDEAS*, 200–209, 2005.
25. N.R. Adam and J.C. Wortmann, "Security-Control Methods for Statistical Databases: A Comparative Study," *ACM Computing Surveys 21(4)*, 515–556, 1989.
26. F.Y. Chin and G. Ozsoyoglu, "Auditing and Inference Control in Statistical Databases," *IEEE Transactions on Software Engineering 8(6)*, 574–582, 1982.
27. J. Schlorer, "Security of Statistical Databases: Multidimensional Transformation," *ACM Transactions on Database Systems 6(1)*, 95–112, 1981.
28. D.E. Denning and J. Schlorer, "Inference Controls for Statistical Databases," *IEEE Computer 16(7)*, 69–82, 1983.
29. N. Zhang, W. Zhao and J. Chen, "Cardinality-Based Inference Control in OLAP Systems: An Information Theoretic Approach," in: *Proceedings of ACM DOLAP*, 59–64, 2004.
30. F.M. Malvestuto, M. Mezzini and M. Moscarini, "Auditing Sum-Queries to Make a Statistical Database Secure," *ACM Transactions on Information and System Security 9(1)*, 31–60, 2006.
31. L. Wang, D. Wijesekera and S. Jajodia, "Cardinality-based Inference Control in Data Cubes," *Journal of Computer Security 12(5)*, 655–692, 2004.
32. L. Wang, S. Jajodia and D. Wijesekera, "Securing OLAP Data Cubes against Privacy Breaches," in: *Proceedings of IEEE SSP*, 161–175, 2004.
33. J. Gray et al., "Data Cube: A Relational Aggregation Operator Generalizing Group-By, Cross-Tab, and Sub-Totals," *Data Mining and Knowledge Discovery 1(1)*, 29–53, 1997.
34. M. Hua et al., "FMC: An Approach for Privacy Preserving OLAP," in: *Proceedings of the 7th International Conference on Data Warehousing and Knowledge Discovery, LNCS Vol. 3589*, 408–417, 2005.

35. R. Agrawal, R. Srikant and D. Thomas, "Privacy-Preserving OLAP," in: *Proceedings of the 2005 ACM International Conference on Management of Data*, 251–262, 2005.
36. J. Vaidya and C. Clifton, "Privacy Preserving Association Rule Mining in Vertically Partitioned Data," in: *Proceedings of the 8th ACM International Conference on Knowledge Discovery and Data Mining*, 639–644, 2002.
37. M. Kantarcioglu and C. Clifton, "Privacy-Preserving Distributed Mining of Association Rules on Horizontally Partitioned Data," *IEEE Transactions on Knowledge and Data Engineering 16(9)*, 1026–1037, 2004.
38. J. Vaidya and C. Clifton, "Privacy-Preserving K-Means Clustering over Vertically Partitioned Data," in: *Proceedings of the 9th ACM International Conference on Knowledge Discovery and Data Mining*, 206–215, 2003.
39. G. Jagannathan, K. Pillaipakkamnatt and R. Wright, "A New Privacy-Preserving Distributed K-Clustering Algorithm," in: *Proceedings of the 2006 SIAM International Conference on Data Mining*, 492–496, 2006.
40. G. Jagannathan and R. Wright, "Privacy-Preserving Distributed K-Means Clustering over Arbitrarily Partitioned Data," in: *Proceedings of the 11th ACM International Conference on Knowledge Discovery and Data Mining*, 593–599, 2002.
41. C. Clifton et al., "Tools for Privacy Preserving Distributed Data Mining," *SIGKDD Explorations 4(2)*, 28–34, 2002.
42. Y. Tong et al., "Privacy-Preserving OLAP based on Output Perturbation Across Multiple Sites," in: *Proceedings of the 2006 International Conference on Privacy, Security and Trust, AICPS Vol. 380*, 46, 2006.
43. W. He et al., "PDA: Privacy-Preserving Data Aggregation in Wireless Sensor Networks," in: *Proceedings of the 26th IEEE Annual Conference on Computer Communications*, 2045–2053, 2007.
44. A. Cuzzocrea, "Accuracy Control in Compressed Multidimensional Data Cubes for Quality of Answer-based OLAP Tools," in: *Proceedings of the 18th IEEE International Conference on Scientific and Statistical Database Management*, 301–310, 2006.
45. A. Cuzzocrea, "Improving Range-Sum Query Evaluation on Data Cubes via Polynomial Approximation," *Data & Knowledge Engineering 56(2)*, 85–121, 2006.
46. A. Stuart and K.J. Ord, *Kendall's Advanced Theory of Statistics, Vol. 1: Distribution Theory*, 6th ed., Oxford University Press, New York City, 1998.
47. S. Agarwal et al., "On the Computation of Multidimensional Aggregates," in: *Proceedings of VLDB*, 506–521, 1996.
48. G.K. Zipf, *Human Behaviour and the Principle of Least Effort: An Introduction to Human Ecology*, Addison-Wesley, Boston, MA, 1949.
49. Transaction Processing Council, TPC Benchmark H, available at http://www.tpc.org/tpch/.
50. UCI KDD Archive, The Forest CoverType Data Set, available at http://kdd.ics.uci.edu /databases/covertype/covertype.html.
51. K. Beyer et al., "Extending XQuery for Analytics," in: *Proceedings of the 2005 ACM International Conference on Management of Data*, 503–514, 2005.
52. R.R. Bordawekar and C.A. Lang, "Analytical Processing of XML Documents: Opportunities and Challenges," *SIGMOD Record 34(2)*, 27–32, 2005.

# V

# Big Data Applications

CHAPTER 17

# Big Data in Finance

Taruna Seth and Vipin Chaudhary

## CONTENTS



## BACKGROUND

The financial industry has always been driven by data. Today, Big Data is prevalent at various levels of this field, ranging from the financial services sector to capital markets. The availability of Big Data in this domain has opened up new avenues for innovation and has offered immense opportunities for growth and sustainability. At the same time, it has presented several new challenges that must be overcome to gain the maximum value out of it. This chapter considers the impact and applications of Big Data in the financial domain. It examines some of the key advancements and transformations driven by Big Data in this field. The chapter also highlights important Big Data challenges that remain to be addressed in the financial domain.

## 17.1 INTRODUCTION

In recent years, the financial industry has seen an upsurge of interest in Big Data. This comes as no surprise to finance experts, who understand the potential value of data in this field and are aware that no industry can benefit more from Big Data than the financial services industry. After all, the industry not only is driven by data but also thrives on data. Today, the data, characterized by the four *V*s, which refer to volume, variety, velocity, and veracity, are prevalent at various levels of this field, ranging from capital markets to the financial services industry. In recent years, capital markets have gone through an unprecedented change, resulting in the generation of massive amounts of high-velocity and heterogeneous data. For instance, about 70% of the US equity trades today are generated by high-frequency trades (HFTs) and are machine driven [1]. The prevalence of electronic trading has spurred up growth in trading activity and HFT, which, among other factors, have led to the availability of very large-scale ultrahigh-frequency data (UHFD). These high-speed data are already having a huge impact in the field in several areas ranging from risk assessment and management to business intelligence (BI). For example, the availability of UHFD is forcing the market participants to rethink the traditional ways of risk assessment and bringing up attention to more accurate, short-term risk assessment measures. Similar trends can be observed in the financial services sector, where Big Data is increasingly becoming the most significant, promising, and differentiating asset for the financial services companies. For instance, today, customers expect more personalized banking services, and to remain competitive as well as comply with the increased regulatory surveillance, the banking services sector is under tremendous pressure to best utilize the breadth and depth of the available data. In recent years, firms have already started using the information obtained from the vast oceans of available data to gain customer knowledge, anticipate market conditions, and better gauge customer preferences and behavior ahead of time, so as to offer highly personalized customer-centric products and services to their customers, such as sentiment analysis–enabled brand strategy management and real-time location-based product offerings as opposed to the historically offered product-centric services. Moreover, events like the credit crisis of 2008 have further shifted the focus of such financial entities towards Big Data as a strategic imperative for dealing with the acute stresses of renewed economic uncertainty, systemic monitoring, increasing regulatory pressure, and banking sector reforms. Unarguably, similar developments can be seen in other areas like asset management and insurance.

Clearly, such examples are indicative of the transformations ensuing in the finance sector, whereby more and more financial institutions are resorting to Big Data to strategize their business decisions based on reliable factual insights supported by real data rather than just intuition. Additionally, Big Data is now playing a critical role in several areas like investment analysis, econometrics, risk assessment, fraud detection, trading, customer interactions analysis, and behavior modeling.

In this digital era, we create approximately 2.5 quintillion bytes of data every day, and 90% of the data in the world today have been created in the last 2 years alone. The Big Data market is estimated to be at $5.1 billion this year and is expected to grow to $32.1 billion by 2015 and to $53.4 billion by the year 2017 [2]. Today, almost all sectors of the financial field are inundated with data generated from a myriad number of heterogeneous sources, such as hundreds of millions of transactions conducted daily, ultrahigh-frequency trading activities, news, social media, and logs. A recent survey shows that around 62% of companies recognize the ability of Big Data to gain competitive edge [2], and there is no doubt that the prevalent Big Data offers immense potential and opportunity in the finance sector. However, the enormously large financial data volumes, high generation speeds, and heterogeneity associated with the relevant financial domain data, along with its susceptibility to errors, make the ingestion, processing, and timely analysis of such vast volumes of often heterogeneous data very challenging.

There is no clear consensus among and within financial institutions today on the best strategies to harvest and leverage the available Big Data to actionable knowledge. This can be attributed to the fact that a single solution is unlikely to cater to the growing needs of different businesses within the financial domain spectrum. Today, many financial organizations are exploring and adopting customized Big Data solutions that are specifically tailored for their domain-specific needs. Section 17.2 presents an in-depth view of the financial domain with details on the historical trends in this sector and innovations in the field as a result of Big Data. The section will also cover the three key elements involved in financial domain dynamics, namely, Big Data sources in finance, information flow, and data analytics.

## 17.2 FINANCIAL DOMAIN DYNAMICS

### 17.2.1 Historical Landscape versus Emerging Trends

From the past several years, data warehouse systems primarily based on relational database management systems (RDBMSs) have been serving as the front-runners in providing access to the business community with necessary intelligence in the field of finance. These systems are mostly constructed out of quantitative data from operational systems, and BI tools are used to access the mostly well-understood operational data in the data warehouses [3]. Such systems are still widely used when it comes to simple analytical jobs or tasks like online analytical processing (OLAP), but their usage is restricted to small-scaled, well-defined, and structured data sets. However, the well-defined boundaries that once existed between the operational and decision-making tasks handled by such systems are increasingly becoming fuzzy in the financial world today. Although these systems were the norm in the early 90s, they are slowly losing their precedence in the area of BI and, hence, decision making due to their limited data handling and analytical capabilities.

In recent years, financial organizations have started to rethink and restructure the way they do business. This change is driven by the confluence of several factors like escalating regulatory pressures, ever-increasing compliance requirements, regulatory oversight, global economic instability, increasing competition in the global markets, growing business demands, need to optimize capital and liquidity, need to improve product and customer relationship margins, and so forth. For instance, regulations such as the European Market Infrastructure Regulation (EMIR) and the Financial Stability Oversight council's

Frank–Dodd Act have by themselves added hundreds of new rules affecting banking and securities industries. These directives for greater transparency are leading to enormous increases in the data volumes across such industries and forcing them to redesign their services infrastructure to cater to the new demands. Similar data growth trends can be seen in other parts of the financial domain. For example, there was a tenfold increase in market data volumes between 2008 and 2011, and and the data volumes are growing stronger in all areas of the financial domain; for example, some of the top European insurers reported a sixfold increase in the amount of data and analytic reporting required by just the first pillar of the Solvency II insurance reform regulation [4]. The New York Stock Exchange (NYSE) by itself creates about several terabytes of market and reference data per day covering the use and exchange of financial instruments, whereas Twitter feeds, often analyzed for sentiment analysis in the financial domain, generate about 8 terabytes of data per day of social interactions [4]. There are around 10,000 payment card transactions executed per second across the globe; there were about 210 billion electronic payments generated worldwide in 2010, and the number is expected to double by the end of the decade [4]. Various other developments in the financial system are also contributing enormously to the overall volume of the data in the system. One such example of this shift can be explained by the emergence of the originate-to-distribute model that has broken down the traditional origination process into a sequence of highly specialized transactions and has led to an increase in the volume of the data in this domain. In this model, financial products like mortgages are systematically securitized and then structured, repackaged, and distributed again, so the loan details that traditionally might have been recorded only by the original lender and the borrower are now shared across multiple, diverse entities such as the originating bank, borrower, loan servicer, securitization trust and bondholders, as well as buyers and sellers of credit protection derivatives [5]. Besides contributing to the data volume, each new entity in the system adds to the complexity of the involved data. The digital universe is expected to grow nearly 20-fold, to approximately 35 zettabytes of data, by the year 2020 [6].

Traditional data management practices in finance can no longer effectively cope with the ever-increasing, huge, and rapid influx of heterogeneous (structured, semistructured, unstructured) data originating from a wide range of internal processes and external sources, including social media, blogs, audio, video, and so forth. Conventional data management technologies are destined to fail with such growing data volumes, which far exceed the storage and analysis capabilities of many traditional systems, and they have in many instances. For instance, with regard to the volume and complexity of created data, back offices of trading firms have failed to keep up with their own front office processes as well as the emerging data management practices adopted in other industries to handle growing data volumes and a multitude of diverse data types [5]. Moreover, traditional systems are not equipped to handle the wide variety of data, especially unstructured data, from social media, like news, Twitter, blogs, videos, and so forth, that is needed to gain insights about businesses processes (e.g., risk analysis, trading predictions) and keep up with the evolving needs of the customer in the financial services industry. Such systems often fail when it comes to integration of heterogeneous data or even real-time processing of structured data. In fact, this is becoming a bottleneck for many top-tier global banking systems since the introduction of new regulations that require the banks to provide a complete horizontal view of risk within their trading arms. This task entails the integration of data from different schemas unique to each of the trade capture systems into a central repository for positions, counterparty information, and trades. Extraction, transformation, cleansing, and integration of such data via traditional extract, transform, load (ETL)–based approaches, coupled with samplings, often span several days and are not very accurate in the scenarios where only a sample of the data is used for analysis. New regulations, however, demand that this entire pipeline be executed several times a day, a feat clearly infeasible using the conventional approach. These regulations are similarly applicable to the capital markets where the regulations necessitate an accurate view of different risk exposures across asset classes and lines of businesses and firms to better estimate and manage systemic interplays. These tasks require simulations of a variety of risk situations, which can result in the generation of terabytes of additional data per day [7].

In recent years, it is becoming increasingly important for financial firms to adopt a data-centric perspective to handle the mounting regulatory pressures and succeed in today's digital, global marketplace. In the past, financial organizations collected large amounts of data. However, these institutions depended primarily on the conventional ETL framework and lacked the ability to process the data and produce actionable knowledge out of it within realistic time frames. This approach prevented them from gaining a full perspective of their business insights and made it difficult for them to anticipate and respond to changing market conditions, business needs, and emerging opportunities, a few must-haves essential to thrive in today's dynamic business environment. As a result, the firms relying on the traditional schemes have started to address the limitations inherent in their conventional systems. Today, a growing number of financial institutions are exploring new ways of unlocking the potential of available data to gain insights that can help them improve their performance and gain competitive advantage through factual and actionable knowledge, timely decisions, risk management and mitigation, and efficient operations in highly complex and often volatile business environments. Figure 17.1 highlights the importance and applicability of Big Data in the financial domain. The figure illustrates the key sectors of finance in which the power of Big Data is being harnessed to address critical business needs ranging from product innovation and fact-driven strategic decision making to the development of novel and intelligent business solutions.

The financial industry has always been one of the most data-driven industries. In the past few years, the prevalence of Big Data has opened up new horizons in the financial fields. Several industries have already started exploiting the value out of Big Data for information discovery in areas like predictive analytics based on social behavior mining, deep analytics, fraud detection, and risk management. For example, most of the credit card companies mine several thousands to millions of records, aggregated from customer transactional data (structured), call records (unstructured), e-mails (semistructured), and claims data (unstructured), to proactively anticipate future risks, accurately predict customer card switching behavior, and devise measures to improve customer relationships based on such behavioral modeling. Likewise, several firms involved in financial risk management perform risk assessment by integrating large volumes of transactional data with

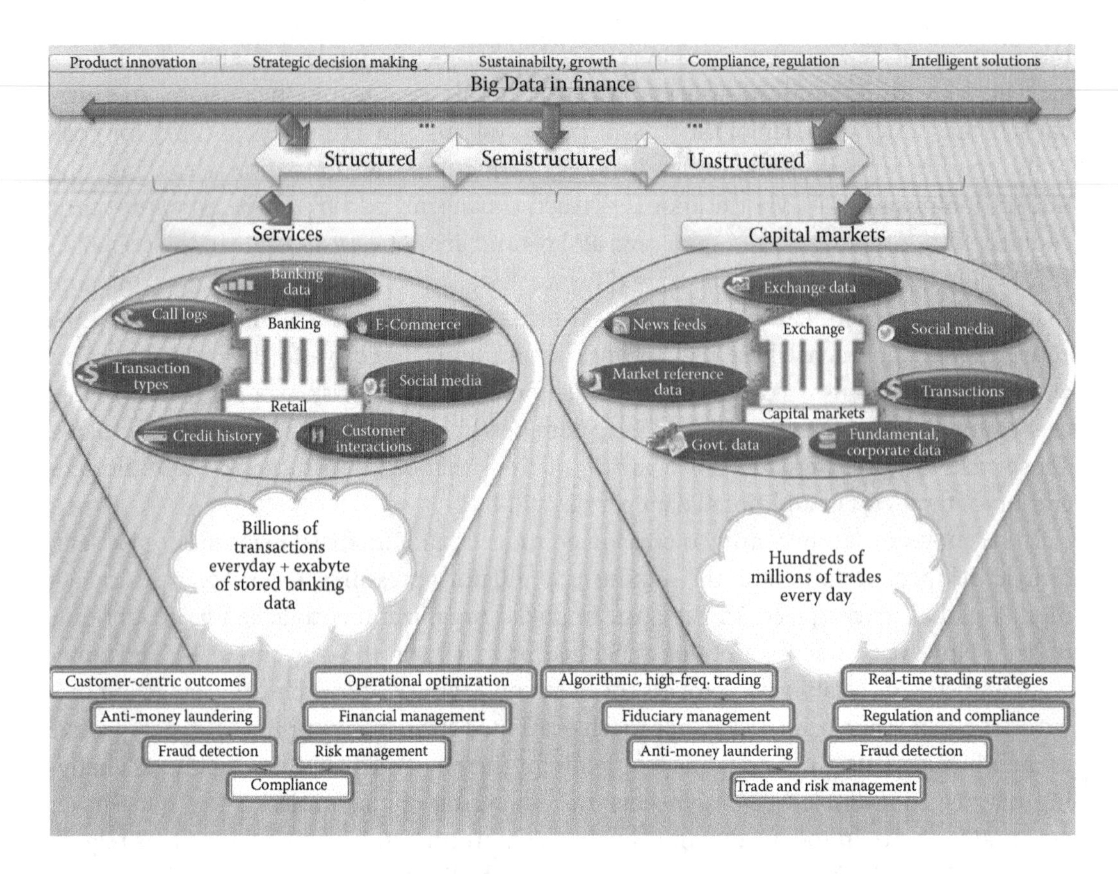


FIGURE 17.1 Big Data applications in key financial domain sectors.

data from other sources like the real-time market data as well as aggregate global positions data, pricing calculations, and the value-at-risk (VaR) crossing capacity data of current systems [2]. In the past couple of years, Big Data has begun to evolve as a front-runner for financial institutions interested in improving their performance and gaining a competitive edge. Section 17.3 exemplifies some of the Big Data innovations in the financial domain and highlights different key players involved in the financial dynamics pipeline of the capital markets.

## 17.3 FINANCIAL CAPITAL MARKET DOMAIN: IN-DEPTH VIEW

### 17.3.1 Big Data Origins

Nowadays, the financial markets are getting inundated with a flurry of data that is increasing in complexity as well as size. Today, the stock markets encompass a wide variety of data from diverse sources such as market data, which include orders and trades; reference data comprising data related to ticker symbols, exchanges, security descriptions, corporate actions, and so forth; and fundamental data, including corporate financials, analyst reports, filings, news (including earnings reports, economic news, etc.), and social media data (comprising blogs, Twitter feeds, etc.) [8,9]. These data sources provide a variety of information in different, structured, semistructured, and unstructured, formats, thereby adding to the heterogeneity of the data. Moreover, the data generated and utilized by these markets are highly voluminous and complex. The reason for the complexity in financial data is that agents acting in markets trade increasingly faster and in more numerous and complex financial instruments, and have better information-acquiring tools than ever before. The reason for heterogeneity stems not just from the fact that agents and regulations across the globe are themselves heterogeneous but also from the more mundane fact that people report information in ways that are not standardized. For example, investment managers reporting holdings to the Securities and Exchange Commission (SEC) often err regarding the proper unit of measurement (thousands or units); performance data are often "dressed" to appear more attractive (most often, smoother); and so on.

Recent projections by the Options Pricing Reporting Authority for the years 2014–2015 estimate a total of 26.9 to 28.7 billion messages per day, 17.7 to 19 million messages per second, and a maximum output rate of at least 1 million messages per second [10]. For example, the data covering the quotes and transactions from the major US exchanges (trade and quote database [TAQ]) grow exponentially, now at a rate of hundreds of terabytes per year. Analysis of the operational structures of the underlying data generation entities partially explains the enormity and complexity of the massive data sets. For instance, unlike the traditional days of specialists and natural price discovery, today, the US stock market structure comprises an aggregation of different exchanges, broker-sponsored execution venues, and alternative trading systems, each of which contributes differently to the market data and volumes [11]. Specifically, around 14 exchanges, approximately 50 dark pools, and more than 200 international platforms or venues contribute around 66%, 13%, and 21% volume, respectively [8]. Orders are submitted through more than 2000 broker deals, and the system is governed by various regulatory agencies including SECs, SROs, and so forth [8]. The estimated average trading volumes for the market include about $50- to $100-billion-value trades, at least 2 billion order submissions, and 5 billion share trades [8,12]. The dynamic system environment comprising complex trade work flows (e.g., billions of trades or order submissions, price matches, executions, rejections, modifications, acknowledgments, etc.) and the changing market trading practices such as HFT further contribute to the volume and complexity of the generated data. Figure 17.2 portrays an example showing the high-level view of a typical automated electronic trading system. The key blocks shown in the figure are representative of extremely intricate models, strategies, and data, among other factors, which add to the complexity of the overall system. In the United States, the HFTs were estimated to account for more than 70% of equity trades in the year 2010 [11]. Besides adding to the volume, such evolving trading techniques are resulting in very high data generation speeds. For example, order matching and subsequent trade execution can now be accomplished in less than 100 μs via a colocated server in the exchange, and algorithmic trading can now be done within microseconds [11].

### 17.3.2 Information Flow

Financial markets offer comprehensive platforms that facilitate complex interplays among different market participants; support large-scale information ingestion and aggregation for price discovery; and provide liquidity for uninformed, liquidity-seeking

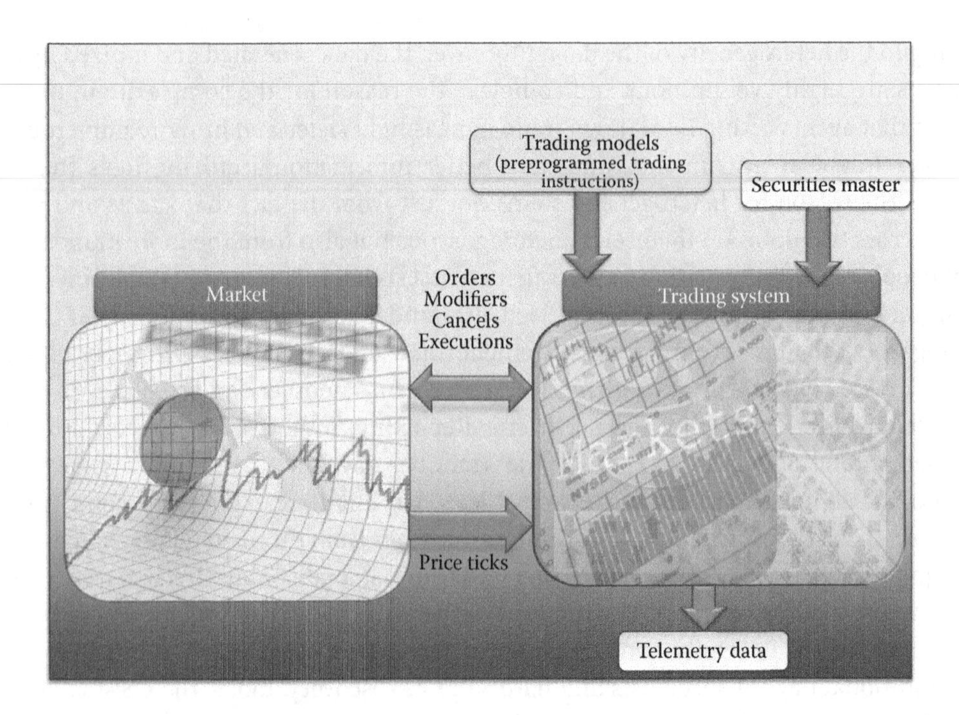


FIGURE 17.2 A typical automated electronic trading system.

order flows [13]. The data generated through the individual market platforms are disseminated to the market participants through different channels. High-speed market data are directly delivered to some entities through the principal electronic communication networks (ECNs) such as INET, whereas the data are delivered to a majority of other entities through distribution channels like the National Association of Securities Dealers Automated Quotations (NASDAQs) dissemination or the Consolidated Trade System (CTS), which directly or indirectly collects all US trades, and the Consolidated Quote System (CQS) [14]. The latter dissemination channel, however, collects data at a much slower pace compared to the speed at which the data are generated [14]. Different market participants often require and utilize data with different price granularities, and hence precision, depending upon their diverse trading objectives. For instance, high-frequency traders and market makers like the NYSE dedicated market makers (DMMs) generally utilize UHFD (tick data), are highly sensitive to small price changes, and deal with several thousands of orders per day. In contrast, investors like the pension funds investors normally base their investment decisions on low-frequency or aggregated data, are not too sensitive to small price changes (e.g., at the intraday level), and usually deal with no more than a few hundreds of orders per day. Unregulated investors like the hedge funds and other speculators like the day traders, on the other hand, generally fall somewhere in between the investors and market makers. Figure 17.3 further exemplifies such market dynamics that exist among different market participants. All these different players and the inherent intricacies of the underlying processes complicate the order or other information flows. For example, about one-third of price discovery nowadays occurs in dark

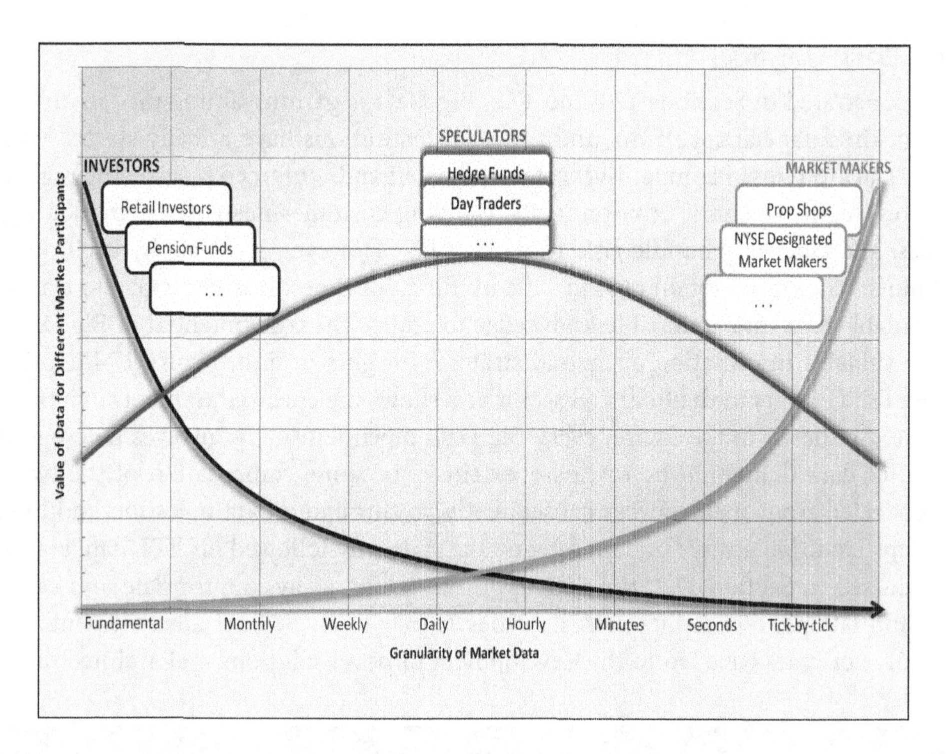


FIGURE 17.3 Value of data for different market participants.

pools, and even today, these pools remain largely unobservable. Such factors add to the complexity of the market structures and make it difficult to understand the transformations of orders into trades and how they drive the price discovery process.

Market participants usually deploy different trading strategies in line with their trading objectives. Typically, the raw market data, collected by these market agents, are refined, aggregated, and analyzed. Today, many market participants are resorting to novel ways of information discovery and incorporating additional information in their trading strategies, which involve integration of traditional data (e.g., orders and trades) with nontraditional data (e.g., sentiment data from social networks, trending over time, news, exploratory and deep analysis of the available data through efficient interactive or ad hoc queries). There is no doubt that the vast amounts of information that is generated by such trading systems along with the information that exists in the complex networks of legal and business relationships that define the modern financial system hold all the answers required to understand and accurately predict unexpected market events as well as address the most demanding questions that plague the financial systems and, hence, the regulators. In recent years, much of the focus has shifted to finding ways to extract all the necessary answers from these ever-growing financial data within realistic time frames. However, to date, the problem remains challenging for many in this field not only because of the Big Data constraints but also due to factors like the lack of transparency; absence of standardized communication protocols; and ill-defined work flows among different data generation systems, processes, and organizations.

### 17.3.3 Data Analytics

As demonstrated in Sections 17.1 and 17.2, Big Data is gaining abundance in almost all levels of the financial spectrum, and financial institutions have already started leveraging Big Data to remain competitive; cater to the demands enforced by growing regulatory pressures, highly dynamic environments, evolving customer needs; seize market opportunities; and efficiently handle risk, to name a few. However, Big Data by itself does not hold much value, and not all of it may be useful at all times. It is necessary to transform the available data into actionable knowledge to realize the true potential of Big Data and extract valuable information or factual strategic insights from it. Figure 17.4 depicts the key Big Data drivers in the financial sector and shows the essential elements of a Big Data pipeline. As shown in the figure, every Big Data pipeline usually involves heterogeneous sources of data that could be intrinsic, extrinsic, or some combination of the two. The data collected from such sources subsequently go through a data ingestion and integration step. Traditional ways of data integration primarily followed an ETL data approach. As discussed in Section 17.2, the ETL approach is not always appropriate and can have many limitations, especially when it comes to Big Data. Several advancements in the last couple of years have led to the development of novel solutions and architectures that

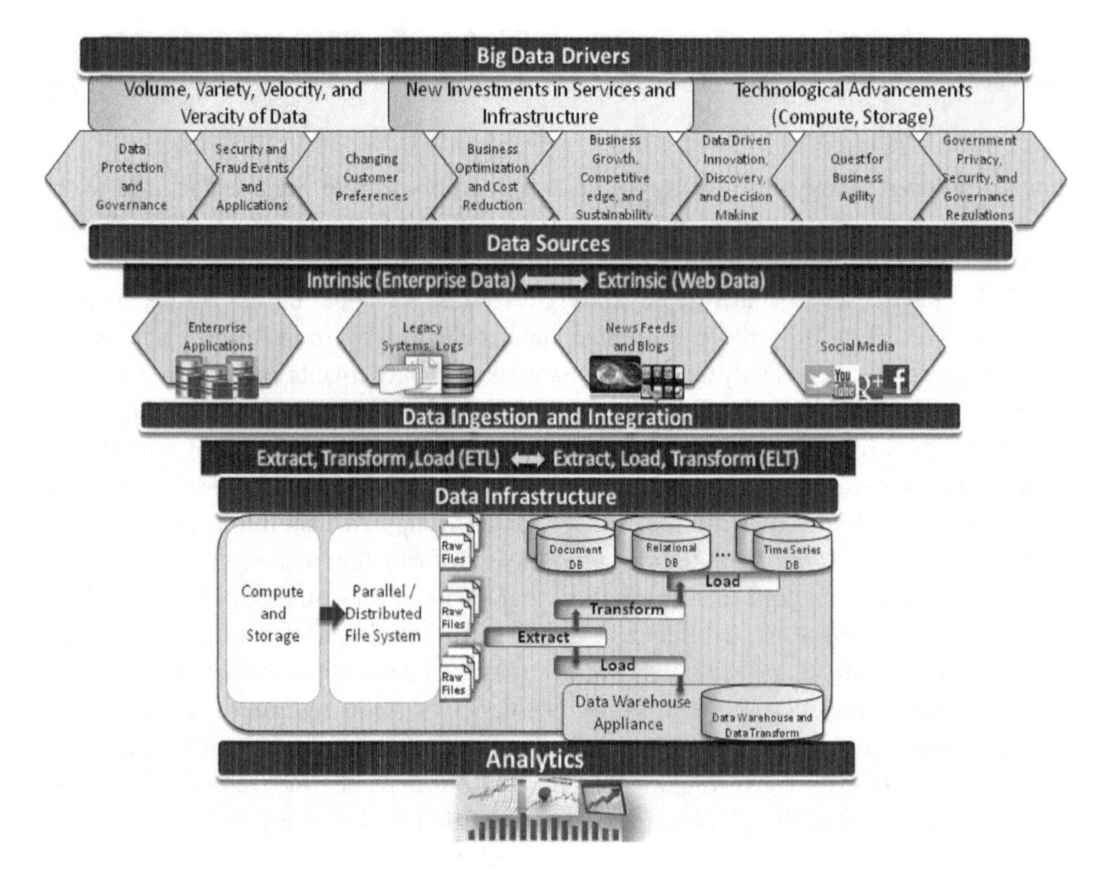


FIGURE 17.4 Big Data ecosystem in finance.

address some of the limitations inherent in the traditional systems and offer unique capabilities to efficiently handle today's Big Data needs. The ultimate goal of a Big Data pipeline is to facilitate analytics on the available data. Big Data analytics provides the ability to infer actionable insights from massive amounts of data and can assist with information discovery during the process. It has become a core component that is being deployed and used by entities operating at various spheres of the financial field. For example, predictive analytics tools are increasingly being deployed by the banks to predict and prevent fraud in real time. Predictive analytics applies techniques from data mining, data modeling, and statistics to identify relevant factors or interactions and predict future outcomes based on such interactions. Predictive analytics tools are also increasingly being used by market participants for tasks like decision making, improving trading strategies, and maximizing return on equities.

Big Data analytics is also being used in the capital markets for governance-based activities such as detection of illegal trading patterns and risk management. For example, NYSE Euronext has deployed a market surveillance platform that employs Big Data analytics to efficiently analyze billions of trades to detect new patterns of illegal trading within realistic time frames. The analytics platform allows Euronext to process approximately 2 terabytes of data volume everyday, and this volume is expected to exceed 10 petabytes a day by 2015. The deployed infrastructure has been reported to decrease the time required to run market surveillance algorithms by more than 99% and improve the ability of regulators or compliance personnel to detect suspicious or illegal patterns in trading activities, allowing them to take proactive investigative action to mitigate risks [15]. Similarly, another market information data analytics system (MIDAS) went online at the SEC in January 2013. MIDAS focuses on business data sources in the financial domain, with particular emphasis on the filings periodically required by the companies to be made with the SEC and Federal Deposit Insurance Corporation (FDIC). The system provides valuable insights about financial institutions at systemic and individual company levels by harnessing the value out of market data as well as data archived by the SEC and FDIC. It processes about 1 terabyte of stocks, options, and futures data per day and millions of messages per second. The analytics system is being utilized by subscribers to analyze mini flash crashes, assess impacts of rule changes, and detect abnormal patterns in the captured data [8].

## 17.4 EMERGING BIG DATA LANDSCAPE IN FINANCE

As discussed in Sections 17.1–17.3, Big Data has already initiated several innovations in the field of finance and has started to reshape some of the traditional ways of doing business. For instance, in recent years, capital markets have evolved from low-frequency trading involving simple trading strategies, like 1980s-paired models, to ultrahigh-frequency trading and the intricate gaming strategies of today. An ever-increasing need to efficiently and effectively exploit Big Data is driving the development of a new generation of technologies and architectures that allow real-time information extraction and discovery by facilitating high-speed data ingestion, data integration, and analysis of massively large and heterogeneous data sets within realistic time frames. Nonetheless, the development and adoption of Big Data–centric technologies and architectures remain challenging. Section 17.4.1 discusses some of these challenges in the financial domain.

### 17.4.1 Challenges

Despite being aware of the significant promise that Big Data holds, many companies still have not started to make any investments to reap its benefits. A lot of companies today waste more than half of the data they already hold. If we assess the value of these data based on the Pareto principle, that is, 80% of the value comes from 20% of the data, then clearly, a lot of value is getting lost [6]. There is no doubt that the Big Data domain is rapidly evolving, but the domain is still premature in the financial field, with several factors slowing its growth and adoption in this domain. Data management and sharing has been a difficult problem for capital market firms for decades. The recent financial collapse and the mortgage/credit crisis of 2008 have uncovered some of the bottlenecks and inadequacies inherent in the information structure of the US financial systems. Factors like the lack of standardized data communication protocols, lack of transparency, complex interactions, and data quality gaps across different financial units in the system make it extremely difficult to unravel and connect different systems, processes, and organizations for any kind of analysis within realistic time frames. Many of these limiting factors represent an evolutionary outcome of the years of mergers, internal fragmentations, and realignments within the financial institutions and have been made worse by the business silos and inflexible Information Technology (IT) infrastructures. To date, the work flows in the system largely remain ill defined, and data reside in unconnected databases and/or spreadsheets, resulting in multiple formats and inconsistent definitions across different organizational units. Data integration remains point to point and occurs tactically in response to emergencies [9]. Convergence of such factors makes it almost infeasible to ingest, integrate, and analyze large-scale, heterogeneous data efficiently across different financial entities within a financial organization. The lack of best practices, data sharing procedures, quality metrics, mathematical modeling, and fact-based reasoning has left even the federal regulatory agencies unable to ingest market information in a timely manner, permit a proactive response, or even determine what information might be missing [9].

Financial institutions have historically spent vast sums on gathering, organizing, storing, analyzing, and reporting data through traditional data management approaches. As demonstrated in Section 17.2, the conventional data management structures are no longer sufficient to handle the massively large, high-velocity, heterogeneous financial data. Today, it is critical to deploy new supporting infrastructure components, data ingestion and integration platforms, as well as Big Data analytics and reporting tools, to handle extremely large-scale, often real-time, heterogeneous data sets. However, due to the varied data and business requirements of different organizational units, it is not realistic to expect a single solution to fit or even be equally applicable to all the financial entities. Moreover, due to the lack of well-defined solutions, an exploratory, incremental, and possibly iterative approach would be needed to devise customized and efficient Big Data solutions. Therefore, it is important for the financial organizations to devise their Big Data investment strategies with focus on their business-specific goals, like what is needed for risk management, product innovation, risk and market intelligence, cost reduction, services, operations, and so forth. For instance, risk analytics and reporting requirements would necessitate a platform that could help consolidate risk measure and provide powerful risk management and reporting capabilities. A market data management system, on the other hand, would require a highly scalable platform that could store, process, and analyze massive amounts of heterogeneous data sets in real time.

### 17.4.2 New Models of Computation and Novel Architectures

Real-time Big Data platforms entail a multitiered architecture mainly comprising data ingestion, integration, analytics, visualization/reporting, and decision-making components. A lot of technological transformations are underway towards the development of such platforms in the financial sector to efficiently manage and utilize Big Data. For instance, to store and manage extremely large and growing volumes of data, the focus has shifted from the scale-up storage solutions to the scale-out storage solutions wherein large shared pools of storage devices are used to generate more capacity while reducing operational costs. Similarly, new data-intensive supercomputing (DISC) appliances are increasingly being deployed to overcome the limitations inherent in the traditional financial data management systems and perform complex analytics on the massively large data sets. Specifically, NASDAQ, NYSE Euronext, and Lucera have started to utilize systems like Netezza [16], Greenplum [17], and Scalable Informatics' Cadence and Resonance [18] for their Big Data and analytics needs. DISC appliances like Netezza represent a new paradigm in data-intensive computing wherein the processing is done at the location of the data. Technologies like complex event processors (CEPs), in-memory databases, and data grids are also being used for real-time analytics in the financial industry. CEP engines enable analytics on streaming data and are used by front offices for algorithmic trading and so forth. They are used by the services industry to detect real-time events such as generation and monitoring of payment transactions or other core banking transactions [7].

Data grids facilitate distributed processing and storage of data in memory and are often used to complement CEP engines. To handle unstructured data, solutions based on open-source frameworks like Hadoop and MapReduce are commonly used in the industry today. These technologies offer high scalability and performance via parallelism and data locality–aware processing [7]. Besides Hadoop, other NoSQL (interpreted as Not only SQL) databases such as graph databases and document databases are also in use today for handling nontraditional, semistructured and unstructured, financial data sets. These Big Data management systems are primarily deployed based on the Brewer's theorem, also known as CAP theorem, which states that the systems have to pick two out of the three properties, consistency, availability, and partition tolerance. Most of the solutions developed using these technologies are tailored as basically available, soft-state, eventually consistent (BASE) systems, unlike the strictly atomicity, consistency, isolation, durability (ACID) compliant traditional data management systems. Big Data technologies like Hadoop and NoSQL working in conjunction with event-stream processing systems like Storm are already playing a key role in financial market operations [19]. Many financial organizations have already started to reap the benefits of their Big Data forays in the field using a combination of existing and new Big Data technologies. More innovations and technological improvements in this field are expected to follow in the near future as more financial units begin to incorporate Big Data solutions.

## 17.5 IMPACT ON FINANCIAL RESEARCH AND EMERGING RESEARCH LANDSCAPE

### 17.5.1 Background

Financial markets are complex systems driven by the influx of exogenous information coming from sources external to the system and regulated by an internal dynamics [20]. In recent years, the availability of UHFD has opened up new horizons in the field. UHFD refer to a financial market data set that comprises of all the transactions or tick-level activity [21,22]. Nowadays, millions of data points, representing the tick- or transaction-level activity, stream out of the global financial markets, driving decision strategies of the various market participants. The data hold high practical relevance because a rising number of market participants today execute trades based on high-frequency strategies and are thus exposed to high-frequency market risks [23]. UHFD serve as the basic building blocks for such analysis. Consequently, analysis and modeling of UHFD has gained a lot of momentum in the academic sector and has become one of the key focal points of research in the fields of finance, financial econometrics, and statistics. Most of the financial studies in these areas are primarily concentrated to two fields, time-series analysis of fixed-resolution UHFD and inferences based on diffusion processes [24]. However, active research in these fields spans several areas like trading, microstructure theory, option pricing, risk management, and trading strategies like statistical arbitrage [25].

In the past several years, much of the research in the financial field has focused on measuring and forecasting volatility, and even today, volatility of asset returns remains one of the most important elements in finance. In the financial jargon, volatility is commonly defined as the dispersion in asset price movements over a certain time period. Measuring and forecasting volatility of financial asset returns plays a crucial role in several areas, including risk management, asset allocation, options, derivatives pricing, investment analysis, and portfolio management. In addition, financial market volatility has a direct impact on policy making, facilitated primarily due to its ability to gauge market and economic vulnerability, and plays a vital role in seizing profitable investment opportunities with optimal to high returns on risk trade-offs [26,27]. For instance, different market participants usually have different market expectations in terms of expected returns on investments and risks. Their investment decisions mainly revolve around the strategies that maximize their returns on investments while minimizing the associated risks. Among other market participants, equity and derivative traders use several volatility measures and forecasts, usually based on historical data, as proxies to assess risks associated with the different asset classes [28]. Today, volatility unarguably serves as one of the key measures of financial risk and drives the hedging and pricing of options or other securities and construction of optimal portfolios [27]. Correct estimation or assessment of financial risks helps reduce the probability of failures during extended periods of financial distress and is critical for the viability and stability of the financial system [29]. Accurate financial market risk assessments are much more important during periods of financial turmoil, and hence high volatility, due to the extensive risk of global financial instability [30]. During recent crisis events in the financial markets, such as the US subprime mortgage crisis and

Europe's ongoing sovereign debt fiasco, a large number of financial organizations did not successfully enforce the Basel Committee of Banking Supervision mandates with respect to their VaR estimations [31]. These adverse financial episodes further underscore the significance of extreme asset price movements, and hence, accurate volatility forecasts, for efficient risk mitigation through appropriate risk measurement and management measures that can adapt in accordance with the changing market environments. Due to these reasons, it is not surprising that volatility estimation and inference have drawn a lot of attention in recent years. Also, another reason that makes volatility a more popular measure comes from the fact that unlike daily returns, which offer very little explanatory power and hence are difficult to predict, volatility of daily returns, due to their relatively high persistence and conditional dependence, is nonstationary and predictable [32,33]. Volatility clustering effects were first reported by Mandelbrot [34], who observed that periods of low volatility were followed by periods of low volatility and vice versa. These factors explain the large number of contributions and research efforts dedicated to the measurement and prediction of volatility.

Despite several advancements, accurate measurement of ex post volatility remains nontrivial largely due to the fact that volatility is unobservable and cannot be directly observed from the data [27]. In the past, several models have been developed to forecast volatility. The very first popular volatility model was the autoregressive conditional heteroskedasticity (ARCH) model introduced by Engle [35]. Subsequently, Bollerslev [36] proposed a more generalized representation of the ARCH model, namely, a generalized autoregressive conditional heteroskedasticity (GARCH) model. These models have been followed by a large number of models based on different variations of the original ARCH model. These include parametric models like the Glosten–Jagannathan–Runkle GARCH (GJR-GARCH), fractionally integrated GARCH (FIGARCH), component GARCH (CGARCH), stochastic volatility (SV), and other models [37–43]. The ARCH class of models incorporates time variation in the conditional distribution mainly through the conditional variance and is geared towards capturing the heavy-tails and long-memory effects in volatility. The stochastic models usually based on the assumption that the repeated low-frequency observations of assets return patterns are generated by an underlying but unknown stochastic process [44]. These models have been successful in explaining several empirical features of the financial return series, such as heavy-tails and long-memory effects in volatility. Since their introduction, an extensive literature has been developed for modeling the conditional distribution of stock prices, interest rates, and exchange rates [45]. This class of models has been extensively used in the literature to capture the dynamics of the volatility process. Until recent years, the ARCH model and its variants had been used by many in the field to model asset return volatility dynamics for daily, weekly, or higher-interval data across multiple asset classes and institutional settings [37,46]; however, despite their provably good in-sample forecasting performance, their out-of-sample forecasting performance remains questionable. Many studies in the past have reported insignificant forecasting ability of this class of models [47–50] based on low correlation coefficients in the assessments. These models usually forecast volatility based on low-frequency asset return data at daily or longer time horizons. The use of low-frequency observations in these models results in a significant amount of information loss, which, for some asset classes like equities, can account for more than 99% of the available data [51]. Also, such models usually use daily squared returns as the highest order of granularity to compute estimates of "true volatility." Since daily squared returns are calculated from closing prices, they fail to capture the intraday price fluctuations or interdaily volatility movements [52]. This results in less accurate volatility estimates due to reduced statistical efficiency. Moreover, models based on low-frequency data are often severely impacted by structural breaks and hence cannot appropriately adjust to the changing drifts in financial markets [53]. Most models in the past had been developed and tested by employing data sets that represented only a small subset of the available trading data [51]. In addition, today, many market participants, such as the high-frequency traders, noise traders, and speculators in future markets, deploy trading strategies that are often very sensitive to even the slightest changes in the asset prices. For these traders, even a minor intraday price fluctuation can result in significant trading volumes. Clearly, low-frequency asset returns–based volatility or risk models lack a tremendous amount of information central to the trading strategies of these market participants.

### 17.5.2 UHFD (Big Data)–Driven Research

In recent years, the increased availability of ultrahigh-frequency trading data has spurred strong growth in the development of techniques that exploit tick-level intraday price data to better estimate volatility forecasts and overcome at least some of the limitations inherent in the traditional low-frequency–based systems. UHFD carry a lot more information compared to their low-frequency counterpart and are not only useful for measuring volatility but also vital for model estimation and forecast evaluation [54]. Availability of high-frequency data coupled with recent technological advancements has made analysis of large-scale trading data more accessible to market participants [25]. A lot of research efforts today are dedicated to the use of high-frequency data for measuring volatility [44,55]. Access to UHFD within academia has led to significant recent progress in the field of econometrics of financial volatility [56]. Recently, much of the interest has shifted from low-frequency data–based volatility modeling to UHFD-based volatility models and even model-free approaches. Among other things, UHFD provide an appropriate basis for empirical tests of market microstructure theories, improve statistical efficiency, and have been shown to be extremely significant in the meaningful ex post evaluation of daily volatility forecasts [57,58].

In the last few years, modeling of financial data observed at high frequencies has become one of the key research areas in the field of financial econometrics. Consequently, several volatility measures and models have been proposed that exploit the power of UHFD for improved volatility forecasts. The most popular measure in this group is the realized volatility estimator [53,59,60] that utilizes intraday returns to measure the true volatility, thereby allowing the measurement of a latent variance process. Realized volatility is defined as the sum of squared intraday returns. It has been shown theoretically that realized volatility converges to the true integrated variance at ultrahigh frequencies, that is, as the length of the intraday intervals approaches zero [61]. Specifically, it has been shown empirically that the sum of squared high-frequency intraday returns of an asset can be used as an approximation to the daily volatility [52,53,61–67]. This quadratic variation is known as the estimator of the daily integrated volatility. Moreover, the forecasting performance of this estimator has been shown to be superior compared to the performance of standard ARCH-type models [52]. Based on the theoretical and empirical properties of realized volatility, several other studies also confirm that precise volatility forecasts can be obtained using UHFD [61,63,68,69]. However, the experimental results do not exactly match with the empirical or theoretical justifications, which mainly rely on the limit theory and suggest that by increasing the observation frequency of asset returns and representing the true integrated volatility of the underlying returns process via realized volatility, one can obtain more efficient, less noisy estimates in comparison to the estimate obtained using low-frequency data, such as the daily data [52,64,70–72]. This discrepancy can be primarily attributed to the presence of market microstructure noise. Market microstructure noise collectively refers to the vast array of frictions inherent in the trading process. These imperfections in the trading process arise due to several factors, such as the bid–ask bounce, infrequent trading, discreteness of price changes, varied informational content of price changes, and so forth. For example, changes in the market prices occur in discrete units. Specifically, the prices usually fluctuate between the bid and the ask prices (the bid–ask spread) and multiple prices may be quoted simultaneously by competitive market participants due to the heterogeneous market hypothesis [73], thereby resulting in market microstructure frictions. At ultrahigh frequencies, the problem gets worse because the volatility of the true price process shrinks with the time interval, while the volatility of the noise components remains largely unchanged [74]. Consequently, at extremely high frequencies, the observed market prices reflect values contributed largely by the noise component compared to the unobserved true price component. As a result, the realized volatility measures become biased and thus are not robust at ultrahigh frequencies when the price is contaminated with noise and the bias is not accounted for during subsequent evaluations [52,75,76]. Market microstructure noise effects and their impact in the high-frequency scenarios have been discussed and analyzed in several studies [74–81].

Since the inception of the high-frequency–based realized volatility measure, several alternative volatility estimation measures have evolved in an attempt to improve upon the basic high-frequency–based realized volatility measure and cater to the growing demands imposed by the availability of trading data at ultrahigh frequencies. For instance, in recent years, many new methods have been developed to estimate spot or instantaneous volatility, for example, volatility per time unit or per transaction, using high-frequency data. Spot volatility is particularly useful to high-frequency traders, who often strategize using the finest granularity of available tick data and to whose operations immediate revelation of sudden price movements is critical. Various high-frequency data–based methods for spot volatility estimation have been proposed in the literature, ranging from nonlinear state space–based models to nonlinear market microstructure noise–based and particle filter–based models [82–88].

Similarly, in an attempt to overcome the limitations inherent in the high-frequency data–based realized volatility measure, several high-frequency–based alternative measures have been proposed.

Many recent studies model the daily volatility of the underlying returns process using parametric models incorporating the realized volatility measure [89–92]. Different high-frequency variants of the realized volatility measure have been used to address a number of other problems. For example, the variants have been shown to be useful for reduced-form volatility forecasting [62,93,94] and formulation of highly informative and directly assessable distributional implications for asset returns [95]. Recently, a heterogeneous autoregressive model (HAR) was proposed [96]. This model estimates volatility by autoregressing volatilities realized over varying interval sizes (daily, weekly, monthly). Since its introduction, several variants of the base model have been developed, for example, HAR realized volatility model with jumps (HARRVJ) and HAR realized volatility model with discrete and continuous jump components (HARRVCJ). The latter variations of the model are based on the indications that relatively frequent jumps could be significant in the evolution of the price process and the volatility originating from jumps in the price level is less persistent than that generated by the continuous component of the realized volatility measure. The high-frequency–based volatility (HEAVY) method represents yet another example in this category. This method incorporates momentum and mean reversion effects of volatility and can adapt to structural breaks in the volatility process. It focuses on two main measures, namely, close-to-close conditional variance and conditional expectation of the open-to-close variation. It estimates both of these measures using the Gaussian quasi-maximum likelihood approach [97]. A realized range–based estimator is another example of HEAVY estimators. It is similar to the realized volatility estimator but uses squared intraperiod high–low price ranges instead to mitigate the effects of bid–ask bias.

Notably, in the last couple of years, due to the ill effects caused by the market microstructure noise at ultrahigh frequencies, when the prices are contaminated by market microstructure noise, academic interest has largely shifted towards the development of robust realized volatility measures that are immune to such noise. A number of variants of the realized volatility measure have been proposed, ranging from optimized sampling frequency–based estimators [75,98], subsampling-based methods [76,77], and Fourier volatility–based methods to realized kernel–based estimators [78], wavelet realized volatility methods [99], and preaveraging methods [100,101]. Most of the methods can be coarsely classified into two main groups. The first group primarily focuses on determining the optimal sampling frequency for volatility estimation in the presence of noise, and the second group comprises bias correction methods that accommodate the microstructure noise effects and validate the use of higher sampling frequencies. Specifically, in the former approach, the microstructure noise effects are alleviated by sampling prices sparsely at coarser intervals. With high-frequency data, even sampling intervals of 5 min can result in a significant amount of potentially useful data points getting discarded, especially for highly traded assets. This results in discretization error and reduced statistical efficiency [65,102,103]. The latter group focuses on utilizing all the available high-frequency data. For example, the two- and multi-time-scale realized volatility estimators utilize two or more time scales respectively, to estimate integrated volatility. These estimators have been shown to be robust to time-series–dependent noise [14] and better than the classic realized volatility measure. Other estimators like the kernel-based estimators have also been shown to capture important characteristics of market microstructure noise and outperform the classic realized volatility measure [80]. UHFD also allow learning about jumps in the price process. Jumps represent discontinuous variations or movements in the price process that are totally incompatible with the observed volatility [104]. In the last couple of years, several parametric and nonparametric methods have been developed that distinguish between jumps and the continuous price process to estimate time-varying volatility robustly to jumps [105–110]. To date, bipower variation (BPV) or its variants remain to be among the dominant methods used to model jumps. These methods are based on the sums of powers and products of powers of absolute returns and have been shown to be robust to rare jumps in the log-price process [56,109,110].

All the high-frequency data–based methods discussed are mostly applicable to the univariate or single-asset volatility estimation scenarios. Nowadays, volatility estimation of the multiple asset scenarios is becoming increasingly important. Also, precise estimation of the covariance matrix of multiple asset returns is central to many issues in finance such as portfolio risk assessment and asset pricing. The availability of UHFD has spurred its use in many recent covariance asset estimation methods. This is mainly because UHFD better reflect the underlying assets returns processes because of better statistical efficiency and can greatly improve the accuracy of the covariance matrix estimates. However, as discussed earlier in this section, the use of UHFD can introduce noise as a result of market microstructure frictions. Estimation methods in multiple asset scenarios face additional difficulties due to nonsynchronous trading issues. These issues arise because the transactions for different assets occur at different points in time, are random, and are thus nonsynchronous. Due to this mismatch in the time points of the recorded transactions, returns sampled at regular intervals in calendar time will correlate with the previous or successive returns on other assets even in the absence of any underlying correlation structure [111,112]. This, called the Epps effect, causes the covariance estimator to be biased towards zero with increasing sampling frequencies [113]. Two key approaches have generally been used in the past to address these issues. One attempts to reduce the microstructure noise effects through the use of lead and lag autocovariance terms in the realized covariance estimator based on synchronized returns, whereas the other produces unbiased estimates of the covariance matrix by using the cross-product of all fully and partially overlapping event-time returns [114]. In recent years, several high-frequency data–based volatility estimation methods have been proposed for multiple asset scenarios, starting with the realized covariance estimator that is basically the sum of cross-products of intraday returns [115]. Like the realized volatility measure, this measure also suffers from the impact of microstructure noise [114,116–120]. Since then, many methods have been developed to deal with the inherent problems in the multiple asset settings, like the market microstructure noise and bias due to nonsynchronicity [114,116,118,119,121–123]. Many market participants often need to estimate matrices comprising a large number of assets using high-frequency data. However, many existing estimators can only be used for a small number of asset classes and become inconsistent as the size of the matrix becomes closer to or exceeds the sample size [124]. A few recent methods suggest different ways to estimate large volatility matrices, some of which are robust to the presence of microstructure noise and nonsynchronicity effects. They incorporate different schemes, ranging from the use of factor models and low-frequency dynamic models to pairwise and all-refresh time schemes [112,124–126].

### 17.5.3 UHFD (Big Data) Implications

UHFD-based estimators have important implications in many areas of finance, especially risk assessment and management or single assets and portfolios. For example, several studies have utilized the realized volatility estimators based on UHFD for VaR forecasting [127–131]. Realized volatility–based estimators have also been shown to explain variations in the cross-sectional and temporal behavior of risk premiums, when used in conjunction with the implied volatility measures [70]. Forecasting performance of realized volatility–based measures and its variants, such as the realized range–based estimators, has been shown to be favorable in many other risk measurement studies involving real high-frequency data from the NYSE and the Standards and Poors (S&P) 500 stock index [55,90,132,133]. Other studies based on high-frequency measures, such as BPV, reveal the significance of jumps in the price process evolution of commonly held assets [109,134]. Evidently, the growing prevalence of HFT has necessitated the utilization of UHFD for information gains in areas like short-term risk assessments. Unarguably, compared to the low-frequency–based risk models, high-frequency data yield relatively more precise and more adaptive short-term risk models. Besides benefiting high-frequency traders, UHFD have also been shown to benefit low-frequency traders like the traditional low-frequency risk and portfolio managers, who can benefit by utilizing high-frequency dynamic factor exposure estimates to hedge short-term risk factor exposures [135]. The use of UHFD is also being increasingly explored in other areas like asset pricing and governance [136–139].

### 17.5.4 UHFD (Big Data) Challenges

Clearly, the availability of UHFD has led to significant advancements in the field of finance and has made it possible for empirical researchers to address problems that cannot be handled using data collected at lower frequencies. However, various factors, such as the underlying market structure, market dynamics, internal process flows, trading frequency, and so forth, present several difficulties in the effective utilization of the vast amounts of UHFD available today. Moreover, the storage, analysis, and management of near-continuous data present significant new challenges. For instance, a highly traded stock can easily result in several millions of data points per year and require several hundreds of gigabytes of storage for a 12-month period [139]. Recorded trade data at ultrahigh frequencies generally have numerous data errors. Usually, the reason for the higher number of errors is attributed to the large volumes of trading. Data errors could originate from various sources. For example, HFT data could have isolated bad ticks, multiple bad ticks in succession, wrong ticks, decimal errors, transposition errors, typing errors, and reporting errors like duplicate trades or delayed trades [25,140]. These errors necessitate the need for accurate data filtering mechanisms that can identify and resolve such errors and can help convert the data into a usable form [25,140]. Also, as previously discussed in Sections 17.5.1–17.5.4, the usage of UHFD generally requires a trade-off between precision of the estimation technique and bias induced by the market microstructure noise effects. The bias increases with the increase in the sampling frequencies, which renders the estimator less accurate at high sampling frequencies. Organizational structure and institutional evolution of the equity markets further exacerbate such errors in the UHFD [14]. These issues represent some of the challenges that are being faced by both the researchers and practitioners with regard to UHFD. Despite several advancements in this field, a lot more remains to be done to efficiently utilize the available UHFD and extract the best value out of them.

## 17.6 SUMMARY

The financial industry has always been a data-intensive industry. Recent technological advancements coupled with several other factors like changing customer preferences and changing business needs have led to the generation and consumption of prolific amounts of data. Several changes in the last couple of years, driven by the confluence of factors like the escalating regulatory pressures, ever-increasing compliance requirements, regulatory oversight, global economic instability, increasing competition in the global markets, growing business demands, growing pressures to optimize capital and liquidity, and so forth, are forcing the financial organizations to rethink and restructure the way they do business. Also, traditional data management practices prevalent in finance can no longer effectively cope with the ever-increasing, huge, and rapid influx of heterogeneous data originating from a wide range of internal processes and external sources, including social media, blogs, audio, and video. Consequently, a growing number of financial institutions are resorting to Big Data to strategize their business decisions based on reliable factual insights supported by real data rather than just intuition. Increasingly, Big Data is being utilized in several areas such as investment analysis, econometrics, risk assessment, fraud detection, trading, customer interactions analysis, and behavior modeling. Efficient utilization of Big Data has become essential to the progress and success of many in this data-driven industry. However, Big Data by itself does not hold much value, and not all of it may be useful at all times. To gain relevant insights from the data, it is very important to deploy efficient solutions that can help analyze, manage, and utilize data. Many solutions have already been deployed in the financial domain to manage relevant data and perform analytics on them. Despite such advancements, many in the industry still lack the ability to address their Big Data needs. This is most likely due to the fact that different organizational units usually have different domain-specific requirements and, hence, solution specifications. So, it is highly unlikely that a solution deployed by one unit would be equally useful to others. The chapter described the impact of Big Data on the financial industry and presented some of the key transformations being driven by the data today. The availability of UHFD has resulted in significant advancements in the field of finance and has made it possible for empirical researchers to address problems that could not be handled using data collected at lower frequencies. Besides showcasing the impact of Big Data in the industrial sector, the chapter also highlighted the Big Data–driven progress in research in the fields of finance, financial econometrics, and statistics. The chapter exemplified some of the key developments in this area with special focus on financial risk measurement and management through the use UHDF-based volatility metrics. In recent years, the increased availability of ultrahigh-frequency trading data has spurred strong growth in the development of techniques that exploit tick-level intraday price data to better estimate volatility forecasts and overcome some of the limitations inherent in the traditional low-frequency–based systems. Availability of high-frequency data coupled with recent technological advancements has made analysis of large-scale trading data more accessible to market participants. Ultrahigh-frequency–based estimators have important implications in many areas of finance, specially risk assessment and management or single assets and portfolios. Among other things, such data have been shown to be extremely useful in the meaningful ex post evaluation of daily volatility forecasts. However, various factors, such as the underlying market structure, market dynamics, internal process flows, trading frequency, and so forth, present several difficulties in the effective utilization of the vast amounts of high-frequency data available today. Particularly, the market microstructure noise effects, such as those due to bid–ask bounce and infrequent trading, introduce a significant bias in the estimation procedures based on high-frequency data. The chapter also touched upon the trade-off that is often required between the precision of the estimation technique and bias induced by the market microstructure noise effects. The availability of Big Data in the financial domain has opened up new avenues for innovation and presented immense opportunities for growth and sustainability. Despite significant progress in the development and adoption of Big Data–based solutions in the field, a lot more remains to be done to effectively utilize the relevant data available in this domain and extract insightful information out of it. Compared to the promise Big Data holds in this domain and its potential, the progress in this field is still in its nascent stages, and a lot more growth in this area remains to be seen in the coming years.

## REFERENCES


1. Zervoudakis, F., Lawrence, D., Gontikas, G., and Al Merey, M., Perspectives on high-frequency trading. Available at http://www0.cs.ucl.ac.uk/staff/f.zervoudakis/documents/Perspectives_on_High-Frequency_Trading.pdf (accessed February 2014).
2. Connors, S., Courbe, J., and Waishampayan, V., Where have you been all my life? How the financial services industry can unlock the value in Big Data. PwC FS Viewpoint, October 2013.
3. Gutierrez, D., Why Big Data matters to finance, 2013. Available at http://www.inside-bigdata.com (accessed February 2014).
4. Versace, M., and Massey, K., The case for Big Data in the financial services industry. IDC Financial Insights, White paper, September 2012.
5. Flood, M., Mendelowitz, A., and Nichols, W., Monitoring financial stability in a complex world. In: Lemieux, V. (Ed.), *Financial Analysis and Risk Management: Data Governance, Analytics and Life Cycle Management.* Heidelberg: Springer Berlin, pp. 15–45, 2013.
6. Brett, L., Love, R., and Lewis, H., Big Data: Time for a lean approach in financial services. A Deloitte Analytics paper, 2012.
7. Oracle Corporation, Financial services data management: Big Data technology in financial services. *Oracle Financial Services*, An Oracle White paper, June 2012.
8. Rauchman, M., and Nazaruk, A., Big Data in Capital Markets. *Proceedings of the 2013 International Conference on Management of Data.* New York: ACM, 2013.
9. Jagadish, H.V., Kyle, A., and Raschid, L., Envisioning the next generation financial cyberinfrastructure: Transforming the monitoring and regulation of systemic risk. Available at ftp://ftp.umiacs.umd.edu/incoming/louiqa/PUB2012/HBR-NextGen_v4.pdf (last accessed 2013).

10. Available at http://www.opradata.com/specs/Traffic_Projections.pdf (accessed February 2014).
11. RBC Global Asset Management (U.S.) Inc., U.S. market structure: Is this what we asked for? *RBC Global Asset Management*, White paper, February 2012.
12. [BATS Global Markets]. Available at http://batstrading.com/market_summary/[BATS Global Markets]/ (accessed February 2014).
13. Adamic, L., Brunetti, C., Harris, J., and Kirilenko, A., Trading networks. Manuscript, University of Michagan, Johns Hopkins University, University of Delaware, and Commodity Futures Trading Commission, 2010.
14. Ait Sahalia, Y., Mykland, P.A., and Zhang, L., Ultra high frequency volatility estimation with dependent microstructure noise. *Journal of Econometrics*, 160(1), 160, 2011.
15. IBM Corp., NYSE Euronext: Adapting to market changes with near-real-time insight into information. IBM Case Study, March 2013. Available at http://www-01.ibm.com/software/success/cssdb.nsf/CS/JHUN-95XMPN?OpenDocument&Site=default&cty=en_us.
16. Available at http://www-01.ibm.com/software/data/netezza/ (accessed February 2014).
17. Available at http://www.gopivotal.com/ (accessed February 2014).
18. Available at https://scalableinformatics.com/ (accessed February 2014).
19. Bertolucci, J., Big Data opens new doors for financial analysts. *Information Week*. Available at http://www.informationweek.com/big-data/big-data-analytics/big-data-opens-new-doors-for-financial-analysts/d/d-id/1111460? (accessed February 2014).
20. Bartiromo, R., Maximum entropy distribution of stock price fluctuations. *Physica A: Statistical Mechanics and Its Applications*, 392(7), 1638, 2013.
21. Verousis, T., and Gwilym, O., An improved algorithm for cleaning ultra high-frequency data. *Journal of Derivatives and Hedge Funds*, 15(4), 323–340, 2010.
22. Engle, R.F., The econometrics of ultra high frequency data. *Econometrica*, 68(1), 1–22, 2000.
23. Dahlhaus, R., and Neddermeyer, J.C., Online spot volatility-estimation and decomposition with nonlinear market microstructure noise models. *Journal of Financial Econometrics*, 12(1), 174–212, 2013.
24. Miao, H., Potential applications of function data analysis in high-frequency financial research. *Journal of Business and Financial Affairs*, 2, e125, 2013.
25. Falkenberry, T.N., High frequency data filtering. White paper, 2002. Available at http://www.tickdata.com/pdf/Tick_Data_Filtering_White_Paper.pdf.
26. Poon, S.H., and Granger, C., Forecasting financial market volatility: A review. *Journal of Economic Literature*, 41(2), 478–539, 2003.
27. Ait-Sahalia, Y., and Yu, J., High frequency market microstructure noise estimates and liquidity measures. *Annals of Applied Statistics*, 3(1), 422–457, 2009.
28. Tripathy, T., and Gil-Alana, L.A., Suitability of volatility models for forecasting stock market returns: A study on the Indian National Stock Exchange. *American Journal of Applied Sciences*, 7(11), 1487–1494, 2010.
29. Louzis, D.P., Xanthopoulos-Sisinis, S., and Refenes, A.P., The role of high-frequency intra-daily data, daily range and implied volatility in multi-period value-at-risk forecasting. *Journal of Forecasting*, 32(6), 561–576, 2013.
30. Drakos, A.A., Kouretas, G.P., and Zarangas, L.P., Forecasting financial volatility of the Athens stock exchange daily returns: An application of the asymmetric normal mixture GARCH model. *International Journal of Finance and Economics*, 15, 331–350, 2010.
31. Campel, A., and Chen, X.L., The year of living riskily. *Risk*, 21, 28–32, 2008.
32. Bannough, K., Measuring and forecasting financial market volatility using high-frequency data. Thesis, Erasmus Research Institute of Management (ERIM), Erasmus University Rotterdam, Netherlands, January 2013.
33. Webb, R.I., Muthuswamy, J., and Segara, R., Market microstructure effects on volatility at the TAIFEX. *Journal of Futures Markets*, 27(12), 1219–1243, 2007.

34. Mandelbrot, B., The variation of certain speculative prices. *Journal of Business*, 36(4), 394–419, 1963.
35. Engle, R., Autoregressive conditional heteroscedasticity with estimates of the variance of United Kingdom inflation. *Econometrica*, 50(4), 987–1007, 1982.
36. Bollerslev, T., Generalized autoregressive conditional heteroscedasticity. *Journal of Econometrics*, 31(3), 307–327, 1986.
37. Bollerslev, T., Chou, R.Y., and Kroner, K.F., ARCH modeling in finance: A review of the theory and empirical evidence. *Journal of Econometrics*, 52, 5–59, 1992.
38. Gourieroux, C., *ARCH Models and Financial Applications*. New York: Springer, 1997.
39. Shephard, N., Statistical aspects of ARCH and stochastic volatility. In: Cox, D.R., Hinkley, D.V., and Barndorff-Nielsen, O.E. (Eds.), *Time Series Models in Econometrics, Finance, and Other Fields*. London: Chapman and Hall, pp. 1–67, 1996.
40. Wang, Y., Asymptotic nonequivalence of ARCH models and diffusions. *The Annals of Statistics*, 30, 754–783, 2002.
41. Glosten, L.R., Jagannathan, R., and Runkle, D.E., On the relation between the expected value and the volatility of the nominal excess return on stocks. *Journal of Finance*, 48, 1779–1801, 1993.
42. Engle, R.F., and Lee, G.G., A permanent and transitory component model of stock return volatility. In: Engle, R.F., and White, H. (Eds.), *Cointegation, Casuality and Forecasting: A Festschrift in Honour of Clive W. J. Granger*. Oxford: Oxford University Press, pp. 475–497, 1999.
43. Baillie, R.T., Bollerslev, T., and Mikkelsen, H.O., Fractionally integrated generalized autoregressive conditional heteroskedasticity. *Journal of Econometrics*, 73, 3–20, 1996.
44. Müller, H.-G., Sen, R., and Stadtmller, U., Functional data analysis for volatility. *Journal of Econometrics*, 165(2), 233–245, 2011.
45. Ghysels, E., and Sinko, A., Volatility forecasting and microstructure noise. *Journal of Econometrics*, 160(1), 257–271, 2011.
46. Bollerslev, T., Engle, R., and Nelson, D., ARCH models. In: Engle, R., and McFadden, D. (Eds.), *Handbook of Econometrics*, Vol. IV. Amsterdam: Elsevier, 1994.
47. Akgiray, V., Conditional heteroskedasticity in time series of stock returns: Evidence and forecasts. *Journal of Business*, 62, 55–80, 1989.
48. Brailsford, T.F., and Faff, R.W., An evaluation of volatility forecasting techniques. *Journal of Banking and Finance*, 20, 419–438, 1996.
49. Figlewski, S., Forecasting volatility. *Financial Markets, Institutions and Instruments*, 6, 1–88, 1997.
50. Frances, P.H., and Van Dijk, D., Forecasting stock market volatility using (non-linear) GARCH models. *Journal of Forecasting*, 15, 229–235, 1995.
51. Cartea, Á., and Karyampas, D., Volatility and covariation of financial assets: A high-frequency analysis. *Journal of Banking and Finance*, 35(12), 3319–3334, 2011.
52. Andersen, T.G., Bollerslev, T., Diebold, F.X., and Labys, P., Modeling and forecasting realized volatility. *Econometrica*, 71, 579–625, 2003.
53. Bardorff-Nielsen, O., and Shephard, N., Non-Gaussian Ornstein-Uhlenbeck based models and some of their applications in financial economics. *Journal of the Royal Statistical Society B*, 63, 167–241, 2001.
54. Chortareas, G., Jiang, Y., and Nankervis, J.C., Forecasting exchange rate volatility using high-frequency data: Is the euro different? *International Journal of Forecasting*, 27(4), 1089–1107, 2011.
55. Martens, M., and van Dijk, D., Measuring volatility with the realized range. *Journal of Econometrics*, 138(1), 181–207, 2007.
56. Barndorff-Nielsen, O.E., and Shephard, N., Power and bipower variation with stochastic volatility and jumps. *Journal of Financial Econometrics*, 2(1), 1–48, 2004.
57. Grammig, J., and Wellner, M., Modeling the interdependence of volatility and inter-transaction duration processes. *Journal of Econometrics*, 106(2), 369–400, 2002.

58. Bollerslev, T., and Wright, J.H., High-frequency data, frequency domain inference, and volatility forecasting. *Review of Economics and Statistics*, 83(4), 596–602, 2001.
59. Comte, F., and Renault, E., Long-memory in continuous-time stochastic volatility models. *Mathematical Finance*, 8, 291–323, 1998.
60. McAller, M., and Medeiros, M., Realized volatility: A review. *Econometric Reviews*, 27, 10–45, 2008.
61. Andersen, T.G., Bollerslev, T., Diebold, F.X., and Labys, P., The distribution of realized exchange rate volatility. *Journal of the American Statistical Association*, 96, 42–55, 2001.
62. Andersen, T.G., and Bollerslev, T., Answering the skeptics: Yes, standard volatility models do provide accurate forecasts. *International Economic Review*, 39, 885–905, 1998.
63. Andersen, T.G., and Bollerslev, T., Deutsche Mark–dollar volatility: Intraday activity patterns, macroeconomic announcements, and longer run dependencies. *Journal of Finance*, 53, 219–265, 1998.
64. Andersen, T.G., Bollerslev, T., Diebold, F.X., and Ebens, H., The distribution of realized stock return volatility. *Journal of Financial Economics*, 61, 43–76, 2001.
65. Barndorff-Nielsen, O.E., and Shephard, N., Econometric analysis of realized volatility and its use in estimating stochastic volatility models. *Journal of the Royal Statistical Society B*, 64, 253–280, 2002.
66. Barndorff-Nielsen, O.L., and Shephard, N., Estimating quadratic variation using realized variance. *Journal of Applied Econometrics*, 17, 457–477, 2002.
67. Andreou, E., and Ghysels, E., Rolling-sample volatility estimators: Some new theoretical, simulation and empirical results. *Journal of Business and Economic Statistics*, 20, 363–376, 2002.
68. Andersen, T., Bollerslev, T., and Meddahi, N., Correcting the errors: Volatility forecast evaluation using high-frequency data and realized volatilities. *Econometrica*, 73, 279–296, 2005.
69. Koopman, S., Jungbacker, B., and Hol, E., Forecasting daily variability of the S&P100 stock index using historical, realised and implied volatility measurements. *Journal of Empirical Finance*, 12, 445–475, 2005.
70. Bollerslev, T., Tauchen, G., and Zhou, H., Expected stock returns and variance risk premia. *Review of Financial Studies*, 22, 4463–4492, 2009.
71. Bollerslev, T., Gibson, M., and Zhou, H., Dynamic estimation of volatility risk premia and investor risk aversion from option implied and realized volatilities. *Journal of Econometrics*, 160, 235–245, 2011.
72. Ghysels, E., and Sinko, A., Comment. *Journal of Business and Economic Statistics*, 24, 192–194, 2006.
73. Müller, U.A., Dacorogna, M.M., Davé, R.D., Pictet, O.V., Olsen, R.B., and Ward, J.R., Fractals and intrinsic time—A challenge to econometricians. In: *International AEA Conference on Real Time Econometrics*, Luxembourg, October 14–15, 1993.
74. Ait-Sahalia, Y., Mykland, P.A., and Zhang, L., How often to sample a continuous-time process in the presence of market microstructure noise. *Review of Financial Studies*, 18, 351–416, 2005.
75. Bandi, F.M., and Russell, J.R., Microstructure noise, realized variance, and optimal sampling. *Review of Economic Studies*, 75, 339–369, 2008.
76. Zhang, L., Mykland, P., and Aït-Sahalia, Y., A tale of two time scales: Determining integrated volatility with noisy high frequency data. *Journal of the American Statistical Association*, 100, 1394–1411, 2005.
77. Zhang, L., Efficient estimation of stochastic volatility using noisy observations: A multi-scale approach. *Bernoulli*, 12, 1019–1043, 2006.
78. Barndorff-Nielsen, O.E., Hansen, P.R., Lunde, A., and Shephard, N., Designing realised kernels to measure the ex-post variation of equity prices in the presence of noise. *Econometrica*, 76, 1481–1536, 2008.
79. Barndorff-Neilsen, O.E., Hansen, P.R., Lunde, A., and Shephard, N., Subsampling realised kernels. *Journal of Econometrics*, 160, 204–219, 2011.

80. Hansen, P.R., and Lunde, A., Realized variance and market microstructure noise (with comments and rejoinder). *Journal of Business and Economic Statistics*, 24, 127–218, 2006.
81. Barucci, E., Magno, D., and Mancino, M., Fourier volatility forecasting with high-frequency data and microstructure noise. *Quantitative Finance*, 12(2), 281–293, 2012.
82. Harris, L., Estimation of stock price variances and serial covariances from discrete observations. *Journal of Financial and Quantitative Analysis*, 25, 291–306, 1990.
83. Zeng, Y., A partially observed model for micro movement of asset prices with Bayes estimation via filtering. *Mathematical Finance*, 13, 411–444, 2003.
84. Fan, J., and Wang, Y., Spot volatility estimation for high-frequency data. *Statistics and Its Interface*, 1, 279–288, 2008.
85. Bos, C.S., Janus, P., and Koopman, S.J., Spot variance path estimation and its application to high frequency jump testing. Discussion Paper TI 2009-110/4, Tinbergen Institute, 2009.
86. Kristensen, D., Nonparametric filtering of the realized spot volatility: A kernel-based approach. *Econometric Theory*, 26, 60–93, 2010.
87. Munk, A., and Schmidt-Hieber, J., Nonparametric estimation of the volatility function in a high-frequency model corrupted by noise. Unpublished manuscript, 2009.
88. Zu, Y., and Boswijk, P., Estimating spot volatility with high frequency financial data. Preprint, University of Amsterdam, 2010.
89. Shephard, N., and Sheppard, K., Realising the future: Forecasting with high frequency-based volatility (HEAVY) models. *Journal of Applied Econometrics*, 25, 197–231, 2010.
90. Brownlees, C.T., and Gallo, G.M., Comparison of volatility measures: A risk management perspective. *Journal of Financial Econometrics*, 8, 29–56, 2010.
91. Maheu, J.M., and McCurdy, T.J., Do high-frequency measures of volatility improve forecasts of return distributions? *Journal of Econometrics*, 160, 69–76, 2011.
92. Hansen, P.R., Huang, Z., and Shek, H.H., Realized GARCH: A joint model for returns and realized measures of volatility. *Journal of Applied Econometrics*, 27, 877–906, 2012.
93. Andersen, T.G., Bollerslev, T., and Meddahi, N., Analytical evaluation of volatility forecasts. *International Economic Review*, 45, 1079–1110, 2004.
94. Ghysels, E., Santa-Clara, P., and Valkanov, R., Predicting volatility: Getting the most out of return data sampled at different frequencies. *Journal of Econometrics*, 131, 59–95, 2006.
95. Andersen, T.G., Bollerslev, T., Frederiksen, P.H., and Nielsen, M.Ø., Continuous-time models, realized volatilities, and testable distributional implications for daily stock returns. Working paper, Northwestern University, 2006.
96. Corsi, F., A simple approximate long-memory model of realized volatility. *Journal of Financial Econometrics*, 7(2), 174–196, 2009.
97. Boudt, K., Cornelissen, J., and Payseur, S., High-frequency: Toolkit for the analysis of high frequency financial data in R. Available at http://highfrequency.herokuapp.com (accessed August 2013).
98. Andersen, T.G., Bollerslev, T., Diebold, F.X., and Labys, P., Great realizations. *Risk*, 13, 105–108, 2000.
99. Hafner, C.M., Cross-correlating wavelet coefficients with applications to high-frequency financial time series. *Journal of Applied Statistics*, 39(6), 1363–1379, 2012.
100. Podolskij, M., and Vetter, M., Estimation of volatility functionals in the simultaneous presence of microstructure noise and jumps. *Bernoulli*, 15(3), 634–658, 2009.
101. Jacod, J., Li, Y., Mykland, P.A., Podolskij, M., and Vetter, M., Microstructure noise in the continuous case: The pre-averaging approach. *Stochastic Processes and Their Applications*, 119(7), 2249–2276, 2009.
102. Jacod, J., and Protter, P., Asymptotic error distributions for the Euler method for stochastic differential equations. *The Annals of Probability*, 26, 267–307, 1998.

103. Meddahi, N., A theoretical comparison between integrated and realized volatility. *Journal of Applied Econometrics*, 17, 479–508, 2002.
104. Roberto, R., Jump-diffusion models: Including discontinuous variation. Lecture Notes, November 2011. Available at http://www.econ-pol.unisi.it/fm20/jump_diffusion_notes.pdf (accessed May 2013).
105. Mancini, C., Estimation of the characteristics of jump of a general Poisson-diffusion process. *Scandinavian Actuarial Journal*, 1, 42–52, 2004.
106. Aït-Sahalia, Y., and Jacod, J., Volatility estimators for discretely sampled Lévy processes. *Annals of Statistics*, 35, 335–392, 2007.
107. Jacod, J., Asymptotic properties of realized power variations and related functionals of semi-martingales. *Stochastic Processes and Their Applications*, 118, 517–559, 2008.
108. Jacod, J., Statistics and high frequency data. In: Kessler, M., Lindner, A., and Sorensen, M. (Eds.), *Statistical Methods for Stochastic Differential Equations*. Boca Raton, FL: Chapman and Hall, pp. 191–310, 2012.
109. Barndorff-Nielsen, O.E., and Shephard, N., Econometrics using bipower variation. *Journal of Financial Econometrics*, 4, 1–30, 2006.
110. Barndorff-Nielsen, O.E., Graversen, S.E., Jacod, J., Podolskij, M., and Shephard, N., A central limit theorem for realised power and bipower variations of continuous semimartingales. In: Kabanov, Y., Lipster, R., and Stoyanov, J. (Eds.), *From Stochastic Analysis to Mathematical Finance, Festschrift for Albert Shiryaev*. New York: Springer, pp. 33–68, 2006.
111. Fisher, L., Some new stock-market indexes. *Journal of Business*, 39(1–2), 191–225, 1966.
112. Tao, M., Large volatility matrix inference via combining low frequency and high-frequency approaches. *Journal of the American Statistical Association*, 106(495), 1025–1040, 2011.
113. Epps, T.W., Co-movements in stock prices in the very short run. *Journal of the American Statistical Association*, 74(366), 291–298, 1979.
114. Griffin, J.E., and Oomen, R.A., Covariance measurement in the presence of non-synchronous trading and market microstructure noise. *Journal of Econometrics*, 160(1), 58–68, 2011.
115. Barndorff-Nielsen, O.E., and Shephard, N., Econometric analysis of realized covariation: High frequency based covariance, regression, and correlation in financial economics. *Econometrica*, 72, 885–925, 2004.
116. Hayashi, T., and Yoshida, N., On covariance estimation of non-synchronously observed diffusion processes. *Bernoulli*, 11, 359–379, 2005.
117. Sheppard, K., Realized covariance and scrambling. Unpublished manuscript, 30, 2006.
118. Zhang, L., Estimating covariation: Epps effect and microstructure noise. *Journal of Econometrics*, 160(1), 33–47, 2011.
119. Voev, V., and Lunde, A., Integrated covariance estimation using high-frequency data in the presence of noise. *Journal of Financial Econometrics*, 5(1), 68–104, 2007.
120. Bandi, F.M., Russell, J.R., and Zhu, Y., Using high-frequency data in dynamic portfolio choice. *Econometric Reviews*, 27, 163–198, 2008.
121. Bandi, F.M., and Russell, J.R., Realized covariation, realized beta and microstructure noise. Working paper, Graduate School of Business, University of Chicago, 2005.
122. Martens, M., Estimating unbiased and precise realized covariances. In: *EFA 2004 Maastricht Meetings Paper No. 4299*, June 2004.
123. Hautsch, N., Kyj, L., and Oomen, R., A blocking and regularization approach to high dimensional realized covariance estimation. *Journal of Applied Econometrics*, 27, 625–645, 2012.
124. Tao, M., Wang, Y., and Chen, X., Fast convergence rates in estimating large volatility matrices using high-frequency financial data. *Econometric Theory*, 29(4), 838–856, 2013.
125. Noureldin, D., Shephard, N., and Sheppard, K., Multivariate high-frequency-based volatility (HEAVY) models. *Journal of Applied Econometrics*, 27(6), 907–933, 2012.

126. Fan, J., Li, Y., and Yu, K., Vast volatility matrix estimation using high-frequency data for portfolio selection. *Journal of the American Statistical Association*, 107(497), 412–428, 2012.
127. Giot, P., and Laurent, S., Modeling daily value-at-risk using realized volatility and ARCH type models. *Journal of Empirical Finance*, 11, 379–398, 2004.
128. Beltratti, A., and Morana, C., Statistical benefits of value-at-risk with long memory. *Journal of Risk*, 7, 21–45, 2005.
129. Angelidis, T., and Degiannakis, S., Volatility forecasting: Intra-day versus inter-day models. *Journal of International Financial Markets, Institutions and Money*, 18, 449–465, 2008.
130. Martens, M., van Dijk, D., and Pooter, M., Forecasting S&P 500 volatility: Long memory, level shifts, leverage effects, day of the week seasonality and macroeconomic announcements. *International Journal of Forecasting*, 25, 282–303, 2009.
131. Louzis, D.P., Xanthopoulos-Sisinis, S., and Refenes, A., Are realized volatility models good candidates for alternative value at risk prediction strategies? Germany: University Library of Munich, 2011.
132. Shao, X.D., Lian, Y.J., and Yin, L.Q., Forecasting value-at-risk using high frequency data: The realized range model. *Global Finance Journal*, 20, 128–136, 2009.
133. Christensen, K., and Podolskij, M., Realized range-based estimation of integrated variance. *Journal of Econometrics*, 141, 323–349, 2007.
134. Huang, X., and Tauchen, G., The relative contribution of jumps to total price variation. *Journal of Financial Econometrics*, 3, 456–499, 2005.
135. De Rossi, G., Zhang, H., Jessop, D., and Jones, C., High frequency data for low frequency managers: Hedging market exposure. White paper, UBS Investment Research, 2012.
136. Aït-Sahalia, Y., and Saglam, M., High frequency traders: Taking advantage of speed (No. w19531). National Bureau of Economic Research, 2013.
137. Brogaard, J., Hendershott, T., and Riordan, R., High frequency trading and price discovery, April 2013. Available at http://ssrn.com/abstract=1928510.
138. Bollerslev, T., and Todorov, V., Tails, fears and risk premia. *Journal of Finance*, 66(6), 2165–2211, 2011.
139. Todorov, V., and Bollerslev, T., Jumps and betas: A new framework for disentangling and estimating systematic risks. *Journal of Econometrics*, 157(2), 220–235, 2010.
140. Brownlees, C.T., and Gallo, G.M., Financial econometric analysis at ultra-high frequency: Data handling concerns. *Computational Statistics and Data Analysis*, 51(4), 2232–2245, 2006.

CHAPTER 18

# Semantic-Based Heterogeneous Multimedia Big Data Retrieval

Kehua Guo and Jianhua Ma

CONTENTS



ABSTRACT

Nowadays, data heterogeneity is one of the most critical features for multimedia Big Data; searching heterogeneous multimedia documents reflecting users' query intent from a Big Data environment is a difficult task in information retrieval and pattern recognition. This chapter proposes a heterogeneous multimedia Big Data retrieval framework that can achieve good retrieval accuracy and performance. The chapter is organized as follows. In Section 18.1, we address the particularity of heterogeneous multimedia retrieval in a Big Data environment and introduce the background of the topic. Then literatures related to current multimedia retrieval approaches are briefly reviewed, and the general concept of the proposed framework is introduced briefly in Section 18.2. In Section 18.3, the description of this framework is given in detail including semantic information extraction, representation, storage, and multimedia Big Data retrieval. The performance evaluations are shown in Section 18.4, and finally, we conclude the chapter in Section 18.5.

## 18.1 INTRODUCTION

Multimedia retrieval is an important technology in many applications such as web-scale multimedia search engines, mobile multimedia search, remote video surveillance, automation creation, and e-government [1]. With the widespread use of multimedia documents, our world will be swamped with multimedia content such as massive images, videos, audios, and other content. Therefore, traditional multimedia retrieval has been switching into a Big Data environment, and the research into solving some problems according to the features of multimedia Big Data retrieval attracts considerable attention.

At present, multimedia retrieval from Big Data environments is facing two problems: The first is document type heterogeneity. The multimedia content generated from various applications may be huge and unstructured. For example, the different types of multimedia services will generate images, videos, audios, graphics, or text documents. Such heterogeneity makes it difficult to execute a heterogeneous search. The second is intent expression. In multimedia retrieval, the query intent generally can be represented by text. However, the text can only express very limited query intent. Users do not want to enter too many key words, but short text may lead to ambiguity. In many cases, the query intent may be described by content (for example, we can search some portraits that are similar to an uploaded image), but content-based retrieval may ignore personal understanding because semantic information cannot be described by physical features such as color, texture, shape, and so forth. Therefore, the returned results may be far from satisfying users' search intent.

These features have become new challenges in the research of Big Data–based multimedia information retrieval. Therefore, how to simultaneously consider the type heterogeneity and user intent in order to guarantee good retrieval performance and economical efficiency has been an important issue.

Traditional multimedia retrieval approaches can be divided into three classes: text-based, content-based, and semantic-based retrieval [1,2]. The text-based approach has been widely used in many commercial Internet-scale multimedia search engines. In this approach, a user types some key words, and then the search engine searches for the text in a host document and returns the multimedia files whose surrounding text contains the key words. This approach has the following disadvantages: (1) Users can only type short or ambiguous key words because the users' query intent usually cannot be correctly described by text. For example, when a user inputs the key word "Apple," the result may contain fruit, a logo, and an Apple phone. (2) To a multimedia database that stores only multimedia documents and in which the surrounding text does not exist, the text-based approach will be useless. For example, if a user stores many animation videos in the database, he/she can only query the videos based on simple information (e.g., filename), because the text related to them is limited.

Another multimedia retrieval approach is content based. In this approach, a user uploads a file (e.g., an image); the search engine searches for documents that are similar to it using content-based approaches. This approach will suffer from three disadvantages: (1) This approach cannot support heterogeneous search (e.g., upload an image to search for audio). (2) The search engine will ignore the users' query intent and cannot get similar results satisfying users' search intent. (3) The computation time of feature attraction will cost much computation resources.

The third approach is semantic based. In this approach, semantic features of multimedia documents are described by an ontology method and stored in the server's knowledge base; when the match requirement arrives, the server will execute retrieval in the knowledge base. However, if the multimedia document leaves the knowledge base, the retrieval process cannot be executed unless the semantic information is rebuilt.

This chapter is summarized from our recent work [1,2]. In this chapter, we use the semantic-based approach to represent users' intent and propose a storage and retrieval framework supporting heterogeneous multimedia Big Data retrieval. The characteristics of this framework are as follows: (1) It supports heterogeneous multimedia retrieval. We can upload a multimedia document with any multimedia type (such as image, video, or audio) to obtain suitable documents with various types. (2) There is convenience in interaction. This framework provides retrieval interfaces similar to traditional commercial search engines for convenient retrieval. (3) It saves data space. We store the text-represented semantic information in the database and then provide links to the real multimedia documents instead of directly processing multimedia data with large size. (4) There is efficiency with economic processing. We use a NoSQL database on some inexpensive computers to store the semantic information and use an open-sourced Hadoop tool to process the retrieval. The experiment results show that this framework can effectively search the heterogeneous multimedia documents reflecting users' query intent.

## 18.2 RELATED WORK

In the past decades, heterogeneous multimedia retrieval was done mainly using text-based approaches. But, in fact, this retrieval is only based on the text surrounding multimedia documents in host files (e.g., web pages). Although we can type key words to obtain various types of multimedia documents, these methods do not intrinsically support heterogeneous retrieval. If the text does not exist, the retrieval cannot be executed.

In a Big Data environment, information retrieval encounters some particular problems because of the data complexity, uncertainty, and emergence [3]. At present, research is focusing on the massive data capture, indexing optimization, and improvement of retrieval [4]. In a Big Data process, NoSQL technology is useful to store the information, which can be represented in map format. Apache HBase is a typical database to realize the NoSQL idea that simplifies the design, provides horizontal scaling, and gives finer control over availability. The features of HBase are outlined in the original Google File System [5] and

BigTable [6]. Tables in HBase can serve as the input and output for MapReduce [7] jobs running in Hadoop [8] and may be accessed through some typical application program interfaces (APIs), such as Java [9].

The research on heterogeneous multimedia retrieval approaches has mainly concentrated on how to combine the text-based approach with other methods [10,11]. Although it is more convenient for users to type text key words, content-based retrieval (e.g., images) has been widely used in some commercial search engines (e.g., Google Images search). However, it is very difficult to execute heterogeneous retrieval based on multimedia content [12,13]. For example, given a video and audio document relevant to the same artist, it is difficult to identify the artist or extract other similar features from the binary data of the two documents because their data formats are different. Therefore, full heterogeneous multimedia retrieval has not been achieved.

The approach proposed in this chapter mainly uses semantic information to support heterogeneous multimedia retrieval. Feature recognition is based on not only the low-level visual features, such as color, texture, and shape, but also full consideration of the semantic information, such as event, experience, and sensibility [14,15]. At present, the semantic information extraction approaches generally use the model of text semantic analysis, which constructs the relation between a text phrase and visual description in latent space or a random field. For an image, a bag-of-words model [16] is widely used to express visual words. Objects can be described by visual words to realize semantic learning. In addition, a topic model is widely used for the semantic extraction. The typical topic models are probabilistic latent semantic analysis (PLSA) [17] and latent Dirichlet allocation (LDA) [18]. Based on these models, many semantic learning approaches have been proposed. A study [19] proposed an algorithm to improve the training effect of image classification in high-dimension feature space. Previous works [20,21] have proposed some multiple-class annotation approaches based on supervision.

In the field of unsupervised learning, a study [22] has proposed a normalized cut clustering algorithm. Another study [23] presented an improved PLSA model. In addition, some researchers have combined relative feedback and machine learning. One study [24] used feedback and obtained a model collaborative method to extract the semantic information and get recognition results with higher precision. Another [25] used a support vector machine (SVM) active learning algorithm to handle feedback information; the users could choose the image reflecting the query intent.

In the field of system development, one study [26] used the hidden Markov model (HMM) to construct an automatic semantic image retrieval system. This approach could express the relation of image features after being given a semantic class. A Digital Shape Workbench [27–29] provided an approach to realize sharing and reuse of design resources and used ontology technology to describe the resources and high-level semantic information of three-dimensional models. In this system, ontology-driven annotation of the Shapes method and ShapeAnnotator were used for user interaction. Based on this work, another study [30] investigated the ontology expression to virtual humans, covering the features, functions, and skills of a human. Purdue University is responsible for the conception of the Engineering Ontology and Engineering Lexicon and proposed a calculable search framework [31]. A study [32] used the Semantic Web Rule Language to define the semantic matching. Another [33] proposed a hierarchical ontology-based knowledge representation model.

In current approaches, semantic features and multimedia documents are stored in the server's rational database, and we have to purchase expensive servers to process the retrieval. On the one hand, if the multimedia document is not stored in the knowledge base, the retrieval process cannot be conducted unless the semantic information is rebuilt. On the other hand, multimedia data with large size will cost much storage space. So, our framework presents an effective and economical framework that uses an inexpensive investment to store and retrieve the semantic information from heterogeneous multimedia data. In this framework, we do not directly process multimedia data with large size; in HBase, we only store the ontology-represented semantic information, which can be parallel-processed in distributed nodes with a MapReduce-based retrieval algorithm.

In our framework, we apply some valuable assets to facilitate the intent-reflective retrieval of heterogeneous multimedia Big Data. Big Data processing tools (e.g., Hadoop) are open source and convenient in that they can be freely obtained. For example, Hadoop only provides a programming model to perform distributed computing, and we can use the traditional retrieval algorithm after designing the computing model that is suitable for the MapReduce programming specification. In addition, the semantic information of the multimedia documents can also be easily obtained and saved because of the existence of many computing models. For example, we can provide the semantic information to the multimedia documents through annotating by social user and describe the semantic information by ontology technology. These technologies have been widely used in various fields.

## 18.3 PROPOSED FRAMEWORK

### 18.3.1 Overview

This section will present our framework on how to combine the semantic information and multimedia documents to perform Big Data retrieval. Our framework adopts a four-step architecture, shown in Figure 18.1. The architecture mainly consists of semantic

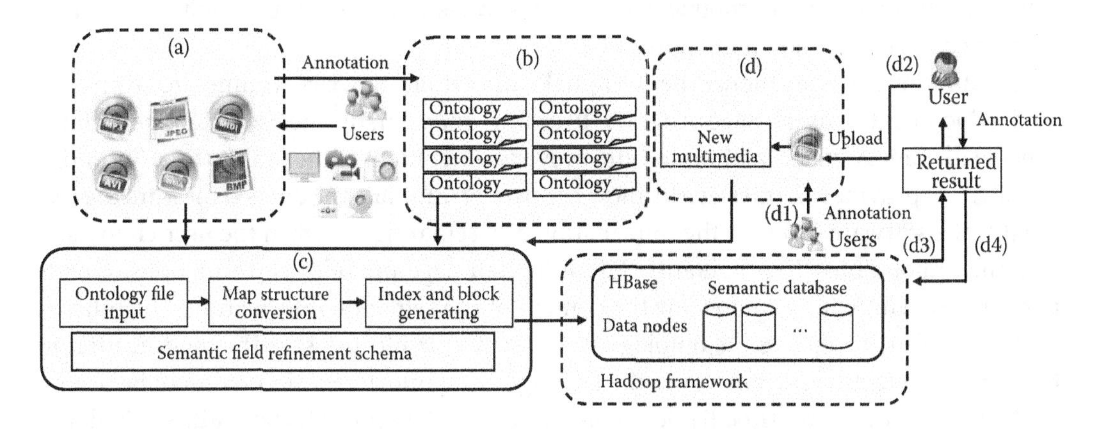


FIGURE 18.1 Structure of the framework.

annotation (Figure 18.1a), semantic representation (Figure 18.1b), NoSQL-based semantic storage (Figure 18.1c), and heterogeneous multimedia retrieval (Figure 18.1d) steps.

In the semantic annotation step, each multimedia document will be annotated by social users according to their personal understanding. The multimedia types may include images, videos, audio, and text with various formats. The multimedia content will be obtained from various generating sources, such as web crawling, sensor collection, user generation, and so forth. After the semantic extraction, the text annotations provided by users will be represented by ontology in the second step. A weight adjustment schema is used to adjust the weight of every semantic field in the ontology.

In the third step, the ontology file will be saved into a NoSQL-based storage linked to the real multimedia data with the location information. In order to better adapt to the NoSQL-based Big Data processing tool, we will use the map (key/value) structure conversion process to normalize the storage format of the multimedia location and its corresponding semantic information. After the new structures are generated, we will generate the index and storage block, according to which the ontology will be saved into the NoSQL-based distributed semantic database managed by HBase.

The fourth step is the retrieval process. In this step, the user can upload an annotated multimedia document with arbitrarily arbitrary format to execute heterogeneous multimedia retrieval (Figure 18.1d1–d2). In this case, the engine will return the results by matching the annotations of an uploaded document with semantic information in the database (Figure 18.1d3). Finally, users will be asked to give additional annotations to the multimedia document they selected (Figure 18.1d4) to make the annotations more abundant and accurate.

### 18.3.2 Semantic Annotation

In this framework, multimedia documents will be semantically annotated by several users, and the semantic information will be saved into HBase. The users can annotate a multimedia document in the interfaces provided by software tools. All the annotations will be described by text.

For annotation users, we have developed software named SemanticMultimediaViewer (SMV) [1], which can read, analyze, and annotate semantic multimedia documents. This software, running on PC or mobile devices, can save new documents to a multimedia database. The interface is shown in Figure 18.2.

In SMV, the users can choose the "Original Annotation" tab page to annotate any multimedia document. This tab page provides three functions. Firstly, the users can open a multimedia file (image, video, audio, etc.) and preview the content of the multimedia. Secondly, the tab page provides an interface to allow the user to annotate the opened multimedia file. Finally, the software can save the information as a semantic file when the user chooses to save the annotations. The "Semantic File Edit" tab page provides two functions. One is to open a semantic file and preview the content of the semantic information. The other is to insert, delete, or update the ontology in the semantic file and save the modification to the semantic file. In our framework, the tab page "Semantic Operation" will not be used.

We define $m$ as a multimedia document and $C$ as the set of all the multimedia documents (including images, videos, audio, etc.) satisfying $C = \{m_1, m_2,\ldots,m_N\}$ (where $N$ is the

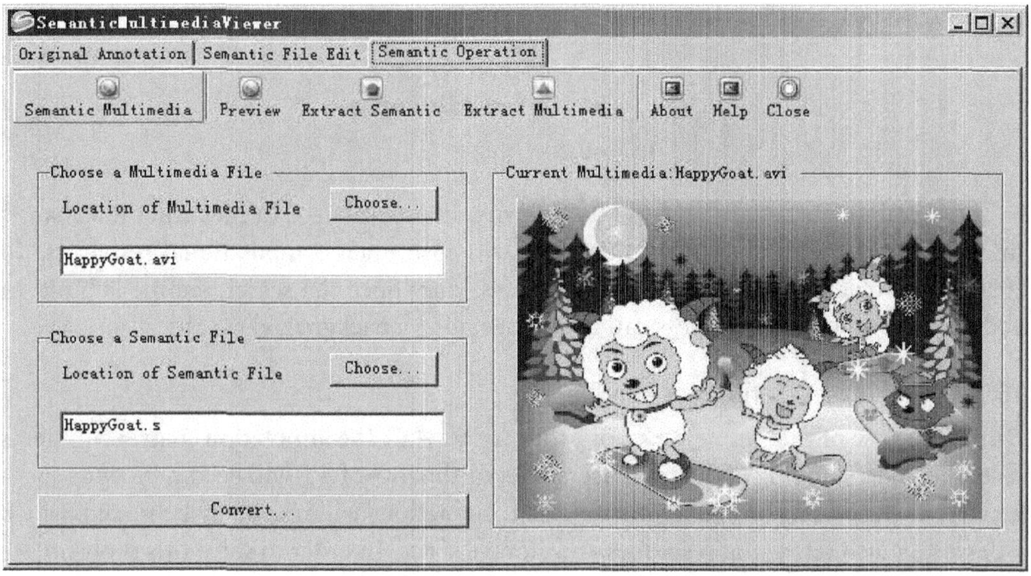

FIGURE 18.2 The interface of SMV. (From Guo, K. et al., *Wirel. Pers. Commun.*, 72, 2013; Guo, K. and Zhang, S., *Comput. Math. Methods Med.*, 2013, 2013.)

number of multimedia documents). For arbitrary $m_i \in C$, $m_i$ will be saved in a hard disk. The location information of $m_i$ is saved in HBase linked to the real file. Semantic annotations will provide meaningful key words reflecting users' personal understanding of $m_i$.

Define set $A_{mi}$ as the annotation set of $m_i$ satisfying $A_{mi} = \{a_1, a_2, \ldots, a_n\}$ (where $n$ is the number of annotations for $m_i$). For arbitrary $m_i \in C$, users will give many annotations. However, not all the annotations can accurately represent the semantic information of $m_i$. Therefore, we assign every $a_i \in A_{mi}$ a weight. Therefore, for arbitrary $m_i \in C$, the final annotation matrix of $A_{mi}$ will be defined as

$$A_{mi} = \begin{bmatrix} a_1, a_2, \ldots, a_n \\ w_1, w_2, \ldots, w_n \end{bmatrix}^T \tag{18.1}$$

where $a_i$ is the $i$th annotation and $w_i$ is the corresponding weight. Therefore, all the annotation matrices for the multimedia documents can be defined as $A = \{A_{m1}, A_{m2}, \ldots, A_{mN}\}$. After the semantic annotation, for arbitrary $m_i \in C$, we assign the initial value of $w_i$ as $\frac{1}{n}$.

It is evident that $w_i$ for every annotation could not be constant after retrieval. Obviously, more frequently used annotations during the retrieval process can better express semantic information; they should be assigned a greater weight. We design an adjustment schema as follows:

$$w_i = w_i + k_i \times \frac{1}{n} \tag{18.2}$$

The value of $k_i$ satisfies the following:

$$k_i = \begin{cases} 1 & m_i \text{ is retrieved based on } a_i \\ 0 & \text{others} \end{cases} \tag{18.3}$$

The initial weight assignment and the adjustment process need to check all the semantic information in the database, and this work will cost much computational resources. To solve this problem, we can execute this process only once the search engine is built. In addition, the adjustment process can be performed in a background thread.

### 18.3.3 Optimization and User Feedback

During the use of this framework, for arbitrary $m_i \in C$, the annotation matrix $A_{mi}$ stems from the understanding of different users. The cardinality of $A_{mi}$ will become progressively greater. In $A_{mi}$, wrong or less frequently used annotations will inevitably be mixed, which will waste much retrieving resources and storage space. In order to solve this problem, we define an optimization approach to eliminate the annotations that may be useless. We use a weight adjustment schema to adjust the weight of every semantic field.

This process is called annotation refinement. The purpose is to retain most of the high-frequency annotations and eliminate the annotations with less use. For arbitrary $m_i \in C$, we will check $A_{mi}$ and remove the $i$th row when $a_i$ satisfies

$$w_i < \frac{1}{n}\sum_{i=1}^{n} w_i \tag{18.4}$$

The initial weight assignment and the adjustment schema need to check all the semantic documents in HBase, and this work will cost much computational resources. To solve this problem, the adjustment process can be performed in a background thread.

After retrieval, this framework will return some multimedia documents. This framework supports user feedback, so for a particular returned document, the user can add additional annotations to enrich the semantic information. For these annotations, the initial weight will be $1/n$ too.

In summary, as the retrieval progresses, the annotations will be more and more abundant. But rarely used annotations will also be removed. There will be some new annotations added into the annotation matrix $A_{mi}$ because of the user feedback. Therefore, this framework is a dynamic framework; the longer it is used, the more accurate the results we can obtain.

### 18.3.4 Semantic Representation

This framework uses ontology technology to describe the multimedia semantic information. In the ontology representation, each node describes one certain semantic concept, and the ontology representation satisfies a recursive and hierarchical structure. The ontology nodes at the first level are used to represent the most obvious features. The second and other levels of semantic annotations will be provided based on the previous levels. All the information is annotated by the users in the original annotation or feedback progress. Figure 18.3 shows the annotation structure of an image.

This framework adopts a composite pattern as the data structure to represent the relation of annotations. In a composite pattern, objects can be composed as a tree structure to represent the part and whole hierarchy. This pattern regards simple and complex elements as common elements. A client can use the same method to deal with complex elements as simple elements, so that the internal structure of the complex elements will be independent with the client program. The data structure using composite pattern is shown in Figure 18.4 [1].

In this structure, *OntologyComponent* is a declared object interface in the composition. In many cases, this interface will implement all the default methods that are shared by all the ontologies. *OntologyLeaf* represents the leaf node object in the composition, and these nodes have no children nodes. In *OntologyComposite*, the methods of branch nodes will be

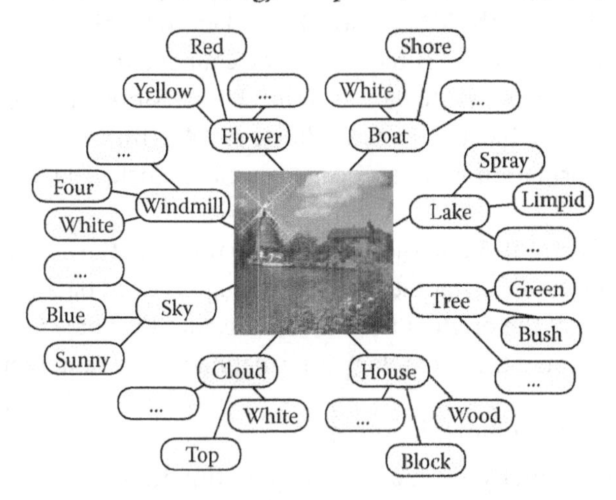


FIGURE 18.3 Annotation structure of an image.

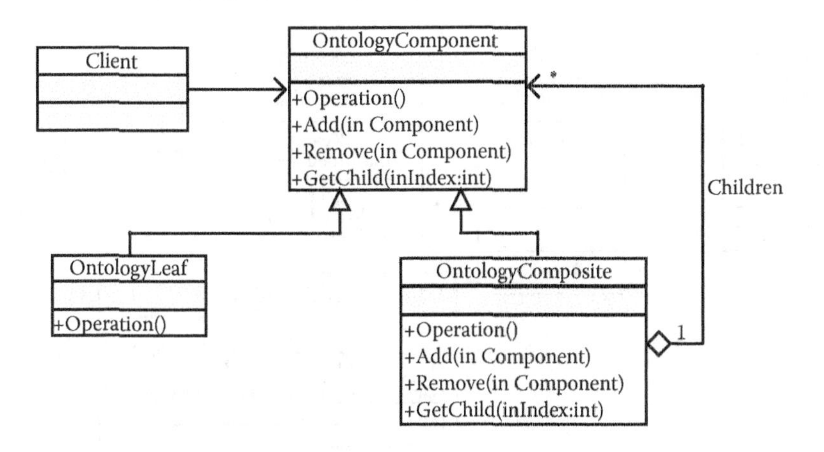


FIGURE 18.4 Ontology structure based on composite pattern. (From Guo, K. et al., *Wirel. Pers. Commun.*, 72, 779–793, 2013; Guo, K. and Zhang, S., *Comput. Math. Methods Med.*, 2013, 1–7, 2013.)

defined to store the children components; the operations relative to children components will be implemented in the *OntologyComponent* interface. Therefore, a composite pattern allows the user to use ontology as a consistency method.

### 18.3.5 NoSQL-Based Semantic Storage

NoSQL databases have been widely used in industry, including Big Data and real-time web applications. We use NoSQL technology to store the semantic information and the multimedia location, which are represented in map format. A NoSQL database presents all the semantic information in a highly optimized key/value format. We can store and retrieve data using models, which employs less constrained consistency than traditional relational databases such as Oracle, Microsoft SQL Server, and so forth. In this chapter, when using this mechanism, we select Apache HBase to simplify storage.

In a Big Data environment, the velocity and cost efficiency are the most important factors in processing heterogeneous multimedia. In Hadoop, HBase is regarded as the data storage solution, which uses the idea of NoSQL. To facilitate the following data processing in MapReduce, the data structure is required to be changed into (key, value) pairs because the map function takes a key/value pair as input. HBase stores the files with some blocks in data nodes, the size of a block is a fixed value (e.g., 64 MB), and the corresponding multimedia semantic information will be recorded in each block. The file structure is shown in Figure 18.5 [34].

We can see from Figure 18.5 that the block files do not store the multimedia data, in order to reduce the network load during the job dispatching between data nodes. For the record, the key is the location of the original multimedia file, and the value is the ontology content of the multimedia document. The ontology information will be represented as a byte array.

### 18.3.6 Heterogeneous Multimedia Retrieval

In the retrieval process, firstly, we specify a mapper function, which processes a key/value pair to generate a set of intermediate key/value pairs. Secondly, we design a reducer function that processes intermediate values associated with the same intermediate key.

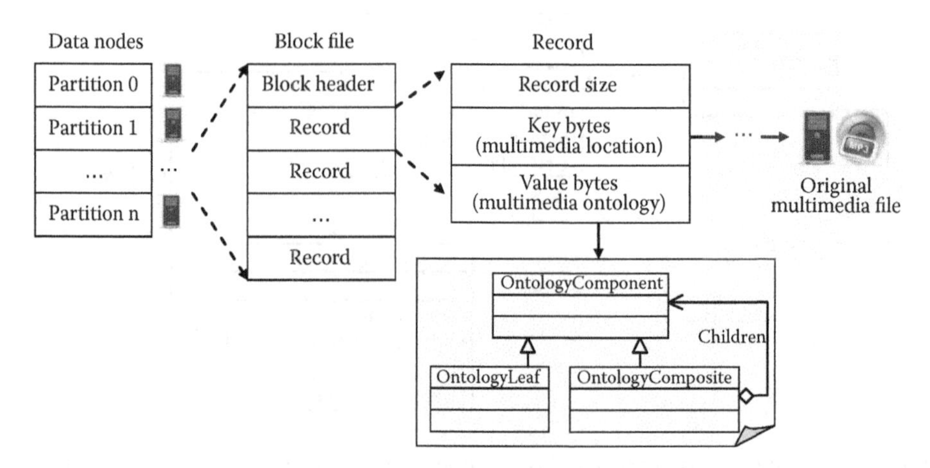


FIGURE 18.5 Map structure of block and record. (From Guo, K. et al., *J. Syst. Software*, 87, 2014.)

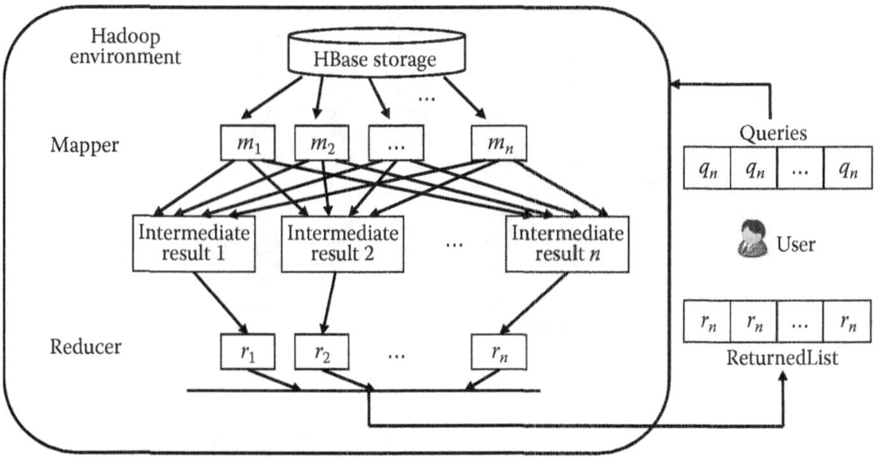

FIGURE 18.6 The process of MapReduce-based retrieval. (From Guo, K. et al., *J. Syst. Software*, 87, 2014.)

The user can upload some multimedia documents to execute the retrieval. Similarly, these documents will be assigned semantic information through social annotation. The semantic information will be represented as an ontology and converted into a map structure. In the queries, we assign every query a QueryId and QueryOntology; the returned result will be formed as a ReturnedList. The process of MapReduce-based retrieval is shown in Figure 18.6 [34].

In Figure 18.6, the queries will be submitted to the Hadoop environment, and then the mapper and reducer will run. All the returned records will be put into a list. In the queries and returned list, the information will be represented as a byte array. The mapper function takes pairs of record keys (multimedia location) and record values (ontology information). For each pair, a retrieval engine runs all queries and outputs for each matching query using a similarity function. The MapReduce tool runs the mapper functions in parallel on each machine. When the map step finishes, the MapReduce tool groups the intermediate output according to every QueryId. For each corresponding QueryId, the reducer function that runs locally on each machine will simply take the result whose similarity is above the average value and output it into the ReturnedList.

## 18.4 PERFORMANCE EVALUATION

### 18.4.1 Running Environment and Software Tools

We developed and deployed our framework on the basis of 10 computers as the data nodes, which can simulate the parallel and distributed system. Every node is a common PC (2.0 GHz CPU, 2 GB RAM), on which are installed Ubuntu Linux and Hadoop 2.0 together with the supporting tools (e.g., Java SDK 6.0, OpenSSH, etc.). We numbered the nodes from 01 to 10 and took node 01 as the master node, and all the nodes were arranged as slave nodes. Therefore, in total, this parallel system has 10 machines, 10 processors, 20 GB memory, 10 disks, and 10 slave data modes. All the experiments were conducted in such a simulated environment. Figure 18.7 shows the running framework in the experiments.

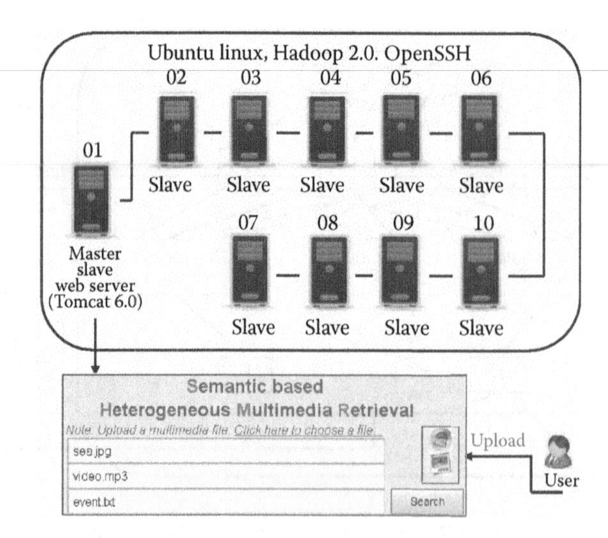


FIGURE 18.7 Running framework of performance evaluation. (From Guo, K. et al., *J. Syst. Software*, 87, 2014.)

In this chapter, we developed some other software tools to verify the effectiveness of our framework. These tools include the following: (1) An annotation interface is used for the users to provide the annotations to the multimedia documents. We developed SMV for PC and mobile device users [1]. (2) Our framework provides a convenient operating interface, which is very similar to the traditional commercial search engines (e.g., Google Images search). Users can upload multimedia documents in the interface and submit the information to the server. The interface was developed using HTML5, and it can run on a typical terminal. (3) For the web server, the search engine was deployed in Tomcat 6.0. In the cluster, the background process was executed every 24 h. Table 18.1 shows the introduction of software tools.

How to construct the data set is an important problem in the experiment. Some general databases have been proposed. However, these databases can only perform the experiments aiming at one particular multimedia type (e.g., image files). Heterogeneous multimedia retrieval requires a wide variety of files such as images, videos, audio, and so forth. So these databases are not suitable for performing the experiments. In this chapter, we have constructed a multimedia database containing various multimedia types including

TABLE 18.1 Introduction of Software Tools

| Software Tool | Development Environment | Running Environment |
|---|---|---|
| SMV | Microsoft Foundation Classes for PC users<br>Android 5.0 for mobile device users | Mobile device: 1.2 GHz CPU, 1 GB RAM<br>PC: 2 GHz CPU, 2 GB RAM |
| User interface | HTML5 for browser users<br>and mobile device users | |
| Search engine | Java Enterprise Edition 5.0 | PC: 2 GHz CPU, 2 GB RAM |
| Server | Tomcat 6.0 | PC: 2 GHz CPU, 2 GB RAM |
| Development tools | MyEclipse 8.5<br>Java Enterprise Edition 5.0 | PC: 2 GHz CPU, 2 GB RAM |

images, videos, audio, and text documents. In the experiment, we used a multimedia database containing 50,000 multimedia documents, including 20,000 images, 10,000 videos, 10,000 audio files, and 10,000 text documents. All the semantic information of the multimedia documents were provided through the users manually annotating and analyzing the text from the host file (e.g., web page) where the documents are downloaded.

### 18.4.2 Performance Evaluation Model

In this section, a performance evaluation model is designed to measure the performance of our framework. These models are based on the following three criteria: precision ratio, time cost, and storage cost.

1. *Precision ratio.* Precision ratio is a very common measurement for evaluating retrieval performance. In the experiment, we slightly modify the traditional definition of precision ratio. For each retrieval process, we let the user choose multimedia documents that reflect his/her query intent. We define the set of retrieved results as $R_t = \{M_1, M_2, \ldots, M_t\}$ (where $t$ is the number of retrieved multimedia documents) and define the set of all the multimedia documents reflecting users' intent as $R_l = \{M_1, M_2, \ldots, M_l\}$ (where $l$ is the number of relevant documents).

   The precision ratio is computed by the proportion of retrieved relevant documents to total retrieved documents. Therefore, the modified precision ratio $p$ can be defined as follows:

$$p = \frac{\#|R_l|}{\#|R_t|} \tag{18.5}$$

   The way to determine whether an image reflects the query intent is important. In this chapter, we make the judgment based on the users. However, because the database is large, the user can select results in the first few pages to evaluate the precision ratio.

2. *Time cost.* Time cost includes two factors. The first factor is data process time cost. In our framework, several background processes will cost time. We define the background process time as follows:

$$t_b = t_{pre} + t_{ref} \tag{18.6}$$

   where $t_{pre}$ is the preprocess time (convert the semantic and multimedia location to map structure) and satisfies

$$t_{pre} = \sum_{i=1}^{N} t_{pre}^{i} \tag{18.7}$$

   $t_{ref}$ represents the annotation refinement time (eliminate the redundant or error semantic information and add the new semantic information in the feedback).

The second factor is retrieval time. We define $t_r$ as the time cost for retrieval. In fact, $t_r$ includes extraction time (extract the semantic information from the HBase) and the matching time (match the semantic similarity of the sample document with the stored documents).

3. *Storage cost.* Because the HBase will store the map information, the storage cost will be taken into consideration. The rate of increase for storage $p_s$ is defined as follows:

$$p_s = s_{ont}/s_{org} \tag{18.8}$$

where $s_{ont}$ is the size of semantic information of the multimedia documents and $s_{org}$ is the size of original multimedia documents,

$$s_{ont} = \sum_{i=1}^{N} s_{ont}^{i}, \quad s_{org} = \sum_{i=1}^{N} s_{org}^{i} \tag{18.9}$$

### 18.4.3 Precision Ratio Evaluation

In the experiment, we first upload a multimedia document to the search engine to search for multimedia documents similar to it. The uploaded file will be annotated by some other users. After uploading the file in the interface, the system will search for all the records whose semantic information is similar to that of the sample document. The time cost includes extracting semantic information from the sample file and matching the semantic information with multimedia documents in HBase.

To measure the performance, we use the images, videos, audio, and text as the sample documents to execute the retrieval. For every sample type, we perform 10 different retrievals using 10 different documents, which are numbered from 01 to 10. The precision ratios are listed in Figure 18.8.

We can see from Figure 18.8 that our framework achieves good retrieval precision ratios. In order to demonstrate the performance of the heterogeneous retrieval, we specially record the precision ratios of using one type to search for the four types (e.g., use an image to search for images, videos, audio, and text documents). For every document type, we perform 10 different retrievals using 10 different sample documents and compute the average precision ratios. The average precision ratios are illustrated in Figure 18.9.

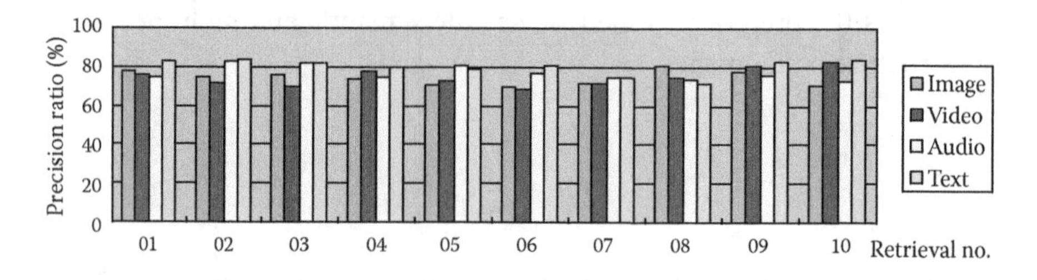


FIGURE 18.8 Precision ratios of 40 retrievals.

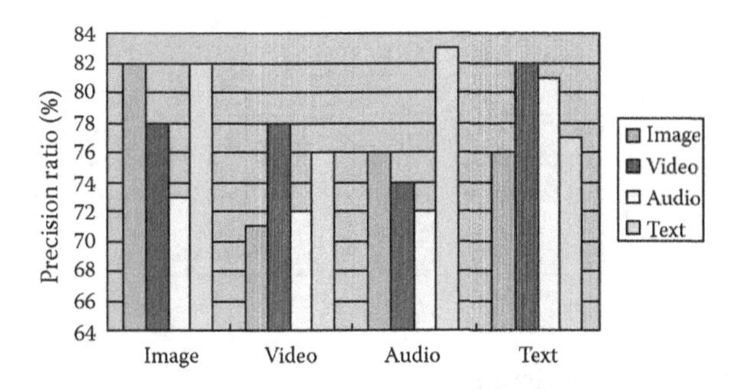


FIGURE 18.9 Average precision ratios of heterogeneous retrievals.

Figure 18.9 indicates that even in the retrieval process between different multimedia types, the precision ratios are not reduced. This is because this framework completely abandons physical feature extraction and executes the retrieval process based only on semantic information.

However, we cannot ignore that some traditional technologies supporting content-based retrieval have good performances too. For example, Google Images search actually can get perfect results reflecting users' intent in content-based image retrieval. However, it has the following disadvantages in comparison with this framework: (1) This search pattern cannot support heterogeneous retrieval, because of the physical feature extraction. (2) Compared with physical features, annotations can better represent the users' query intent, so our framework can get more accurate results in case of the semantic multimedia retrieval. If the documents contain more abundant annotations, the retrieval performance will be better. In addition, our framework has the advantage of good speed because of skipping the physical feature extraction.

### 18.4.4 Time and Storage Cost

In order to carry out the retrieval process, we have to perform several background processes whose time cost is $t_b$, which includes $t_{pre}$ and $t_{ref}$. Table 18.2 shows the time cost of the background processes.

We can see from Table 18.2 that $t_{pre}$, $t_{ref}$ will cost some seconds ($t_b$ costs about 132 s for image type, 91 s for video type, 71 s for audio type, and 102 s for text type). However, the background processes are not always executed. In the server, the background process will be executed every 24 h in background thread, so this time cost can be acceptable.

Now we will measure the retrieval time $t_r$. We specially record the time cost of 16 retrieval processes. For every document type (image, video, audio, and text), we perform four different retrievals (the samples are numbered from 01 to 04). In every retrieval, $t_r$

TABLE 18.2 Time Cost of Background Processes (s)

| Multimedia Type | $t_{pre}$ | $t_{ref}$ | $t_b$ |
|---|---|---|---|
| Image | 93 | 39 | 132 |
| Video | 63 | 28 | 91 |
| Audio | 52 | 19 | 71 |
| Text | 78 | 24 | 102 |

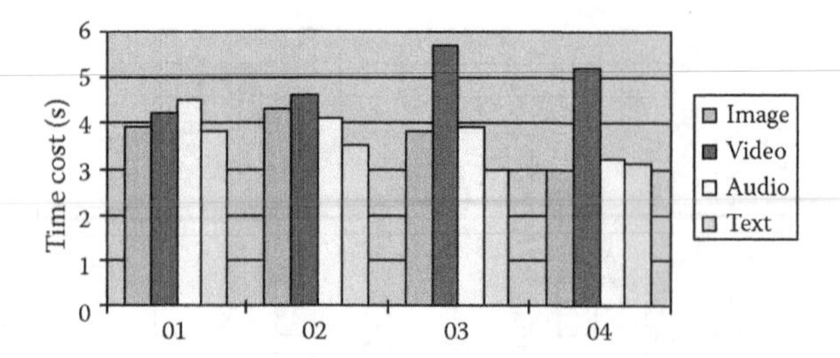


FIGURE 18.10 Time cost of 16 retrievals.

TABLE 18.3 Storage Space Cost

| Multimedia Type | $p_s$ (%) |
|---|---|
| Image | 1.56 |
| Video | 0.18 |
| Audio | 0.32 |
| Text | 15.23 |

will be recorded respectively. The detailed time cost of 16 retrieval experiments is listed in Figure 18.10.

Figure 18.10 shows that the semantic information extraction only costs a very short time; this is because we only need to read the sample document and directly extract the semantic segment from it. After the extraction, the retrieval process will be similar to the text-based retrieval, and Table 18.2 indicates that this process can be executed in acceptable time.

The storage cost will be taken into consideration because the HBase will store the semantic information. Table 18.3 shows the storage space cost.

We can see from Table 18.3 that the semantic information file size has almost not increased for image, video, and audio ($p_s$ is about 1.56% for image, 0.18% for video, and 0.32% for audio type); this is because the semantic information is represented as text and the size of semantic files is small. However, the semantic information file size occupies 15.23% for text type; this is because the semantic information in a text file is abundant.

## 18.5 DISCUSSIONS AND CONCLUSIONS

In this chapter, a new approach supporting heterogeneous multimedia retrieval and reflecting users' retrieval intent has been proposed. We designed the framework and described semantic annotation, ontology representation, ontology storage, a MapReduce-based retrieval process, and a performance evaluation model. Experiment results show that this framework can achieve good performance in heterogeneous multimedia retrieval reflecting the users' intent.

In this framework, we only need to purchase some cheap computers to perform the semantic storage and retrieval process. Because investment in high-performance server hardware is more expensive than organizing cheap computers into a distributed computing environment, our framework can help us to reduce the cost of hardware investment.

Also, in our framework, we use some open-source tools such as Ubuntu Linux, Java SDK, and Hadoop tools. The operating system and software tools can be freely downloaded from the corresponding websites. This will save the investment of software. In addition, Apache Hadoop provides simplified programming models for reliable, scalable, distributed computing. It allows the distributed processing of large data sets across clusters of computers using simple programming models; the programmer can use Hadoop and MapReduce with low learning cost. Therefore, the investment cost is very economical for heterogeneous multimedia Big Data retrieval using our framework.

## ACKNOWLEDGMENTS

This work is supported by the Natural Science Foundation of China (61202341, 61103203) and the China Scholarship (201308430049).

## REFERENCES

1. K. Guo, J. Ma, G. Duan. DHSR: A novel semantic retrieval approach for ubiquitous multimedia. *Wireless Personal Communications*, 2013, 72(4): 779–793.
2. K. Guo, S. Zhang. A semantic medical multimedia retrieval approach using ontology information hiding. *Computational and Mathematical Methods in Medicine*, 2013, 2013(407917): 1–7.
3. C. Liu, J. Chen, L. Yang et al. Authorized public auditing of dynamic Big Data storage on cloud with efficient verifiable fine-grained updates. *IEEE Transactions on Parallel and Distributed Systems*, 2014, 25(9): 2234–2244.
4. J.R. Smith. Minding the gap. *IEEE MultiMedia*, 2012, 19(2): 2–3.
5. S. Ghemawat, H. Gobioff, S.T. Leung. The Google file system. *ACM SIGOPS Operating Systems Review. ACM*, 2003, 37(5): 29–43.
6. F. Chang, J. Dean, S. Ghemawat et al. Bigtable: A distributed storage system for structured data. *ACM Transactions on Computer Systems*, 2008, 26(2): 1–26, Article no. 4.
7. J. Dean, S. Ghemawat. MapReduce: Simplified data processing on large clusters. *Communications of the ACM*, 2008, 51(1): 107–113.
8. Apache Hadoop. Available at http://hadoop.apache.org/.
9. Apache Hbase. Available at http://en.wikipedia.org/wiki/HBase.
10. R. Zhao, W.I. Grosky. Narrowing the semantic gap-improved text-based web document retrieval using visual features. *IEEE Transactions on Multimedia*, 2002, 4(2): 189–200.
11. Y. Yang, F. Nie, D. Xu et al. A multimedia retrieval architecture based on semi-supervised ranking and relevance feedback. *IEEE Transactions on Pattern Analysis and Machine Intelligence*, 2012, 34(4): 723–742.
12. A. Smeulders, M. Worring, S. Santini et al. Content-based image retrieval at the end of the early years. *IEEE Transactions Pattern Analysis and Machine Intelligence*, 2000, 22(12): 1349–1380.
13. G. Zhou, K. Ting, F. Liu, Y. Yin. Relevance feature mapping for content-based multimedia information retrieval. *Pattern Recognition*, 2012, 45(4): 1707–1720.
14. R.C.F. Wong, C.H.C. Leung. Automatic semantic annotation of real-world web images. *IEEE Transactions on Pattern Analysis and Machine Intelligence*, 2008, 30(11): 1933–1944.
15. A. Gijsenij, T. Gevers. Color constancy using natural image statistics and scene semantics. *IEEE Transactions on Pattern Analysis and Machine Intelligence*, 2010, 33(4): 687–698.
16. W. Lei, S.C.H. Hoi, Y. Nenghai. Semantics-preserving bag-of-words models and applications. *IEEE Transactions on Image Processing*, 2010, 19(7): 1908–1920.
17. T. Hofmann. Unsupervised learning by probabilistic latent semantic analysis. *Machine Learning*, 2001, 42(1–2): 177–196.

18. D.M. Blei, A.Y. Ng, M.I. Jordan. Latent Dirichlet allocation. *Journal of Machine Learning Research*, 2003, 3(1): 993–1022.
19. Y. Gao, J. Fan, H. Luo et al. Automatic image annotation by incorporating feature hierarchy and boosting to scale up SVM classifiers. In *Proceedings of the 14th ACM International Conference on Multimedia*, Santa Barbara, CA, 2006: 901–910.
20. G. Carneiro, A. Chan, P. Moreno, N. Vasconcelos. Supervised learning of semantic classes for image annotation and retrieval. *IEEE Transactions on Pattern Analysis and Machine Intelligence*, 2007, 29(3): 394–410.
21. N. Rasiwasia, P.J. Moreno, N. Vasconcelos. Bridging the gap: Query by semantic example. *IEEE Transactions on Multimedia*, 2007, 9(5): 923–938.
22. J. Shi, J. Malik. Normalized cuts and image segmentation. *IEEE Transactions on Pattern Analysis and Machine Intelligence*, 2000, 22(8): 888–905.
23. F. Monay, D. Gatica-Perez. Modeling semantic aspects for cross-media image indexing. *IEEE Transactions on Pattern Analysis and Machine Intelligence*, 2007, 29(10): 1802–1817.
24. D. Djordjevic, E. Izquierdo. An object and user driven system for semantic-based image annotation and retrieval. *IEEE Transactions on Circuits and Systems for Video Technology*, 2007, 17(3): 313–323.
25. S.C.H. Hoi, J. Rong, J. Zhu, M.R. Lyu. Semi-supervised SVM batch mode active learning for image retrieval. In *Proceedings of 2008 IEEE Conference on Computer Vision and Pattern Recognition*, Anchorage, AK, IEEE Computer Society, 2008: 1–7.
26. J. Li, J.Z. Wang. Automatic linguistic indexing of pictures by a statistical modeling approach. *IEEE Transactions on Pattern Analysis and Machine Intelligence*, 2003, 25(9): 1075–1088.
27. R. Albertoni, R. Papaleo, M. Pitikakis. Ontology-based searching framework for digital shapes. *Lecture Notes in Computer Science*, 2005, 3762: 896–905.
28. M. Attene, F. Robbiano, M. Spagnuolo. Part-based annotation of virtual 3d shapes. In *Proceedings of International Conference on Cyberworlds*, Hannover, Germany, IEEE Computer Society Press, 2007: 427–436.
29. M. Attene, F. Robbiano, M. Spagnuolo, B. Falcidieno. Semantic annotation of 3d surface meshes based on feature characterization. *Lecture Notes in Computer Science*, 2009, 4816: 126–139.
30. M. Gutiérrez, A. García-Rojas, D. Thalmann. An ontology of virtual humans incorporating semantics into human shapes. *The Visual Computer*, 2007, 23(3): 207–218.
31. Z.J. Li, V. Raskinm, K. Ramani. Developing ontologies for engineering information retrieval. *IASME Transactions Journal of Computing and Information Science in Engineering*, 2008, 8(1): 1–13.
32. X.Y. Wang, T.Y. Lv, S.S. Wang. An ontology and swrl based 3d model retrieval system. *Lecture Notes in Computer Science*, 2008, 4993: 335–344.
33. D. Yang, M. Dong, R. Miao. Development of a product configuration system with an ontology-based approach. *Computer-Aided Design*, 2008, 40(8): 863–878.
34. K. Guo, W. Pan, M. Lu, X. Zhou, J. Ma. An effective and economical architecture for semantic-based heterogeneous multimedia big data retrieval. *Journal of Systems and Software*, 2014, 87 pp.

CHAPTER 19

# Topic Modeling for Large-Scale Multimedia Analysis and Retrieval

Juan Hu, Yi Fang, Nam Ling, and Li Song

CONTENTS



ABSTRACT

The explosion of multimedia data in social media raises a great demand for developing effective and efficient computational tools to facilitate producing, analyzing, and retrieving large-scale multimedia content. Probabilistic topic models prove to be an effective way to organize large volumes of text documents, while much fewer related models are proposed for other types of unstructured data such as multimedia content, partly due to the high computational cost. With the emergence of cloud computing, topic models are expected to become increasingly applicable to multimedia data. Furthermore, the growing demand for a deep understanding of multimedia data on the web drives the development of sophisticated machine learning methods. Thus, it is greatly desirable to develop topic modeling approaches to multimedia applications that are consistently effective, highly efficient, and easily scalable. In this chapter, we present a review of topic models for large-scale multimedia analysis. Our goal is to show the current challenges from various perspectives and to present a comprehensive overview of related work that addresses these challenges. We will also discuss several research directions in the field.

## 19.1 INTRODUCTION

With the arrival of the Big Data era, recent years have witnessed an exponential growth of multimedia data, thanks to the rapid increase of processor speed, cheaper data storage, prevalence of digital content capture devices, as well as the flooding of social media like Facebook and YouTube. New data generated each day have reached 2.5 quintillion bytes as of 2012 (Dean and Ghemawat 2008). Particularly, more than 10 h of videos are uploaded onto YouTube every minute, and millions of photos are available online every week. The explosion of multimedia data in social media raises great demand in developing effective and efficient computational tools to facilitate producing, analyzing, and retrieving large-scale multimedia content. Big Data analysis for basic tasks such as classification, retrieval, and prediction has become ever popular for multimedia sources in the form of text, graphics, images, audio, and video. The data set is so large and noisy that the scalability of the traditional data mining algorithms needs to be improved. The MapReduce framework designed by Google is very simple to implement and very flexible in that it can be extended for various large-scale data processing functions. This framework is a powerful tool to develop scalable parallel applications to process Big Data on large clusters of commodity machines. The equivalent open-source Hadoop MapReduce developed by Yahoo is now very popular in both the academic community and industry.

In the past decade, much effort has been made in the information retrieval (IR) field to find lower-dimensional representation of the original high-dimensional data, which enables efficient processing of a massive data set while preserving essential features. Probabilistic topic models proved to be an effective way to organize large volumes of text documents. In natural language processing, a topic model refers to a type of statistical model for representing a collection of documents by discovering abstract topics. At the early stage, a generative probabilistic model for text corpora is developed to address the issues of term frequency and inverse document frequency (TF-IDF), namely, that the dimension reduction effect using TF-IDF is rather small (Papadimitriou et al. 1998). Later, another important topic model named probabilistic latent semantic indexing (PLSI) was created by Thomas Hofmann in 1999. Essentially, PLSI is a two-level hierarchical Bayesian model where each word is generated from a single topic and each document is reduced to a probability distribution of a fixed set of topics. Latent Dirichlet allocation (LDA) (Blei et al. 2003) is a generalization of PLSI developed by providing a probabilistic model at the level of documents, which avoids the serious overfitting problem as the number of parameters in the model does not grow linearly with the size of the corpus. LDA is now the most common topic model, and many topic models are generally an extension of LDA by relaxing some of the statistical assumptions. The probabilistic topic model exemplified by LDA aims to discover the hidden themes running through the words that can help us organize and understand the vast information conveyed by massive data sets.

To improve the scalability of a topic model for Big Data analysis, much effort has been put into large-scale topic modeling. Parallel LDA (PLDA) was designed by distributing

Gibbs sampling for LDA on multiple machines (Wang et al. 2009). Another flexible large-scale topic modeling package named Mr.LDA is implemented in MapReduce, where model parameters are estimated by variational inference (Zhai et al. 2012). A novel architecture for a parallel topic model is demonstrated to yield better performance (Smola and Narayanamurthy 2010).

While topic models are proving to be effective methods for corpus analysis, much fewer related models have been proposed for other types of unstructured data such as multimedia content, partly due to high computational cost. With the emergence of cloud computing, topic models are expected to become increasingly applicable to multimedia data. Furthermore, the growing demand for a deep understanding of multimedia data on the web drives the development of sophisticated machine learning methods. Thus, it is greatly desirable to develop topic modeling approaches for large-scale multimedia applications that are consistently effective, highly efficient, and easily scalable. In this chapter, we present a review of topic models for large-scale multimedia analysis.

The chapter is organized as follows. Section 19.2 gives an overview of several distributed large-scale computing frameworks followed by a detailed introduction of the MapReduce framework. In Section 19.3, topic models are introduced, and we present LDA as a typical example of a topic model as well as inference techniques such as Gibbs sampling and variational inference. More advanced topic models are also discussed in this section. A review of recent work on large-scale topic modeling, topic modeling for multimedia analysis, and large-scale multimedia analysis is presented in Section 19.4. Section 19.5 demonstrates recent efforts done in large-scale topic modeling for multimedia retrieval and analysis. Finally, Section 19.6 concludes this chapter.

## 19.2 LARGE-SCALE COMPUTING FRAMEWORKS

The prevalence of large-scale computing drives parallel processing with thousands of computer nodes compared to most computing in the past done on a single processor. Academic research areas and industries have been strongly connected in this area as lots of companies such as Google, Microsoft, Yahoo, Twitter, and Facebook face the technical challenge of efficiently and reliably dealing with ever-growing data sets. These companies, in turn, created many innovative large-scale computing frameworks. For example, Google's MapReduce (Dean and Ghemawat 2008) is perhaps the most popular distributed framework, and the open-source version of MapReduce created by Yahoo with Apache Hadoop can be seen anywhere nowadays. There are also many other successful frameworks such as Microsoft's Dryad (Isard et al. 2007), Yahoo's S4 for unbounded streaming data, Storm used by Twitter, and so on. As most of these frameworks can be considered as extensions of MapReduce with enhancements for specific applications, in the rest of the section, the large-scale computing framework is exemplified and illustrated by MapReduce.

MapReduce is a framework for processing massive data sets in a parallel diagram across a large number of computer nodes. The computer nodes can be a cluster where they are on the same local network and use similar hardware, or a grid where computer nodes are shared across geographically and administratively distributed systems and use more heterogeneous hardware. In the parallel computing architecture, computer nodes are stored on racks, roughly 8 to 64 for each rack. The computer nodes on the same rack are connected by gigabyte Ethernet, while the racks are interconnected by a switch.

There are two kinds of computer nodes in this framework, the master node and the worker node. The master node basically assigns tasks for worker nodes and keeps track of the status of the worker nodes, which can be idle, executing a particular task, or finished completing it. Thus, the master node takes central control role of the whole process, and failure at the master node can be disastrous. The entire process could be down, and all the tasks need to be restarted. On the other hand, failure in the worker node can be detected by the master node as it periodically pings the worker processes. The master node can manage the failures at the worker nodes, and all the computing tasks will be complete eventually.

The name of MapReduce comes naturally from the essential two functions under this framework, that is, the Map step and the Reduce step, as described in Figure 19.1. The input of the computing process is chunks of data, which can be any type, such as documents or tuples. The Map function converts input data into key–value pairs. Then the master controller chooses a hash function that is applied to keys in the Map step and produces a bucket number, which is the total number of Reduce tasks defined by a user. Each key in the Map task is hashed, and its key–value pair is put in the local buckets by grouping and aggregation, each of which is destined for one of the Reduce tasks. The Reduce function takes pairs consisting of a key and its list of associated values combined in a user-defined way. The output of a Reduce task is a sequence of key–value pairs, which consists of the key received from the Map task and the combined value constructed from the list of values that the Reduce task received along with the key. Finally, the outputs from all the Reduce tasks are merged into a single file.

The MapReduce computation framework can be best illustrated with a classic example of word count. Word count is very important as it is exactly the term frequency in the IR model. The input file for the framework is a corpus of many documents. The words are the keys of the Map function, and the count of occurrences of each word is the value corresponding to the key. If a word $w$ appears $m$ times among all the documents assigned to that process, in the Reduce task, we simply add up all values so that after grouping and aggregation, the output is a sequence of pairs $(w, m)$.

In the real MapReduce execution, as shown in Figure 19.1, a worker node can handle either a *Map* task or a *Reduce* task but will be assigned only one task at a time. It is

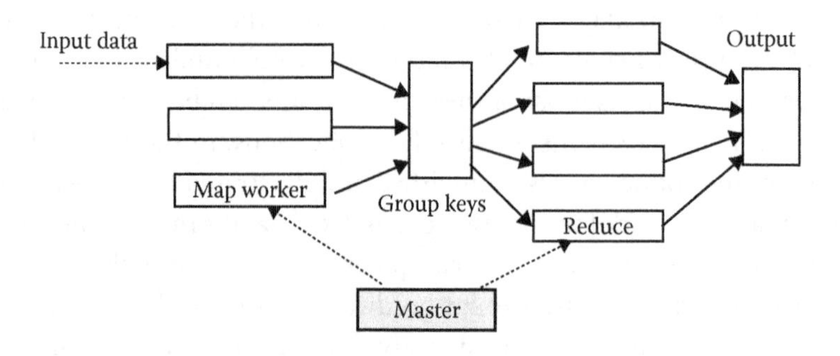


FIGURE 19.1 Schematic of MapReduce computation.

reasonable to have a smaller number of *Reduce* tasks compared to *Map* tasks as it is necessary for each *Map* task to create an intermediate file for each *Reduce* task and if there are too may *Reduce* tasks, the number of intermediate files explodes. The original implementation of the MapReduce framework was done by Google via Google File System. The open-source implementation by Yahoo was called Hadoop, which can be downloaded along with the Hadoop file system, from the Apache foundation. Computational processing can occur on data stored either in a distributed file system or in a database. MapReduce can take advantage of locality of data, processing data on or near the storage assets to decrease transmission of data with a tolerance of hardware failure.

MapReduce allows for distributed processing of the map and reduction operations. Provided each mapping operation is independent of the others, all maps can be performed in parallel, though in practice, it is limited by the number of independent data sources and the number of CPUs near each source. Similarly, a set of "reducers" can perform the reduction tasks. All outputs of the map operation that share the same key are presented to the same reducer at the same time. While this process can often appear inefficient compared to algorithms that are more sequential, MapReduce can be applied to significantly larger data sets than "commodity" servers can handle. A large server farm can use MapReduce to sort petabytes of data in only a few hours. The parallelism also offers some possibility of recovering from partial failure of servers or storage during the operation: If one mapper or reducer fails, the work can be rescheduled as long as the input data are still available.

While the MapReduce framework is very simple, flexible, and powerful, the data-flow model is ideally suited for batch processing of on-disk data, where the latency can be very poor and its scalability to real-time computation is limited. To solve these issues, Facebook's Puma and Yahoo's S4 are proposed for real-time aggregation of unbounded streaming data. A novel columnar storage representation for nested records was proposed in the Dremel framework (Melnik et al. 2010), which improves the efficiency of MapReduce. In-memory computation was allowed in another data-centric programming model. Piccolo (Powell and Li 2010) and Spark, which were created by the Berkeley AMPLab, have been demonstrated to be able to deliver much faster performance. Pregel was proposed for large-scale graphical processing (Malewicz et al. 2010).

## 19.3 PROBABILISTIC TOPIC MODELING

The motivation of topic models comes from intuition in real life. Suppose you want to find books about tennis at a public library. You would directly go to the bookshelves labeled "sports." You can do this because the librarian has arranged books into categories by the topics inside the books. However, with the exponentially growing information available online, it is impossible to hire enough human power to read and label web pages, images, or videos with their topics; it is highly desirable to label the large-scale information automatically so that we can easily find what we are looking for. Probabilistic topic models provide powerful tools to discover the hidden themes running through large-scale texts, images, and videos without any prior annotations or label information.

Latent semantic indexing (LSI) utilizes the singular value decomposition of the document term frequency (Deerwester et al. 1990), which provides a solution for high-dimensional

Big Data sets by giving shorter and lower-dimensional representation of original documents. PLSI took a remarkable step forward based on LSI (Hofmann 1999). PLSI is essentially a generative probabilistic topic model where each word in the document is considered to be a sample from a mixture model, and the mixture topic is multinomial distribution. As shown in Figure 19.2a, for each document in the corpus, each word $w$ is generated from the latent topic $x$ with the probability $p(w|d)$, where $x$ is from the topic distribution $p(x|d)$ of the document $d$. PLSI models the probability of occurrence of a word in a document as the following mixture of conditionally independent multinomial distributions:

$$p(w,d) = p(d)\, p(w|d)$$

where

$$p(w|d) = \sum_{x} p(w|x)p(x|d)$$

Although PLSI proves to be useful, it has an issue in that there is no probabilistic generative model at a document level to generate the proportions of different topics. LDA was proposed to fix this issue by constructing a three-level hierarchical Bayesian probabilistic topic model (Blei et al. 2003), which is a generative probabilistic model of a corpus. Probabilistic topic models are exemplified by LDA. In the next paragraph, we are going to introduce LDA in the environment of natural language processing, where we are dealing with text documents.

A topic is defined as a distribution over a fixed vocabulary. Intuitively, documents can exhibit multiple topics with different proportions. For example, a book on topic models can have several topics, such as probabilistic general models, LSI, and LDA, with LDA having the largest proportion. We assume that the topics are specified before the documents are generated. Thus, all the documents in the corpus share the same set of topics but with different distribution. In the generative probabilistic model, data arise from the generative process with hidden variables.

In LDA, documents are represented as random mixtures over latent topics, and the prior distribution of the hidden topic structure is described by Dirichlet distribution. Each word

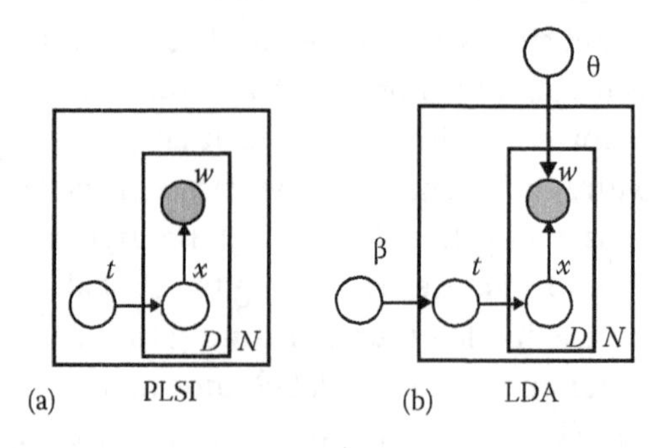


FIGURE 19.2 Graph model representation of PLSI (a) and LDA (b).

is generated by looking up a topic in the document it refers to and finding out the probability of the word within that topic, where we treat each topic as a bag of words and apply the language model. As depicted in Figure 19.2, LDA is a three-level hierarchical probabilistic model, while the PLSI has only two levels. LDA takes the advantage of defining document distribution over multiple hidden topics, compared to PLSI, where the number of topics is fixed.

The three-level LDA model is described as a probabilistic graphical model in Figure 19.2, where a corpus consists of $N$ documents and each document has a sequence of $D$ words. The parameter $\theta$ describes the topic distribution over the vocabulary, assuming that the dimensionality of the topic variable $x$ is known and fixed. The parameter $t$ describes the topic mixture at the document level, which is multinomially distributed over all the topics: $Dir$ ($\beta$). Remember that the topics are generated before the documents, so both $\beta$ and $\theta$ are corpus-level parameters. The posterior distribution of the hidden variables given a document, which is the key problem we need to solve LDA, is described in the following equation:

$$p(t,x\,|\,w,\theta,\beta)=\frac{p(t,x,w\,|\,\theta,\beta)}{p(w\,|\,\theta,\beta)}$$

where

$$p(t,x,w\,|\,\theta,\beta)=p(t\,|\,\beta)\prod_{n=1}^{D}p(x_n\,|\,t)p(w_n\,|\,x_n,\theta)$$

The topic structure is the hidden variable, and our goal is to discover the topic structure automatically with the observed documents. All we need is a posterior distribution, the conditional distribution of the hidden topic structure given all the documents in the corpus. A joint distribution over both the documents and the topic structure is defined to compute this posterior distribution. Unfortunately, the posterior distribution is intractable for exact inference. However, there are a wide variety of approximate inference algorithms developed for LDA, including sampling-based algorithms such as Gibbs sampling and variational algorithms. The sampling-based algorithms approximate the posterior with samples from an empirical distribution. In Gibbs sampling, a Markov chain is defined on the hidden topic variables for the corpus. The Monte Carlo method is used to run the chain for a long time and collect samples and then approximate the posterior distribution with these samples as the posterior distribution is the limiting distribution of the Markov chain. On the other hand, variational inference is a deterministic method to approximate the posterior distribution. The basic idea is to find the nearest lower band by optimization. The optimal lower band is the approximated posterior distribution.

There are many other probabilistic topic models, which are more sophisticated than LDA and usually obtained by relaxing the assumptions of LDA. A topic model relaxes the assumption of bags of words by generating words conditionally on the previous words (Wallach 2006). A dynamic topic model relaxes the assumption by respecting the ordering of the documents (Blei and Lafferty 2006). The third assumption that the number of topics is fixed is relaxed by the Bayesian nonparametric (BNP) topic model (Teh et al. 2006). The correlated topic model (Blei and Lafferty 2007) captures the insight that topics are not independent of each other. Instead, topics might provide evidence for each other. Other more advanced topic models, for example, the author–topic model (Rosen-Zvi et al. 2004) and relational topic model (Chang and Blei 2010), consider additional information such as authors of documents and links between the documents.

## 19.4 COUPLINGS AMONG TOPIC MODELS, CLOUD COMPUTING, AND MULTIMEDIA ANALYSIS

### 19.4.1 Large-Scale Topic Modeling

Topic modeling such as LDA enables us to discover the hidden topic structure running through the data without any supervision, which is quite useful for handling today's Big Data at a scale that we can hardly annotate with human power. However, the real-world applications of topic modeling are limited due to scalability issues. When the number of documents and the vocabulary size reach the million or even the billion level, which is quite common now in practical processing, we need to boost the scalability of topic modeling to very large data sets by using a distributed computing framework. The broadly applicable MapReduce framework proves to be powerful for large-scale data processing. Thus, it is highly desirable to implement parallelized topic modeling algorithms in the MapReduce programming framework.

The challenge of scaling topic modeling to very large document collections has driven lots of effort in this research area to establish distributed or parallelized topic models. A distributed learning algorithm for LDA was proposed to improve the scalability by modifying the existing Gibbs sampling inference approach (Newman et al. 2007). The learning process is distributed on multiple processors, and each processor performs a local Gibbs sampling iteration. This approximate distributed LDA (AD-LDA) successfully opens the door for parallel implementation of topic models for large-scale computing. PLDA is based on AD-LDA, implemented on message passing interface (MPI) and MapReduce (Wang et al. 2009). Similarly, PLDA distributes all input documents over multiple processors. Each processor maintains a copy of word-topic counts, recomputes the word-topic counts after each local Gibbs sampling iteration, and then broadcasts the new word-topic count to all other processors via communication. In the MapReduce framework, each Gibbs sampling iteration is modeled as a Map–Reduce task. In the Map step, map workers do the Gibbs sampling task, while in the Reduce step, reduce workers update the topic assignments. All workers, both in the Map step and in the Reduce step, work in parallel, and the master processor constantly checks the worker status to manage failure at worker processors. Thus, PLDA boosted by MapReduce provides very robust and efficient data processing.

However, Gibbs sampling has its drawbacks for large-scale computing. First, the sampling algorithms can be slow for high-dimensional models. It has been demonstrated that the PLDA can yield very good performance with a small number (millions) of documents with a large cluster of computers. However, it might not be able to process tens to hundreds of millions of documents, while today's Internet data, such as Yahoo and Facebook user profiles, can easily beat this level. To handle hundreds of millions of user profiles, scalable distributed inference of dynamic user interests was designed for a time-varying user model (Ahmed et al. 2011). User interests can change over time, which is very valuable for prediction of user purchase intent. Besides, the topics can also vary over time. Thus, the topic model for user behavior targeting should be dynamic and updated online. However, it is very challenging for inference of a topic model for millions of users over several months. Even for an approximate inference, for example, Gibbs sampling, it is also computationally infeasible. While sequential Monte Carlo (SMC) could be a possible solution, the SMC estimator can quickly become too heavy in long-range dependence, which makes it also infeasible to transfer and update so many computers. Instead, only forward sampling is utilized for inference so that we do not need to go back to look at old data anymore. The new inference method has been demonstrated to be very efficient for analyzing Yahoo user profiles, which provides powerful tools for web applications such as online advertising targeting, content personalization, and social recommendations.

Second, the Gibbs sampler is highly tuned for LDA, which makes it very hard to extend for other applications. To address this issue, a flexible large-scale topic modeling package in MapReduce (Mr.LDA) was proposed by using the variational inference method as an alternative (Zhai et al. 2012). Compared to random Gibbs sampling, variational inference is deterministic given an initialization, which ensures the same output of each step no matter where or when the implementation is running. This uniformity is important for a MapReduce system to check the failure at a worker node and thus have greater fault tolerance, while for random Gibbs sampling, this is very hard. Thus, variational inference is more suitable for a distributed computing framework such as MapReduce.

Variational inference is meant to find variational parameters to minimize the Kullback–Leibler divergence between the variational distribution and the posterior distribution, which is the target of the LDA problem, as mentioned earlier. The variational distribution is carefully chosen in such a way that the probabilistic dependencies from the true distribution are well maintained. This independence results in natural palatalization of computing across multiple processors. Furthermore, it only takes dozens of iterations to converge for variational inference, while it might take thousands for Gibbs sampling. Most significantly, Mr.LDA is very flexible and can be easily extended to efficiently handle online updates.

While distributed LDA has proven to be efficient to solve a large-scale topic modeling problem, regularized latent semantic indexing (RLSI) (Wang et al. 2011) is another smart design for topic modeling parallelization. The text corpus is represented as a term–document matrix, which is then approximated by two matrices: a term–topic matrix and a topic–document matrix. The RLSI is meant to solve the optimal problem that minimizes a quadratic loss function on term–document occurrence with regularization. Specifically, the topics are regularized with $l_1$ norm and $l_2$ for the documents. The formulation of RLSI is carefully chosen so that the inference process can be decomposed into many subproblems. This is the key ingredient that RLSI can scale up via parallelization. With smooth regularization, the overfitting problem is effectively solved for RLSI.

The most distinguishing advantage of RLSI is that it can scale to very large data sets without reducing vocabulary. Most techniques reduce the number of terms to improve the efficiency when the matrix becomes very large, which improves scalability at the cost of decreasing learning accuracy.

### 19.4.2 Topic Modeling for Multimedia

The richness of multimedia information in the form of text, graphics, images, audio, and video has driven the traditional IR algorithms for text documents to be extended to multimedia computing. While topic modeling proves to be an effective method for organizing text documents, it is highly desirable to propose probabilistic topic models for multimedia IR. Indeed, in recent years, topic models have been successfully employed to solve many computer vision problems such as image classification and annotations (Barnard et al. 2003; Blei and Jordan 2003; Duygulu et al. 2002; Li and Perona 2005; Sivic et al. 2006; Wang et al. 2009).

Similar to text retrieval, image files are growing exponentially so that it is infeasible to annotate images with human power. Automatic image annotation as well as automatic image region annotation is very important to efficiently process image files. Other tasks, such as content-based image retrieval and text-based image retrieval, are quite-popular research problems. The former retrieves a matched image with a query image, which can be a sketch or simple graph, while the latter finds corresponding image results by text query. Obviously, the latter is harder as we need to establish the relationship between the image and text data, while the former can be done with visual image features. These image retrieval tasks as a whole, however, are much more difficult compared to text retrieval. Words for text document retrieval can be indexed in vocabulary, while most images are based on creation, which is rather abstract, and it is very hard to teach a machine to recognize the theme of an image (Datta et al. 2005). In the past decade, a lot of research efforts have been done in this area to make computers organize and annotate images.

Correspondence LDA (Corr-LDA) (Blei and Jordan 2003) models annotate data where one type of data is a description of another, for example, a caption and an image. The model establishes the conditional relationship between the image regions considered as latent variables and sets of words. In the generative process, the image is first generated, which consists of regions, like the topics for text documents. Then each caption word is generated from a selected region. Thus, the words of captions must be conditional on factors presented in the image. The correspondence, however, need not be one to one. Multiple caption words can come from the same region. With typical variational inference, the Corr-LDA model is demonstrated to provide much better performance for the tasks of automatic annotation and text-based image retrieval compared to the ordinary Gaussian mixture model and Gaussian-multinomial LDA.

So far, the dominant topic model LDA has been demonstrated to be very powerful in discovering hidden topic structure as lower-dimensional latent representation for originally high-dimensional features, and Corr-LDA is very effective in fulfilling tasks in multimedia retrieval. Essentially, a topic vector is a random point in a topic simplex parameterized by a multinomial distribution, and topic mixing is established by drawing each word repeatedly from the topics of this multinomial. While this Bayesian network framework does make sampling rather easy, the conditional dependencies between hidden variables become the bottleneck of efficient learning for the LDA models. A multiwing harmonium (MWH) was proposed for captioned images by combining a multivariate Poisson distribution for word count in a caption and a multivariate Gaussian for a color histogram of the image (Xiang et al. 2005). Unlike all the aforementioned models, MWH can be considered as an undirected topic model because of the bidirectional relationship between the latent variables and the inputs: The hidden topic structure can be viewed as predictors from a discriminative model taking the inputs, while it also describes how the input is generated. This formalism enjoys a significant advantage for fast inference as it maintains the conditional independence between the hidden variables. MWH has multiple harmoniums, which group observations from all sources using a shared array of hidden variables to model the theme of all sources. This is consistent with the fact in real applications that the input data unnecessarily come from a single source. In Xiang et al.'s (2005) work, a contrastive divergence and variational learning algorithms are both designed and evaluated on a dual-wing harmonium (DWH) model for the tasks of image annotation and retrieval on news video collections. DWH is demonstrated to be robust and efficient with both learning algorithms. It is worth noting that while the conditional independence enables fast learning for MWH, it also makes learning more difficult. This trade-off can be acceptable for off-line learning, but in other cases, it might need further investigation.

While image classification is an independent problem from image annotations, the two tasks can be connected as they provide evidence for each other and share the goal of automatically organizing images. A single coherent topic model was proposed for simultaneous image classification and annotation (Wang et al. 2009). The typical supervised topic modeling supervised LDA (Blei and McAuliffe 2007) for image classification was extended to multiclasses and embedded with a probabilistic model of image annotation followed by substantial development of inference algorithms based on a variational method. The coherent model, examined on real-world image data sets, provides comparable annotation performance and better than state-of-the-art classification performance.

All these probabilistic topic models estimate the joint distribution of caption words and regional image features to discover the statistical relationship between visual image features and words in an unsupervised way. While the results from these models are encouraging, simpler topic models such as latent semantic analysis (LSA) and direct image matching can also achieve good performance for automatic image annotation through supervised learning where image classes associated with a set of words are predefined. A comparison study between LSA and probabilistic LSA was presented with application on the Coral database (Monay and Gatica-Perez 2003). Surprisingly, the result has shown that simple LSA based on annotation by propagation outperformed probabilistic LSA on this application.

### 19.4.3 Large-Scale Computing in Multimedia

The amount of photos and videos on popular websites such as Facebook, Twitter, and YouTube is now on the scale of tens of billions. This real-life scenario is twofold. For one thing, with such huge multimedia data sets, it is even much more necessary for strategies such as topic models to automatically organize the multimedia Big Data by capturing the intrinsic topic structure, which makes good sense in simplifying the Big Data analysis. For another, it is prohibitively challenging to handle such large-scale multimedia content as images, audio, and videos. Cloud computing provides a new generation of computing infrastructure to manage and process Big Data by distribution and parallelization. Multimedia cloud computing has become an emerging technology for providing multimedia services and applications.

In cloud-based multimedia computing, the multimedia application data can be stored and accessed in a distributed manner, unlike traditional multimedia processing, which is done on client or server ends. One of the big challenges with multimedia data is associated with the heterogeneous data types and services such as video conferencing and photo sharing. Another challenge is how to make the multimedia cloud provide distributed parallel processing services (Zhu et al. 2011). Correspondingly, parallel algorithms such as parallel spectral clustering (Chen et al. 2011), parallel support vector machines (SVMs) (Chang et al. 2007), and the aforementioned PLDA provide important tools for mining large-scale rich-media data with cloud service. More recently, several large-scale multimedia data storage and processing methods were proposed based on the Hadoop platform (Lai et al. 2013; Kim et al. 2014). Python-based content analysis using specialization (PyCASP), a Python-based framework, is also presented for multimedia content analysis by automatically mapping computation onto parallel platforms from Python application code to a variety of parallel platforms (Gonina et al. 2014).

## 19.5 LARGE-SCALE TOPIC MODELING FOR MULTIMEDIA RETRIEVAL AND ANALYSIS

As mentioned in Section 19.4.3, it is highly desirable to develop efficient algorithms for multimedia retrieval and analysis at a very large scale, and it is quite a natural idea to develop topic models given the success of large-scale topic modeling and topic modeling for multimedia analysis, which is demonstrated in Sections 19.4.1 and 19.4.2. However, it is not an easy task to build effective topic models and develop efficient inference algorithms for large-scale multimedia analysis, due to the exponential growth of computing complexity.

The web image collection is perhaps the largest photo source. The current web image search service provided by Google or Yahoo mostly employs text-based image retrieval without using any actual image content, that is, only analyzing images based on the surrounding text features. This is very applicable as the HTML code of web pages contains rich context information for embedded images' annotation. There is, correspondingly, another class of algorithms, named content-based image retrieval, which tries to retrieve images based on analysis of image regional content. Instead of purely considering the surrounding text or the visual features of the image content, such as the color and texture, it is increasingly important to develop a semantic concept model for semantic-based applications such as images and video retrieval as well as effective and efficient management of a massive amount of multimedia data.

When the amount of data scales to the tens of billions, the computational cost for feature extraction and machine learning can be disastrous. For example, some kernel machines, such as SVMs, have computing complexity at least quadratic to the amount of data, which makes it infeasible to track ever-growing data at the level of terabytes or petabytes. To solve the scalability issues, data-intensive scalable computing (DISC) was specially designed as a new computing diagram for large-scale data analysis (Bryant 2007). In this computing diagram, data itself is more emphasized, and it takes special care to consider constantly growing and changing data collections besides performing large-scale computations. The aforementioned MapReduce framework is a popular application built on top of this diagram. The intuition for MapReduce is very simple: If we have a very large set of tasks, we can easily tackle the problem by hiring more workers, distributing the tasks to these workers, and finally, grouping or combining all the results from the parallelized tasks. MapReduce is simple to understand; however, the key problem is how to build semantic concept modeling and develop efficient learning algorithms that can be scalable to the distributed platforms.

In recent years, the implementation of topic modeling of multimedia data on the MapReduce framework has provided a good solution for large-scale multimedia retrieval and analysis. An overview of MapReduce for multimedia data mining was presented for its application in web-scale computer vision (White et al. 2010). An efficient semantic concept named robust subspace bagging (RSB) was proposed, combining random subspace bagging and forward model selection (Yan et al. 2009). Consider the common problem of overfitting due to the high dimensionality of multimedia data; this semantic concept model has remarkably reduced the risk of overfitting by iteratively searching for the best models in the forward model selection step based on validation collection. MapReduce implementation was also presented in Yan et al. (2009) and tested on standard image and video data sets. Normally, tasks in the same category of either a Map task or a Reduce task would be assumed to require roughly the same execution time so that longer tasks would not slow down the distributed computing process. However, the features of multimedia data are heterogeneous with various dimensions; thus, the tasks cannot be guaranteed to have similar execution times. Therefore, extra effort should be made to organize these unbalanced tasks. A task scheduling algorithm was specially designed in Yan et al. (2009) to estimate the running time of each task and optimize task placement for heterogeneous tasks, which achieved significantly improved performance compared to the baseline SVM scheduler.

Most of the multimedia data we talked are about text and images. Actually, large-scale topic modeling also finds success in video retrieval and analysis. Scene understanding is an established field in computer vision for automatic video surveillance. The two challenges are the robustness of the features and the computing complexity for scalability to massive data sets, especially for real-world data. The real-world data streams are unbounded, and the number of motion patterns is unknown in advance. This makes traditional LDA inapplicable as the number of topics of LDA is usually fixed before the documents are generated. Instead, BNP models are more appealing to discover unknown patterns (Teh et al. 2006). The selection of features and the incremental inference corresponding to a continuous stream are crucial to enable the scalability of the models for large-scale multimedia data.

A BNP model was designed for large-scale statistical modeling of motion patterns (Rana and Venkatesh 2012). In the data preprocessing stage, coarse-scale low-level features are selected as a trade-off between model and complexity; that is, a sophisticated model can define a finer pattern but sacrifice computational complexity. A robust and efficient principle component analysis (PCA) was utilized to extract sparse foregrounds in video. To avoid the costly singular vector decomposition operation in each iteration, rank 1 constraint due to the almost-unchanged background during a short period was used for PCA. Then the Dirichlet process mixture was employed to model the sparse components extracted from PCA. Similar to the generative model for text documents, in the bag-of-event model, each feature vector was considered as a sample from motion pattern. The combination of multinomial distribution and mixture of Gaussians defines the location and quantity of the motion. A decayed Markov chain Monte Carlo (MCMC) incremental inference was developed for fixed-cost update in the online setting. The posterior can be guaranteed to converge to the true value on the condition that any point in past would be selected at nonzero probability. A traditional decay function such as exponential decay only depends on time, while for the scene-understanding problem, the probability distribution also depends on data in the clustering space, which is the distance between the past observations and the current ones. To improve the scalability to large-scale settings, the distance between clusters is measured to represent the distance between samples in each cluster. The framework was tested on a 140-h-long video, whose size is quite large, and it was demonstrated to provide comparable pattern-discovering performance with existing scene-understanding algorithms. This work enhances our ability to effectively and efficiently discover unknown patterns from unbounded video streams, providing a very promising framework for large-scale topic modeling of multimedia data.

## 19.6 CONCLUSIONS AND FUTURE DIRECTIONS

The Internet and social media platforms such as Facebook, Twitter, and YouTube have been so prevalent that in every minute, millions and billions of new media data such as photos and videos are generated. Users are overwhelmed by so much information, and it becomes rather difficult and time consuming to digest a vast amount of information. Probabilistic topic models from the IR field are highly desirable to effectively organize large volumes of multimedia data by discovering the hidden topic structure. On the other hand, traditional topic models such as PLSI and LDA are hard to scale to massive data sets at the level of terabytes or petabytes due to the prohibitive computational complexity, which is also the curse for many traditional machine learning and data mining algorithms. With the emergence of cloud computing, topic models are expected to become increasingly applicable to multimedia data. Google's MapReduce and Yahoo's Hadoop are notable examples of large-scale data processing. These techniques have demonstrated great effectiveness and efficiency for Big Data analysis tasks.

While the success of topic models has been well established for text documents, much effort has been recently devoted to their application to large-scale multimedia analysis, where the multimedia data can be in the form of text, images, audio, and video. It is evident from Sections 19.3 through 19.5 that topic modeling is a very useful and powerful technique to address multimedia cloud computing at a very large scale (Barnard et al. 2003; Blei and Jordan 2003; Duygulu et al. 2002; Li and Perona 2005; Sivic et al. 2006; Wang et al. 2009). For example, comparative studies between an ordinary Gaussian mixture model and Corr-LDA have demonstrated that Corr-LDA is significantly better for the tasks of image automatic annotation and text-based image retrieval. Various distributed and parallel computing frameworks such as PLDA and parallel support vector machine (PSVM) (Chang et al. 2007) have been proposed to solve scalability issues with multimedia computing. The DISC diagram is another popular design for large-scale data analysis (Bryant 2007). The popular Big Data processing framework MapReduce also find its success in multimedia cloud computing applications such as web-scale computer vision (White et al. 2010). Another major issue with high dimensionality of large-scale data is overfitting. A novel semantic concept model was proposed (Yan et al. 2009) to address the overfitting issue, and the results have demonstrated the effectiveness of the method.

In future work, scalability will be a more prominent issue with the ever-growing data size. Web-based multimedia data mining (White et al. 2010) demonstrates the power to implement computer vision algorithms such as training, clustering, and background subtraction in the MapReduce framework. Cloud-based computing will remain a very active research area for large-scale multimedia analysis. Moreover, for cloud-based large-scale multimedia computing, it is very desirable to visualize high-dimensional data with easy interpretation at the user interface. The current method mainly displays topics with term frequency (Blei 2012), which is limited as hidden topic structure also connects different documents. The relations among multimedia data and the underlying topics are even more complex. To visualize these complex relations will significantly help people consume multimedia content. In addition, topic models are usually trained in an off-line fashion, where the whole batch of data is used once to construct the model. This process does not suit the online setting, where data streams continuously arrive with time. With online learning (Barbara and Domeniconi 2008), we do not need to rebuild the whole topic model when new data arrive but just incrementally update the parameters based on the new data. Thus, online learning could provide a much more efficient solution to multimedia analysis as some multimedia content is streaming in nature. Furthermore, feature selection and feature engineering are crucial for multimedia analysis and retrieval. With the recent advances in Deep Learning (Hinton et al. 2006; Hinton and Salakhutdinov 2006), learning a compact representation and features of multimedia data will become an important research topic.

## REFERENCES

Ahmed, A., Low, Y., and Smola, A. Scalable Distributed Inference of Dynamic User Interests for Behavioral Targeting. *ACM SIGKDD Conference on Knowledge Discovery from Data*, 2011.

Barbara, D., and Domeniconi, C. On-line LDA: Adaptive Topic Models for Mining Text Streams with Applications to Topic Detection and Tracking. *IEEE International Conference on Data Mining*, 2008.

Barnard, K., Duygulu, P., Forsyth, D., De Freitas, N., Blei, D., and Jordan, M. Matching Words and Pictures. *Journal of Machine Learning Research*, 3, 1107–1135, 2003.

Blei, D. Probabilistic Topic Models. *Communications of the ACM*, 55(4), 77–84, 2012.

Blei, D., and Jordan, M. Modeling Annotated Data. *ACM SIGIR Conference on Research and Development in Information Retrieval*, 127–134, 2003.

Blei, D., and Lafferty, J. Dynamic Topic Models. *International Conference on Machine Learning*, 113–120, 2006.

Blei, D., and Lafferty, J. A Correlated Topic Model of Science. *Annals of Applied Statistics*, 1(1), 17–35, 2007.

Blei, D., and McAuliffe, J. Supervised Topic Models. *Neural Information Processing Systems*, 2007.

Blei, D., Ng, A., and Jordan, M. Latent Dirichlet Allocation. *Journal of Machine Learning Research*, 3, 993–1022, 2003.

Bryant, R.E. Data-Intensive Supercomputing: The Case for Disc. Technical Report, School of Computer Science, Carnegie Mellon University, Pittsburgh, PA, 2007.

Chang, E.Y., Zhu, K., Wang, H., and Bai, H. PSVM: Parallelizing Support Vector Machines on distributed computers. *Neural Information Process System*, 2007.

Chang, J., and Blei, D. Hierarchical Relational Models for Document Networks. *Annals of Applied Statistics*, 4(1), 124–150, 2010.

Chen, Y.W., Song, Y., Bai, H., Lin, C. J., and Chang, E.Y. Parallel Spectral Clustering in Distributed Systems. *IEEE Transactions on Pattern Analysis and Machine Intelligence*, 33(3), 568–586, 2011.

Datta, R., Li, J., and Wang, J. Content-based Image Retrieval: Approaches and Trends of the New Age. *ACM SIGMM International Workshop on Multimedia Information Retrieval*, 253–262, 2005.

Dean, J., and Ghemawat, S. MapReduce: Simplified Data Processing on Large Clusters. *Communications of ACM*, 2008.

Deerwester, S., Dumais, S., Landauer, T., Furnas, G., and Harshman, R. Indexing by latent semantic analysis. *Journal of the American Society of Information Science*, 41(6), 391–407, 1990.

Duygulu, P., Barnard, K., de Freitas, N., and Forsyth, D.A. Object Recognition as Machine Translation: Learning a Lexicon for a Fixed Image Vocabulary. *European Conference on Computer Vision*, 97–112, 2002.

Gonina, E., Friedland, G., Battenberg, E., Koanantakool, P., Driscoll, M., Georganas, E., and Keutzer, K. Scalable Multimedia Content Analysis on Parallel Platforms using Python. *ACM Transactions on Multimedia Computing, Communications, and Applications (TOMCCAP)*, 10(2), 18–38, 2014.

Hinton, G.E., Osindero, S., and Teh, Y. A Fast Learning Algorithm for Deep Belief Nets. *Neural Computation*, 18, 1527–1554, 2006.

Hinton, G.E., and Salakhutdinov, R.R. Reducing the dimensionality of data with neural networks. *Science*, 313(5786), 504–507, 2006.

Hofmann, T. Probabilistic Latent Semantic Indexing. *ACM SIGIR Conference on Research and Development in Information Retrieval*, 50–57, 1999.

Isard, M., Budiu, M., Yu, Y., Birrell, A., and Fetterly, D. Drad: Distributed Data-Parallel Programs from Sequential Building Blocks. *European Conference on Computer Systems (EuroSys)*, 2007.

Kim, M., Han, S., Jung, J., Lee, H., and Choi, O. A Robust Cloud-based Service Architecture for Multimedia Streaming Using Hadoop. *Lecture Notes in Electrical Engineering*, 274, 365–370, 2014.

Lai, W.K., Chen, Y.U., Wu, T.Y., and Obaidat, M.S. Towards a Framework for Large-scale Multimedia Data Storage and Processing on Hadoop Platform. *The Journal of Supercomputing*, 68(1), 488–507, 2013.

Li, F.F., and Perona, P. A Bayesian Hierarchical Model for Learning Natural Scene Categories. *IEEE Conference on Computer Vision and Pattern Recognition*, 524–531, 2005.

Malewicz, G., Austern, M., and Czajkowski, G. Pregel: A System for Large-Scale Graph Processing. *ACM SIGMOD Conference*, 2010.

Melnik, S., Gubarev, A., Long, J.J., Romer, G., Shivakumar, S., Tolton, M., and Vassilakis, T. Dremel: Interactive Analysis of Web-scale Datasets. *International Conference on Very Large Data Bases*, 330–339, 2010.

Monay, F., and Gatica-Perez, D. On Image Auto-annotation with Latent Space Models. *ACM International Conference on Multimedia*, 275–278, 2003.

Newman, D., Asuncion, A., Smyth, P., and Welling, M. Distributed Inference for Latent Dirichlet Allocation. *Neural Information Process System*, 1081–1088, 2007.

Papadimitriou, C.H., Raghavan, P., Tamaki, H., and Vempala, S. Latent Semantic Indexing: A Probabilistic Analysis. *ACM SIGACT-SIGMOD-SIGART Symposium on Principles of Database Systems*, 159–168, 1998.

Powell, R., and Li, J. Piccolo: Building Fast, Distributed Programs with Partitioned Tables. *USENIX Conference on Operating Systems Design and Implementation*, 2010.

Rana, S., and Venkatesh, S. Large-Scale Statistical Modeling of Motion Patterns: A Bayesian Nonparametric Approach. *Indian Conference on Computer Vision, Graphics and Image Processing*, 2012.

Rosen-Zvi, M., Griffiths, T., Steyvers, M., and Smith, P. The Author-Topic Model for Authors and Documents. *Uncertainty in Artificial Intelligence*, 2004.

Sivic, J., Russell, B., Zisserman, A., Freeman, W., and Efros, A. Unsupervised Discovery of Visual Object Class Hierarchies. *IEEE Conference on Computer Vision and Pattern Recognition*, 2008.

Smola, A., and Narayanamurthy, S. An Architecture for Parallel Topic Model. *Very Large Database (VLDB)*, 3(1), 703–710, 2010.

Teh, Y., Jordan, M., Beal, M., and Blei, D. Hierarchical Dirichlet Processes. *Journal of American Statistical Association*, 101(476), 1566–1581, 2006.

Wallach, H. Topic Modeling: Beyond Bag of Words. *International Conference on Machine Learning*, 2006.

Wang, Q., Xu, J., Li, H., and Craswell, N. Regularized Latent Semantic Indexing. *ACM SIGIR Conference on Research and Development in Information Retrieval*, 2011.

Wang, Y., Bai, H., Stanton, M., Chen, M., and Chang, E.Y. PLDA: Parallel Latent Dirichlet Allocation for Large-Scale Applications. *Conference on Algorithmic Aspects in Information and Management*, 301–314, 2009.

White, B., Yeh, T., Lin, J., and Davis, L. Web-Scale Computer Vision using MapReduce for Multimedia Data Mining. *ACM KDD Conference Multimedia Data Mining Workshop*, 2010.

Xiang, E., Yan, R., and Hauptmann, A. Mining Associated Text and Images with Dual-Wing Harmoniums. *Uncertainty in Artificial Intelligence*, 2005.

Yan, R., Fleury, M., and Smith, J., Large-Scale Multimedia Semantic Concept Modeling Using Robust Subspace Bagging and MapReduce. *ACM Workshop on Large-Scale Multimedia Retrieval and Mining*, 2009.

Zhai, K., Graber, J., Asadi, N., and Alkhouja, M. Mr.LDA: A Flexible Large Scale Topic Modeling Package Using Variational Inference in MapReduce. *ACM Conference World Wide Web*, 2012.

Zhu, W., Luo, C., Wang, J., and Li, S. Multimedia Cloud Computing. *IEEE Signal Processing Magazine*, 28(3), 59–69, 2011.

CHAPTER 20

# Big Data Biometrics Processing

## *A Case Study of an Iris Matching Algorithm on Intel Xeon Phi*

Xueyan Li and Chen Liu

CONTENTS



ABSTRACT

With the rapidly expanded biometric data collected by various sources for identification and verification purposes, how to manage and process such Big Data draws great concern. On one hand, biometric applications normally involve comparing a huge amount of samples and templates, which has strict requirements on the computational capability of the underlying hardware platform. On the other hand, the number of cores and associated threads that hardware can support has increased greatly; an example is the newly released Intel Xeon Phi coprocessor. Hence, Big Data biometrics processing demands the execution of the applications at a higher parallelism level. Taking an iris matching algorithm as a case study, we implemented an open multi-processing (OpenMP) version of the algorithm to examine its performance on the Intel Xeon Phi coprocessor. Our target is to evaluate our parallelization approach and the influence from the optimal number of threads, the impact of thread-to-core affinity, and the impact of the hardware vector engine.

## 20.1 INTRODUCTION

With the drive towards achieving higher computation capability, the most advanced computing systems have been adopting alternatives from the traditional general purpose processors (GPPs) as their main components to better prepare for Big Data processing. NVIDIA's graphic processing units (GPUs) have powered many of the top-ranked supercomputer systems since 2008. In the latest list published by Top500.org, two systems with Intel Xeon Phi coprocessors have claimed positions 1 and 7 [1].

While it is clear that the need to improve efficiency for Big Data processing will continuously drive changes in hardware, it is important to understand that these new systems have their own advantages as well as limitations. The required effort from the researchers to port their codes onto the new platforms is also of great significance. Unlike other coprocessors and accelerators, the Intel Xeon Phi coprocessor does not require learning a new programming language or new parallelization techniques. It presents an opportunity for the researchers to share parallel programming with the GPP. This platform follows the standard parallel programming model, which is familiar to developers who already work with x86-based parallel systems.

This does not mean that achieving good performance on this platform is simple. The hardware, while presenting many similarities with other existing multicore systems, has its own characteristics and unique features. In order to port the code in an efficient way, those aspects must be understood.

The specific application that we focus on in this parallelization study is biometrics. The importance of biometrics and other identification technology has increased significantly, and we can observe biometric systems being deployed everywhere in our daily lives. This has dramatically expanded the amount of biometric data collected and examined everyday in health care, employment, law enforcement, and security systems. There can be up to petabytes of biometric data from hundreds of millions of identities that need to be accessed in real time. For example, the Department of Homeland Security (DHS) Automated Biometric Identification System (IDENT) database held 110 million identities and enrolled or verified over 125,000 individuals per day in 2010 [2]. India has their unique identification (UID) program with over 100 million people to date and expected to cover more than 1 billion individuals. This Big Data processing poses a great computational challenge. In addition, running biometric applications with sizable data sets consumes large amounts of power and energy. Therefore, efficient power management is another challenge we are facing.

As a result, we propose to execute biometric applications on many-core architecture with great energy efficiency while achieving superb performance at the same time. In this study, we want to analyze the performance of an OpenMP implementation of an iris recognition algorithm on the new Intel Xeon Phi coprocessor platform. The focus is to understand the workload characteristics, analyze the results from specific features of the hardware platform, and discuss aspects that could help improve overall performance on Intel Xeon Phi.

## 20.2 BACKGROUND

In this section, we will introduce the Xeon Phi coprocessor, the iris matching algorithm, the OpenMP programming model, and the Intel VTune performance analyzer, all serving as background knowledge for this research.

### 20.2.1 Intel Xeon Phi

The Intel Xeon Phi coprocessor (code-named *Knight Corner*) is the first commercial product employing Intel's Many Integrated Core (MIC) architecture. The Xeon Phi coprocessor is implemented as a Peripheral Component Interconnect Express (PCIe) form-factor add-in card. It supports the x86 memory model, Institute of Electrical and Electronics Engineers (IEEE) 754 floating-point arithmetic, and applications written in industry-standard programming languages such as FORTRAN, C, and C++ [3]. A rich development environment supports the coprocessor, which includes compilers, numerous libraries such as threading libraries and high-performance math libraries, performance characterizing and tuning tools, and debuggers.

The cores of Xeon Phi run independently of each other and have very powerful vector units. Each core has a vector processing unit (VPU) for fused multiply–add (FMA) operations. The vector width is 512 bits, and the VPU may operate on 8 double-precision (or 16 single-precision) data elements [4]. The instructions are pipelined, and after every cycle, a vector result is produced once the pipeline is filled. Additionally, Xeon Phi coprocessors include memory controllers that support the double data rate type five synchronous graphics random access memory (GDDR5) specification and special-function devices such as the PCIe interface. The memory subsystem provides full cache coherence. Cores and other components of Xeon Phi are connected via a ring interconnect.

From the software point of view, the Xeon Phi coprocessor is of the shared-memory computing domain, which is loosely coupled with the computing domain of the host. The host software stack is based on a standard Linux kernel. In comparison with that, the Xeon Phi software stack is based on a modified Linux kernel. In fact, the operating system on the Xeon Phi coprocessor is an embedded Linux environment, which provides basic functionality such as process creation, scheduling, or memory management [5].

A Xeon Phi coprocessor comprises more than 50 Intel architecture (IA) cores, each with four-way Hyper-Threading (HT), to produce a total of over 200 logical cores. Because Xeon Phi supports four hardware threads per core, users need to determine how threads are used. The software environment allows the user to control thread affinity in two modes: compact and scatter. Compact affinity means using all threads on one core before using other cores; scatter affinity means scattering threads across cores before using multiple threads on a given core.

The single-threaded, scalar (i.e., nonvectorial) code performance on Xeon Phi is fairly low, for example, when compared with that of an Intel Sandy Bridge central processing unit (CPU) [6,7]. Xeon Phi makes up for this deficit by offering more cores and wider vectors. Since a single thread can issue a vector instruction only every other cycle on Xeon Phi, at least two threads per core must be used to achieve full utilization. Therefore, the programs need to be highly parallel and vectorial in order to optimize the performance on this platform. The developers can maximize parallel performance on Xeon Phi in four ways:

- Increasing concurrency in parallel regions
- Reduction of serial code
- Minimizing the number of critical sections
- Improvement of load balance

They should also vectorize applications to maximize single instruction multiple data (SIMD) parallelism by including vectorial directives and applying loop transformations as needed, in case the code cannot be automatically vectorized by the compiler [5].

Xeon Phi can be used in two different modes: native and offload. In our experiments, we focused on the offload mode. There are two problems we need to handle in the offload mode: One is managing the data transfer between the host and the coprocessor; the other is launching the execution of functions (or loops) on the coprocessor. For the offload mode, all the data transfer between the host and the coprocessor goes through the PCIe bus. Therefore, an attempt should be made to exploit data reuse. This is also a consideration for applications written using OpenMP that span both the host and the coprocessor. Getting the benefit from both the host and Xeon Phi is our goal.

### 20.2.2 Iris Matching Algorithm

An iris matching algorithm has been recognized as one of the most reliable methods for biometric identification. The algorithm we used in this study is based on Daugman's iris matching algorithm [8], the most representative iris matching method. The procedure of an iris matching algorithm can be summarized as follows:

1. Generate a template along with a mask for the iris sample.
2. Compare two iris templates using Hamming distances.
3. Shift Hamming distance calculation to counter rotational inconsistencies.
4. Get the sample pair with the lowest Hamming distance reading.

After sampling the iris region (pupil and limbus), the iris sample is transformed into a two-dimensional (2-D) image. Then Gabor filtering will extract the most discriminating information from the iris pattern as a biometric template. Each iris sample is represented with two 2-D matrices of bits (20 rows by 480 columns). The first one is the template matrix representing the iris pattern. The second one is the mask matrix, representing aspects such as eyelid, eyelash, and noise (e.g., spectacular reflections), which may occlude parts of the iris. Two iris samples are compared based on the Hamming distance calculation, with both matrices from each sample being used. The calculated matching score is normalized between 0 and 1, where a lower score indicates that the two samples are more similar.

The main body of the algorithm is a 17-round for-loop structure with the index shifting from −8 to 8, performing bit-wise logical operations (AND, OR, XOR) on the matrices. We suppose that the matrices of the first iris sample are named as *template1* and *mask1* and the matrices of the second iris sample are named as *template2* and *mask2*. In each round, three steps are needed to finish the procedure. Firstly, *template1* and *mask1* are rotated for 2 × |*shift*| columns to form two intermediate matrices, named *template1s* and *mask1s*. They will be left-rotated if shift is less than 0 or right-rotated if shift is larger than 0. Secondly, we will XOR *template1s* with *template2* into *temp* and OR *mask1s* with *mask2* into *mask*. Finally, Matrix *temp* will be ANDed with Matrix *mask* to form Matrix *result*. So the Hamming distance from this round is calculated by dividing the number of 1s of Matrix *result* by the total number of the 2-D matrix entries (which is 20 × 480) minus the number of 1s in Matrix *temp* [9]. The minimum-valued Hamming distance of all 17 iterations shows the two samples' similarity, which is returned as the matching score of the two samples.

### 20.2.3 OpenMP

OpenMP [10] is an application programming interface (API) that supports multiplatform shared-memory parallel programming in FORTRAN, C, and C++. It consists of a set of compiler directives and library routines, in which runtime behavior can be controlled by environment variables [11]. Annotating loop bodies and sections of the codes, compiler directives are used for parallel execution and marking variables as local or shared (global). Certain constructs exist for critical sections, completely independent tasks, or reductions on variables [12].

When a for loop is declared to be parallel, the iterations of the loop are executed concurrently. The iterations of the loop can be allocated to the working threads according to three scheduling policies: static, dynamic, and guided. In static scheduling, the iterations are either partitioned in as many intervals as the number of threads or partitioned in chunks that are allocated to the threads in a round-robin fashion. Dynamic scheduling partitions the iterations in chunks. Those chunks are dynamically allocated to the threads using a first-come first-served policy. Finally, the guided scheduling policy tries to reduce the scheduling overhead by allocating first a large amount of work to each thread. It geometrically decreases the amount of work allocated to the thread (up to a given minimum chunk size) in order to optimize the load balance [12]. In this study, we use the default static scheduling policy. In addition, the *#pragma omp parallel* directive can be used to mark the parallel section, and the *#pragma simd* directive can be used to perform the vectorization. In this study, we employed both.

### 20.2.4 Intel VTune Amplifier

In order to get an insight into the software behavior, we employed the Intel VTune Amplifier XE 2013 [13] (in short, VTune) to perform the workload characterization of the iris matching algorithm. VTune analyzes the software performance on IA-32– and Intel 64–based machines, and the performance data are displayed in an interactive view. At the granularity-of-function level, VTune breaks down the total execution time and shows the time spent on each function, so the function consuming most of the execution time is identified as the "hot-spot function." At a finer granularity level, it can also show the time spent on a basic block, several lines of code, or even a single line of the source code inside a function. In this way, we can identify the performance-intensive function (hot-spot function) or lines of code (hot block) in the software algorithm [14].

## 20.3 EXPERIMENTS

In this section, we first present how we set up the experiments. Then we present the results from our experimental data in detail.

### 20.3.1 Experiment Setup

The experiments have been conducted on BEACON, an energy-efficient cluster that utilizes Intel Xeon Phi coprocessors. BEACON is funded by the National Science Foundation (NSF) to port and optimize scientific codes to the coprocessors based on Intel's MIC architecture [15]. By June 2013, BEACON was listed at the no. 3 position among the most energy-efficient supercomputers in the world on the Green500 list [16] and at the no. 397 position among the most powerful supercomputers in the world on the TOP500 List [17]. BEACON currently has 48 compute nodes. Each compute node houses two 8-core Intel Xeon E5-2670 processors, 256 GB of memory, and four Intel Xeon Phi 5110P coprocessors. This specific Xeon Phi model has 60 cores and 8 GB of memory [15]. To compile the code, we use icc version 14.0.2 and OpenMP version 4.0.

To obtain the results, we conducted this experiment using the following steps:

1. Write the source code in C code on BEACON and then verify the correctness and validate against the original results obtained from Microsoft Visual Studio 2010.
2. Study the workload characterization of the iris matching algorithm by using Intel VTune.
3. Based on the results from previous steps, rewrite this program by adding the parallelism and vectorization features.
4. Compare all the results; get the optimal number of threads, the influence of different affinities, as well as vectorization on the performance.

### 20.3.2 Workload Characteristics

The workload used in the experiment is from the Quality in Face and Iris Research Ensemble (Q-FIRE) data set [18] and Q-FIRE II Unconstrained data set, where we used a total number of 574 iris samples. The iris matching process is to match every iris sample against the entire data set, resulting in a total of 164,451 comparisons. In this section, we will demonstrate the results of the performance analysis.

In our iris matching algorithm, each sample needs to be compared against all the other samples in the database, so the main function will have a two-level nested loop for comparison. Each iteration will load two samples (both template and mask) for comparison and then calculate the Hamming distance to generate the matching score [14]. VTune

TABLE 20.1 Experimental Environment for Intel VTune

| CPU | I-Cache and D-Cache | L2 Cache | L3 Cache | Memory Size | Operating System | Compiler |
|---|---|---|---|---|---|---|
| Intel Xeon E5-2630 2.3 GHz 6-core | 192 KB per core | 1.5 MB | 15 MB | 16 GB | Windows 7 64-bit | Visual Studio 2010 |

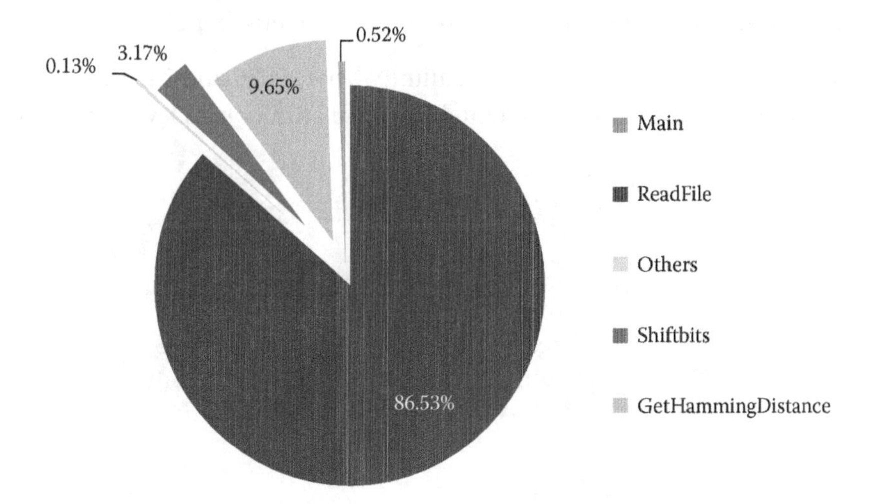


FIGURE 20.1 Percentage of execution time of the iris matching algorithm functions.

performs the function breakdown in order to identify the software performance bottleneck. Table 20.1 shows the environment in which we used VTune to get the characteristics of the iris matching algorithm on an Intel Xeon E5-2630 CPU.

Figure 20.1 shows the functions of the iris matching algorithm along with their consumed percentage of total execution time. Note that the time recorded for each function does not include the time spent on subfunction calls [14]. For example, the time for function GetHammingDistance does not include the time spent on function shiftbits. The results show that the ReadFile function, which loads the samples into memory, occupies the majority of execution time. It means that the data transfer is the bottleneck of the iris matching algorithm.

### 20.3.3 Impact of Different Affinity

To study the impact of thread-to-core affinity, the experiment was conducted on Xeon Phi in both compact and scatter modes. The compact affinity assigns OpenMP thread $n + 1$ to a free thread context closely to another thread context where the $n$th OpenMP thread was placed. The scatter affinity distributes the threads evenly across the entire system.

Figures 20.2 through 20.4 provide the comparison in execution time between scatter affinity and compact affinity, with the light-colored bar representing the compact affinity and the dark-colored bar representing the scatter affinity, the $x$-axis representing the number of threads, and the $y$-axis representing the execution time in seconds. The difference between the two affinities can be contrasted in three parts:

1. Compact affinity is more effective than scatter affinity when only a small number of threads are needed, as shown in Figure 20.2. When the number of threads is less than eight, compact affinity only uses a small number of cores (one to two cores), and hence, the communication overhead between the threads is small; on the other hand, scatter affinity uses the same number of cores as the number of threads.

2. Scatter affinity is faster than compact affinity between 16 and 240 threads, as shown in Figure 20.3. In this case, even though compact affinity uses fewer cores, every core

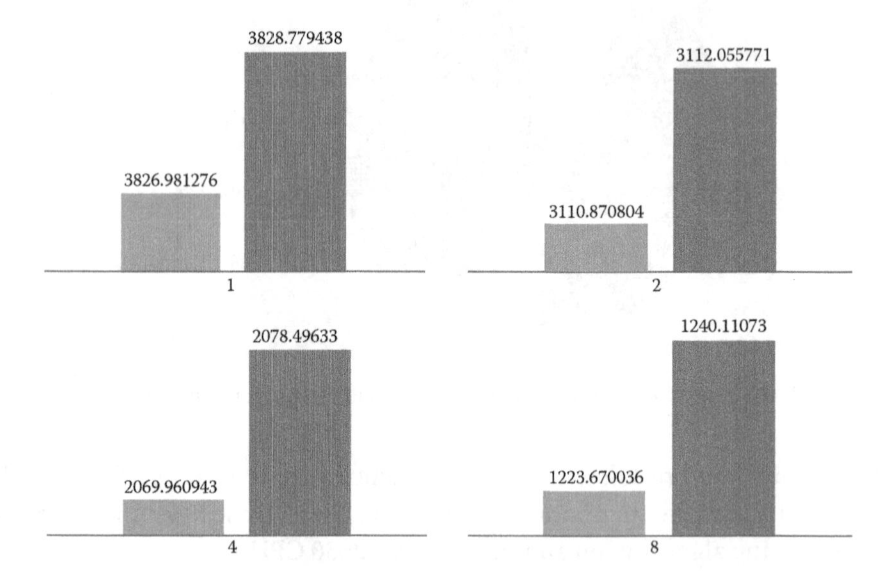


FIGURE 20.2 Difference between scatter and compact affinity for one to two cores.

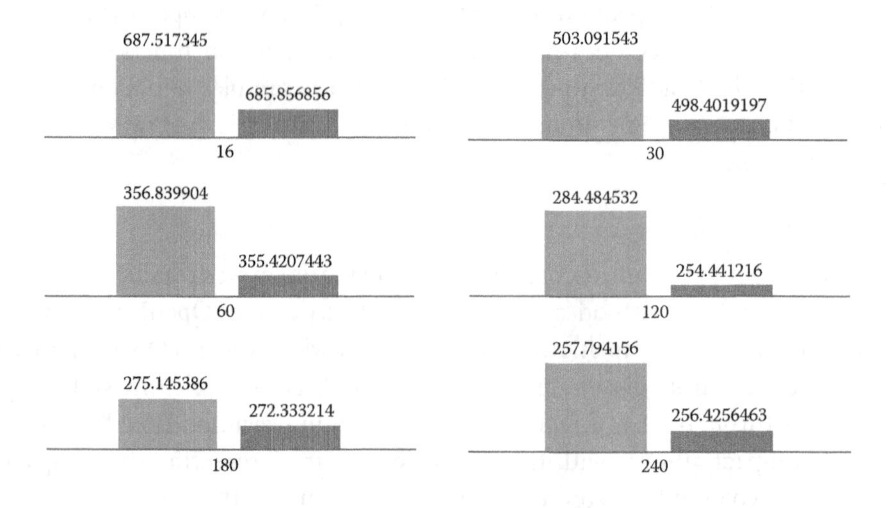


FIGURE 20.3 Difference between scatter and compact affinity for more than two cores.

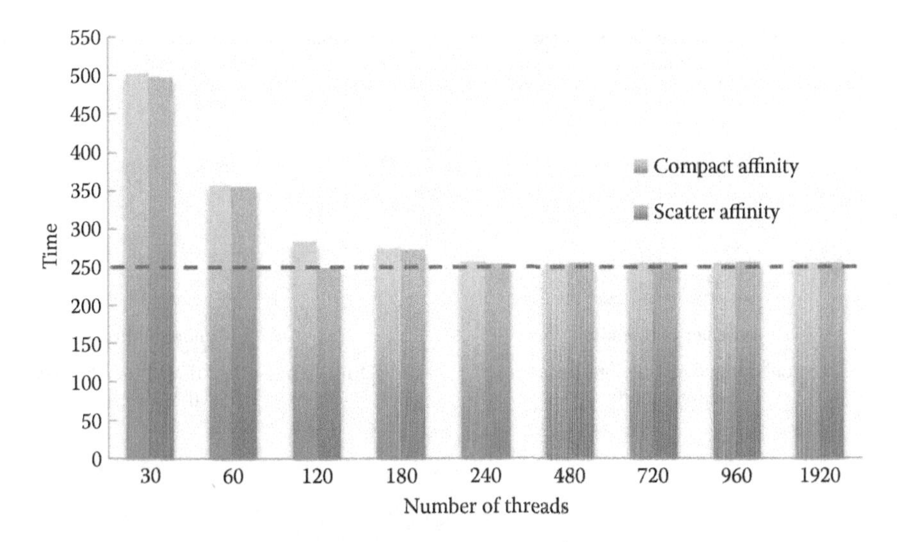


FIGURE 20.4 Difference between scatter and compact affinity.

is fully loaded with four threads; they will compete for the hardware resource for their execution and slow down the individual execution, what we call "interthread interference," which outweighs the benefit they get from saved communication overhead. On the other hand, in scattered affinity, the cores are not that fully loaded, which shows the benefit.

3. When the number of threads is large (more than 240 threads), the effect of two affinities is almost the same, as shown in Figure 20.4. In this case, the number of threads is over the hardware capability (60 cores with 4 threads each, resulting in a total of 240 threads), so all the cores are fully loaded in either affinity.

Overall, when we run a program on Xeon Phi, the suitable affinity must be considered based on the number of threads and the correspondingly parallelism level.

### 20.3.4 Optimal Number of Threads

After discussing the effect of affinity, finding the optimal number of threads for the execution is another objective. We try to identify the optimal number of threads for both compact affinity and scatter affinity. The results from Figure 20.5 show that the optimal number of threads is 120 when using scatter affinity, corresponding to using all the 60 cores on Xeon Phi, with 2 threads per core; the optimal number of threads is 240 when using compact affinity, corresponding to using all the 60 cores on Xeon Phi, with 4 threads per core.

### 20.3.5 Vectorization

As we can see from Figure 20.1, function GetHammingDistance takes the second longest time, only next to the ReadFile function. Since Xeon Phi is equipped with a very powerful vector engine [19], our target is to utilize the vector engine to improve the overall performance. As a result, we rewrite the GetHammingDistance function by adding parallelism

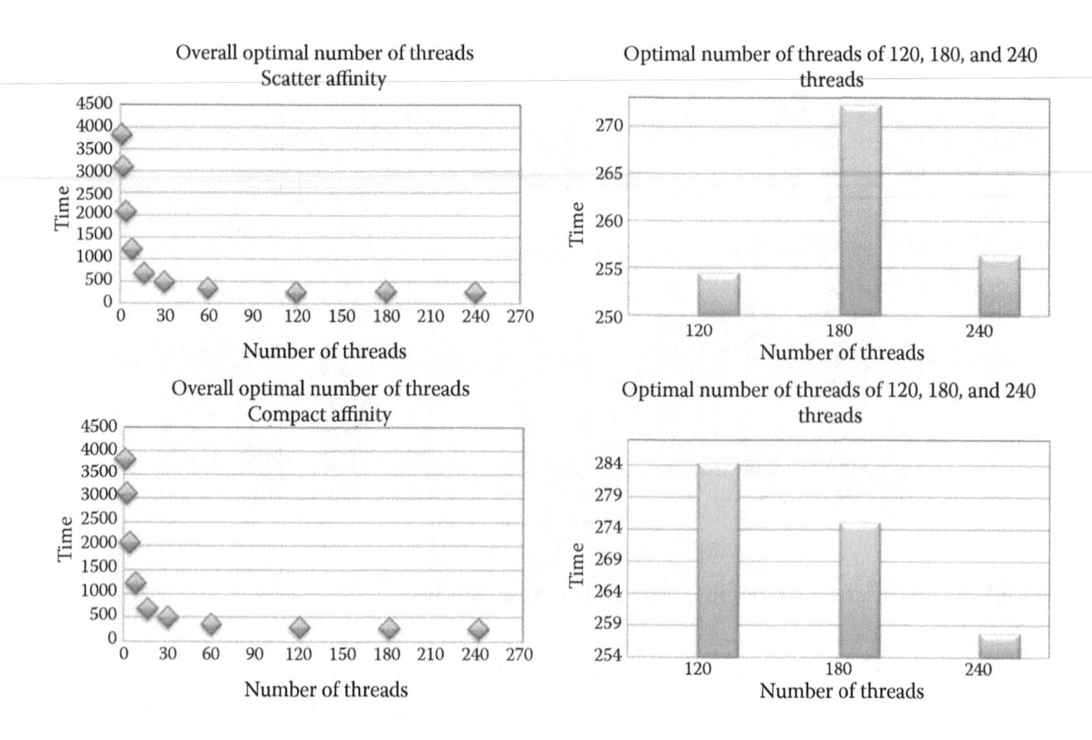


FIGURE 20.5 Optimal number of threads.

and vectorization. Since we have discussed the optimal number of threads and the suitable affinity, in this case, we run the program with scatter affinity. Figure 20.6 shows the execution time of running the iris matching algorithm on Xeon Phi in the offload mode with and without vectorization. Please note that the $y$-axis is in log scale, so basically, 2 means 100 s, 3 means 1000 s, and so on and so forth. Clearly, in all cases, vectorization improves the performance. The best time occurs when we use the optimal number of threads of 120.

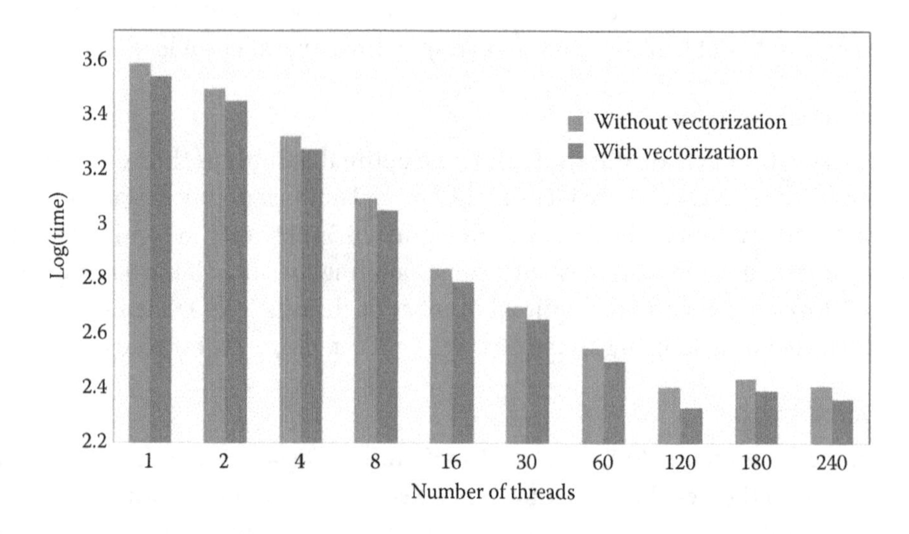


FIGURE 20.6 Difference between multithreading with vectorization and without vectorization.

## 20.4 CONCLUSIONS

Big Data biometrics processing poses a great computational need, while in the meantime, it is embarrassingly parallel by nature. In this research, as a case study, we implemented an OpenMP version of the iris matching algorithm and evaluated its performance on the innovative Intel Xeon Phi coprocessor platform. We performed workload characterization of the iris matching algorithm, analyzed the impact of different thread-to-core affinity settings on the performance, identified the optimal number of threads to run the workload, and most important of all, vectorized the code to take advantage of the powerful vector engines that come with Xeon Phi to improve the performance. Based on our experience, even though the emerging many-core platform is able to provide adequate parallelism, redesigning the application to take advantage of the specific features of the hardware platform is very important in order to achieve the optimal performance for Big Data processing.

## ACKNOWLEDGMENTS

We thank the National Institute for Computational Sciences (NICS) for providing us access to the BEACON supercomputer to conduct this research. We would also like to thank Gildo Torres and Guthrie Cordone for their feedback that greatly improved the results of this experiment and the quality of this study. This work is supported by the National Science Foundation under grant number IIP-1332046. Any opinions, findings, and conclusions or recommendations expressed in this material are those of the authors and do not necessarily reflect the views of the National Science Foundation.

## REFERENCES

1. Top 500 Supercomputer Sites, November 2013. Available at http://www.top500.org/.
2. A. Sussman, "Biometrics and Cloud Computing," presented at the *Biometrics Consortium Conference 2012*, September 19, 2012.
3. L. Meadows, "Experiments with WRF on Intel Many Integrated Core (Intel MIC) Architecture," in *Proceedings of the 8th International Conference on OpenMP in a Heterogeneous World, Ser. IWOMP'12*, Springer-Verlag, Berlin, Heidelberg, pp. 130–139, 2012.
4. J. Jeffers and J. Reinders, *Intel Xeon Phi Coprocessor High Performance Programming*, 1st ed., Morgan Kaufmann, Boston, 2013.
5. J. Dokulil, E. Bajrovic, S. Benkner, S. Pllana and M. Sandrieser. "Efficient Hybrid Execution of C++ Applications using Intel Xeon PhiTM Coprocessor," Research Group Scientific Computing, University of Vienna, Austria, November 2012.
6. K. Krommydas, T. R. W. Scogland and W.-C. Feng, "On the Programmability and Performance of Heterogeneous Platforms," in *Parallel and Distributed Systems (ICPADS), 2013 International Conference on*, pp. 224–231, December 15–18, 2013.
7. P. Sutton, Benchmarks: Intel Xeon Phi vs. Sandy Bridge, 2013. Available at http://blog.xcelerit.com/benchmarks-intel-xeon-phi-vs-intel-sandybridge/.
8. J. G. Daugman, "High Confidence Visual Recognition of Persons by a Test of Statistical Independence," *Pattern Analysis and Machine Intelligence, IEEE Transactions on*, vol. 15, no. 11, pp. 1148–1161, 1993.
9. G. Torres, J. K.-T. Chang, F. Hua, C. Liu and S. Schuckers, "A Power-Aware Study of Iris Matching Algorithms on Intel's SCC," in *2013 International Workshop on Embedded Multicore Systems (ICPP-EMS 2013)*, in conjunction with ICPP 2013, Lyon, France, October 1–4, 2013.
10. OpenMP. Available at http://openmp.org/wp/.

11. E. Saule and U. V. Catalyurek, "An Early Evaluation of the Scalability of Graph Algorithms on the Intel MIC Architecture," in *26th International Symposium on Parallel and Distributed Processing, Workshops and PhD Forum (IPDPSW), Workshop on Multithreaded Architectures and Applications (MTAAP)*, 2012.
12. G. Gan, "Programming Model and Execution Model for OpenMP on the Cyclops-64 Manycore Processor," Ph.D. Dissertation, University of Delaware, Newark, DE, 2010.
13. Intel® VTune™ Amplifier XE 2013. Available at https://software.intel.com/en-us/intel-vtune-amplifier-xe.
14. J. K.-T. Chang, F. Hua, G. Torres, C. Liu and S. Schuckers, "Workload Characteristics for Iris Matching Algorithm: A Case Study," in *13th Annual IEEE Conference on Technologies for Homeland Security (HST'13)*, Boston, MA, November 12–14, 2013.
15. Beacon. Available at http://www.nics.tennessee.edu/beacon.
16. The Green500 List, June 2013. Available at http://www.green500.org/lists/green201306.
17. Beacon on Top500. Available at http://www.top500.org/system/177997#.U3MMkl4dufR.
18. P. A. Johnson, P. Lopez-Meyer, N. Sazonova, F. Hua and S. Schuckers, "Quality in Face and Iris Research Ensemble (Q-FIRE)," in *Fourth IEEE International Conference on Biometrics: Theory Applications and Systems (BTAS)*, pp. 1–6, 2010.
19. E. Saule, K. Kaya and U. V. Catalyurek, "Performance Evaluation of Sparse Matrix Multiplication Kernels on Intel Xeon Phi," arXiv preprint arXiv:1302.1078, 2013.

CHAPTER 21

# Storing, Managing, and Analyzing Big Satellite Data

## *Experiences and Lessons Learned from a Real-World Application*

Ziliang Zong

CONTENTS



## 21.1 INTRODUCTION

Big Data has shown great capability in yielding extremely useful information and extraordinary potential in revolutionizing scientific discoveries and traditional commercial models. Numerous corporations have started to utilize Big Data to understand their customers' behavior at a fine-grained level, rethink their business process work flow, and increase their productivity and competitiveness. Scientists are using Big Data to make new discoveries that were not possible before. As the volume, velocity, variety, and veracity of Big Data keep increasing, we are facing significant challenges with respect to innovative Big Data management, efficient Big Data analytics, and low-cost Big Data storage solutions.

In this chapter, we will (1) provide a case study on how the big satellite data (at the petabyte level) of the world's largest satellite imagery distribution system is captured, stored, and managed by the National Aeronautics and Space Administration (NASA) and the US Geological Survey (USGS); (2) provide a unique example of how a changed policy could significantly affect the traditional ways of storing, managing, and distributing Big Data, which will be quite different from typical commercial cases driven by sales; (3) discuss how does the USGS Earth Resources Observation and Science (EROS) center swiftly overcome the challenges from serving few government laboratories to hundreds of thousands of global users; (4) discuss how are data visualization and data mining techniques used to analyze the characteristics of millions of requests and how can they be used to improve the performance, cost, and energy efficiency of the EROS system; and (5) summarize the experiences and lessons that we learned from conducting this Big Data project in the past 4 years.

The Big Data project we discuss in this chapter has several unique features compared to other typical Big Data projects:

- It is not driven by sales or other commercial benefits. Instead, it is driven by the promise of high-quality service to global researchers and users.
- The Big Data already existed in the previous system, which did not pose any challenges before. However, a simple policy change brought numerous new challenges to the existing system, which is unusual in most Big Data projects.
- The characteristics of users and their behaviors changed dramatically after the new policy was announced, which is not common in real-world Big Data projects.
- Although the previous system has been designed to store a massive volume of data, the new system designs still involve almost every aspect of the challenges faced by any Big Data project (e.g., system architecture, data storage, work flow design, data mining, data visualization, etc.).
- We share not only our experiences but also several lessons that we learned from this Big Data project, which probably are more valuable to readers.

The case study we conduct in this chapter will provide new perspectives for readers to think wider and deeper about the challenges we are facing in today's Big Data projects as well as possible solutions that can help us transit smoothly toward the exciting Big Data era.

The remainder of the chapter is organized as follows. Section 21.2 provides a brief background about the NASA/USGS Landsat program. Section 21.3 will discuss how did a new policy change almost everything (work flow, system architecture, system hardware and software, users, and user behaviors) of the conventional system and how does USGS EROS swiftly overcome the new challenges with great agility. In Section 21.4, we will present our previous research efforts on how to utilize data visualization and data mining techniques to improve the performance and reduce the operation cost of the newly designed satellite imagery distribution system. Section 21.5 concludes our study by summarizing our successful experiences and hard-learned lessons.

## 21.2 THE LANDSAT PROGRAM

The Landsat program [1,2] is the longest continuous human effort in recording the Earth's surface. Since the early 1970s, a number of Landsat satellites have continuously and consistently captured and archived petabytes of visually stunning and scientifically valuable images of our planet. The Landsat program is jointly managed by NASA and USGS. In this joint mission, NASA is responsible for the development and launch of satellites as well as the development of the ground system, while USGS is responsible for storing, managing, and distributing the captured massive satellite data. Figure 21.1a demonstrates a Landsat satellite orbiting the Earth taking images, and Figure 21.1b shows the huge antenna in front of the USGS EROS center, which is used to periodically receive the captured raw satellite images from a satellite on service. Inside the EROS building, there is a large-scale data center that stores, processes, and distributes the big satellite data (more details will be discussed in Section 21.3).

For over 40 years, the Landsat program has created a historical archive that no other system can match in terms of quality, detail, coverage, and length. This unparalleled satellite data archive has provided scientists the ability to assess changes in the Earth's landscape in a wide variety of research domains (e.g., hydrology, agriculture, atmospheric science, natural hazards, and global climate change). It is worth noting that the Landsat sensors only have a moderate spatial resolution. You cannot see individual houses on a Landsat image (like you can see from Google Satellite Images), but you can see large geographical items such as lakes or rivers and man-made objects such as highways and cities. This is an important spatial resolution decision because it is coarse enough for allowing global coverage with manageable data volume. At the same time, it is detailed enough to identify landscape changes and characterize human-scale processes such as urban growth. Figure

FIGURE 21.1 (a) A Landsat satellite capturing images. (b) USGS Earth Resources Observation and Science (EROS) Center, Sioux Falls, South Dakota.

FIGURE 21.2 (a) An example of using Landsat satellite images to study global climate change. (b) An example of using Landsat satellite images to study the urban growth of Las Vegas, NV.

21.2a shows an example of using Landsat satellite images to study the ice melting speed of Antarctica, and Figure 21.2b shows an example of using Landsat satellite images to study the urban growth of Las Vegas. All images shown in Figures 21.1 and 21.2 are provided by the USGS and NASA.

## 21.3 NEW CHALLENGES AND SOLUTIONS

### 21.3.1 The Conventional Satellite Imagery Distribution System

Before 2008, EROS has offered the satellite distribution service for many years using the conventional distribution system. Generally, users send specific requests to the USGS EROS, which includes the row (latitude), path (longitude), and time (when the images are captured and by which satellite) information of the satellite imagery. Once the payment is received, the USGS EROS will process the requests, burn the produced satellite images to CDs or DVDs, and send them to users through snail mails. Although the USGS EROS has successfully managed their workload and distributed a large amount of valuable satellite data using this traditional way for decades, only a few research groups could afford to access these expensive satellite images. Therefore, the majority of raw satellite data were stored on tape systems to reduce cost, and a moderate size of storage system was sufficient for processing the limited number of user requests.

### 21.3.2 The New Satellite Data Distribution Policy

On April 21, 2008, the USGS made an important "Imagery for Everyone" announcement that the entire USGS Landsat archive would be released to public *at no charge* by February 2009 [3]. This announcement also provided a detailed timeline for the availability of different satellite data. Based on the announced timeline, the full archive of historical Enhanced Thematic Mapper Plus (ETM+) data acquired by the Landsat 7 satellite, which was launched in 1999, would become available for free downloading by the end of September 2008. At that time, all Landsat 7 data purchasing options from the USGS, wherein users paid for on-demand processing to various parameters, would be discontinued. This new policy generated a significant impact on almost every component of the conventional EROS satellite imagery distribution system.

### 21.3.3 Impact on the Data Process Work Flow

The conventional work flow is simple and straightforward. Users contact EROS directly about their requests. Once the payment is confirmed, EROS will process the requested satellite images, which typically include the following steps: (1) the raw data is located from the tape system; (2) the raw data is loaded to a server for image optimization; (3) the optimized satellite images are burned to CDs or DVDs; and (4) the CDs and DVDs are delivered to users via snail mail. A large portion of the conventional data process work flow was handled manually, and the end users only expected their requests to be processed in weeks or even months. The major challenge of the new policy is that the satellite imagery distribution service must be available 24/7 and that users will expect their requests to be processed almost immediately (in seconds or minutes). Therefore, the new data process work flow must be completely automatic. Manual intervention should only be allowed in rare circumstances (e.g., when a user requests an extremely large number of satellite images).

### 21.3.4 Impact on the System Architecture, Hardware, and Software

In order to automate the entire data process work flow and ensure the quality of the satellite imagery distribution service, EROS has no other choice but to rethink and redesign the whole system. From the system architecture perspective, the new architecture must be highly reliable and provide clear interfaces between users, storage subsystems, and processing subsystems. In addition, the new architecture has to be flexible and scalable enough to accommodate possible future hardware/software changes and significant growth of global users. From the hardware perspective, the trade-offs between hardware performance, power consumption, reliability, and cost need to be considered altogether. In addition, the new hardware system should be able to reuse hardware components of the previous system whenever it is possible to reduce cost. From the software perspective, the primary challenge is to integrate the existing software components (e.g., satellite image processing) as well as newly developed software components (e.g., download web portal, database searching, request scheduling, prefetching) into the redesigned software stack of the new architecture. In addition, some of the previous software needs to be modified to fully take advantage of the advanced hardware. For example, the satellite image processing software needs to be parallelized to better utilize multicore CPUs.

### 21.3.5 Impact on the Characteristics of Users and Their Behaviors

In the conventional system, user behavior was more conservative and cautious because of the high cost of requesting satellite images. For example, all users tried to avoid redundant requests because they did not want to pay for the same satellite images more than once. Therefore, users carefully tracked what satellite images they have requested. Meanwhile, users would not request satellite images without a specific research purpose, because the requested images were usually paid out of a research grant and the payment needed to be approved by a project manager. Additionally, users were unlikely to request a substantially large number of satellite images due to the budget limitation. However, in the new system, with the "Imagery for Everyone" policy, there is no cost for users requesting satellite images. This greatly influences the characteristics of users and their behaviors. First, the number of users grows significantly. Before the new policy was announced, there were only hundreds of users, and most of them were from the United States. The new EROS system serves hundreds of thousands of global users now. Figure 21.3 shows the rapid growth of monthly download requests since September 2008. It can be observed that the workload of the EROS system (in terms of the number of requests) has been quadrupled in the first 29 months after applying the free download policy, and this growth is expected to be continuous in the future. Second, users behave rather aggressively and randomly in the new system. Figure 21.4 shows the analysis results of more than 60,000 users. We find that the number of requests sent by the top nine users account for 17.5% of the total number of requests. These users can be classified as aggressive users. On the other hand, the majority

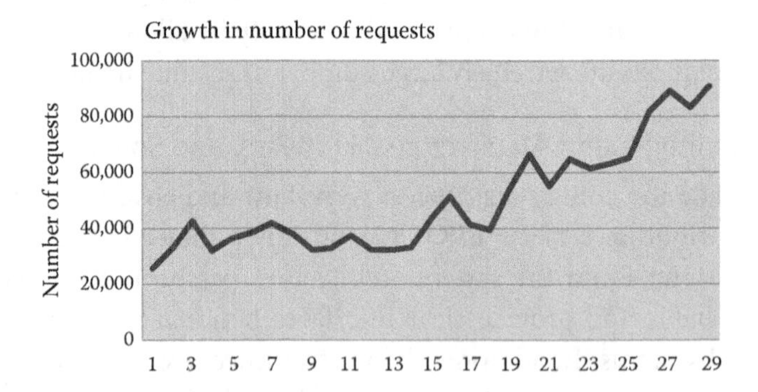


FIGURE 21.3 The number of monthly requests grew significantly in the first 29 months.

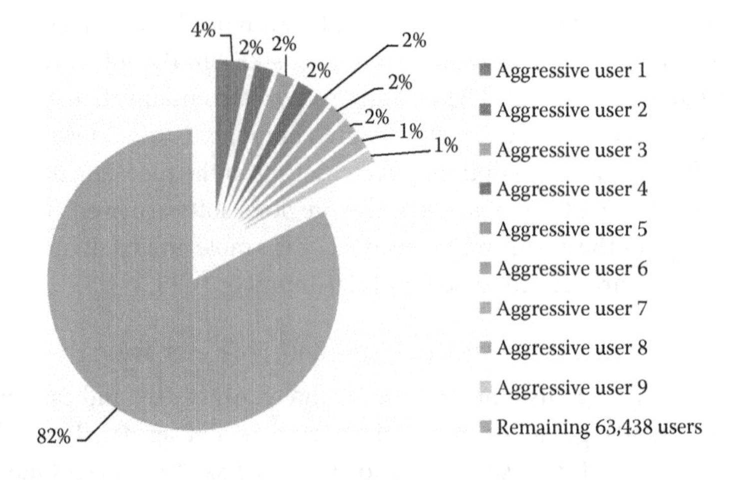


FIGURE 21.4 The user download behaviors tend to be more aggressive and random.

of users only request one or two satellite images, and they rarely come back to request again. These users can be classified as random users. In addition, we observe that among the 2,875,548 total requests, 27.7% of them are duplicated requests. Users send redundant requests because there is no cost on their side and they do not need to track what satellite images have been requested and what have not.

### 21.3.6 The New System Architecture

In order to provide a high-quality and no-cost satellite imagery distribution service to global users, EROS designed the following new system architecture (as shown in Figure 21.5), which addresses most of the aforementioned challenges. This new architecture contains four major modules: the User Request Handler; the Work Order Handler; the Hybrid Storage Subsystem (including the FTP server, the Storage Area Network or SAN for short, and tapes); and the Satellite Image Processing Subsystem.

*User Request Handler and Workorder Handler*: Global users are able to send their satellite image requests via the EarthExplorer [4] or the Global Visualization Viewer [5] web portal. The user requests will be forwarded to the Workorder Handler. The Workorder Handler will determine where are the requested satellite images. If the requested satellite images are present in the FTP server, the User Request Handler will send an e-mail to users with the download links to the requested satellite images. If, however, a satellite image cannot be found in the FTP server, a "work order" is generated and sent to the Satellite Image Processing Subsystem, which must fetch the corresponding raw data from the magnetic tapes to the SAN and apply appropriate filtering and processing algorithms to create the download-ready

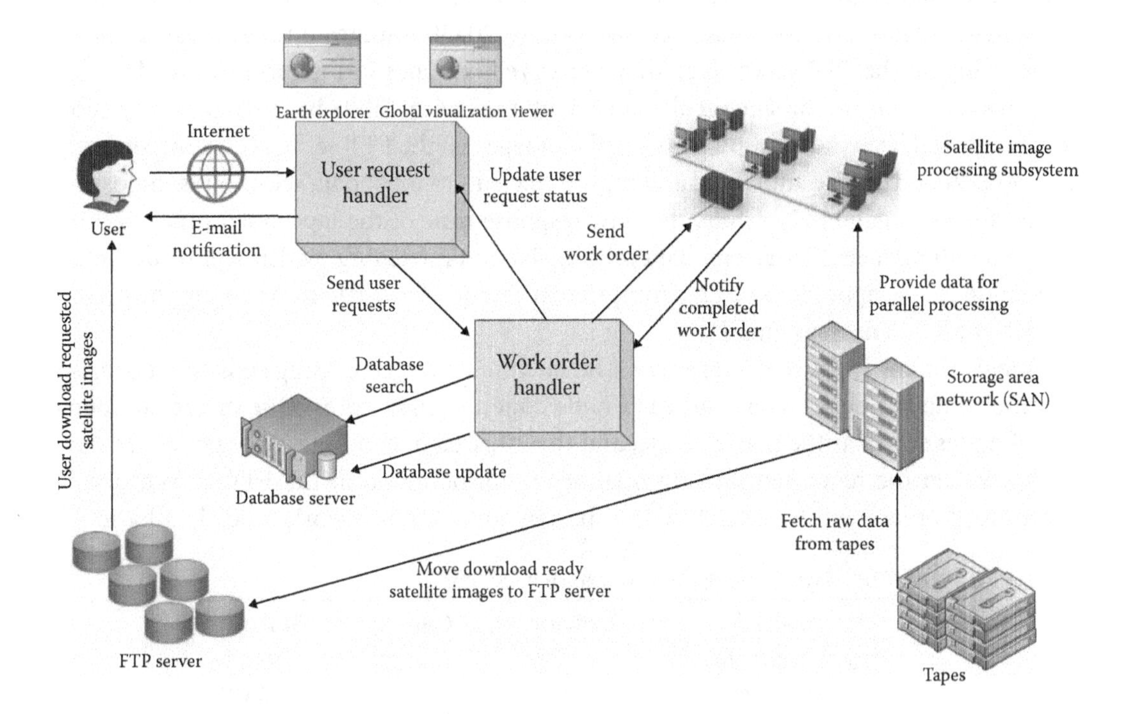


FIGURE 21.5 The new architecture of the EROS satellite imagery distribution system.

satellite images. Once a work order is completed, the download-ready satellite images will be moved to the FTP server, and a notification e-mail (with download links) will be sent to the user. After receiving the e-mail, users will be given at least 7 days to download their requested images. In other words, the system will not delete a satellite image that is not older than 7 days even when the FTP server is full. Since the size of each requested satellite image is generally greater than 200 MB, this asynchronous e-mail notification mechanism successfully improved the usability of the EROS system because users do not have to worry about not being able to download the requested satellite images immediately.

*Hybrid Storage Subsystem*: In the new architecture, the USGS EROS stores its massive satellite images in long-term archived (LTA) format using a tiered storage subsystem. Table 21.1 outlines the tiered storage components in the USGS EROS satellite imagery distribution system. The raw data of all satellite images are first stored in the tape system. Once a satellite image is requested, its raw data will be found in the tape system and then loaded to the SAN system composed of hard drive disks (HDDs) for image processing and optimization. The processed images will then be fetched to the FTP server for downloading.

Since the EROS system is highly data intensive and the majority of requests will be "read" operations, the use of solid state disks (SSDs) in the FTP server layer can significantly reduce the data transfer time once the satellite images are ready. However, the high cost of SSDs also sets the capacity limitation of FTP servers. Currently, the total amount of satellite data of EROS is at the petabyte level, while the current storage capacity of the FTP server is designed at the terabyte level. Therefore, only a small portion of satellite data can reside in the FTP server, and the large chunk of data have to be stored in the low-cost tape system. If the requested satellite images reside on the FTP server, users can download them almost immediately. However, users may need to wait for 20–30 minutes if the requested images are missing on the FTP server (i.e., they are stored on tapes in the form of raw data). To improve the quality of the satellite data distribution service, EROS has to make every effort to identify and keep the most popular satellite images on the FTP server. Most importantly, this process needs to be automatic and highly efficient. Manual data selection is impossible for petabytes of data. Meanwhile, the slow response time of the tape system may become the system bottleneck. Therefore, data mining–based prefetching, which will be discussed in Section 21.4.2, appears to be an appropriate technique for further improving the performance of EROS's new system.

*Satellite Image Processing Subsystem*: The satellite image processing subsystem consists of nine computational nodes, and each node contains eight cores. One of the computational nodes serves as the master node, and the other eight computational nodes are slave nodes. When the requested satellite images cannot be found at the FTP server, one or more work orders will be issued by the master node. Here, a work order is a basic job

TABLE 21.1 Hybrid Storage Subsystem Architecture

| Tier | Model | Hardware | Capacity | Bus Interface |
|---|---|---|---|---|
| 1 | Sun/Oracle F5100 | SSD | 100 TB | SAS/FC |
| 2 | IBM DS3400 | HDD | 1 PB | SATA |
| 3 | Sun/Oracle T10K | Tape | 10 PB | Infiniband |

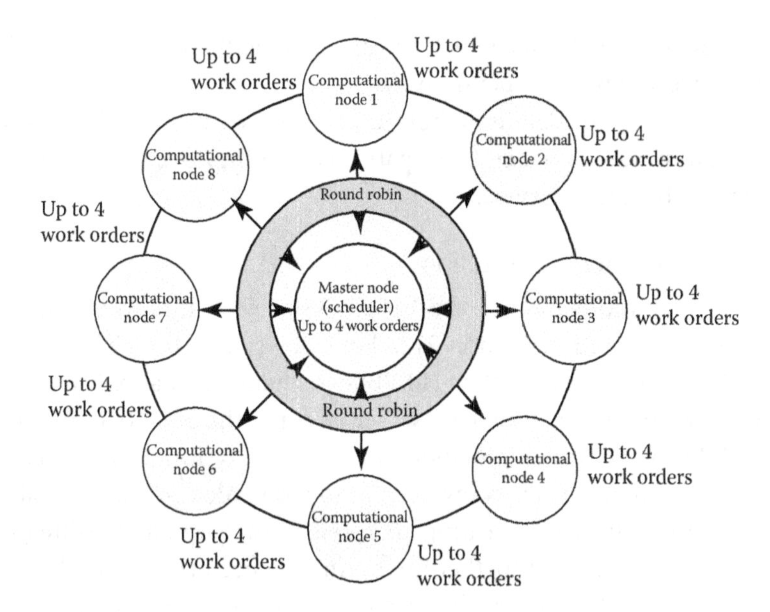


FIGURE 21.6 The round-robin work order scheduling algorithm.

executed by a core, which contains 15 different algorithms for raw satellite data preprocessing and image optimization. The master node decides how to schedule work orders to slave nodes based on the round-robin scheduling algorithm, which is shown in Figure 21.6. Specifically, the scheduler will calculate the number of work orders being executed on each slave node in a round-robin way. The next work order will be allocated to the first identified slave node that runs less than four work orders. If all slave nodes are busy (four work orders are running simultaneously in each slave node), the work order will be allocated to the master node if it runs less than four work orders. Otherwise, the work orders will be put in the waiting queue. Did you notice in Figure 21.6 that only half of the available cores are utilized and wonder why not utilize all the eight cores of each node? This question will be answered in Section 21.5 as one of the lessons we learned. It is also worth noting that the work orders running on different cores have no knowledge about other corunning work orders, even for those work orders scheduled on the same node. In other words, the execution flow and data flow of different work orders are completely isolated and work in an arbitrary way, which may cause potential cache and memory contention problems as discussed in Section 21.5.

## 21.4 USING BIG DATA ANALYTICS TO IMPROVE PERFORMANCE AND REDUCE OPERATION COST

Although the new architecture successfully automates the satellite data processing work flow and handles significantly increased workload, EROS still needs to continuously identify system bottlenecks and improve the efficiency of its system because the user space as well as the data volume keep growing. The good news is that EROS has collected millions of download requests from its global users after the new policy was applied in September 2008. These data have great potential (if analyzed correctly) to help EROS better understand user behaviors and download patterns, which in turn can facilitate EROS to further improve system performance and reduce operation cost.

In this section, we provide details on how to find the user download patterns by analyzing historical requests and how can these patterns help improve the quality of the EROS satellite imagery distribution service.

### 21.4.1 Vis-EROS: Big Data Visualization

Before we discuss the motivation of the Vis-EROS project, it is worth restating several facts that we have discussed in Section 21.3. First, the size of the FTP server (100 TB) is merely 1% of the size of the tape system (10 PB), which means that only a small portion of the satellite images can reside at the FTP server. Second, the penalty of missing a satellite image in the FTP server is very high (approximately 20–30 minutes) because finding and processing a satellite image from its raw format to a download-ready format is very time consuming. Therefore, effectively caching and prefetching popular satellite images on the FTP server will be paramount in further improving the performance of the EROS system. However, there are several questions that need to be answered before designing any caching or prefetching algorithms: (1) Do popular images really exist? (2) How long will they stay popular? (3) How can we find them easily? The primary goal of the Vis-EROS project [6] is to find out answers to these questions.

Since each satellite image has a unique geographical ID (i.e., the combination of row and path), we realized that existing data visualization techniques can be leveraged to visually represent and identify the "hot spots" (i.e., popular satellite images) on Earth. Compared to plain text data, visualized data communicates information much more clearly and effectively. There are two free tool kits that we can choose to visualize the EROS download patterns. The first one is Google Earth [7], which is a widely used visualization tool kit that provides an exploratory interface to a rich series of spatial data sets. Google Earth displays information specified by the Keyhole Markup Language (KML) [8], which is an extensible markup language (XML) grammar and file format for modeling and storing geographic

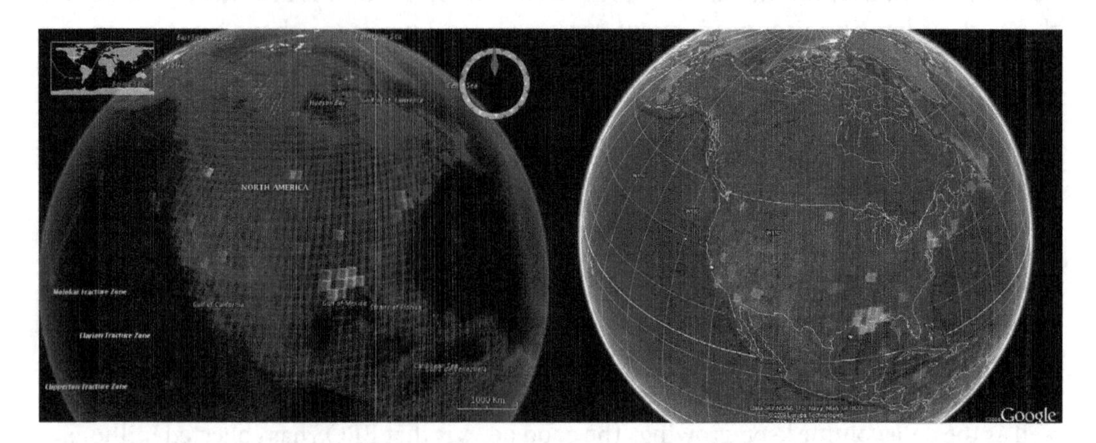


FIGURE 21.7 Visualization of EROS global download requests in NASA World Wind (left) and Google Earth (right). The bright color indicates a large number of requests for this area.

features such as points, lines, images, and polygons. The second choice is NASA World Wind [9], which was developed by NASA using the Java software development kit (SDK). It uses Java OpenGL (JOGL) [10] to render the globe and display visualization results.

At the beginning, we were not sure which visualization tool kit would be more appropriate for our needs. Therefore, we started two separate projects focusing on each of them. Figure 21.7 illustrates the first generation of visualization results of EROS data using Google Earth (right) and NASA World Wind (left). Although NASA World Wind has its advantages (e.g., fine-grained control when rendering the globe), it turns out that Google Earth is easier to use and program. Therefore, in the second generation, we only optimized the Google Earth version, which includes a better color rendering algorithm and the capability of displaying visualization results of a smaller region. Note that the satellite images marked in brighter color indicate that they are more popular. Figure 21.8 shows the visualization results of North America and South America, and Figure 21.9 shows the visualization results of the United States. Figure 21.10 demonstrates the hot spots of Haiti after the 2010 earthquake. Here, we do not include detailed information about how we visualized the EROS data using Google Earth and NASA World Wind, simply because visualization techniques are not the focus of our discussion here. Readers can visit the Vis-EROS website [6] and refer to our papers [11,12] for implementation details. Our focus in this chapter is to discuss how visualization can help with Big Data analytics.

The Vis-EROS project clearly answered the aforementioned questions. Popular images do exist, and they can be identified without sophisticated algorithms. The popularity of images evolves over time. For example, before the Haiti earthquake happened, no satellite images of Haiti were considered popular. However, the location of the earthquake did become popular after the event. The Vis-EROS project provides strong evidence that caching and prefetching techniques can improve the performance of the EROS system by preprocessing, loading, and keeping popular images in the FTP servers.

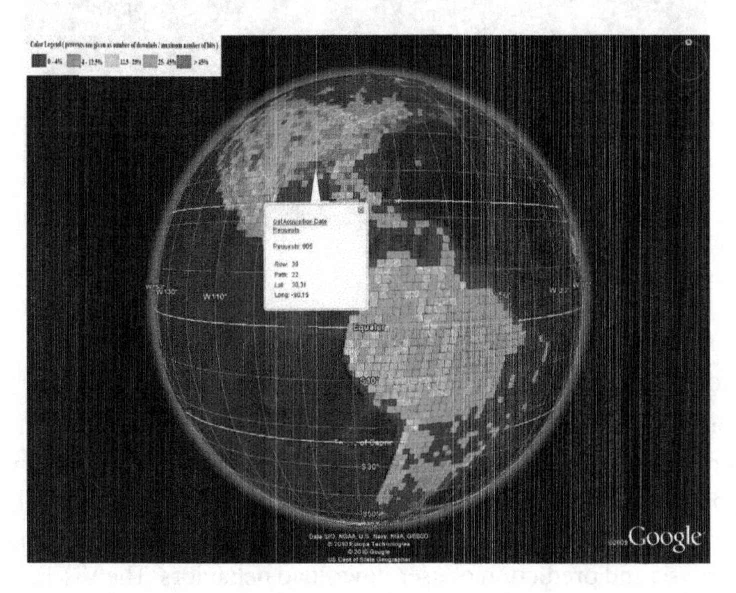


FIGURE 21.8 Visualization results of North America and South America.

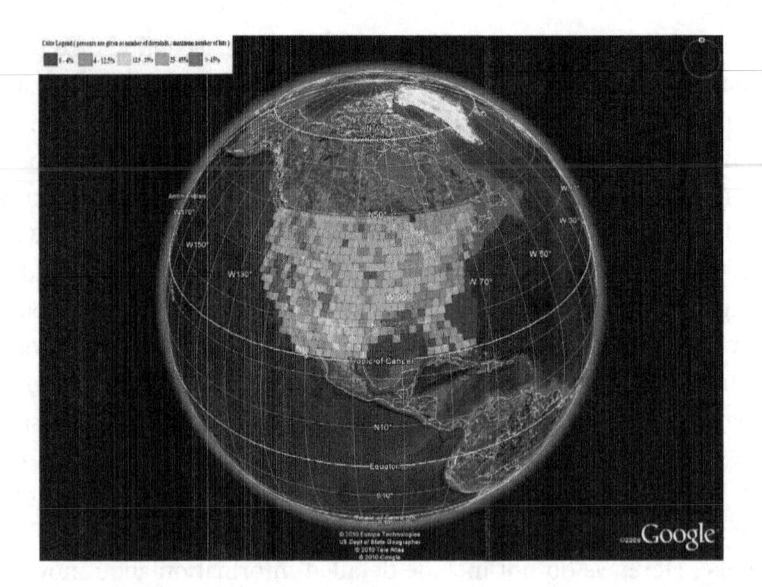


FIGURE 21.9 Visualization results of the United States.

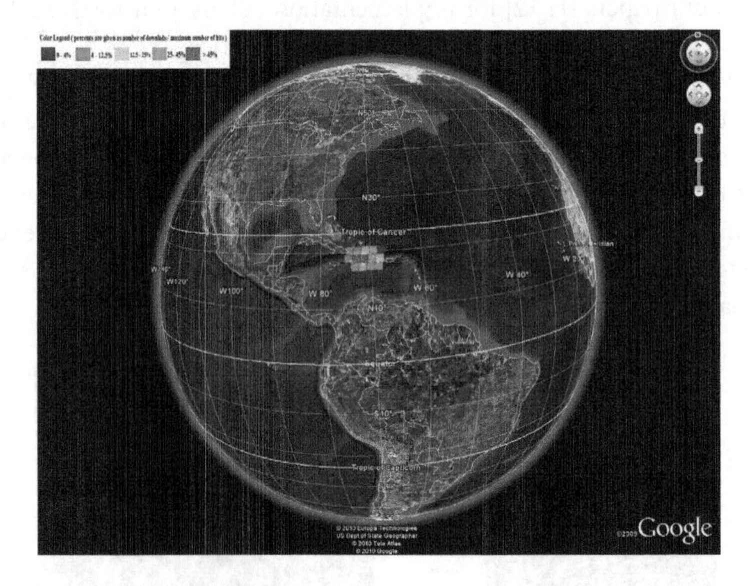


FIGURE 21.10 Visualization results of Haiti.

### 21.4.2 FastStor: Data Mining-Based Multilayer Prefetching

The FastStor project is conducted after the Vis-EROS project is completed. The primary goal of the FastStor project is to develop effective caching and prefetching algorithms that can preload the popular satellite images before the arrival of user requests and keep the popular satellite images in the cache (i.e., the FTP server) as long as possible. This relies heavily on accurate analysis and prediction of user download behaviors. The Vis-EROS project laid a good foundation for that purpose, and it provided a list of valuable observations:

- Few users request many images, while many users only request a few.
- Very few images are very popular, while most of them are unpopular.
- The popularity of images evolves over time. Newly captured images tend to be more popular than old images in general.
- Some image requests are triggered right after important global events (e.g., earthquake, tsunami, and forest fire).
- Images of extreme historical events stay popular (e.g., UFO crashing and Chernobyl nuclear power plant explosion) for many years.

More details about our analysis report can be found in Reference 13. With these observations in mind, we started searching for techniques that can achieve high performance of the EROS system. During this process, we encountered a series of challenges and pertinent problems, several of which are listed as follows.

- Will cache optimization techniques that are typically applied to very small caches still remain effective when the cache is orders of magnitude larger?
- New satellite data are added to the EROS system every week, and user download patterns evolve over time. What techniques can effectively catch newly emerged patterns with a continually increasing set of information?
- EROS currently has a large number of global users. Some users show aggressive behavior by requesting a large number of images frequently, while others only download very few images and never come back. Will it be possible to identify patterns of each user, and can user-specific algorithms improve system performance?

We decided to address these problems one by one. First, we narrowed down our focus to the caching algorithms. We studied the impact of three widely used cache replacement algorithms, namely, First In First Out (FIFO), Least Recently Used (LRU), and Least Frequently Used (LFU), on system performance using historical user download requests. These algorithms determine which images will be evicted from the FTP servers when the maximum capacity is reached. Specifically, FIFO evicts the earliest entry in cache when cache replacement happens. No action is taken on cache hits (i.e., if an entry in a cache gets requested again, its position does not change). The LRU algorithm removes the least recently used entry in a cache when it is full. LFU exploits the overall popularity of entries rather than their recency. LFU sorts entries by popularity. The least popular item is always chosen for eviction.

After observing and analyzing the results generated from the real-world data provided by EROS [14], we concluded that traditional approaches to caching can be used to successfully improve performance in environments with large-scale data storage systems. Throughout the course of evaluation, the FIFO cache replacement algorithm frequently resulted in a much lower cache hit rate than either LRU or LFU. The LFU and LRU algorithms result in similar cache hit rates, but the LFU algorithm is more difficult to implement. Overall, we found LRU to be the best caching algorithm with the consideration of both performance and ease of implementation.

Prefetching is another technology that has the potential to significantly increase cache performance. Prefetching offsets high-latency input/output (I/O) accesses by predicting user access patterns and preloading data that will be requested before the actual request arrives. In fact, previous studies have shown that an effective prefetching algorithm can improve cache hit ratio by as much as 50% [15].

Initially, we proposed a data mining–based multilayer prefetching framework (as shown in Figure 21.11), which contains two engines, the offline pattern mining engine and the online pattern matching engine. More specifically, the offline pattern mining engine contains three modules: the pattern discovery module, the pattern activation module, and the pattern amendment module. The pattern discovery module takes the user request history table, applies pattern discovery strategy, and generates the candidate pattern table. The candidate patterns will not take effect until the pattern activation module activates them. Based on the priority or urgency of requested files, the candidate patterns will be categorized into urgent patterns and nonurgent patterns. The activated urgent patterns and nonurgent patterns will be sent to the Urgent Pattern Table and Nonurgent Pattern Table, respectively. The upper-level prefetching will prefetch the satellite images generated

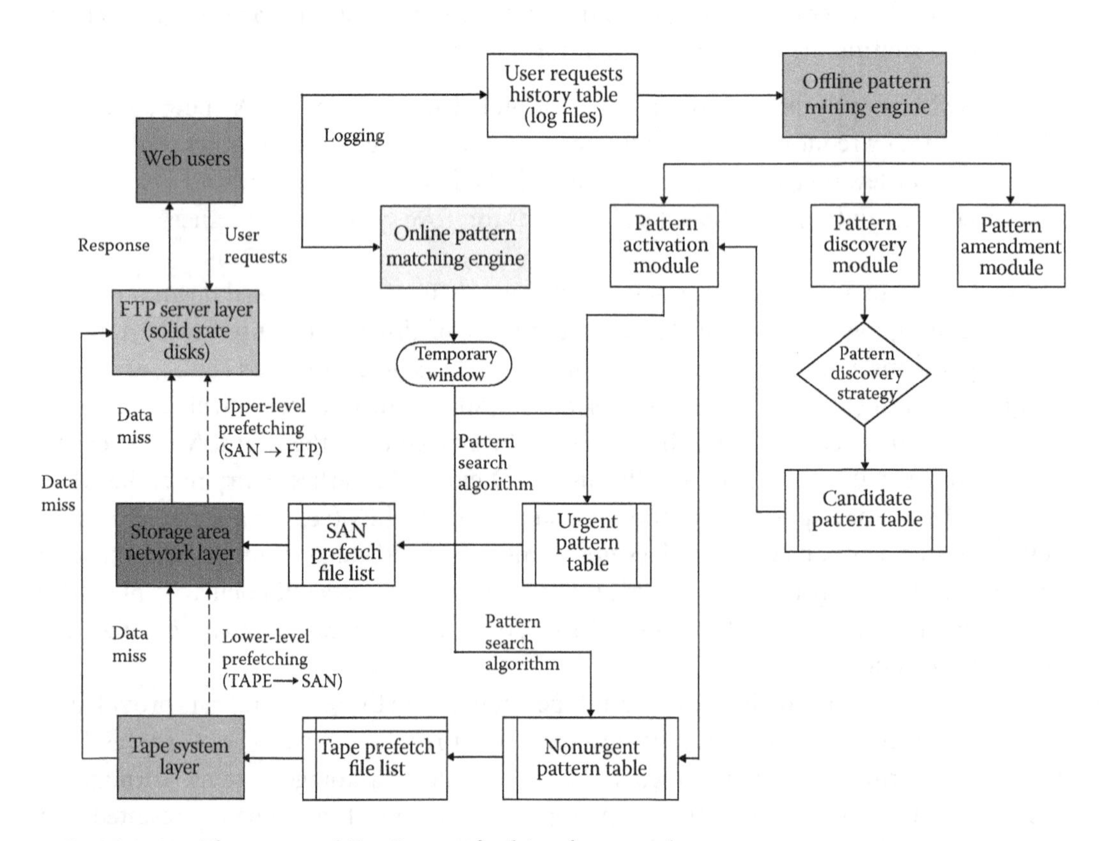


FIGURE 21.11 The proposed FastStor prefetching framework.

by the pattern search algorithm based on the Urgent Pattern Table, and the lower-level prefetching file list should be generated from the Nonurgent Pattern Table. This framework appeared to be overly complicated when we tried to implement it. More importantly, it missed some critical factors (e.g., evolvement of popularity and consideration of aggressive users) that need to be considered. In the real implementation, we did not differentiate patterns based on urgency, because it is very hard to determine which pattern is urgent and which is not. Also, we dropped the online pattern matching engine because it usually does not have a sufficient number of requests to analyze within the given window and it is fairly expensive to keep it running all the time.

Meanwhile, we implemented two specific prefetching algorithms in our emulator, which are called Popularity-Based Prefetching and User-Based Prefetching. In the Popularity-Based Prefetching algorithm, we distinguished archived popular images from current popular images. An archival popular image usually owes its popularity to historically significant events that occurred in the past. For example, the satellite image of Chernobyl falls into this category. A current popular image is popular because users are interested in the newly captured scene. For example, research monitoring the effects of global warming may request the newest images of Greenland as soon as they become available. In this case, it would be beneficial to prefetch new satellite images of Greenland once they are available.

The User-Based Prefetching algorithm is designed based on the important observation that EROS users behave significantly differently. Some users downloaded thousands of satellite images while others only sent requests once and never came back again. Therefore, prefetching rules generated based on one user's pattern may not be suitable for another user at all. With this in mind, we designed the User-Based Prefetching algorithm that generates prefetching rules based solely on that user's historical request pattern. Although this will result in more accurate prefetching results, generating separate rules for each user is a very ambitious goal because of the nature of EROS data. There are six attributes in each request that together uniquely identify a satellite image. These attributes are, in order, satellite number, satellite sensor, row, path, acquisition year, and acquisition day of year. Users can switch their download pattern in any of these dimensions between two requests, which makes it really hard to establish the pattern library for each user. Additionally, this process must be completed in $O(n)$ time, considering the size of the EROS user space.

Our solution is to represent each important attribute of a request as an integer. Then, we concatenate all six attribute integers to create a single long integer that represents a unique satellite image. Having a long integer representation of an image where each set of digits corresponds to a different attribute allows for a simple subtraction of two images that captures the movement in the multidimensional space. Then, we can treat these movements as prefetching rules. For example, if a user requests two images 7-1-100-100-2000-100 (an image in row 100 and path 100, captured on the 100th day of year 2000 by the first sensor of Landsat 7) and 7-1-101-099-2000-100, we subtract the second image from the first one and get the following movement: 711010992000100 – 711001002000100 = 9990000000. The difference, 9990000000, uniquely represents the movement of +1 path and –1 row because every other attribute stays the same. This simple digitalization solution reduces the time complexity of the prefetching algorithms to $O(n)$, which allows us to effectively

TABLE 21.2 Summary of Average Monthly Performance, Power Consumption, and Electricity Cost

| Cache Configuration | Avg. Monthly Hit Rate | Avg. Monthly Power (watts) | Avg. Monthly Cost ($) |
|---|---|---|---|
| 100 TB LRU, no prefetching (current) | 65.86% | 13,908,620 | 1127 |
| 100 TB LRU with prefetching | 70.26% | 13,748,657 | 1114 |
| 200 TB LRU, no prefetching | 69.58% | 20,152,915 | 1632 |
| 200 TB LRU with prefetching | 71.83% | 17,052,688 | 1381 |

evaluate the impact of our proposed prefetching algorithms on performance. To further reduce the overhead caused by prefetching, we set three thresholds to limit the number of the prefetched images. The user popularity (UP) threshold filters out users that only request very few satellite images. In other words, prefetching should apply to users who have sent a sufficient number of requests. The window size (WS) threshold sets the number of recent requests considered for rule generation. WS also ensures that prefetching rules are generated from recent patterns only. The rule confidence (RC) threshold is set to filter out infrequent rules in the window. Our experimental results presented in Reference 16 (and shown in Table 21.2) demonstrate that the EROS system can maintain the same level of performance with 31.78% power savings and $8 million reduction in capital cost (for purchasing another 100 TB of SSDs) by using our proposed popularity-based and user-specific prefetching algorithms, compared to doubling the size of the FTP server farm.

## 21.5 CONCLUSIONS: EXPERIENCES AND LESSONS LEARNED

The forthcoming era of Big Data will bring significant challenges to conventional ways of capturing, storing, analyzing, and managing information. In this chapter, we conducted an in-depth case study on how USGS EROS handles its big satellite data for the world's largest satellite imagery distribution system. We also provided a unique example of how could a changed policy significantly affect the traditional ways of Big Data storage and distribution and discussed how the USGS EROS swiftly overcame the new challenges. More importantly, we revealed in detail how can data visualization and data mining techniques be used to analyze the characteristics of millions of requests and improve the performance, cost, and energy efficiency of the EROS system.

During the course of conducting this Big Data project, we have obtained numerous valuable experiences, which are listed as follows:

- Experience #1: Big Data is not scary. Even for a large-scale system like EROS, it can adapt to the new challenges rather quickly via careful planning and well-thought designing.
- Experience #2: Big Data analytics is worth exploring because it has great potential to improve the quality of service and reduce cost.
- Experience #3: Data visualization techniques sometimes can be very helpful in finding hidden patterns, and many freely available visualization tool kits are available.

- Experience #4: Big Data analytics is complicated. It will take substantial efforts to figure out a working solution. Do not be overly aggressive at the beginning. Take one step at a time. For example, in our case, the Vis-EROS project served as a pilot project that provided support for further efforts and created excitement.
- Experience #5: Having a comprehensive understanding of user behavior or system workload will help greatly in proposing suitable algorithms for a large-scale system like EROS. Our proposed Popularity-Based Prefetching and User-Based Prefetching algorithms are both inspired by our user behavior analysis and workload characterization results.
- Experience #6: If there does not exist a solution that can significantly improve system performance, combining different methods that individually have small improvements on the system can add up to make a noticeable overall increase in system performance. For example, we achieved about 70% hit rate on the FTP server with LRU alone. By applying the current Popularity-Based Prefetching algorithm, we were able to achieve an extra 2% improvement. Finally, we improved the FTP server hit rate to more than 76% by combining the LRU, Popularity-Based Prefetching, and User-Based Prefetching algorithms altogether.
- Experience #7: The proposed algorithms for large-scale systems must be highly efficient. An algorithm with a time complexity larger than $O(n^2)$ will not be feasible to implement. Fast but simple solutions are always preferred (e.g., representing a user request using a long integer significantly reduced the time needed to complete User-Based Prefetching).

Meanwhile, we also learned several lessons from our study, listed as follows:

- Lesson #1: Think twice before physically expanding the existing system. The cost of maintaining a Big Data system is fairly expensive. Do not scale the system up when the number of users grows rapidly, although it is the easiest solution. It is wise to span the entire hardware and software stack to see if performance can be improved before investing on new hardware. For example, the EROS user space grew nearly 400% in the first 3 years. To guarantee performance, the managers can choose to double the size of the FTP servers. However, our experiments showed that the EROS system could maintain the same performance with over 30% energy savings as well as $8 million direct cost reduction (for purchasing an additional 100 TB of SSDs) by utilizing our proposed prefetching algorithms, compared to the alternative solution of doubling the size of the current FTP server farm.
- Lesson #2: Understanding the workload will help in reducing hardware cost and improve hardware utilization. Recall that in Section 21.3, we mentioned that only four cores of the eight-core node can be used simultaneously; this is because the typical data file required by each work order is several hundred megabytes. Cores within the same socket could race for cache resources. Additionally, cores within the same

node may also fight for memory resources if too many work orders are assigned to the node. Even worse, cache contention could further cause a thread thrashing problem, which may lead to unpredicted server downtime. The current solution is to limit the number of work orders simultaneously running on each computational node to four, which will degrade the potential peak performance by almost 50%. EROS would be able to purchase more nodes with fewer cores if this problem was identified earlier. These problems will still exist or even get worse when the application is ported to many-core systems with Intel Many Integrated Core (MIC) coprocessors or NVIDIA GPU accelerators. Fortunately, EROS did not invest on the coprocessors/accelerators.

- Lesson #3: The volume and variety of Big Data prefers the simplicity of any implemented framework or algorithms. A complicated framework (e.g., the original data mining–based prefetching framework presented in Figure 21.11) probably needs to be simplified before it can be deployed at Big Data systems.
- Lesson #4: It is possible that no existing solutions can be applied directly to solve your specific problem. Before we started developing our own prefetching algorithms, we spent substantial amount of time on evaluating several well-known data mining and machine learning algorithms, which include market basket analysis (MBA) [17], C4.5 [18], KNN [19], naive Bayes [20], Bayes networks [21], and support vector machines [22]. However, no performance improvement is achieved when using all possible combinations of input features. These widely used techniques failed because of the unique characteristics of EROS data and user behaviors.

We hope that readers can benefit from our previous experiences and lessons and have a successful Big Data practice in their own organizations.

## ACKNOWLEDGMENTS

The work reported in this chapter is supported by the US National Science Foundation under grant nos. CNS-1212535 and CNS-0915762. We also gratefully acknowledge the support from the US Geological Survey (USGS) Earth Resources Observation and Science (EROS) Center, who provided us the global user download requests. We also thank the reviewers for their helpful comments. Any opinions, findings, and conclusions or recommendations expressed in this study are those of the author and do not necessarily reflect the views of NSF or USGS.

## REFERENCES

1. National Aeronautics and Space Administration (NASA) Landsat Program Introduction. Available at http://landsat.gsfc.nasa.gov/.
2. U.S. Geological Survey (USGS) Landsat Program Introduction. Available at http://landsat handbook.gsfc.nasa.gov/handbook/handbook_htmls/chapter1/chapter1.html.
3. Available at http://landsat.usgs.gov/documents/USGS_Landsat_Imagery_Release.pdf.
4. USGS Global Visualization Viewer. Available at http://glovis.usgs.gov/.
5. Earth Explorer. Available at http://earthexplorer.usgs.gov/.
6. Vis-EROS. Available at http://cs.txstate.edu/~zz11/viseros/.

7. Google Earth. Available at http://www.google.com/earth/.
8. KML. Available at https://developers.google.com/kml/.
9. NASA World Wind. Available at http://worldwind.arc.nasa.gov/java/.
10. Java OpenGL. Available at http://en.wikipedia.org/wiki/Java_OpenGL.
11. Z. L. Zong, J. Job, X. S. Zhang, M. Nijim, and X. Qin, "Case Study of Visualizing Global User Download Patterns Using Google Earth and NASA World Wind," *Journal of Applied Remote Sensing*, vol. 6, no. 1, p. 061703, 2012. doi: 10.1117/1.JRS.6.061703.
12. G. D. Standart, K. R. Stulken, X. S. Zhang, and Z. L. Zong, "Vis-EROS: Geospatial Visualization of Global Satellite Images Download Requests in Google Earth," *Journal of Environmental Modelling and Software (ENVSOFT)*, vol. 26, no. 7, pp. 980–982, 2011.
13. B. Romoser, R. Fares, P. Janovics, X. J. Ruan, X. Qin, and Z. L. Zong, "Global Workload Characterization of A Large Scale Satellite Image Distribution System," In *Proceedings of the 2012 IEEE International Performance Computing and Communications Conference (IPCCC12)*, Austin, TX, December 2012.
14. R. Fares, B. Romoser, Z. L. Zong, M. Nijim, and X. Qin, "Performance Evaluation of Traditional Caching Policies on A Large System with Petabytes of Data," In *Proceedings of the 7th IEEE International Conference on Networking, Architecture, and Storage (NAS 2012)*, Xiamen, China, June 2012.
15. A. Nanopoulos, D. Katsaros, and Y. Manolopoulos, "A Data Mining Algorithm for Generalized Web Prefetching," *IEEE Transactions on Knowledge and Data Engineering*, vol. 15, no. 5, pp. 1155–1169, 2003.
16. B. Romoser, Z. L. Zong, R. Fares, J. Wood, and R. Ge, "Using Intelligent Prefetching to Reduce the Energy Consumption of A Large-Scale Storage System," In *Proceedings of the IEEE International Performance Computing and Communications Conference (IPCCC 2013)*, San Diego, CA, December 2013.
17. Available at http://en.wikipedia.org/wiki/Market_basket_analysis.
18. Available at http://en.wikipedia.org/wiki/C4.5_algorithm.
19. Available at http://en.wikipedia.org/wiki/K-nearest_neighbor_algorithm.
20. Available at http://en.wikipedia.org/wiki/Naive_Bayes_classifier.
21. Available at http://en.wikipedia.org/wiki/Bayesian_network.
22. Available at http://en.wikipedia.org/wiki/Support_vector_machine.

CHAPTER 22

# Barriers to the Adoption of Big Data Applications in the Social Sector

Elena Strange

CONTENTS



## 22.1 INTRODUCTION

Effectively working with and leveraging Big Data has the potential to change the world. Indeed, in many ways, it already has [1,2]. Summarizing today's conventional wisdom, in *PANEL Crowds, Clouds, and Algorithms*, Doan et al. [1] credit the pervasiveness of Big Data to the "connectivity of billions of device-enabled people to massive cloud computing infrastructure, [which] has created a new dynamic that is moving data to the forefront of many human endeavors."

If there is a ceiling on realizing the benefits of Big Data algorithms, applications, and techniques, we have not yet reached it. The research field, initially coined "Big Data" in 2001 [1], is maturing rapidly. No longer are we seeking to understand what Big Data is and whether it is useful. No longer is Big Data processing the province of niche computer science research. Rather, the concept of Big Data has been widely accepted as important and inexorable, and the buzzwords "Big Data" have found their way beyond computer science into the essential tools of business, government, and media.

Tools and algorithms to leverage Big Data have been increasingly democratized over the last 10 years [1,3]. By 2010, over 100 organizations reported using the distributed file system and framework Hadoop [4]. Early adopters leveraged Hadoop on in-house Beowulf clusters to process tremendous amounts of data. Today, well over 1000 organizations use Hadoop. That number is climbing [5] and now includes companies with a range of technical competencies and those with and without access to internal clusters and other tools.

Whereas Big Data processing once belonged to specialized parallel and distributed programmers, it eventually reached programmers and computer scientists of all subfields and specialties. Today, even nonprogrammers who can navigate a simple web interface have access to all that Big Data has to offer. Foster et al. [6] highlight the accessibility of cloud computing via "[g]ateways [that] provide access to a variety of capabilities including workflows, visualization, resource discovery and job execution services through a browser-based user interface (which can arguably hide much of the complexities)."

Yet, the benefits of Big Data have not been fully realized by businesses, governments, and particularly the social sector. The remainder of this chapter will describe the impact of this gap on the social sector and the broader implications engendered by the sector in a broader context. Section 22.2 highlights the opportunity gap: the unrealized potential of Big Data in the social sector. Section 22.3 lays out the channels through which the social sector has access to Big Data. Section 22.5 describes the current perceptions of and reactions to Big Data algorithms and applications in the social sector. Section 22.6 offers some recommendations to accelerate the adoption of Big Data. Finally, Section 22.7 offers some concluding remarks.

## 22.2 THE POTENTIAL OF BIG DATA: BENEFITS TO THE SOCIAL SECTOR—FROM BUSINESS TO SOCIAL ENTERPRISE TO NGO

The social sector—the world of nonprofit organizations (NPOs) and nongovernmental organizations (NGOs)—has much to gain from leveraging Big Data. (Throughout this chapter, we will refer to this group of organizations as "NGOs" or the "social sector.") With Big Data applications, NGOs can better understand the behavior of their constituents; design and collect data sets to reveal relevant patterns; and work more effectively with partner technology companies, businesses, and governments.

Like businesses, government, and academia, NGOs are an essential component of the world's economic and social engines. The sector's ability to leverage Big Data has the potential to benefit the organizations themselves, their constituents, their donors and supporters, and their employees. As long as a gap in knowledge and understanding exists that prevents NGOs from leveraging Big Data applications and algorithms, the deficit is as far-reaching and meaningful as the omission of Big Data would be in the business world. When the gap is eventually bridged, the reward will be as far-reaching and meaningful as it has been for the business world over the last several years.

Barriers to the social sector realizing Big Data potential remain staunchly in place, however. These inhibitors are not about technology, applications, or the data themselves. Rather, NGOs are inhibited by their perception of risk and reward; they have little incentive to overcome a natural mistrust in new technologies, including those associated with Big Data.

Unlike for-profit businesses, NGOs seemingly have little to gain from the use of Big Data; because they do not operate in a market-driven reality, no financial payoff awaits them for successfully utilizing Big Data applications and techniques. And so, the perceived risk of adopting new algorithms and methods often outweighs the perceived benefit of using them well. The benefits that can be brought by Big Data can be as significant for NGOs as for businesses, but they are far more abstract. First, they can improve services to constituents, but this is a metric hard to quantify for many NGOs. Constituents, after all, do not demonstrate support by "voting with their wallet" the way that customers of businesses do. Second, they can improve their own operational efficiencies, but they do not benefit by doing so. In fact, many NGOs find operational improvements to be disadvantageous when it comes to requesting additional funding and support from their foundations and other benefactors.

As we will see later in this chapter, the NGO/for-profit hybrid organizations called *social enterprises* represent a construct better suited to leveraging Big Data. These organizations have social missions like NGOs do, but they also have a fiduciary responsibility to generate revenue, providing a catalyst for them to make the most of the data they capture and collect. Many are also web-based businesses, more naturally inclined to using current technologies to achieve their missions.

Since 2006, Big Data has had an increasingly strong and pervasive impact on businesses, governments, and scientific endeavors. The impact is such that the United States will need as many at 140,000 data-specialized experts and 1.5 million data-literate managers [7]. It has changed the way business and organizations operate, evaluate profit strategies, and contextualize their data assets.

Meanwhile, the social sector has lagged behind; indeed, they are "losing ground because they fear what might happen if they open themselves up to this new world" [8]. Collectively, NGOs have not benefited from advances in Big Data algorithms and applications the way other industries have.

The significance of this deficiency is substantial, both for individual organizations and, more broadly, for the sector in aggregate. Organizations face an opportunity cost when they fail to leverage Big Data applications and algorithms to better achieve their missions. Moreover, the social sector runs a risk over time of becoming altogether ineffectual as Big Data is increasingly and more eagerly embraced by businesses, governments, and individuals.

At the organizational level, by passing over the benefits of Big Data, an NGO misses out on opportunities to leverage Big Data to improve services and offerings, increase memberships and donations, and streamline operational efficiencies.

Beyond this, the implications in the aggregate are even more significant. The organizations that comprise the social sector are tasked with solving the world's problems: feeding the hungry, saving the environment, curing disease, rescuing animals, and the like. NGOs help children, the poor, and the disenfranchised. When the organizations fail to embrace potentially beneficial technologies such as Big Data, the impact resonates among all these constituent groups and beneficiaries.

As the for-profit sector is learning, the effective utility of Big Data is no longer restricted to the technical realm. Businesses of all kinds, including those with little technical competency, are able to leverage applications and knowledge: As described by Null to his business readers, "Chances are, you're already processing Big Data, even if you aren't aware of it" [9]. Businesses use Big Data tools, including mobile apps and websites, to understand traffic and user behavior that impacts their bottom line.

Consider a use case as simple as the Google Analytics tool [10], which has made interpretation of website statistics a layman application. Plaza [11] describes a use case of a small Spain-based NGO that leverages Google Analytics to understand how users are attracted to research work generated by the Guggenheim Museum. In addition to gleaning important information from the collected data, the organization uses the tables and charts created by Google Analytics to better position its materials, with little scientific or technical requirements on the part of the organization itself. As Big Data becomes increasingly democratized, organizations like this NGO can not only use these tools but also communicate the results and impact more effectively; today, the readers of the NGO's reports will understand language around site traffic patterns and user behavior, and immediately comprehend the significance of the Big Data tools and information, with little or no convincing needed.

Tools such as Google Analytics do not create data where there was none but, rather, make existing data more accessible. Without an accessible tool such as Google Analytics, only a highly technical, strongly motivated business owner would be inclined to follow a traditional, work-intensive approach to learning about site traffic: by manually downloading their server logs and meticulously identifying visitation patterns to the company website. After the tool's wide launch in 2006, however, every business and NGO, including those with no technical background, has been able to set up a Google ID on their website and intuitively navigate through the Google Analytics interface to understand and interpret the implications of traffic on their business.

The analysis provided by tools such as Google Analytics, true to the nature of Big Data, provides insight proportional to the quantity of information you have. With the web-based tool, anyone can glean more insight about which pages are relevant, where conversions happen, and how they make money from the website—and they learn more the more site visitors they have. This is the power of Big Data for businesses; it has become so accessible that more data make it deceptively easier to understand the implications of user behavior and spot relevant patterns.

In between the for-profit businesses who have eagerly embraced Big Data and NGOs who are lagging behind sits the *social enterprise*. Social enterprises provide a more fitting example of a sector actively using and benefitting from Big Data.

Unlike NGOs, social enterprises must generate revenue and maintain profitability in order to be successful. As a result, they see the potential benefits of Big Data far more clearly than non-market-driven NGOs do. The remainder of this section explores the role of the social enterprise within the social sector.

The social enterprise, a concept relatively new to both the business world and the social sector, fits squarely between both. Loosely defined, a social enterprise is an NGO that prioritizes generating revenue or a for-profit business that maximizes revenue for a social goal. The social enterprise represents "a new entrepreneurial spirit focused on social aims," yet it remains "a subdivision of the third sector.... In this sense they reflect a trend, a groundswell involving the whole of the third sector" [12]. These social enterprises, increasingly a large and key component of the social sector, are also well positioned to compete with traditional for-profit businesses [13].

It is social enterprises, more than traditional NGOs, who have led the way in adopting Big Data trends and algorithms. They will continue to do so, with traditional NGOs lagging behind.

Consider the use case of social enterprise and travel site Couchsurfing, a peer-to-peer portal where travelers seeking places to stay connect with local hosts seeking to forge new connections with travelers. No money changes hands between host and visitor. Visitors benefit from the accommodations and local connection to a new city, whereas hosts benefit from discovering new friendships and building their online and offline reputations. Couchsurfing itself has a social mission: It is a community-building site that is cultivating a new kind of culture in the digital age.

Couchsurfing was launched as a not-for-profit organization in 2004 and eventually reestablished as a "B-corporation" company: a for-profit business with a social mission. Although some controversy ensued when they gave up their not-for-profit status, this organization represents a key example of a social enterprise that has made tremendous use of Big Data to both increase profits and achieve their social mission.

In addition to its need to turn a profit, the key distinction of Couchsurfing, relative to other social-mission organizations, lies in its origins as a technical organization. As a community-building site, it assumed the use of Big Data to drive decision making from its first inception. Like many web start-ups, Couchsurfing relies on Big Data to create and continually improve the user experience on its site. As a traveler, you can use Couchsurfing to find compatible hosts, locations, and events. With over 6 million members, it must use data mining and patterns to connect guests with appropriate hosts and, more importantly, to establish and maintain hosts' and guests' reputations [14].

Big Data cannot be decoupled from a site like Couchsurfing. Patterned after for-profit web businesses such as Airbnb, Facebook, LinkedIn, and Quora, Couchsurfing cannot possibly keep tabs reliably on all of its hosts, nor would it be beneficial for the organization to do so. There is no central system to vet the people offering a place for travelers to stay, yet these hosts must be reliable and safe in order for Couchsurfing to stay in business. The reputations of its hosts and travelers are paramount to the stability and reputability of the site itself. As in other community sites, these reputations are best managed and engender confidence when they are built and reinforced by community members.

Moreover, the Couchsurfing user experience must be able to reliably connect travelers with hosts based not only on individual preferences but also on patterns of use. It must be able to filter through the millions of hosts in order to provide each traveler with recommendations that suit their needs, or the traveler will not participate.

Straddling the line between for-profit and nonprofit, Couchsurfing has built its platform like any for-profit technical business would, relying on Big Data algorithms and applications to make the most of user experience both on and off the site. All the while, they have established themselves and seek to maintain their reputation as a community-minded enterprise with the best interests of the community at heart. They invest in the engineering and research capabilities that enable them to leverage Big Data because it will benefit both their profits and their social mission.

In fact, it is this latter need that drove Couchsurfing to transition to a B-corporation. As an NGO, they were unable to raise the capital needed to invest in the infrastructure and core competencies needed to effectively leverage Big Data. When they were able to turn to investors and venture capitalists for an infusion of resources, rather than foundations and individual donors, they found a ready audience to accept and embrace their high-growth, data-reliant approach to community building.

NGOs may not yet have caught up with Big Data, but social enterprises such as Couchsurfing are well on their way. It is these midway organizations, between for-profit and NGO, that light the way for the adoption of Big Data tools and techniques throughout the social sector.

## 22.3 HOW NGOs CAN LEVERAGE BIG DATA TO ACHIEVE THEIR MISSIONS

Every NGO has a mission: a cause the organization was conceived to fight for or against. Many interdependent factors determine whether and how well a given NGO achieves its mission. This is the opportunity cost for Big Data: the ways in which an NGO can achieve their mission more quickly or more fully.

The remainder of this section explores ways in which the social sector can leverage Big Data, including the following, both direct and indirect entry points:

1. Improve its services
2. Improve its offerings
3. Increase donations and memberships
4. Streamline operational efficiencies
5. Package mission-oriented results for foundations and supporters

The first two leverage points directly impact the efficacy of an organization's achieving a social mission. The latter three are indirect: They impact an organization's structure, efficiency, and even size. When these indirect factors are improved, a given NGO is better positioned to achieve any of the goals associated with its mission.

First, let us consider how an NGO can improve its services. An NGO that leverages Big Data can improve the services they make available to constituents and other end users and beneficiaries. In the social sector, a "service" is a core competency of the NGO made available to constituents for the purpose of achieving its mission. It might be a medical service, a food service, or a logistical service. Service-based NGOs are constantly trying to assess the services they offer and how and whether they might be improved. To do so, they need information.

Consider a use case, such as Doctors Without Borders, an NGO that "provides urgent medical care in countries to victims of war and disaster regardless of race, religion, or politics" [15]. This type of NGO must understand how effective their services are, whether those who partake of the service benefit, and how services can be improved. They must understand the impact their services have at the individual level, such as the level of care rendered to a patient. They must also understand the organization's impact at a broader level: the effectiveness they impart at an aggregate level among their wide group of constituents.

These needed evaluations even lend themselves to data mining metrics and language. Doctors Without Borders needs to understand the effectiveness of their services in terms of precision (percentage of constituents served who are in their target market) and recall (percentage of affected victims in a given region the organization was able to reach).

Without Big Data systems or applications, NGOs such as Doctors Without Borders tend to rely on surveys to gather and analyze information. Though useful, survey-based data alone are insufficient for an NGO to fully understand the scope and impact of their services, particularly as the number of constituents reached and number of services delivered grow. Implicit data, captured and analyzed by Big Data applications and algorithms, is an effective companion of explicit data captured through surveys.

Second, let us consider how an NGO can improve its offerings. In the social sector, an NGO's "offering" is a product or service that is sold for a fee, such as a T-shirt with the NGO's logo on it. The revenue captured from these transactions is poured back into the NGO's operational budget. Even NGOs that do not describe themselves as social enterprises often need to generate revenue in order to maintain their sustainability over time. They collect money from end users for products and services rendered—for example, an NGO with a medical mission might charge a patient for a check-up or other medical service.

How does Big Data impact upon an NGO's offerings? In the ways it makes such offerings available. Big Data is tremendously influential in creating web platforms—particularly retail sites—that are user friendly and navigable. In a retail context, Big Data algorithms ensure that users find what they are looking for and enjoy a smooth discovery process.

Users' expectations have changed in the retail context. In traditional terms, end users tend to be forgiving of NGOs whose missions they support. If the organization has a strong and relevant social mission, users will traditionally wade through irrelevant products and slog through a time-consuming purchasing process. Today, however, users' expectations are higher. Thanks to the prevalence of retail platforms that use Big Data to create consumer-friendly experiences, users have to demand these experiences in all of their online interactions. NGOs are less likely to provide these experiences without the help of Big Data shaping their platforms and interactions models.

Data mining techniques have long been applied in traditional marketing contexts as well. In their seminal data mining paper, Agarwal et al. [16] introduce the paradigm of data mining as a technique to leverage businesses' large data sets. For example, they describe data mining as a tool to determine "what products may be impacted if the store stops selling bagels" and "what to put on sale and how to design coupons" [16].

Retail leaders such as Target pioneered the use of Big Data and data mining techniques to drive sales. Target employed its first statisticians to do so as far back as 2002 [17]. As Duhigg explains, "[f]or decades, Target has collected vast amounts of data on every person who regularly walks into one of its stores… demographic information like your age, whether you are married and have kids, which part of town you live in, how long it takes you to drive to the store, your estimated salary, whether you've moved recently, what credit cards you carry in your wallet and what Web sites you visit. All that information is meaningless, however, without someone to analyze and make sense of it" [17].

Fourth, NGOs can leverage Big Data to streamline operational efficiencies. One of the significant by-products of Big Data is the infrastructure of cloud computing technologies used to support Big Data applications. Cloud computing tools are available—and increasingly accessible—to almost everyone. Even NGOs that do not need Big Data applications per se can use the infrastructure that has been built to support Big Data activities to host their web applications, maintain their databases, and engage in other activities common to even nontechnical organizations.

Like many for-profit businesses, especially smaller ones, many NGOs tend to maintain their information technology (IT) departments in-house: Mail servers, networks, and other essential tools are constructed within their own four walls. As an indirect result of the reach and scope of Big Data, NGOs can now take advantage of the proliferation of cloud computing and online tools to maintain their infrastructure.

Fifth and finally, NGOs can leverage Big Data to package their results to funders and supporters. Stories are well known to be "a fundamental part of communication and a powerful part of persuasion" [18, p. 1]. Storytelling is a popular and meaningful way to make a case for financial and emotional support.

## 22.4 HISTORICAL LIMITATIONS AND CONSIDERATIONS

Initially coined in 2001 [1], the concepts and realities of Big Data have been adopted in a relatively short period of time. Many tech-savvy businesses swarmed to Big Data applications almost as soon as technologies became available, and "Big Data" quickly became a part of the lexicon. Following the early adopters, nontechnical businesses were slower to embrace Big Data applications but eventually came to use them as well.

At a high level, solutions around cloud computing came to mean a combination of data mining algorithms (software) and cloud computing (hardware). The two together constitute Big Data solutions. Why? Because in order for an organization to effectively make use of the Big Data available to it, both software and hardware are necessary components of the solution. At the software level, Big Data applications and algorithms harness the power of collected data; in short, they turn flat data into useful information. At the hardware level, systems are needed to manage and process the vast amounts of data that provide meaningful information.

Throughout the remainder of this chapter, we refer to data mining as the software component of Big Data and cloud computing as the hardware. Although additional terms and approaches are included in Big Data solutions, these are simple generalizations that provide the greatest impact for the organizations that are the subject of this chapter.

Among other portals, the rise of Big Data is reflected in the jobs and skills that for-profit businesses are hiring for. For example, the job title "data scientist" came into widespread use only in 2008, and the number of workers operating under that title skyrocketed between 2011 and 2012 [19]. Whereas Big Data applications and algorithms emerged from many IT departments, it has become professionalized over time. Businesses are willing to invest time and resources into roles specifically aimed at Big Data.

In its earliest incarnations, Big Data applications and algorithms were strictly the province of specialized computer scientists. Programmers who specialized in parallel and distributed computing learned how to manipulate memory, processors, and data in order to create tapestries of programs that could work with more data than memory.

Prior to the introduction of the World Wide Web, any enterprise with enough data to constitute Big Data was within a specialized cohort of businesses and researchers. The urgency to develop applications to manage large amounts of data faded as the size of available physical memory increased in most computer systems from the late 1980s and on. For a time, while the capacity of memory available grew, the amount of data that necessitated processing did not. It seemed as if bigger memory and faster processors would be able to quickly manage the data captured by an organization, and certainly enough for any individual.

Then came the World Wide Web, shepherding a new, accessible entry point to the Internet for organizations. New data were created and processed in two significant ways: First, websites started popping up. The number of pages on the Internet grew from under 1 million to an estimated 1 trillion in a short 15-year time span [20].

The growth of web pages was just the beginning, however. More significantly, people began interacting on the web. End users passed all manner of direct data and metadata back and forth to companies, including e-commerce orders, personal information, medical histories, and online messages. The number of web pages paled in comparison to the amount of data captured in Internet transactions.

We began to see that, no matter how large RAM became and no matter how fast processors became, personal computers and even supercomputers would never be able to keep up with the growth of data fueled by the Internet.

This trend toward Big Data necessitated changes in computing tools and programming techniques. Programming languages and techniques were developed to serve these specializations, including open multi-processor (OpenMP) [21] and Erlang [22], as well as standards that evolved, such as portable operating system interface (POSIX) threads. A programmer writing parallel code that worked with massive data had to learn a relevant programming paradigm in order to write programs effectively.

Over time and in parallel to the growth of Internet-fueled data, innovations such as Hadoop [4] and framework generator (FG) [23] became accessible to traditional computer scientists, the vast majority of whom had no specialized training in parallel programming. These middleware frameworks were responsible for the "glue" code that is common to many parallel programs, regardless of the task the specific application sets out to solve.

With these developments, parallel programming and working with massive data sets became increasingly more available to computer scientists of all stripes.

Still, the field of parallel computing belonged to computer scientists and programmers, not to organizations or individuals. Over time, these paradigms evolved still further, to the point where nonprogrammers could not only see the power of but also make use of Big Data applications by using the tools and consumer applications becoming increasingly accessible to them.

Systems like Amazon Web Services (AWS), introduced in 2006, provided access to powerful clusters perfectly suited to operate on massive amounts of data. At launch, AWS offered a command-line interface, limited to highly technical entry points that were best suited to highly technical users. AWS coupled with Hadoop made parallel programming more accessible than ever.

AWS and similar offerings (e.g., Rattle [24]) slowly developed graphical interfaces that made their powerful cloud computing systems more widely available. Slowly, technically minded individuals who were not expert programmers have been able to adopt the intuitive tools that enable them to create their own applications on the bank of computers in the cloud.

When cloud computing and data mining applications eventually become available and accessible to any end user of any technical skill—like the consumer applications we have come to rely upon in our everyday lives, such as the layman's tool Google Analytics—for-profit companies will surely make use of Big Data algorithms and techniques in increasing breadth and depth.

There remains a gap, however, between the user accessibility of these highly technical interfaces and a true for-everyone consumer application. Cloud computing and data mining applications are following the same trajectory as enterprise applications: In *5 Ways Consumer Apps are Driving the Enterprise Web*, Svane asks, "Since the software I use every day at home and on my phone are so friendly and easy to use, why is my expensive business application so cumbersome and stodgy?" [25].

Like enterprise applications, cloud computing and data mining applications remain seemingly impenetrable to end users who have come to expect "one-click easy" interfaces. Still, both enterprise and Big Data applications are undeniably on the same forward trajectory toward becoming more accessible over time. In the meantime, organizations rely on their technical workers and data scientists to collect and leverage the Big Data intrinsic to their businesses.

## 22.5 THE GAP IN UNDERSTANDING WITHIN THE SOCIAL SECTOR

Fundamental barriers inhibit NGOs from leveraging the benefits of Big Data applications and algorithms. NGOs are not, as one might assume, limited by the accessibility or relevance of Big Data—neither in the data themselves nor in accessibility to the applications and algorithms associated with Big Data. Indeed, as we have seen, over the last 10 years, Big Data entry points have become increasingly accessible to all manner of professionals, and NGOs were certainly included in the new wave.

Rather, the issue at hand is the lack of incentive to overcome an innate mistrust of Big Data. New technologies are always slow to be adopted, to be sure, but this is particularly true in the social sector. In the current state, an NGO intent on using Big Data is put into a position where they must outlay the appropriate resources to hire people who can work with their data, and they are simply not incented to do so.

This reality has many underlying reasons: First, the risk of adopting Big Data is magnified due to the limited payoffs in a nonmarket reality. Second, NGOs do not have access to the resources and knowledge required to leverage Big Data while its applications remain out of reach for the everyday user. Third, NGOs tend to enjoy less turnover than for-profit businesses, engendering a corporate culture strikingly resistant to change. In this section, we will examine each of these underlying reasons in turn.

First, the benefits of Big Data are seen in the zeitgeist as primarily financial (although this perception is misleading): It provides a competitive advantage to businesses that know how to leverage the data they collect. In their seminal data mining paper, Agarwal et al. [16] argue for relevance of their work as a key way to improve "business decisions that the management of [a] supermarket has to make, includ[ing] what to put on sale, how to design coupons, how to place merchandise on shelves in order to maximize the profit, etc." All of these decisions, made better and more effective by data mining techniques, are in the service of increasing profits. In the sphere of for-profit businesses, this limited focus is both a necessary and sufficient incentive to adopt and invest in a new technology.

NGOs, on the other hand, have no such driver. They rely upon foundation and government grants, individual donations, memberships, and even some revenue-generating products, but they do not carry a primary responsibility to sustain profitability. Therefore, the potential upside of increasing profits by leveraging Big Data is greatly diminished, and the related risks are magnified.

To be sure, Big Data can be useful to NGOs. As we have seen in this chapter, Big Data can help NGOs serve their constituents better and improve their offerings. To be sure, every NGO has a mission that they strive for and constituents to serve. They must manage their income—generated from foundation grants, individual donations, members, and social revenue—well and responsibly. They risk their organizational reputations if they fail to carry out their social and fiduciary missions. Still, the benefits of Big Data are seemingly indirect and out of reach relative to the risk and resource outlay required to make use of the data in the first place.

Second, NGOs often lack the resources needed to capture, manage, and analyze Big Data. As we have seen, for-profit businesses place cloud computing and data mining applications in the hands of their data scientists and highly technical workers. As these applications become more democratized, end users of all kinds, working at all types of businesses, will take on the roles currently held by data scientists. Until then, for-profit businesses see enough financial benefit in Big Data that they are willing and able to hire data scientists and other dedicated roles to manage their data applications.

Nonprofits, however, are less willing and able to invest in these roles due to the lack of potential payoff from leveraging their data. Projects such as Data Without Borders and DataKind [26] are bringing data scientists into the social sector via collaborations and internships, as a way to bridge the gap of these roles within the social sector.

Lacking the resources of these technical roles and their associated knowledge of Big Data, NGOs face perceived risk when it comes to security and privacy associated with

Big Data applications. For a typical organization working with Big Data, it is unlikely to be profitable or beneficial for them to acquire and maintain a cluster of machines to manipulate their data.

Finally, in general terms, NGOs enjoy less turnover and greater longevity among their employees than the for-profit business sector. One of the downsides of this reality is that long-time employees are less inclined to take risks than young, newer employees. Indeed, many NGOs were slow to adopt technologies such as Twitter and Facebook despite urging from their young staffers [8]. The managers and executives who had been employed by an NGO for a long time were reluctant to deviate from their known path and were slow to trust younger, newer employees.

## 22.6 NEXT STEPS: HOW TO BRIDGE THE GAP

This chapter has described the historical context and current state that underlie the reluctance in the social sector to fully embrace Big Data algorithms and applications. Although this gap is significant, it is not insurmountable. Indeed, it is absolutely inevitable that the gap will close eventually. The time line to close the gap can be accelerated, however, and the remainder of this section suggests some strategies and tactics to do so.

The social sector will come to embrace Big Data in due course. New technologies and new ways of thinking that become commonplace make their way into the social sector in due course, even if this sector is slower to adopt them. Since its formal inception in 2006, Big Data has taken hold not only as an instrument of science but as an instrument of business as well. The social sector will follow.

Big Data is a relatively new concept in the general lexicon, but it is no fad. The amount of data generated in this digital age will only increase, as will the need to transform vast amounts of data into actionable information. Concurrently, the applications and techniques used to leverage Big Data have become increasingly democratized in recent years, and this trend will continue as well. Cloud computing and data mining have moved from the hands of specialized computer scientists to general computer scientists and now to technically minded individuals. Big Data is likely to continue on this trajectory until—like the now-widespread consumer applications on millions of smartphones, computers, and tablets—it is accessible to end users of all kinds.

This transformation will take time on its own, but the time line can be accelerated. The inevitable adoption can be accelerated. We need not leave the whole of change to the steady march of time. The social sector itself bears some responsibility for embracing these new and beneficial techniques. Beyond this, the field of computer science is also uniquely positioned to accelerate the adoption curve.

The remainder of this section describes three distinct ways in which the adoption of Big Data algorithms and techniques in the social sector can be precipitated: through the growing prevalence of social enterprises alongside NGOs; through the improved communication and outreach of data scientists and researchers; and through the increased accessibility of cloud computing and data mining applications.

First, as social enterprises begin to comprise a greater percentage of mission-driven organizations, the landscape of this sector will evolve. The social sector will comprise more than donation- and membership-based nonprofits to include many more social enterprises similar to Couchsurfing. As described in Section 22.2, social enterprises are responsible for generating revenue in addition to carrying out their social missions.

These new types of organizations, many of them highly technical and web based, will lead the way in adopting and using Big Data to their advantage. More traditional NGOs will be inclined to follow suit as they see their sister organizations realize the nonfinancial benefits of leveraging Big Data.

Second, the scientists and researchers in the field of Big Data can improve the way they communicate about advances in the field and related applications. The social sector will be more amenable to Big Data when the applications and benefits are more understandable [18], and it is incumbent upon scientists to translate novel academic and industry research into terms that convey the relevance and real-world applicability of their results.

Scientists are responsible not just for the content of their original research but for the communication layer as well. Big Data research, even the most esoteric results, often has straightforward application for many kinds of end users, including for-profit businesses and NGOs.

Third, cloud computing and data mining applications will become increasingly accessible. These tools started out the sole province of specialized programmers, and they have become democratized to a much wider range of users over time. Still, they currently remain out of reach for those less comfortable with highly technical tools.

Like the social-web consumer applications before them, Big Data tools will grow in accessibility and usability until they become relevant to even the least technically minded users. When the translation and understanding of Big Data is relevant to anyone at any NGO, then organizational Big Data will be truly utilized to its full potential.

## 22.7 CONCLUSION

This chapter has discussed the unmet potential of Big Data algorithms and applications within the social sector. The implications of this gap are far-reaching and impactful, within and beyond the sector itself.

Fundamentally, NGOs are not incented to seek out the resources and individuals they need to make the most of Big Data. Without the financial drive of the market, the benefits of cloud computing and data mining applications stack up poorly against the risk required to outlay resources required to use them.

This dynamic can and will change, however. Social enterprises are leading the way in applying Big Data algorithms and techniques in a mission-driven context. These organizations' leadership, along with the ongoing accessibility of cloud computing and data mining tools, will accelerate the adoption curve in the social sector.

## REFERENCES

1. Doan, A., Kleinberg, J., and Koudas, N. PANEL Crowds, clouds, and algorithms: Exploring the human side of "Big Data" applications. *2010 Special Interest Group on Management of Data (SIGMOD '10)*, June 6–10, 2010.
2. Lui, B., Hsu, W., Han, H.S., and Xia, Y. Mining changes for real-life applications. *2nd International Conference on Data Warehousing and Knowlelge Discovery (DaWaK 2000)*, pp. 337–346, 2000.

3. Dean, J., and Gehmawat, S. MapReduce: Simplified data processing on large clusters. *Proceedings of the 6th Conference on Symposium on Operating Systems Design and Implementation (ODSI 2004)*, pp. 137–150, 2004.
4. Shvachko, K., Hairong, K., Radia, S., and Chansler, R. The Hadoop distributed file system. *Mass Storage Systems and Technologies (MSST 2010)*, pp. 3–7, 2010.
5. Rosenbush, S. More companies, drowning in data, are turning to Hadoop. *Wall Street Journal*, April 14, 2014.
6. Foster, I., Zhao, Y., Raicu, I., and Lu, S. Cloud computing and grid computing 360-degree compared. *Grid Computing Environments Workshop 2008 (GCE '08)*, pp. 12–16, November 2008.
7. Lohr, S. The age of Big Data. The *New York Times*, February 11, 2012.
8. Kanter, B., and Paine, K. *Measuring the Networked Nonprofit: Using Data to Change the World*. Jossey-Bass, New York, 2012.
9. Null, C. How small businesses can mine Big Data. *PC World Magazine*, August 27, 2013.
10. Google, Inc. Available at http://analytics.google.com.
11. Plaza, B. Monitoring web traffic source effectiveness with Google Analytics: An experiment with time series. *Aslib Proceedings*, Vol. 61, Issue 5, pp. 474–482, 2009.
12. DeFourney, J., and Borzaga, C., eds. From third sector to social enterprise. *The Emergence of Social Enterprise*. Routledge, London and New York, pp. 1–18, 2001.
13. Raz, K. Toward an improved legal form for social enterprise. *New York University Review of Law & Social Change*, Vol. 36, Issue 283, pp. 238–308, 2012.
14. Frankel, C., and Bromberger, A. *The Art of Social Enterprise: Business as if People Mattered*. New Society Publishers, New York, 2013.
15. Doctors without Borders. Available at http://www.doctorswithoutborders.org.
16. Agarwal, R., Imieliński, T., and Swami, A. Mining association rules between sets of items in large databases. *Proceedings of the 1993 ACM SIGMOD International Conference on Management of Data*, pp. 207–216, 1993.
17. Duhigg, C. How companies learn your secrets. The *New York Times*, February 16, 2012.
18. Olson, R., Barton, D., and Palermo, B. *Connection: Hollywood Storytelling Meets Critical Thinking*. Prairie Starfish Productions, Los Angeles, 2013.
19. Davenport, T., and Patil, D. Data scientist: The sexiest job of the 21st century. *Harvard Business Review*, October 2012.
20. The Incredible Growth of Web Usage (1984–2013). Available at http://www.whoishostingthis.com/blog/2013/08/21/incredible-growth-web-usage-infographic/.
21. Dagum, L. OpenMP: An industry standard API for shared-memory programming. *Computational Science & Engineering*, Vol. 5, Issue 1, pp. 46–55, 1998.
22. Armstrong, J., Virding, R., Wilkstrom, C., and Williams, M. Concurrent programming in ERLANG, Prentice Hall, Upper Saddle River, NJ, 1993.
23. Cormen, T., and Davidson, E. FG: A framework generator for hiding latency in parallel programs running on clusters. *17th International Conference on Parallel and Distributed Computing Systems (PDCS 2004)*, pp. 127–144, September 2004.
24. Williams, G. Rattle: A data mining GUI for R. *The R Journal*, Vol. 1, Issue 2, pp. 45–55, 2009.
25. Svane, M. 5 ways consumer apps are driving the enterprise web. *Forbes Magazine*, August 2011.
26. Data Kind. Available at http://www.datakind.org.

# Index

Page numbers followed by f and t indicate figures and tables, respectively.

## C

## D

## I

## N

## Q

## R



## S

## T

## U

## V



## W



## X



## Y



## Z